FUNCTIONAL FOODS AND NUTRACEUTICALS

SOURCES AND THEIR DEVELOPMENTAL TECHNIQUES

FUNCTIONAL FOODS AND NUTRACEUTICALS

SOURCES AND THEIR DEVELOPMENTAL TECHNIQUES

C.S. Riar
Associate Professor
Department of Food Engineering & Technology, SLIET
Longowal- 148106, Sangrur, Punjab

D.C. Saxena
Professor & Head
Department of Food Engineering & Technology, SLIET
Longowal- 148106, Sangrur, Punjab

Sukhcharn Singh
Associate Professor
Department of Food Engineering & Technology, SLIET
Longowal- 148106, Sangrur, Punjab

Vikas Nanda
Associate Professor
Department of Food Engineering & Technology, SLIET
Longowal- 148106, Sangrur, Punjab

Navdeep Jindal
Assistant Professor
Department of Food Engineering & Technology, SLIET
Longowal- 148106, Sangrur, Punjab

nipa

NEW INDIA PUBLISHING AGENCY

New Delhi – 110 034

NEW INDIA PUBLISHING AGENCY
101, Vikas Surya Plaza, CU Block, LSC Market
Pitam Pura, New Delhi 110 034, India
Phone: + 91 (11)27 34 17 17 Fax: + 91(11) 27 34 16 16
Email: info@nipabooks.com
Web: www.nipabooks.com

Feedback at feedbacks@nipabooks.com

ISBN: 978-93-83305-96-4

Composed and Designed by NIPA

Foreword

Nutraceuticals represent foods that help in prevention and treatment of diseases. These contain wide array of multi-activity bio-active compounds. Historically, several food items have been known for their disease preventive and curative properties. With advanced scientific understanding of foods, nutrition and human biology, a large number of products have emerged in the market to meet large variety of nutritional as well as medicinal requirements of the consumers.

Globally, functional food industry including nutraceutical products has been experiencing rapid growth. It has been estimated that the market of global functional food industry composed of foods, beverages and supplements would reach US $ 176.7 billion by 2015 with a compounding annual growth rate (CAGR) of 7.4%. Broadly, consumers are increasingly concerned about diets and their nutritional and medicinal characteristics, safety, convenience, rising costs of health-care etc. These concerns are driving a growing demand and robust market for functional foods and natural health products. In several economies including India, the terms of trade are not yet favorable to the farmer producers. Post harvest losses are still very large. Rising costs of farm inputs, uncertain quality, weather dependence, seasonality, price fluctuations of farm produce and depressed prices at the time of harvest remain problem issues for the farmers. Large heaps of fruits and vegetables, food grains and other produces lying by the road sides during harvesting seasons are common sites. These include heaps of harvested paddy, potatoes, citrus fruits, bananas, mangoes, guavas, grapes, tomatoes, cabbage, cucumbers, carrots, peas, green leafy vegetables, turmeric, coriander, other spices and condiments, aromatic produces and so on. Scientific handling of the produce, primary processing at farm level, and appropriate processing and value addition is of great importance for loss prevention and protecting the quality. Vertical linkages of the production sector and diversification to nutraceuticals and functional foods would provide much needed support to farm production sector in augmenting its income and balancing the terms of trade to its favour.

India has the potential to be the food producer for the world. Food processing industry in India has been termed as the Sun-rise Sector that would play very important role in the overall national economy. Increasing globalization and technological advancements have further opened new avenues for food processing sector and development of functional foods.

Consumers today have been looking forward for foods and beverages with specific characteristics. These include weaning foods, foods for growing children, high energy foods, protein/ vitamin/ mineral rich diets and supplements, high fibre foods, foods for aged persons, foods for diabetics, cancer patients, heart patients, partially digested foods etc. There are concerns for maintain youthfulness, strengthening immune system, and attaining longevity. Increased side effects of modern medicines, attraction to alternate/ holistic system of treatments, economy are yet other issues resulting in growing demand for nutraceutical products. A look into the ancient food culture of India reveals great understanding and appreciation for foods with medicinal / healing properties. Foods were also classified into *satwik* and *tamsi* classes resulting in specific behavior anticipated and exhibited by the user.

The nutraceuticals containing phytosterols are effective in lowering LDL cholesterol. Proteins and non-proteins of bovine milk-fat globule membrane are potent nutraceuticals that are considered effective as anticancer, anti-cholesterolemic, and good for coronary heart diseases. Functional foods could be derived from natural sources are processed in order to exhibit certain desirable characteristics and contain desired biologically-active compounds. These may also be of great health benefits. In future, nutraceuticals of plant origin may evolve to be considered a vital aspect of dietary disease-preventive food components. Noteworthy is the recent establishment of Centers for Study on Alternative Medicines around the world. Studies are also in progress on various phyto-nutrients for their role in the prevention of chronic degenerative diseases.

The sources of nutraceuticals include ginseng, spirulina, ginkgo biloba, amino acids, glucosamine, chondroitin and Aegle marmelos. They require to be processed, formulated and consumed or administered internally under the supervision of a qualified medical practitioner. These are intended for specific dietary management of a disease or condition with distinctive nutritional requirements, based on recognized scientific principles. Research would further be needed for prescribing personalized diets in order to maintain good health, improve the quality of life and enhance the lifespan while understanding the role of nutrients at the molecular level. Continued R&D is also needed for identification and classification of such foods, sourcing, their composition, bio-active compounds, efficacy, mechanism of action, handling and processing, formulations, packaging, presentation, safety aspects etc. It is an emerging challenging area encompassing different disciplines of science and technology and integrative approach.

The book focuses on current advances in the area of nutraceuticals and functional foods. The articles also highlight the future role of such foods in maintaining good health and preventing diseases. The book comprises of different chapters written by eminent scholars. Many of them have based their papers on their own long drawn observations and experiences. I congratulate the authors. Also, sincere appreciations are due to the editors for their valuable contribution made in bringing out this book for the benefit of the society. I also congratulate the Sant Longowal Institute of Engineering and Technology, Sangrur for supporting publication of this book. I hope that this book would be of great help to the students, researchers, medical fraternity, policy makers and industry in seeking detailed information on nutraceuticals and functional foods.

Prof. (Dr.) B.S. Bisht
Former Vice Chancellor
G.B Pant University of Agriculture & Technology, Pantnagar
Former Assistant Director General (HRD)
Indian Council of Agricultural Research , New Delhi
Director, Birla Institute of Applied Sciences, Bhimtal
bsbisht@gmail.com

Preface

Consumers are increasingly interested in the health benefits of foods and have begun to look beyond the basic nutritional benefits to the potential disease prevention and health enhancing compounds contained in many foods. Now, the state-of-the-art research in this area at academic, government and industry levels is focusing on product development, bio-actives' standardization and consumer's interest as well as regulatory efforts. Requirement for shaping the future direction for recommending personalized diets for optimum maintenance of health, improving life quality and increasing lifespan while understanding the role of nutrients at the molecular level are needed.

The functional food industry, consisting of food, beverage and supplement sectors, is one of the several areas of the food industry that is experiencing fast growth in recent years. It is estimated by BCC Research that the global market of functional food industry reachs 176.7 billion in 2013 with a compound annual growth rate (CAGR) of 7.4%. Specifically, the functional food sector will experience 6.9% CAGR, the supplement sector will rise by 3.8% and the functional beverage sector will be the fastest growing segment with 10.8% CAGR. This kind of growth is fueled not only by industrial innovation and development of new products that satisfy the demand of health conscious consumers but also by health claims covering a wide range of health issues. Yet, consumer skepticism persists mainly due to the fact that benefits associated with consuming the products may be difficult to be detected. The industry suggests the establishment of a health claim regulating agency, which may increase consumer confidence. Strict examination of some of the functional food claims may discourage some companies from launching their products.

Credible scientific research indicates there are many clinically proven and potential health benefits from food components. Claims on food packages are just one vehicle for informing consumers about these diet and health relationships. Many academic, scientific and regulatory bodies have developed, or are developing, guidelines to establish the scientific evidence base needed to support and further validate claims for functional components or the foods containing them. FDA regulates food products according to their intended use and the nature of claims made on the package.

The papers included in this book have a broad coverage of the topics related to new technologies in functional foods and nutraceuticals, F&Vs and their by-products as valuable ingredients for functional foods and nutraceuticals, potential bioactive components from various food sources, trends and development of nutraceuticals and functional foods as well as functional food and nutraceuticals as ingredients in the value addition for health promotion, standardization and quality control. We are thankful to the contributors who deserve credit for the contents in this book and deliberated excellent and informative presentations on various topics dealt with functional foods and nutraceuticals. This compilation helps to overcome the problems faced in exploring the potential of nutraceuticals in naturopathy and device strategies to encounter such problems.

Editors

Contents

Unit - II: Functional Foods and Nutraceuticals Their Health Benefits

Unit-III: Functional Foods and Nutraceuticals Their Development Techniques

Unit-IV: Cereals and other Food Grains as a Source of Functional Foods and Nutraceuticals

List of Contributors

Aabida Jabeen
Department of Food Technology, Islamic University of Science and Technology, Awantipora Pulwama, Kashmir, (J&K)

Aamir Hussain
Department of Food Technology, Islamic University of Science and Technology, Awantipora Pulwama, Kashmir, (J&K)

Aasima Rafiq
Department of Food Science and Technology, PAU, Ludhiana 141004, Punjab

Aditya Madan
Department of Food Process Engineering, IICPT, MFPI, GOI, Pudukkottai Road, Thanjavur - 613 005, Tamil Nadu

Amarjeet Kaur
Department of Food Science and Technology, PAU, Ludhiana 141004, Punjab

Arti Chauhan
Department of Food Engineering and Technology, SLIET, Longowal 148106, Punjab

Ashok Kumar
Additional Director Research (FSNE), PAU, Ludhiana 141004, Punjab

Bade P.V.
Department of Food Technology, SM College of Food Technology, MPKV, Rahuri 413705, Maharashtra

Baljit Singh
Department of Food Science and Technology, PAU, Ludhiana 141004, Punjab

Basharat Yousuf
Department of Post Harvest Engineering and Technology, Faculty of Agricultural Sciences, AMU, Aligarh, Uttar Pradesh

Bhatia Aruna
Faculty of Life Sciences, Punjabi University, Patiala 147002

Bhupendar Singh Khatkar
Department of Food Technology, Guru Jambheshwar University of Science and Tech. Hisar 125001, Haryana

Binnar A.S.
Department of Food Technology, SM College of Food Technology, MPKV, Rahuri 413705, Maharashtra

Chothe Deepak
Department of Food Science and Technology, SM College of Food Technology, Mahatma Phule Krishi Vidyapeeth, Rahuri 413 705, Maharashtra

Choudhari D.M.
Shramshakti College of Food Technology, Maldad, Tal-Sangamner, Dist- Ahmednagar, Maharashtra

D.S. Sogi
Department of Food Science and Technology, GNDU, Amritsar 143 005, India

D.C. Saxena
Department of Food Engineering and Technology, SLIET, Longowal 148106, Punjab

Davinder Kaur
Department of Food Science and Technology, PAU, Ludhiana 141004, Punjab

Deepika Kamble
Department of Food Technology, SM College of Food Technology, MPKV, Rahuri 413705, Maharashtra

Deshmukh Chaitanya
Department of Food Science and Technology SM College of Food Technology, Mahatma Phule Krishi Vidyapeeth, Rahuri 413 705, Maharashtra

Dnyaneshwar Holkar
Department of Food Technology, SM College of Food Technology, MPKV, Rahuri 413705, Maharashtra

Farhan Mohiuddin Bhat
Department of Food Engineering and Technology, SLIET, Longowal 148106, Punjab

Gaikwad V.G.
Shramshakti College of Food Technology, Maldad, Tal-Sangamner, Dist- Ahmednagar Maharashtra

Gulzar Ahmad Nayik
Department of Food Engineering and Technology, SLIET, Longowal 148106, Punjab

Gurjant Singh
Department of Food Science and Technology, CDLU, Sirsa, Haryana

Gurkirat Kaur
Department of Food Science and Technology, PAU, Ludhiana 141004, Punjab

H. R. Naik
Division of Post Harvest Technology, SKUAST, J&K

H. Rather
Division of Post Harvest Technology, SKUAST, J&K

H.K. Sharma
Department of Food Engineering and Technology SLIET, Longowal 148106, Punjab

Hanuman Bobade
Department of Food Science and Technology, PAU, Ludhiana 141004, Punjab

Ishpreet Kaur Walia
Department of Life Sciences, Dolphin (PG) College of Life Sciences, Chunni Kalan 140307, Punjab

Ishrat Majid
Department of Food Engineering and Technology, SLIET, Longowal 148106, Punjab

Jagrani Minj
Dairy Microbiology Division, NDRI, Karnal 132001, Haryana

Jaspreet Kaur
Department of Food Science and Technology, PAU, Ludhiana 141004, Punjab

Jundhare S.S.
Shramshakti College of Food Technology, Maldad, Tal-Sangamner, Ahmednagar, Maharashtra

Jyoti Jha
Department of Animal Products Technology, College of Animal Sciences, CCSHAU, Hisar 125 004, Haryana

Jyotika Sharma
Department of Food Engineering and Technology, SLIET, Longowal 148106, Punjab

K. Bakhshi
Department of Food Science and Technology, PAU, Ludhiana 141004, Punjab

Kale K.R.
Department of Food Technology, SM College of Food Technology, MPKV, Rahuri 413705, Maharashtra

Khalid Bashir
Department of Basic and Applied Sciences, NIFTEM, Sonipat, Haryana

Khalid Muzaffar
Department of Food Engineering and Technology, SLIET, Longowal 148106, Punjab

Khan Nadiya Jan
Department of Food Engineering and Technology, SLIET, Longowal 148106, Punjab

Kishor Pawar
Department of Food Technology, SM College of Food Technology, MPKV, Rahuri 413705 Maharashtra

Kulsum Jan
Department of Food Engineering and Technology, SLIET, Longowal 148106, Punjab

Mandeep Singh Sibia
Department of Food Engineering and Technology, SLIET, Longowal 148106, Punjab

Manjeet Aggarwal
Department of Basic and Applied Sciences, NIFTEM, Sonipat, Haryana

Manju V. Nehra
Department of Food Science and Technology, CDLU, Sirsa, Haryana

Monika Sharma
CIPHET, PAU Campus, Ludhiana 141004, Punjab

Muneer Ahmad Dar
Department of Food Technology, Islamic University of Science and Technology, Awantipora Pulwama, Kashmir (J&K)

Narender K Chandla
Department of Food Engineering and Technology, SLIET, Longowal 148106, Punjab

Neeraj Gandhi
Department of Food Science and Technology, PAU, Ludhiana 141004, Punjab

Neha Balan
Department of Food Process Engineering, IICPT, MFPI, GOI, Pudukkottai Road, Thanjavur 613 005, Tamil Nadu

P. S. Ranote
Department of Food Science and Technology, PAU, Ludhiana 141004, Punjab

Patil Priyanka
Department of Food Science and Technology, SM College of Food Technology, Mahatma Phule Krishi Vidyapeeth, Rahuri 413 705, Maharashtra

Pawase P.A.
Shramshakti College of Food Technology, Maldad, Tal-Sangamner, Dist- Ahmednagar, Maharashtra

Poonam A.
Department of Food Science and Technology, PAU, Ludhiana 141004, Punjab

Preetinder Kaur
Department of Processing and Food Engineering, PAU, Ludhiana 141004, Punjab

Priyanka Sharma
Department of Food Science and Technology, PAU, Ludhiana 141004, Punjab

R. K. Gupta
Director, CIPHET, PAU Campus, Ludhiana 141004, Punjab

R. S. Dabur
Department of Animal Products Technology, College of Animal Sciences, CCSHAU, Hisar 125 004, Haryana

Rahul Debbarma
Department of Food Process Engineering, IICPT, MFPI, GOI, Pudukkottai Road, Thanjavur 613 005, Tamil Nadu

Rakesh Sharma
Department of Food Science and Technology, Dr. YS Parmar, University of Horticulture and Forestry, Nauni, Solan -173230, Himachal Pradesh

Ranjit Singh
Department of Food Process Engineering, IICPT, MFPI, GOI, Pudukkottai Road, Thanjavur 613 005, Tamil Nadu

Reeba Panesar
Department of Food Engineering and Technology, SLIET, Longowal 148106, Punjab

Riar, C.S.
Department of Food Engineering and Technology, SLIET, Longowal 148106, Punjab

Ritika Oberoi
Department of Food Science and Technology, PAU, Ludhiana 141004, Punjab

RK Malik
Dairy Microbiology Division, NDRI, Karnal 132001, Haryana

Rodge A.B.
Associate Dean and Principal, MGM College of Food Technology, Aurangabad

Romee Jan
Department of Food Engineering and Technology, SLIET, Longowal 148106, Punjab

Sadewale V.S.
Department of Food Technology, SM College of Food Technology, MPKV, Rahuri 413705, Maharashtra

Sakshi
Department of Food Engineering and Technology, SLIET, Longowal 148106, Punjab

Sandeep Singh Rana
Department of Food Process Engineering, IICPT, MFPI, GOI, Pudukkottai Road, Thanjavur 613 005, Tamil Nadu

Sanju B. Dhull
Department of Food Science and Technology, CDLU, Sirsa, Haryana

Savita Sharma
Department of Food Science and Technology, PAU, Ludhiana 141004, Punjab

Seema
Department of Food Engineering and Technology, SLIET, Longowal 148106, Punjab

Shilpa Vij
Dairy Microbiology Division, NDRI, Karnal, 132001, Haryana

Shubhneet Kaur
Department of Food Engineering and Technology, SLIET, Longowal 148106, Punjab

Shumaila Jan
Department of Food Engineering and Technology, SLIET, Longowal 148106, Punjab

Sukhcharn Singh
Department of Food Engineering and Technology, SLIET, Longowal 148106, Punjab

Suresh Bhise
Department of Food Science and Technology, PAU, Ludhiana 141004, Punjab

Swati Kapoor
Department of Food Science and Technology, PAU, Ludhiana 141004, Punjab

Syed Insha Rafiq
Department of Food Engineering and Technology, SLIET, Longowal 148106, Punjab

Syed Zameer Hussain
Division of Post Harvest Technology, SKUAST, J&K

Taranpreet Singh
Department of Food Science and Technology, PAU, Ludhiana 141004, Punjab

Usha Bajwa
Department of Food Science and Technology, PAU, Ludhiana 141004, Punjab

V.K. Joshi
Department of Food Science and Technology, Dr. YS Parmar, University of Horticulture and Forestry, Nauni, Solan 173230, Himachal Pradesh

Vijayta Sharma
Department of Life Sciences, Dolphin (PG) College of Life Sciences, Chunni Kalan, 140307, Punjab

Vikas Nanda
Department of Food Engineering and Technology, SLIET, Longowal 148106, Punjab

Vinod Palve
Department of Food Technology, SM College of Food Technology, MPKV, Rahuri 413705, Maharashtra

Walunj Amol
Department of Food Science and Technology, SM College of Food Technology, Mahatma Phule Krishi Vidyapeeth, Rahuri 413 705, Maharashtra

Unit - I : Functional Foods and Nutraceuticals An-Overview

CHAPTER 1

Functional Food and Nutraceuticals from Fruits and Vegetables

D.S. Sogi

Abstract

Food is basic need of human for its existence. Food is a source of nutrients needed for growth, energy and protection. There are number of chemical compounds not considered as nutrients but has positive effect on health. There is a need to look into the health benefits of the food components. Many lifestyle diseases are caused by the unhealthy food habits. Developments in science have led to establish the link between food intake and healthy living. Non-nutrient compounds of food are also called phyto-chemicals. Functional foods are prepared by increasing a particular nutrient or phyto-chemical to get health benefits. The increased cost of health care and drugs trigger the developments in this area. The approximate global market size is 30 billion US$ while Japan, United States, and Europe are the leaders in functional foods. FICCI has reported that functional food market in India is worth Rs 4,400 crore with growth rate of 21% per annum.

Introduction

The food has been considered as source of nutrients in the past but increasing evidences has proved that particular types of food can help in preventing the diseases. The World Health Organization (WHO) has reported that deaths due to Non-Communicable diseases are increasing in South East Asian counties; where as CMPN (Communicable, Maternity, Prenatal, and Nutrition) are decreasing. The major Non-Communicable diseases are cardiovascular disease and cancers. Research on the dietary habit has revealed that vegetarian diet promotes health and the population is less prone to certain diseases like cardiovascular disease and various types of cancers. It has stimulated the research focusing on various components of the food that are responsible for protecting body against the diseases.

This has led to the concept of 'Nutraceuticals' and 'Functional foods' that are being marketed with health claims to manage various diseases. Generally these foods contain 'Bioactive compounds' which are non-nutrient substances having health promoting properties. There is a wide range of the food sources which contain bioactive compounds but fruits and vegetables are important source.

Fruits and Vegetables are the plant parts that are consumed both as raw and after cooking. These are considered major source of vitamins and mineral, however, fiber and sugars are also present in good amount.

Production: According to Food and Agriculture Organization (FAO) of United Nation, the global production of fruits and vegetables including melons was 638 and 1090 million metric tons while India produced 75 and 106 million metric tons in 2011. In addition to this, the world production of roots and tuber was 810 million metric tons whereas India produced 51 million metric tons in 2011. It indicates that fruits and vegetables form an important part of the diet in the world including India. There is an ample opportunity to manufacture various health promoting products based on fruits and vegetables. A number of Fruits and Vegetable based functional food that can be prepared and presented in (Table 1) indicates the role of fruits and vegetables in the nutraceutical production for their health benefits.

Table 1: Functional Food Based on Fruits and Vegetables

Artichoke Extract	Grape Skin Extract	Grapefruit Extract
Bearberry Extract	Green Olive Leaf Extract	Garlic Extract
Blueberry Leaf Extract	Green Tea Extract	Neohesperidin Dihydrochalcone
Bitter Citrus Extract	Lemon Bioflavonoid Complex	Olive Leaf Extract
Citrus Bioflavonoid Complex	Naringin	Orange Bioflavonoid Complex
Citrus Extract	Neohesperidin	Pomegranate Extract Powder

Phyto-chemicals: The health promoting properties of functional food and nutraceuticals is due to the presence of phytochemicals also known as botanicals or bioactive compounds. No comprehensive list of these compounds is available but can be grouped into following categories:

a. Polyphenols
b. Phytosterols
c. Carotenoids
d. Tocopherols
e. Tocotrienols
f. Organosulfur
g. Fiber
h. Prebiotics

a. **Polyphenols**: These compounds are widely distributed in food products. They contain either one or more benzene rings along with varying number of hydroxyl, carbonyl, and carboxyl groups. Generally found conjugated with one or more sugar residues. Flavonoids are the most common type of polyphenols found in food systems. Other important compounds under this category are catechins, thearubingens, theaflavins, isoflavones etc. Vegetables are one of the important sources of polyphenols. Bravo (1998) reported that polyphenol content of onion was 20 to 2025 microg/g fresh weight bases.

b. **Phytosterols**: The structures of phytosterols are similar to cholesterol but the side-chain contains additional double bonds and methyl and/or ethyl groups. Important phytosterols are beta-sitosterol, campesterol, and stigmasterol. Vegetables are not good sources of phytosterols; however legume vegetables contain these bioactive compounds.

c. **Carotenoids**: These are generally forty carbon compounds with or without oxygenation having an extensive conjugated double bond system. These compounds are lipid soluble and have chromatic properties that give various colors to vegetables. The important carotenoids are beta-carotene, alpha-carotene, lycopene and leutin. These are found either esterified to fatty acids or un-esterified in plant tissue. Vegetables are important source of carotenoids eg carrots, squash, sweet potato, and spinach are good source of beta-and alpha-carotene; leafy vegetables such as spinach, mustard greens and green beans are good sources of lutein; and tomatoes are good source of lycopene.

d. **Tocopherols**: These are non-polar compounds containing a phenolic-chromanol ring linked to an isoprenoid side chain which is saturated. Tocopherols are of four types i.e. alpha, beta, gamma, and delta depending upon the number and position of methyl groups on the phenolic-chromanol ring. Other differences are due to three asymmetrical carbons at position 2, 4, and 8 of the isoprenoid side chain. Among the eight isomeric forms, RRR-a-tocopherol is the most abundant in human blood and tissues having highest bioactivity. Dietary sources include vegetable oils, nuts and germ part of grains.

e. **Tocotrienols**: These compounds are similar to tocopherols but phenolic-chromanol ring linked to an isoprenoid side chain is unsaturated. This compound is also found in germ oil, vegetable oil and nuts.

f. **Organosulfur**: Cruciferous and allium vegetables are rich source of these compounds. Sulfur atoms are bound to a cyanate group or a carbon atom in a cyclic or non-cyclic configuration in these compounds. Organosulfur compounds are obtained by crushing the tissues of the vegetables to facilitate enzyme-substrate interaction. In cruciferous vegetables, the enzyme myrosinase acts on glucosinolyates to yield isothiocyanates [sulforaphane, phenethyl-isothiocyanate, and benzyl isothiocyanate]. In allium vegetables, allinase enzyme converts allicin to diallyl sulfide, diallyl disulfide or diallyl trisulfide. The products from both cruciferous and allium vegetables are bioactive compounds. Important cruciferous vegetables are broccoli, cauliflower and brussels sprout whereas predominant allium vegetables are garlic, leeks and onions.

g. **Fiber**: Dietary fiber is composed of long straight chains of carbohydrate molecules that cannot be broken by human digestive enzymes. There are two types of dietary fibers - insoluble and soluble. Insoluble fiber does not dissolve in water and commonly called roughage. They promote regular bowel movement and prevent constipation, remove toxic waste through colon in less time, help prevent colon cancer by keeping an optimal pH in intestines to prevent microbes from producing cancerous substances. Rich sources of insoluble dietary fiber are edible skins of vegetables, green beans and dark green leafy vegetables.

   Soluble fiber dissolves in water and has great role in the uptake of nutrients. They prolong stomach emptying time so that sugar is released and absorbed more slowly. Thus helps to regulate blood sugar in people with diabetes. Soluble fibers bind with fatty acids and reduce their absorption. They reduce the risk of heart disease by lowering total cholesterol and LDL cholesterol (bad cholesterol) in body. Rich sources of soluble dietary fiber are beans, legumes, carrots, cabbage, okra etc.

h. **Prebiotics**: These are non-digestible food fibers or oligosaccharides which stimulate the growth and activity of beneficial bacteria in the intestines. A prebiotic must escape digestion in the upper gastrointestinal tract so that it can be used by beneficial micro-organisms in the colon, mainly Bifidobacteria and Lactobacilli. Probiotic bacteria do not colonize well in intestine however prebiotic helps them to survive and thrive in the human gut. Probiotic bacteria produce short-chain fatty acids (SCFAs) that restrict the growth and activity of less beneficial species. It promotes better gut integrity and function, increases immune system function, improves calcium absorption and cholesterol maintenance.

The oligosaccharides recognized as prebiotics include Fructo-oligosaccharides (FOS), Inulins, Isomalto-oligosaccharides, Lactilol, Lactosucrose, Lactulose, Oligofructose, Pyrodextrins, Soy oligosaccharides, Transgalacto-oligosaccharides (TOS), and Xylo-oligosaccharides. Vegetables as source of prebiotic are asparagus, garlic, tomatoes, Jerusalem artichoke, onions, spinach, kale, mustard greens and legumes.

Mechanism of Action

Precise information about the mechanism by which bioactive components promotes health has not been established well. It has been observed that multiple bioactive components act simultaneously at different or identical target sites. These compounds have potential to reduce the risk of cancer, cardiovascular disease, osteoporosis inflammation, type-2 diabetes, and other chronic degenerative diseases. Bioactive food components have health-promoting roles at various stages of diseases from initiation to development.

Isoflavones have been reported to reduce the risk of cardiovascular disease through multiple mechanisms. They reduce level of oxidized low-density lipoproteins in the plasma, bind cholesterol in the intestinal tract, enhance bile excretion and modulate arterial elasticity. These actions help in reducing cholesterol absorption and fast metabolism.

Polyphenols, Carotenoids, Tocopherols and Allyl Sulfides quench free radicals and reactive oxygen species having unpaired electron. Lipids, proteins and carbohydrates can convert into free radicals. These compounds act as antioxidants and remove the free radicals. Formation of free radicals within a cell triggers various molecular responses such as induction of apoptosis or cell death and activation of signal transduction between cells and within a cell. Thus regulation of physiological and pathological processes by antioxidants is important for health and disease prevention. Polyphenols also get bound to toxins or carcinogens in the intestinal tract and prevent their absorption.

Dietary fibers and Phytosterol lowers the lipids by sequestering cholesterol in the intestinal tract and reducing cholesterol absorption. The long fibrous structure of dietary fibers entraps harmful toxins and carcinogens in the digestive tract. Soluble dietary fibers bind toxins and cholesterol in the intestinal tract. Insoluble dietary fibers dilute carcinogens and decrease contact of toxins/carcinogens with the intestinal tract.

Isoflavones are structurally similar to estrogens and estradiols therefore they may have estrogen-like biological activities. Isoflavones or Phytoestrogens lower the exposure to estrogen in premenopausal women and reduce breast cancer risk due to antagonist estrogen activity (Shimizu et al.,1990; Cassidy et al., 1994 and 1995). Phytoestrogen reduce the hormone-sensitive increases in plasma cholesterol levels and bone loss in postmenopausal women (Potter et al., 1998; Setchell and Cassidy, 1999).

Antioxidants activate phase II detoxifying enzymes via an antioxidant-responsive element (Mukhtar and Ahmad, 2000). Sulforaphane one of the isothiocyanates also induces phase-II detoxifying enzymes (Zhang et al., 1994). Diallyl sulfides from garlic preparations induce both phase I and phase II detoxifying enzymes (Yang et al., 2001). The induction of these enzyme systems reduces susceptibility to deleterious effects of mutagenic compounds.

Isothiocyanates, Allyl Sulfides, Polyphenols, Tocopherols *and* ***Carotenoids*** reduce the ill effects of reactive oxygen species and free radicals responsible for premature death of immune cells (Brennan et al., 2000). They stimulate synthesis of immune cells and protect against infections.

References

1. Bravo, L. (1998). Polyphenols: Chemistry, Dietary Sources, Metabolism, and Nutritional Significance. Nutrition Reviews, 56:317–333.
2. Brennan, L. A. Morris, G. M. Wasson, G. R. Hannigan, B. M. and Barnett, Y. A. (2000).The Effect of Vitamin C or Vitamin E Supplementation on Basal and H_2O_2-Induced DNA Damage in Human Lymphocytes. British Journal of Nutrition, 84:195–202.
3. Cassidy, A. Bingham, S. and Setchell, K. D. R. (1994). Biological Effects of Isoflavones Present in Soy in Premenopausal Women: Implications for the Prevention of Breast Cancer. American Journal of Clinical Nutrition 60: 330–340.
4. Cassidy, A. Bingham, S. and Setchell, K. D. R. (1995). Biological Effects of Isoflavones in Young Women—Importance of the Chemical Composition of Soya Products. British Journal of Nutrition, 74:587–590.
5. Feinberg, B. (1973). Vegetables. In Food Dehydration, 2nd Edition, Vol. 2. W.B. Van Arsdel, M.J. Copley and A.I. Morgan, Jr.(Editors). AVI Publishing Co., Westport, Conn.
6. Mukhtar, H. and Ahmad, N. (2000). Tea Polyphenols: Prevention of Cancer and Optimizing Health. American Journal of Clinical Nutrition, 71:1698S–1702S.
7. Potter, S. M. Baum, J. A. Teng, H. Stillman, R. J. Shay, N. F. and Erdman, J. W. (1998). Soy Protein and Isoflavones: Their Effects on Blood Lipids and Bone Density in Postmenopausal Women. American Journal of Clinical Nutrition, 68:1325S–1379S.
8. Setchell, K.D.R. and Cassidy, A. (1999). Dietary Isoflavones: Biological Effects and Relevance to Human Health. Journal of Nutrition, 129:758S–767S.
9. Shimizu, H. Ross, R. K. Bernstein, L. Pike, M. C. and Henderson, B. E. (1990). Serum Estrogen-Levels in Postmenopausal Women—Comparison of American Whites and Japanese in Japan. British Journal of Cancer, 62:451–453.
10. Yang, C. S. Chhabra, S. K. Hong, J.Y. and Smith, T. (2001). Mechanisms of Inhibition of Chemical Toxicity and Carcinogenesis by Diallyl Sulfide (DAS) and Related Compounds from Garlic. Journal of Nutrition, 131:1041S–1045S.
11. Zhang, Y. Kensler, T. W. Cho, C.G. Posner, G. H. and Talalay, P. (1994). Anticarcinogenic Activities of Sulforaphane and Structurally Related Synthetic Norbornyl Isothiocyanates. Proceedings of the National Academy of Sciences, 91:3147–3150.

CHAPTER 2

Designer Foods - An Overview

Usha Bajwa

Abstract

Designer foods are known by various names like medical foods, fortified foods, nutritional foods, nutraceutical foods, pharma foods etc. These foods are nutritious, maintain health and prevent diseases. Such foods must possess characteristics like modified composition; limit the presence of certain potentially harmful components and include certain desirable ingredients either naturally or by intentional addition. The term designer food emerged some years ago to describe foods tailored to specific concentration and proportion of nutrients that are important to health. These are designed and produced by using different approaches. The term 'designer foods' is subjected to change based on nutritional labelling regulations, new developments, consumer perceptions and expectations. Grain improvement through biotechnology emphasizes crop upgrading with a specific challenge to manipulate the endosperm composition in order to improve its intrinsic properties for diet and health. Breeding fruits with higher bioactive compounds for disease resistance and improved composition have stimulated interest in the nutraceutical and processing aspects of the fruits Nutritive and genetic interventions to alter milk composition for specific health and processing opportunities are gaining importance in dairy biotechnology. Designer eggs having modified fat provide choices with different nutritional benefits. The options suggested for designer meat and meat products are through production practices, post-harvest procedures and reformulation techniques. All these features of designer foods will be presented with suitable examples.

Introduction

The term designer food emerged some years ago to describe foods tailored to certain specific concentration and proportion of nutrients that are important to health. Some foods are specifically designed to control a number of chronic diseases such as obesity, diabetes, hypertension, cardiovascular diseases, cancer, enteric disorders etc. for a target segment/group of population (Aiyar, 2001). Designer foods are known by various names like medical foods, fortified foods, nutritional foods, nutraceutical foods, pharma foods etc. (Joshua et al., 2006). These foods are nutritious, prevent diseases and maintain health. Such foods must possess characteristics like modified composition; limit the presence of certain potentially harmful components and may include certain desirable ingredients either naturally or by addition (Table1).

Table 1: Some Designer Food Ingredients and their health benefits

Ingredient	Examples	Benefits
Phytochemical	Allicin, Flavanol, Curcumin, Phenolic Acids	Anticarcinogenic, Antioxidant, Antihypertensive, Antibacterial
Prebiotics	Dietary Fibres, Oat Fibre, Oligosaccharide, Inulin	Retain water, Improve sensory properties, Replace fat
Poly unsaturated fatty acids	Omega Fatty Acids, Peanut, Olive, Soybean, Fish, Sunflower Oil	Modified fatty acid profile, Reduce cholesterol and Reduce risk Cardiovascular disease, Source of Antioxidant vitamin
Bioactive ingredients	Ascorbic Acid, Vitamin E, β-Carotene	Reduce risk of free radical damage and related oxidative disorders like Cataract, Cancer, Cardiac disease
Probiotics	Fermented Foods: Lactobacillus (L. caselii, L. acidophilus, L. delbrenekii) Bifidobacterium (B. longum, B. infantis)	Anticarcinogenic activity, Immunostimulation and Antagonistic action against Enteric pathogens

(Anandh et al, 2003).

There is no internationally agreed definition for designer foods. The term 'designer foods' is subjected to change based on nutritional labelling regulations, new developments, consumer perceptions and expectations. Therefore, it can be said that designer food is an evolving concept meant for reducing or delaying the risk of diseases (Pszczola, 1993). The factors responsible for innovations in designer foods are presented in Table-2.

Table 2: Motivating factors for development of designer foods

Factors responsible for innovations	Major concern/reason
1. Health and longevity	• Increasing health costs
2. Quick and convenient solutions	• Greying/ageing population
3. Natural products	• Increase in self health maintenance
4. Preventive measures	• Consumer health awareness and wide choice of foods.

(Bawa et al, 2003; Varshney, 2002)

Types of Designer Foods

A) Designer cereals/flour

Protein modification

Wheat plants of the future might provide grain for designer flours that yield delicious, wholesome new breads, pastas, and other appetizing foods. And giving some of wheat's flour genes to other grains e.g. barley, oats, corn, rye, rice etc could yield innovative, multipurpose versatile flours.

The scientists are investigating high molecular-weight glutenins, which are critical to making strong dough to yield flour that produces stronger dough, larger bread loaf sizes and light finer textured breads.Using biotechnological techniques, wheat glutenin genes inserted into other grains may lead to unique and healthful products, impossible to make otherwise. Moving one or two of wheat's high molecular weight glutenin genes into barley, for example, might open the door to popular new products from barley flour (Anonymous, 1999).

Starch modification

Flour's other main component; starch might also be reworked through genetic engineering into a more marketable product. Wheat starch is composed of molecules known as amylose and amylopectin. Wheat flour low in amylose, for example, is desirable for noodle making. Reduced-amylose flours may also improve dough for frozen foods like pizza crusts or ready-to-bake breads by helping maintain flavour. Scientists suspect that boosting the amount of amylopectin in starch may concurrently reduce the amount of amylose, resulting in value-added, low amylose flour. With the wheat genes now in hand, genetic engineers might soon be able to increase the ratio of amylopectin to amylose (Joshua et al., 2006).

Enhancing specific nutrients

This includes enhancing naturally occurring substances in foods to contain specific added ingredients e.g., varieties of rice that include beta-carotene and iron. Genetic engineering for biofortification of rice could be an important approach to improve the nutritional quality of rice. Genetic engineering of the metabolic pathway for biosynthesis of β-carotene (1.6 mg/g total carotenoids) in the endosperm of japonica-type and indica-type rice cultivars has been demonstrated. In addition to beta-carotene, transgenic rice with ferritin and FRO2 genes have also been developed for stable iron enhancement in rice seeds even after polishing/milling.

B) Designer fruits

Technological innovations and new tools in genetics and breeding can be harnessed to revitalize the commercial tree fruit industry whereby new genetic traits will decrease production costs and increase fruit quality and consumer appeal. Selection and molecular biology now allow genetic engineering tools to be applied to the precision breeding of these crops, resulting in exciting new varieties.

a) Breeding fruits with higher bioactive compounds

Studies showed that antioxidant activity of fruits was correlated with total phenolic content rather than with any individual phenolic compound and in some fruits like strawberries ellagic acid along with some flavonoids (anthocyanins, proanthocyanins, catechin and epicatechin) are the main compounds. These compounds are known to influence quality, acceptability and stability of foods by acting as flavorants, colorants or antioxidants. Antioxidant composition of fruits varies among cultivars and genetics play a major role. Furthermore, the phytochemical content of the fruits not only increases the quality of the fruit but also has a major impact on shelf life and susceptibility to diseases (Khanijadeh et al, 2004).

b) Breeding fruits with improved composition and texture

Breeding fruits for improvement in composition, texture and juiciness is the recent outcome For example, the fruit **peacotum–incorporated with characteristics of peach, plum and apricot.** Another success is the **plout** –a plum-apricot hybrid that is available in purple, yellow or green with red polka dots (Carlye, 2004).

Another example of this category includes enhancing naturally occurring substances in food e.g., producing a tomato that contains more lycopene than usual.

Okanagan Specialty Fruits Inc. ("OSF") is a dynamic Canadian agricultural biotechnology company that is developing new commercial tree fruit varieties e.g. non browning apple varieties, ideally suited to the fresh-cut produce market.

C) Designer meat foods

Designer meat foods include a) Low fat meat products b) Low-sodium meat products c) Nitrite reduced meat products d) Meat products with designer food ingredients (Keeton, 1994).

a) Low fat meat products

Most successful method for altering the fat content in meat products is incorporation of non-meat ingredients as fat replacers (Giese, 1996; Akoh, 1998; Lyons et al., 1999).

b) Low-sodium meat products

Salt is an important additive in our diet. According to standard nutrition labelling regulation, the low sodium foods are designated as sodium free (5mg or less), very low sodium (35-40mg or less) and low sodium (120-140mg) products. The two possible options to reduce the sodium content in meat products:

i) To reduce level of salt added during processing to the minimum level and

ii) To replace all or part of the salt being used in processing with salt substitutes like chlorides, phosphates, lactates, potassium sorbates, glutamate, glutamic acid, calcium ascorbate etc.

c) Nitrite reduced meat products

Due to health concern, nitrite reduced meat products are produced by elimination or reduction of nitrite in meat products and use of nitrosamine inhibitors. The possible alternatives for nitrite in meat products are erythrosine, betalines, n-nitrosocystine, p-hydroxy benzoic acid, EDTA salts, propyl gallate, ascorbyl palmitate, lactic cultures, nisin, acid phosphates, α-tocopherols, ascorbates, erythorbates, sorbic acid etc. Cassens, 1997). However, the search for viable alternative to nitrite is still ongoing. It is impossible to find any single compound capable of replacing the functions of nitrite in meat products. The solution, must therefore, be necessary to combine several compounds, which together have a cumulative effect on colour, flavour and antimicrobial activity.

d) Meat foods with designer food ingredients

Food components designed to modulate the human body and cure or prevent diseases are called designer food ingredients. The well known and developed designer food ingredients fall into five major categories, namely phytochemicals, prebiotics, polyunsaturated fatty acids, bioactive ingredients, and probiotics (Table-3).

e) Designer egg foods

The egg industry is in the process of developing new designer eggs. A small percentage of eggs (3-5%) sold are "Designer eggs" because they contain nutrients or added factors that set them apart from generic eggs. The possibilities for designer eggs are many (Table-4) including

a) Nutritionally Enhanced

b) Value Added

c) Added Processing

Table 3: Some examples of Designer (Low Fat) Meat Products

Meat products (% fat)	Low fat ingredients (% addition)	Low fat meat products (% fat)
Beef Patties (20%)	Carrageenan (0.5%)	
Pork Frankfurter (30%)		812
Bologna (30%)	Oat fiber (3.5%)	10
Beef patties (20%)	Soy isolates (2%)	10
Frankfurter (30%)	Malto-dextrin (2%)	12
Frankfurter (30%)	Olive oil (12%)	10
Smoked sausages (30%)	Whey protein (6%)	6

Nutritionally enhanced

Eggs with modified fat content are very popular and in many parts of the country because these contains modified fat. These are marketed as containing less cholesterol, less saturated fat, higher amounts of omega-3 fatty acids, higher vitamin E content, and high amounts of iodine when compared to generic shell eggs. The reported cholesterol level of modified fat eggs is 190 mg per egg in comparison to 215 mg in generic eggs, and the saturated fat content is 1.0 g compared with 1.5 g in generic eggs. Hens are typically fed all grain diets supplemented with canola oil, bran; kelp and vitamin E (Shallo, 2003).

b) Organic eggs

Organic and free range eggs are marketed as value added eggs and are available in many parts of world. Everything that effect the production of the egg must be certified organic for the egg to be considered organic, from the hen's feed to the amount and type of space the hens live in, to the chemicals in the wash water. Organic eggs have the same nutrient content as generic eggs if the feed is of the same quality. Free range eggs are produced by hens that are not kept in cages. Hens have daily access to outdoors, depending on the climate. The nutrient content of free range eggs is identical to generic eggs if the quality of the feed is equivalent. Free range eggs are not nutritionally enhanced unless the feed is modified (Joshua et al., 2006).

c) Processing

Pasteurization has been used for years for liquid egg products; however, the technology to pasteurize a whole shell egg is relatively new. The process involves heating eggs to specific temperatures in water baths designed to heat the egg enough to destroy pathogenic bacteria but not enough to coagulate the proteins. The eggs are then cooled and packaged.

Table 4: Comparison of fat and antioxidant levels in designer and generic eggs

Component	Generic Egg	Modified (Omega-3) egg
Total fat (g)	4.5	4.5
Saturated fat (g)	1.5	1.5
Linolenic acid (mg)	17	1000
DHA (mg)	18	100-150
Total omega-3 (mg)	33	100-150
Linoleic acid (mg)	500	100
Cholesterol (mg)	213	213
Vitamin EU (IU)	1.1	6.0

(Shallo, 2003)

D) Designer milk

Nutritional and genetic interventions to alter the milk composition for specific health and/or processing opportunities are gaining importance in dairy biotechnology. Altered fatty acid and amino acid profiles, more protein, less lactose and absence of β-lactoglobulin are some challenges of 'designing' milk for human health benefits. Alteration of primary structure of casein and lipid profile, increased protein recovery, milk containing nutraceuticals and replacement for infant formula are some of the processing advantages envisaged (Sabikhi, 2004), (Table-5).

Types of modifications include: Milk with Modified Fat, Milk with Modified Protein, and Milk with Modified Carbohydrate

a) Milk with modified fat

The level of saturation of milk fat can be decreased. By feeding cow with highly unsaturated oils (e.g. soybean oil) caused depression in milk fat but increased the proportion of USFA to SFA in milk.Conjugated linoleic acid (CLA) levels can be increased by feeding cows on a diet rich in linoleic acid and grass. CLAs reportedly suppress carcinogens, inhibiting proliferation of leukemia and cancers of colon, prostate, ovary and breast. They are anti-atherogenic, anti-diabetic, enhance immunity and improve bone mineralization. Cows fed with fish oil, fishmeal or plankton can produce milk rich in omega-3 fatty acids (Bauman et al., 2001and Eynard et al, 2003).

A "designer cow" called Daisy which can produce semi skimmed milk has been bred by Britain's Agricultural Development and Advisory Services (ADAS) in collaboration with Samsbury's, the retail grocery chain (Murthy et al, 2002).

b) Milk with modified protein

Modification of milk composition through transgenesis is a promising way for improving existing products and extending the uses of milk components. A human like milk enriched with human lysozyme and lactoferrin might enhance defense against GI infections and promote iron transport in the digestive tract. It might also be possible to produce antibodies in the mammary gland that are capable of preventing mastitis infection or antibodies that aid in preventing of human diseases. Elimination of β-lacto globulin from cow's milk is unlikely to have any detrimental effects, either on cow or human formula and might actually overcome many of the major allergy problems associated with cow milk (Murthy et al, 2002).

One of the major products of the mammary glands being protein, exciting research and technology opportunities extends the frontiers for better protein supplementation. Improved amino acid profile by the addition of L-leucine, L-taurine and L-phenylalanine offers additional nutritional benefits. High value, low volume therapeutic proteins can be produced in the milk of domestic animals transgenic animals can also secrete proteins such as blood clotting factors needed by human hemophilia-sufferers in their milk.

c) Milk with modified carbohydrate

Some infants and even adults are not able to tolerate milk but can assimilate dahi. This is primarily due to lactose intolerance. To combat this problem, low lactose milk can be produced via transgenesis or hydrolysis. Hydrolyzed low lactose products are developed and use of fermented milk products is promoted for lactose intolerant people (Sabikhi, 2004).

Table 5: Potential Changes In Milk Through Genetic Engineering

Potential Modification	Change in Milk
Increase Casein Content	Increased protein, better cheese manufacturing properties
Engineered Casein	Better cheese manufacturing properties, increase recovery
Remove β-lactalbumin	Decreased milk allergies
Remove fat	Easier to produce low fat milk products, decrease butter surplus
Produce β-galactosidase, antibodies of pathogens	Safer food, mastitis prevention

(Murthy et al, 2002)

Status and Food Safety Laws in India

The world over India is among the fastest growing market, by either introducing new brands or repositioning existing ones with functional enhancement. These include main stream food and drink companies' viz. Kellogg, Unilever, Nestle, Tropicana, Danone; Pharma and dietetic companies viz. Novartis, Abbott, Smith Kline Beecham, and Johnson and Johnson (Aiyar, 2001). The last couple of years have witnessed the introduction of some nutraceuticals as well as functional foods in the Indian market also, either developed or manufactured indigenously or imported from abroad. But lack of scientific substantiation to prove designer food claims poses problem in promotion of this market. Variation in definition and product category and difficulties in food regulation and control are other big challenges.

Although functional or designer foods may indeed play key roles in achieving and maintaining health, it is critical that their manufacturers are allowed to make "well established and justifiable claims" only.

India passed the Food Safety and Standards Act to integrate and streamline many regulations covering nutraceuticals and foods and dietary supplements. At present, there is no comprehensive law and regulation to regulate manufacturing, processing and marketing of designer foods in general. Legal definition and regulation of these categories of foods are yet to be developed (Anandh et al, 2003).

References

1. Aiyar, A.S. (2001). Functional foods –from medicine chest to kitchen cabinet. Indian Food Indstry, 20:8-12.
2. Akoh, C.C. (1998). Fat replacers, Food Technology, 52:47-53.
3. Anonymous (1999) U.S. Patent No. 5650558, Glutenin genes and their uses. Patent application No.08/785,716.
4. Anandh, A. M. Lakshamanan, V. and Anjaneyulu, A. S. R. (2003). Designer meat foods. Indian Food Indstry, 22:40-45.
5. Bauman, D.E. Corl, B. A. Baumgard, L. H. and Griinari, J. M. (2001). Conjugated linoleic acid and the dairy cow. In Recent Advances in Animal Nutrition (eds Garnsworthy P C and Wiseman J). Nottingham University Press, Nottingham, UK, pp. 221-50.
6. Carlye, A. (2004). Designer fruit. Fortune Small Business 12:24-46.
7. Cassens, R. G. (1997). Residual nitrite in cured meat. Food Technology, 51:53-55.
8. Cassens, R. G. Greaser, M. C. Ito, T. and Lee, M. (1979). Reactions of nitrite in meat. Food Technollogy, 33:46-57.
9. Eynard, A. R. and Lopez, C. B. (2003). Conjugated linoleic acid versus saturated fats/ cholesterol: their proportion in fatty and lean meats may affect the risk of developing colon cancer. Lipids Health Dis, 2: 6, www.pubmedcentral.nih.gov.htm
10. Giese, J. (1996). Fat and fat replacers: balancing and health benefits. Food Technology 50: 76-78.
11. Joshua, R. M. Sudheer, K. P. and Sinija, V. R. (2006). Designer foods. Processed Food Industry, 10: 28-36.
12. Khanizadeh, S. Ehsani-Moghaddam, B. Tsao, R. Desjardins, Y. and Gosselin, A. (2004). Strategy for breeding fruits with higher content in bioactive compounds, ISHS Acta Horticulture 744: I International Symposium on Human Health. Effects of Fruits and Vegetables. www.actahort.org/books/744/744_23.htm
13. Lyons, P. H. Kerry, J. F. Morrissey, P. A. and Buckley, D. J. (1999). The influence of added whey protein, carrageenan gels and tapioca starch on the textural properties of low fat pork sausages. Meat Science, 51:43-52.
14. Manjula, K. and Suneetha, C. (2011). Designer foods-their role in preventing lifestyle Disorders. International Journal of J Science and Nature, 2(4):878-882.
15. Murthy, G. L. N. and Kanawjia, S. K. (2002). Designer milk: nutritional and technological significance. Indian Dairyman, 54:49-58.
16. Rajasekaran, A. and Kalaivani, M. (2013). Designer foods and their benefits: A review. Journal of Food Science and Technollogy, 50:1-16.
17. Sabikhi, L, (2004). Designer Milk –an imminent milestone in dairy biotechnology. Current Science, 87:1530-35.
18. Shallo, H. E. (2003). Designer foods: egg products. enc@enc-online.org
19. Suraim P. F. McPhersonm A. Speake. B. K. and Sparksm N. H. C. (2000). Designer egg evaluation in a controlled trial. European Journal of Clinical Nutrition, 54:298-305.
20. Surai, P.F. and Sparks, N.H.C. (2001). Designer eggs: from improvement of egg composition to functional food. Trends in Food Science and Technollogy 12:7-16.
21. Varshney, S.C. (2002). Role of functional foods in diet. Indian Food Industry, 21:41-43.

CHAPTER 3

Utilization of Milk Byproduct (Whey) for Preparation of Beverages: A Review

P.A. Pawase, D.M. Choudhari, V.G. Gaikwad and S.S. Jundhare

Introduction

Whey is the serum or watery part of the milk that remains after separation of curd that result from the acid or proteolytic enzyme mediated co-agulation of milk. It is major by product of dairy industries manufacturing products like channa, paneer, cheese and chakka (Sharma and Bhatia, 1999).The current world production of whey is estimated at about 165 million tones, in that cheese whey accounts for nearly 95 % of total whey. In India near about one million tonne of whey is annually derived, in that major source of whey is from the production of channa, paneer and shrikhand production. This is mostly thrown away as a waste which creates environmental pollution, also in that way about 70, 000 tonnes of Nutritious solids are lost (Puranik, 1999).

Whey represents between 85-90 percent of original milk volume processed and contains nearly 50 percent of total solids. It contains all the continents of milk, expect casein, fat and fat soluble vitamins. It is a source of carbohydrates, minerals, proteins and water soluble vitamins. Whey is also rich in calcium, phosphorus, essential amino acids (Parekh, 1997 and Ghosh et al., 1995). It is a very dilute by-product stream with only 6-7% solids and one-fifth of the initial milk proteins present. It contained most nutritious whey protein viz., α-lactolabumin, β-lactalbumin, serum albumin and immunoglobulin, lactose, milk salts (Zadow, 1986). Whey is received from about 80-92 per cent of the volume of milk used for conversion in to different product. It retained about 45 to 55 per cent of the milk nutrients derived whey (Horton, 1995). Most of the whey solids are usually drained off causing great loss of valuable nutritious solids and creating problem of environmental pollution. Since whey contains 6.0 to 7.0% solids and has a biological oxygen demand (BOD) of over 32,000 mg/liter, it can not be discharged to waste treatment systems used for treating dairy plant wash waters. These limitations beg the question. Therefore, many dairy organizations treat the whey before disposal, which is rather expensive (Gandhi, 1989). To avoid pollution as well as utilization of nutritious solids needs to converts whey in to various beverages.

Nutritional and Therapeutic Importance of Whey

Prendergast (1985) reported following important benefits of whey. The Chemical constituents of different whey system are shown in Table 1. Whey has been an important food. Whey drinks are light are refreshing, less acidic than fruit juice, they provide health and nutritional benefits and good source of carbohydrates, minerals and protein. The father of modern medicine, Hippocrates, the legendary Greek physician insisted on a daily dose of two liter of whey. By the year 1600 whey was used to treat jaundice infected lesions of skin, arthritis, anemia, liver, complaints and epilepsy (Prendergast, 1985). It also posses almost all the electrolytes of Oral Rehydration Solution (ORS), which is invariably used to control dehydration (Anonymous 2006).

Table 1: Chemical constituents of different whey system

Sr. No.	Types of whey	Chemical constituents							References
		Fat	Protein	Acidity	Ash	TS	Lactose		
		(%)	(%)	(%)	(%)	(%)	(%)	pH	
1.	Chakka whey	0.01-0.39	0.35-1.03	0.55-1.13	0.33-0.65	4.98-7.03	3.97-5.26	3.94-5.06	Kulkarni (1984)
2.	Channa whey	0.54	0.37	0.46	0.53	6.9	5.08	-	Singh and Mathur (1973)
3.	Paneer whey	0.25	0.45	0.20	0.63	6.75	5.42	-	Kapil (1991)
4.	Cheese whey	0.1-1.63	0.57-0.92	-	0.46-0.80	-	3.50-5.57	4.6-6.6	Racotta et al. (1979)

Whey Beverages

A. Fermented whey beverages

Fermentation is one of the oldest methods of food preservation and biological upgrading of dairy by-product in to value added products. Whey has been found to be useful as a base for growth of some of the lactic acid bacteria in the preparation of beverage. These beverages are obtained due to lactic acid bacteria fermenting lactose in whey to form mainly lactic acid. Lactic acid imparts fresh flavour and is known suppress/ inhibit the growth of spoilage pathogenic organism. Antibacterial substances produced during fermentation are reported to have curative properties in controlling of the several gastro-intestinal disorders (Parasher and Gandhi 1990).

According to Anonymous (2006), processing of whey into fermented whey beverage having several advantages which are as below:

- The composition of whey makes it a suitable base for fermentation.
- It is a genuine thirst quencher unlike most of the soft drinks.
- It has a drink can replace much of the lost organic and inorganic salts to the extra cellular fluid.
- Whey is rapidly absorbed due to minute quantity of fat emulsion.
- Due to fermentation of whey with LAB, it becomes a suitable drinks lactose intolerant people.

Procedure of fermented whey beverage given by (Gandhi 1984)

Procedure of Acido-whey has been developed at NDRI, Karnal. In this process whey is obtained in the production of cheese, channa, or casein.

Whey
↓
Fat separation/Clarification
↓
Heat treatment (85-90˚C for 10 min.)
↓
Cooling (40˚C)
↓
Inoculation with Starter culture (L. acidophilus)
↓
Incubation (20-22 hr.)
↓
Filtration/Clarification
↓
Adding sweetening agent and flavour (Desirable amount of pine apple flavour or any compatible citrus flavour)
↓
Pasteurization if desire
↓
Packing
↓
Cold storage 4-7˚C

B. Fruit pulp/juice/flavour based whey drinks

Since, last ten years India progress in milk production as well as emerged as the top fruit producer in the world, but distressing aspect is that as much as fruit produced in India get spoiled due to lack of infrastructure facilities as well as appropriate post harvest technology. So formulation of a new product using suitable combination of whey and fruit juices both in concentrated and dried firms would permit economic utilization of whey and value addition to high income. Table 2 represents various whey based fruit drinks.

According to Khamuri and Rajori (1998) the simplest whey based fruit drinks are manufactured by mixing predetermined fruit pulp or juice and minimally processed whey. Beverages of this type have been formulated in different parts of the world. Acid whey is most suitable for blending with fruit juices to produce cloudy, still fruit drink. A pH of less than 4.0 is necessary to prevent protein coagulation during pasteurization. The pH reduction can be achieved by addition of lactic or other organic acid.

Gagarin, et al. (1987) prepared fruit flaoured whey beverages with 0.5 % acidity and 200 brix can be prepared and mixed with orange, pineapple, guava and mango juice at 10, 15, 20 and 25 percent of whey. Higher sensory scores are usually obtained using 10 percent orange, 15 percent pineapple, 25 percent guava and 15 percent mango juice in to whey.

Table 2: Various products from whey by using different fruit pulp/ juices:

Sr. No.	Types of whey	Products	References
1.	Channa whey	Mango RTS Beverage	Sikdar and Ghatok (2003)
2.	Paneer whey	Kokum juice	Rupnar (2006)
		Pineapple juice	Kesarkar (2002)
		Guava whey beverage	Singh and Kapoor (1999)
3.	Cheese whey	Kinnow juice Beverage	Khamrui and Rajorhia (1998)
		Chocolate drink	Sudhir Singh (1991)
4.	Chakka whey	Fruit jelly	Joshi (1985)
		Mango whey beverage	Shende (2006)

The technology developed for incorporation of fruit juices/pulp in whey beverages products by different research workers are shown in process diagram below:

a. Kokam juice

Paneer whey
↓
Cooling at room temperature
↓
Defatting
↓
Deprotenization
↓
Cooling at room temperature
↓
Addition of sugar (15 %)
↓
Addition of jerra powder (0.1 %)
↓
Addition of Kokam juice
↓
Addition of Colour
↓
Mixing and bottling
↓
Refrigeration and storage(5-7 °C)

Rupnar (2006)

b. Mango whey beverage

Chakka whey
↓
Neutralization (0.75 %)
↓
Addition of sugar (14 %)
↓
Pasteurization at 60°C for 30 min
↓
Cooling to room temperature
↓
Addition of mango pulp
↓
Mixing
↓
Filtration
↓
Bottling
↓
Storage

Shende (2006)

c. Kinnow juice

Cheese whey
↓
Filtration
↓
Fat separation at 45°C
↓
Transfer to processing wat
↓
Addition of sugar
↓
Warming and stirring sugar
↓
Kinnow pulp addition
Mixing
↓
Adjustment of pH
↓
Addition of stabilizer (pectin-0.05 and CMC-0.15)
↓
Heating (90°C/10sec)
↓
Cooling to 4°C
↓
Cold storage (4°C)

Khamrui and Rajorhia (1998)

d. Chocolate drink

Cheese whey
↓
Heat treatment
↓
Removal of clear supernatant of whey
↓
Blend of whey and milk (equal volume)
↓
Heating
↓
Homogenization
↓
Mixing sugar (60 g/lit) stabilizer (0.5 g/lit)
↓
Filtration
↓
Addition of coca powder (5 g)
↓
Packaging in bottle pasteurization
↓
Cold storage

Sudhir (1991)

e. Whey based pineapple RTS flow chart (Kumar and Manimegali 2003)

Whey
↓
Dissolving of sugar
↓
Filtering

Mature pineapple
↓
Removing of crown
↓
Peeling of outer skin along with eyes
↓
Cutting in to small pieces
↓
Pulping in mixer 10 min.
↓
Pineapple juice
↓

Blending
(Whey+ pineapple juice 90:10)
↓
Heating the blend 800C
↓
Filling in sterilized bottles (head space 1 inch)
↓
Corking
↓
Pasteurizing in boiling water (30 min)
↓
Cooling
↓
Drying surface
↓
Labeling
↓
Storing (Refrigerator)

C. Whey based sport drink

Wagner, et al., (1975) reported that whey drinks could stabilize the osmolar system in the body and had a thirst quenching effect. Whey being rich in lactose and minerals serve as base for development of sports as well as electrolyte drinks of varying composition. Further more addition of other additives and process of carbonation may provide thirst quenching properties and consumer appeal. According to Beucler, et al., (2005) beverages made with 25-50% substitution of hydrolyzed or un hydrolyzed whey permeate were similar to commercial beverages in visual and flavour characteristics than those made with higher level of substitution. All beverages made with whey permeate were higher in electrolytes (Na, K, Zn, Mg, P) compared with commercially available sport drinks.

Due to high amount of proteins with high nutritional value these beverages are ideal source of energy and nutrients for athletes. Whey proteins are a rich source of branched chain amino acids (BCAA) like isoleucine, leucine and valine. BCAAs unlike other essential amino acids are metabolized directly into the muscle tissue and are first amino acids used during periods of exercise and resistance trainings (Sherwood et al., 2007).

Kumar (2008) prepared the best formulation pineapple sport beverage with 20% pineapple juice, 7.49 % sugar, 0.15% stabilizer mix and 0.12% salt. The stored sport drink was acceptable up to 6 months, and acceptable lemon based sport drink can be made with 7.43% sugar, 301% lemon juice, 0.07% lemon flavour and 0.07% CMC add to whey 3.01%.

Conclusion

Present paper leads to conclude that the whey derived from milk is economically converted in to different beverages like kokam juice, pineapple flavour juice, kinnow RTS beverage, mango based RTS, chocolate drink, cold coffee drink, by addition of different fruits, nuts, flavours etc. Using whey we can save the environmental pollution and nutritious solids losses in whey as waste. The whey with fruit juice combinations seems to hold good promises in the manufacture of value added nutritious beverages. Such beverages have been found to be highly acceptable as refreshing drink.

References

1. Beucler, J. Drake, M. and Foegeding, E.A. (2005). Design of beverage from wheypermeate. Journal of Food Science, 70 (4):5277-5285.
2. Gandhi, D.N. (1989). Whey utilization for Beverage Production Indian Dairyman. 41: 35.
3. Horton, B.S. (1995). Whey processing and utilization. Bull. IDF. 308:2.
4. Joshi, P.S. and Waghmare, P.S. (1985). Utilization of curd whey for the preparation of fruit jelly. Indian Food Packer, 39(2):38-42.
5. Kapil, V. (1991). Manufacture of lactose: Effect of processing parameter on yield and purity. Journal of Food Science, 23(3):167.
6. Kesarkar, T.V. (2002). Studies on preparation pineapple flavoured paneer whey beverage. M.Sc. (Agri.) Thesis, M.P.K.V., Rahuri.
7. Khamrui Rajorhia, G.S. (1998). Formulation of ready to serve whey based kinnow juice beverage. Indian Journal of Dairy Science, 51(6):413-419.
8. Kulkarni, M.B. (1984). Studies on chemical composition and utilization of chakka whey. Ph.D. Thesis submitted to M.P.K.V., Rahuri.
9. Kumar, R.S. and Manimegalai (2003). Astudy of storage behaviour of whey based pine apple juice RTS beverage. Indian Food Maker, March-April pp. 51-55.
10. Parekh, J.V. (1997). Milk processing and downstream products science and technological needs. Beverage and Food World, Nov. 53.
11. Prendergast, K. (1985). Whey Drinks – Technology of Processing and Marketing. Journal of the Society of Dairy Technology, 38: 103.

12. Puranik, D.B. (1999). Whey and whey protein in Geriatic Nutrition. Indian Dairyman, 51(5): 5-10.
13. Racotta, V., Bovrges, H. and Navareete, A. (1979). Use of whey proteins for supplementing tortilla. Journal of Agricultural and Food Chemistry, 27(4):668-672.
14. Rupnar, P.S. (2006). Studies on preparation of paneer whey beverage using Kokam juice. M.Sc. (Agri.) Thesis, M.P.K.V., Rahuri.
15. Singh, A. K. (2008). Technological development in whey based non fermented beverages and soups. Course Compendium on Technological advances in the utilization of dairy byproduct. pp. 91-100.
16. Sharma, A. and Bhatia, K.L. (1999). Functional properties of whey protein concentrate – A Review. Journal of Dairying, Food and Home Science, 18(3and4):141.
17. Shende, D.H. (2006). Studies on preparation of chakka whey beverage using mango pulp. M.Sc. (Agri.) Thesis, M.P.K.V., Rahuri.
18. Sikdar, B. and Ghatak, P.K. (2003). Formulation of mango whey beverage. Beverage and Food World. Dec. 2003. pp. 11-12.
19. Singh, G.P. and Mathur, M.L. (1973). Channa whey its composition. Indian Journal of Dairy Science, 41(3):322-325.
20. Sudhir. S. (1991). Utilization of cheese whey for the preparation of a chocolate Drink: Beverage and Food World, pp. 25.
21. Singh, W. and Kapoor, C.M. (1999). Standardization of Technology for the manufacture of Guava Whey Beverage. Indian Journa od Dairy Science,52(5):1999.
22. Sherwood, S. and Jenkins, D. (2007). US Patent US 2007/0178214 A1.
23. Zadow, J.G (1986). Utilization of milk components: Whey in Modern Dairy Technology. Vol. J.R.K. Robinson (ed.) Elsevier Applied Sci. Publ. London, U.K. pp. 273.
24. Wagner, K.H. Albig, K. and Maicold, K.G (1975). The effect of whey basesd drinks on osmolarity. Milchwissenschatt.

CHAPTER 4

Role and Overview of Strawberry as a Potential Source for the Development of Functional Foods and Nutraceuticals

Aamir Hussain Dar, Muneer Ahmad, H.K. Sharma and Aabida Jabeen

Abstract

Research has demonstrated that nutrition plays a crucial role in the prevention of chronic diseases as most of them can be related to diet. Functional food enters the concept of considering food not only necessary for living but also as a source of mental and physical well-being contributing to the prevention and reduction of risk factors for several diseases or enhancing certain physiological functions. One of the most clear objectives of the study of functional foods is to identify and validate biological markers in order to understand the interaction mechanisms between the food component and a body function. Epidemiologic and laboratory data from previous studies suggests that diets rich in fruits and vegetables are associated with a reduced risk of various types of cancers. The reduction in cancer risk may be attributed to the abundance of a family of specific compounds referred to as anthocyanins, which are found in various berries. Anthocyanins are water soluble pigments responsible for the blue, purple, and red color in many plant tissues Strawberry is one of "super fruits", since they can form a significant component of the diet and are a potential resource of functional foods and nutraceuticals due to their antioxidant capacity and nutritional quality. Strawberries contain high levels of anthocyanins and they are also well liked by consumers. Popularity of strawberries in the consumer market makes them an ideal focus for chemoprevention studies. Strawberries are a rich source of phytochemicals (anthocyanins, catechins, the flavonols quercetin and kaempferol, and ellagic acid) and an important source of essential nutrients (ascorbic acid, potassium, folic acid, carotenoids, B-vitamins). Strawberries are consistently ranked among the top food sources of polyphenols and antioxidant capacity; the preventive and therapeutic health benefits of strawberries may be attributed to the synergistic effects of these bioactives and the nutrients contained in the fruit. These bioactive photochemical contribute the major health-promoting function. Strawberries have shown a remarkably high scavenging activity toward chemically generated radicals, thus making them effective in inhibiting oxidation of human low-density lipoproteins. Antioxidant activity of strawberries could contribute to the prevention of cancer, cardiovascular and other chronic diseases. Hence strawberry can be used as a vital and indispensible functional ingredient for the development of functional foods and nutraceuticals.

Introduction

Functional foods are generally considered as those food products which provide a specific health benefit over and above their basic/traditional nutritional value (Salam, 2010; Buttriss, 2010). Functional foods were originally developed to identify and correct nutritional deficiencies, for example, breakfast cereals with folic acid, but many have now gone further to become food products that improve physical and mental well-being (Siro et al, 2008) and reduce the risk of chronic diseases (Abdel-Salam, 2010) or even manage disease (Hasler et al, 2000). The component that the functional food contains that provides the additional health benefit (the "function") can be a non-nutrient, or a nutrient (FAO, 2007). An example of a non-nutrient would be plant sterols to reduce cholesterol.

In the early twentieth century, nutrition science was preoccupied with preventing deficiencies and supporting body growth. Following World War II, although nutrient sufficiency and basic nutritional adequacy remained important, the economic growth enjoyed by numerous countries fostered a more hedonistic approach to food consumption, with a focus on preferences and aversions (Arai, 2002). This led, in the last third of the century, to the recognition by nutritionists that consumption of some food components could have a negative impact on health, if they were consumed in sufficient quantities (Ashwell, 2003). Today, the focus of nutritional science is shifting toward the concept of optimal nutrition, the objective of which is to optimise the daily diet in terms of nutrients and non-nutrients. The development of functional foods appears to be a long-term trend with important market potential. According to Fitzpatrick (2003), the market for functional foods is expanding by 7-10 per cent per year.

The new concept of functional foods raises concerns about food security and the proper selection of an adequate diet, beyond the classical adequacy in energy, protein, essential fats, vitamins and minerals. Today, it is recognized that foods not only provide basic nutrition, but can also prevent diseases and ensure good health and longevity. This concept is not new to people in some parts of the world – they have always believed that certain foods are beneficial to health and some even therapeutic. But these beliefs were primarily anecdotal, based on centuries of tradition and largely lacking scientific bases. Nutraceuticals was introduced also in 1989 by the U. S. Foundation for Innovation in Medicine to refer to "any substance that is a food or a part of a food and provides medical or health benefits, including the prevention and treatment of disease" (DeFelice, 1995). In 1994, the U. S. Institute of Medicine's Food and Nutrition Board defined functional foods as "any food or food ingredient that may provide a health benefit beyond the traditional nutrients it contains" (Thomas and Earl, 1994).

Health Benefits of Functional Foods and Nutraceuticals

Several factors, scientific advances, consumer demand, increasing health care costs, an aging population, technical advances in the food industry, and changing regulatory environment have stimulated the field of functional foods (Hasler, 1996). Scientific investigations have resulted in the accumulation of scientific evidence supporting the vital role of diet in overall health and well-being. For instance, six of the ten leading causes of death in the United States – cancer, coronary heart disease, stroke, diabetes, atherosclerosis, and liver disease have been increasingly shown to be related to diet. The positive relationship between diet and health has increased consumer demand for more information related to healthy diets, including fruits and vegetables, with functional characteristics that help to delay the aging processes and reduce the risk of various diseases, mainly cardiovascular diseases and cancer, as well as other disorders (Paredes-Lopez et al., 2010). During the last years various studies have targeted the characterization and utilization of bioactive compounds in, in functional foods, like berries, and their benefic properties for human health based on clinical research. The World Health Organization (WHO) emphasizes the importance the bioactive compounds, especially from small colorful fruits for prevention of the most important health problems namely cardiovascular diseases, diabetes, cancer, and obesity (Stapleton et al., 2008) functional characteristics that help to delay the aging processes and reduce the risk of various diseases. These physiologically active compounds, referred to simply as 'phytochemicals', are produced via secondary metabolism in relatively small amounts. Until fairly recently, they have been generally assumed to be irrelevant, sometimes even harmful, to human health. The number of identified phytochemicals has increased dramatically in the last decade. Some groups of phytochemicals which have or appear to have significant health potentials are: carotenoids, phenolic compounds (flavonoids, phytoestrogens, phenolic acids), phytosterols and phytostanols, tocotrienols, organosulfur compounds (allium compounds and glucosinolates), and nondigestible carbohydrates (dietary fiber and prebiotics).

Bioactive Compounds in Berries and their Functional Properties

The metabolic activity of bioactive compounds is manifested mainly by the scavenging ability of reactive oxygen species (Szajdek and Borowska, 2008). Fruits like berries may be an important component of a healthy diet because of their bioactive compounds content. Important berries such as: blackberry (*Rubus* sp.), bilberry (*Vaccinium myrtillus*), blackcurrant (*Ribes rugrum*), blueberry (*V. corymbosum*), chokeberry (*Aronia melanocarpa*), cranberry (*V. macrocarpon*), bayberry (*Myrica* sp.), raspberry (*R. ideaus*), black raspberry (*R. occidentalis*), and strawberry (*Fragaria ananassa*) are important sources of bioactive compounds usually consumed in fresh and processed products in the human diet (Cote et al., 2010).

Among the chemical compounds found in this type of fruits are the phenolic compounds, being the major group of phytochemicals including flavonoids (anthocyanins, flavonols, flavones, flavanones, flavanols, and isoflavonoids), stilbenes, tannins, and phenolic acids (Tulipani et al.,2008). Diverse studies have shown that biotic and abiotic factors play an important role in the levels of bioactive compounds in berries and their antioxidant activity (Zhang et al., 2008). The phenolic compounds contribute to protection against degenerative diseases, and their effects on health have been commonly attributed to their antioxidant properties (Seeram, 2008). Improvement in knowledge on their bioavailability is an important step to understand the possible mechanisms of action and their impact on human health (Szajdek and Borowska, 2008). Strawberries were found to contain a wide variety of phenolics such as ellagic acid, ellagic acid glycosides,ellagitannins, gallotannins, flavonols (quercetin and kaempferol glucuronides, and glycosides), anthocyaninsm(pelargonidin and cyanidin glycosides), flavonols and coumaroyl glycosides (Zhang et al., 2008). Ellagic acid is the predominant acid in strawberries where it accounts for 51% of all acids found in this fruit (Aiyer et al., 2008).

The total content of ellagic acid determined after acid hydrolysis ranged from 25.01 to 56.35 mg/100 g fresh weight (Gülçin et al., 2011). Tannins are an important group in berries. They comprise both condensed tannins, known as proanthocyanidins, and esters of gallic and ellagic acid — defined as hydrolysable tannins (Bautista-Ortín et al., 2012). Hydrolysable tannins are less frequently encountered and have been found in strawberries, raspberries and blackberries (Holt et al., 2008). Small quantities of trans-resveratrol were also found in strawberry. The content of trans-resveratrol in strawberry fruits reaches is up to 3.57 μg/g (Ehala et al., 2005). Berry fruits contain small quantities of carotenoids (Marinova and Ribarova, 2007). The chemical profile of phenolic compounds from strawberries has been studied using liquid chromatography electro spray ionization mass spectrometry (LC-ESI–MS) methods (Cote et al., 2010). Berries are not only delicious, low energy food, but also rich sources of fiber, antioxidant vitamins and various phenolic compounds (flavonoids and phenolic acids).

The main classes of flavonoids in berries are anthocyanins, proanthocyanidins, flavonols and catechins (Wang et al., 1997). Phenolic acids present in berries are hydroxylated derivatives of benzoic acid and cinnamic acid (Torronen, 2000).Various potential health benefits from berries have been attributed to flavonoids and phenolic acids. Strawberry and blackcurrant pomace contain polyphenols, which may inhibit the formation of toxic products of Maillard reaction during baking, e.g., neurotoxic and carcinogenic acrylamid or one of the end-products of advanced glycation N (carboxymethyl) lysine (Hedegaard et al. 2008; Peng et al. 2010). Because of these characteristics, products with the addition of fruit pomace may be classified as functional food, which exerts additional positive influence, except for its standard nutritional role (Siro et al. 2008).While there is a rapid growth in the market of functional products, the baking sector has developed poorly in the last decade. On the other hand, the major contribution to the antioxidant capacity of berries of the genus Rubus and Fragaria are the ellagitannins, which represent between 51% and 88% of all phenolic compounds depending on the maturity stage, in fully ripe red fruit concentration of this compounds can be 50% lower (Table-1) (Castrejón et al., 2008).

Table 1: Bioactive compounds in berries (Castrejon et al., 2008).

Berries	Bioactive compounds	Fresh weight (mg/g)
Bilberry	Phenolic compounds	525
	Flavonoids	44
	Anthocyanins	300
Raspberry	Phenolic compounds	121
	Flavonoids	6
	Anthocyanins	99
Blueberry	Phenolic compounds	585
	Flavonoids	50
	Anthocyanins	495
Strawberry	Phenolic compounds	313
	Flavonoids	–
	Anthocyanins	54

Berries of Vaccinium sp. have been reported to possess a wide range of biological activities. Cranberry and wild blueberry has been shown to prevent urinary tract infections. This protective effect has been attributed to condensed tannins or proanthocyanidins, which are thought to act as anti-adhesive agents preventing bacterial colonization (Howell et al., 1998; Ofek et al., 1996). In wild blueberry, proanthocyanidins have also been identified as the active agents inhibiting the in vitro promotion of chemically-induced carcinogenesis (Bomser et al., 1996). The presence of bioactive compounds such as flavonoids and ellagic acid derivatives makes the consumption of strawberry suitable for potential health benefits. Dietary intake of flavonoids has been associated to lower risk of heart disease as well as cancer, probably related to the anti-oxidative activity of these compounds.

Development of Functional Foods and Nutraceuticals from Strawberry

The bakery industry, production of which has been increasing steadily in the country, is the largest among the processed food industries in India. The two major bakery industries, viz. bread and biscuit/cookies account for about 82% of the total bakery products. The production of bread and biscuits/cookies in the country, both in the organised and unorganised sectors, is estimated to be around 0.44 million tonnes and 11 million tonnes respectively. The per-capita consumption of biscuits/cookies in India is around 900gms as compared to 15-20 kg for developed countries. Cookies are the most popular bakery products consumed by a wide range of populations in Kashmir Province. This is mainly due to its ready-to-eat nature, good nutritional quality and availability in different tastes, longer shelf life and relatively low cost.

The consumption of biscuits/cookies is equally divided between the urban and rural population. Bakery products are an important source of nutrients viz. energy, protein, iron, calcium and several vitamins. Cookies are amongst the lowest cost processed food in the country when compared to other Indian sweets and salted snacks. These are easy to use during travel or at home because of its availability in variety of pack sizes. They also offer substantial energy. Thus cookies have an important role to play as a diet supplement for both adults and children. It is no longer viewed as a luxury tea-time snack but essential daily food component for an average Indian household.

Strawberry contains large amounts of anthocyanins, the major pigment of strawberry. Particularly, pelargonidin 3-o-glucoside (P3G) is predominant and is responsible for the brights red colour of fresh berries, followed by cyanidin 3-glucoside (C3G), pelargonidin 3-rutinoside, pelargonidin 3-succinate (Bakker et al., 1994) pelargonidin 3-galactoside (Kikoku et al., 1995), and pelargonidin

3-malonylglucoside (Tamura et al., 1995) as minor anthocyanins, flavanols (catechins), and flavonols (quercetin and kaempferol glycosides) (Cordenunsi et al., 2005).

As it was evident from the recent studies, that the compounds in strawberries have potential power as follows: lower risk of cardiovascular by inhibition of LDL-cholesterol oxidation, promotion of plaque stability, improved vascular endothelial function, and decreased tendency for thrombosis, modulation of the inflammatory process, blocking initiation of carcinogenesis, suppressing progression and proliferation of tumors (Hannum , 2004) inhibition of proliferation of human liver cancer cells (Meyers et al., 2003), inhibition of esophageal cancer (Stoner et al., 1999), anticarcinogenic activities to breast and cervical cancer cells (Wedge et al., 2001) and inhibition of growth and stimulation of apoptosis of human cancer cells (Seeream et al., 2006). Moreover, the strawberry extracts has shown antithrombotic effect (Naemura et al., 2005), and inhibition of Helicobactor pylori with enhanced susceptibility to clarithromycin (Chatterjee et al., 2004). Cookies have been subjected to different physical, chemical, textural and micro structural studies. It was concluded that the consumer acceptance of cookies remained unaffected up to 4 % level of incorporation of strawberry powder. Thus it would help the manufacturers to develop various functional foods especially snack foods by utilizing strawberry as a main functional ingredient.

Conclusion

The incorporation of strawberry in various food products like cookies, frozen desserts etc can facilitate the development of nutraceuticals and functional foods.

References

1. Abdel-Salam A. M. (2010). Functional foods: Hopefulness to Good Health, American Journal of Food Technology, 5:86-99.
2. Aiyer, H. S. Srinivasan, C. and Gupta, R. C. (2008). Dietary berries and ellagic acid diminish estrogen-mediated mammary tumorigenesis " ACI rats. Nutrition and Cancer, 60(2):227–234.
3. Arai, S. (2002). Global view on functional foods: Asian perspectives, British Journal of Nutrition, 88(2):S139-43.
4. Ashwell, M. (2003). ILSI Europe Concise Monograph on Concepts of Functional Foods, The International Life Sciences Institute, Washington, DC.
5. Bakker, J. Bridle, P. and Bellworthy, S.J. (1994). Strawberry juice colour: A study of the quantitative and qualitative pigment composition of juices from 39 genotypes. J Sci Food Agric, 64:31-37.
6. Bautista-Ortín, A. B. Rodríguez-Rodríguez, P. Gil-Muñoz, R. Jiménez-Pascual, E. Busse-Valverde, N. Martínez-Cutillas, A.(2012). Influence of berry ripeness on concentration,

qualitative composition and extractability of grape seed tannins. Australian Journal of Grape and Wine Research, 18:123–130.

7. Bomser, J. Madhavi, D.L. Singletary, K. Smith, M.A.L. (1996). In vitro anticancer activity of fruit extracts from Vaccinium species. Planta Medica, 62:212-6.
8. Buttriss Judith, L. (2010). Are health claims and functional foods a route to improving the nation's health?, Nutritional Bulletin, 35:87-91.
9. Castrejón, A.D.R. Eichholz, I., Rohn, S. Kroh, L.W. and Huyskens-Keil, S. (2008). Phenolic profile and antioxidant activity of highbush blueberry (Vaccinium corymbosum L.) during fruit maturation and ripening. Food Chemistry, 109:564–572.
10. Chatterjee, A. Yasmin, T. Badchi, D. Stohs, S.J. (2004). Inhibition of Helicobacter pylori in vitro by various berry extracts, with enhanced susceptibility to clarithromycin. Mol Cell Biochem, 265:19-26.
11. Cordenunsi, B.R. Genovese, M.I. Nascimento, J.R.O. Hassimotto, N.M.A. Dos Santos, R.J. and Lajolo, F.M. (2005). Effects of temperature on the chemical composition and antioxidant activity of three strawberry cultivars. Food Chem, 91:113-121.
12. Cote, J. Caillet, S. Doyon, G. Sylvain, J. F. and Lacroix, M. (2010). Analyzing cranberry bioactive compounds. Critical Reviews in Food Science and Nutrition, 50:872–888.
13. DeFelice, S.L. (1995). The nutraceutical revolution: its impact on food and industry R and D. Trends Food Sci Technol, 6:59-61.
14. Ehala, S. Vaher, M. and Kaljurand, M. (2005). Characterization of phenolic profiles of Northern European berries by capillary electrophoresis and determination of their antioxidant activity. Journal of Agricultural and Food Chemistry, 53: 6484–6490.
15. Fitzpatrick, K. (2003). Functional foods and nutraceuticals: exploring the Canadian possibilities, Research Report, University of Manitoba, Winnipeg.
16. Gülçin, Ý. Topal, F. Çakmakçý, R. Bilsel,M. Gören, A. C. and Erdogan, U. (2011). Pomological features, nutritional quality, polyphenol content analysis, and antioxidant properties of domesticated and 3 wild ecotype forms of raspberries (Rubus idaeus L.). Journal of Food Science, 76(4):C585–C593.
17. Hannum, S.M.(2004). Potential impact of strawberries on human health: A review of science. Crit Rev Food Sci Nutr, 44:1-17.
18. Hasler Clare M, Kundrat Susan and Wool Deborah (2000). Functional foods and cardiovascular disease, Current Atherosclerosis Reports, 2(6):467-473.
19. Hasler, C.M. (1996). Functional foods: The Western perspective. Nutritional Review, 54:S6-S10.
20. Hedegaard, R.V., Granby, K., Frandsen, H., Thygesen, J. and Skibsted, L.H. (2008). Acrylamide in bread. Effect of prooxidants and antioxidants. Eur. Food Res. Technol. 227:519–525.
21. Holt, H. E. Francis, I. L. Field, J. J. Herderich, M. J. and Iland, P. G. (2008). Relationships between berry size, berry phenolic composition and wine quality scores for cabernet Sauvignon (Vitis Vinifera L.) from different pruning treatments and different vintages. Australian Journal of Grape and Wine Research, 14(3):191–202.
22. Howell, A.B. Versa, N., der Marderosian, A. Foo, L.Y. (1998). Inhibition of the adherence of P-fimbriated Escherichia coli to uroepithelial-cell surfaces by proanthocyanindin extracts from cranberries. New Eng J Med, 339:1085-6.
23. Kikoku, Y. Fukuhara, K. Saito, I. Oota, H. (1995). Identification and high performance liquid chromatographic determination of strawberry anthocyanin pigments. J Jap Soc Food Sci Technology, 42:118-123 (in Japanese).
24. Marinova, D. and Ribarova, F. (2007). HPLC determination of carotenoids in Bulgarian berries. Journal of Food Composition and Analysis, 20:370–374.

25. Meyers, K.J. Watkins, C.B. Pritts, M.P. Liu, R,H. (2003). Antioxidant and antiproliferative activities of strawberries. J Agric Food Chem, 51:6887-6892.
26. Naemura, A. Mitani, T. Ijiri, Y. Tamura, Y. Yamashita, T. Okimura, M. Yamamoto, J. (2005). Anti-thrombotic effect of strawberries. Blood Coag Fibrin, 16:501-509.
27. Ofek, I. Goldhar, J. Sharon, N. (1996). Anti-Escherichia coli adhesion activity of cranberry and blueberry juices. In: Kahane S, Ofek I (editors). Toward anti-adhesion therapy for microbial diseases. New York: Plenum Press; p. 179-83.
28. Paredes-Lopez, O. Cervantes-Ceja, M. L. Vigna-Pérez, M. and Hérnandez-Pérez, T. (2010). Berries: Improving human health and healthy aging, and promoting quality — A review. Plant Foods for Human Nutrition, 65:299–308.
29. Peng, X. Maa, J. Cheng, K.W. Jiang, Y. Chen, F. and Wang, M. (2010). The effects of grape seed extract fortification on the antioxidant activity and quality attributes of bread. Food Chem. 119:49–53.
30. Seeram, N.P. Adams, L.S. Zhang, Y. Lee, R. Sand, D. Scheuller. H.S. Heber, D. (2006). Blackberry, black raspberry, blueberry, cranberry, red raspberry, and strawberry extracts inhibit growth and stimulate apoptosis of human cancer cells in vitro. J Agric Food Chem, 54:9329-9339.
31. Seeram, N. P. (2008). Berry fruits: Compositional elements biochemical activities, and impact of their intake on human health, performance, and disease. Journal of Agricultural and Food Chemistry, 56:627–629.
32. Siro Istvan, Kapolna Emese, Kapolna Beata, Lugasi Andrea (2008). Functional food. Product development, marketing and consumer acceptance – a review, Appetite, 51:456-467.
33. Stapleton, A.P. James, E. M. Goodwill, G. A. and Frisbee, J. C. (2008). Obesity and vascular dysfunction. Pathophysiology, 15, 79–89.
34. Stoner, G.D. Kresty, L.A. Carlton, P.S. Sliglin, J.C. Morse, M.A. Isothiocyanates and freeze-dried strawberries as inhibitors of esophageal cancer. Toxicol Sci 1999,52 (Supplement):95-100.
35. Szajdek, A. and Borowska, E. J. (2008). Bioactive compounds and health-promoting properties of berry fruits: Review. Plant Foods for Human Nutrition, 63:147–156.
36. Tamura, H. Takada, M. Yoshida, Y. (1995). Pelargonidin 3-o-(6-o-malonyl-β-D-glucoside) in Fragaria x ananassa Duch. Cv. Nyoho. Bioscience Biotechnology Biochemistry, 59:1157-1158.
37. Thomas, P. R. (1994). Earl, R (editors). Enhancing the food supply. In: Opportunities in the nutrition and food sciences. Washington, DC: National Academy Press; p. 98-142.
38. Torronen R. (2000). Flavonols and ellagic acid in berries. Recent Advances in Agriculture and Food Chemistry, 1:31-7.
39. Tulipani, S. Mezzetti, B. Capocasa, F. Bompadre, S. Beekwilder, J. Ric de Vos, C. H. (2008). Antioxidants, phenolic compounds, and nutritional quality of different strawberry genotypes. Journal of Agricultural and Food Chemistry, 56:696–704.
40. Wang, H. Cao, G. Prio,r R.L. (1997). Oxygen radical absorbing capacity of anthocyanin. Journal of Agriculture Food Chemistry, 45:3425-31.
41. Wedge, D.E. Meepagala, K.M. Magee, J.B. Smith, S.H. Huang, G. Larcom, L.L. (2001). Anti-carcinogenic activity of strawberry, blueberry, and raspberry extracts to breast and cervical cancer cells. Journal of Medical Food, 4:49-51.
42. Zhang, Y. Seeram, N. P. Lee, R. Feng, L. and Herber, D. (2008). Isolation and identification of strawberry phenolics with antioxidant and human cancer cell antiproliferative properties. Journal of Agricultural and Food Chemistry, 56: 670–675.

CHAPTER 5

Potential Applications of Biopigments in Food and Feed

Reeba Panesar, Shubhneet Kaur, D.C. Saxena

Abstract

The production of natural food colours through bio-processing is preferred now-a-days due to health related issues associated with their synthetic counterparts. Microorganisms are generally preferred sources as these have high production rate, high growth rate, ease of handling under controlled conditions and absence of seasonal variations. Biopigments can be produced from different genera of bacteria, yeast and mold. Different types and colours of pigments have been produced from microorganisms depending upon their origin, media components and process conditions. The biopigments have variety of applications in food industry such as food additives, food preservatives, biological control agents and nutritional supplements. These have also been associated with therapeutic, and antioxidant properties. Lycopene from B. trispora is used as a food ingredient in dietary supplement. Astaxanthin from X. dendrortious has also been reported to have beneficial effect on human health as it possesses strong antioxidant properties. Arpink red P. oxalicum is used as red colorant in meat products, non-alcoholic drinks, milk products and confectionary. β- carotene is considered as leading food colorant in world and is applied to range of baked, dairy and canned foods to improve their appearance. The antioxidant and precursory properties of β-carotene increase its applications in pet foods. Zeaxanthin is used as additives in poultry fed to intensify the colour of the egg yolk. Monascus pigments are utilized for coloration of sausages, ham, marine products and tomato ketchup. Monascus extract is also marketed as dietetic supplement in Japan. Thus, the biopigments can serve as a chromophore for colorant production with broad spectrum of colours and wide applications.

Key words: Biopigment, applications, food, feed

Introduction

Colour is integral to the interface between consumer and food, which act as indices of good aesthetic and sensory value. Due to consumer's awareness, people have preferred natural colourants incorporated into their food, feed, clothing, cosmetics and medicines. Food colourants from natural sources are preferred over synthetic ones because of the adverse affects in terms of tetragenicity and carcinogenicity (Babitha et al. 2006). Colour is added to food product to improve or compensate losses during storage or processing and it also enhances the visual appeal of uncoloured food (Spears, 1988). Consumers are aware of the harmful effect of synthetic colourants despite to the fact that synthetic colourants are relatively stable and cheap to produce, but natural sources are preferred due to safety (Zheng et al., 2009). Natural pigments also posses' anti-cancer properties, precursor of vitamin A and other desirable properties like stability to light, heat and pH. Plants, animals and microbes are the natural sources of pigments. Colours produced by microbial cells are called microbial pigments. Microbial pigments have the advantage of being climate independent, don't require large area for growth and can be produced in any quantity in short time (Joshi et al., 2003). Some of the microbial pigments can be produced from waste material, which may represent an added value to industry and also an environmental friendly way for waste management (Juzlova et al., 1996). The production of microbial pigments having high growth velocity is expected to give such a production that makes them industrially competitive (Babitha, 2009). Microbial strains differ in the amount and tone of pigments produced. Factors like temperature, pH and humidity influence the micro-organism growth and pigmentation in given media (Carvalho et al., 2005).

Microbial Sources

Microorganisms like bacteria, mold, yeast and fungi produce different types of pigments depending upon their source. Biopigments can be derived from wide range of bacterial strains like *Serratia marcescens*, *Pseudomonas magneslorubra*, *Vibrio psychroerythrous*, *S. rubidaea*, *Vibrio gazogenes*, *Alteromonas rubra*, *Rugamonas rubra*, *Streptoverticillium rubrireticuli* and *Streptomyces longisporus* (Joshi et al., 2003; Dufosse, 2006; Gupta et al., 2011).

Yeast cultures are considered as a good source of microbial pigments due to its unicellular nature and high growth rate by utilizing low cost media (Malisorn and Suntornsuk, 2008) as compared to bacteria and fungi. Media and process parameters play a key role in the production of various carotenoid pigments from yeast strains. Yeast strains like *Rhodotorula glutinis*, *Yarrowia lipolytica*, *Cryptococcus* sp. and *Phaffia rhodozyma* are potential sources of pigments (Vijayalakshmi et al., 2001). Among mold cultures, *Monascus purpureus* strains are most commonly employed for the pigment production through solid state fermentation (SSF).

Monascus sp. produces variety of azaphilone pigments with colour variants, which can be classified into three groups. These include yellow pigments (ankaflavins and monascin), orange pigments (rubropunctatin and monascorubin) and red (rubropuntamine and monascorubramine) pigments (Pattanagul et al., 2007). Besides Monascus, microorganisms belonging to the genera Aspergillus and Penicillium have also been studied as potential producers of natural pigments.

Applications in Food and Feed

Applications of microbial pigments are multifold in nature. They are matter of intense interest in the present decade because of their wide range of potential applications in food and feed industries (Table-1), which are discussed as below:

Food Industries

Pigments from fungus sp. have remarkable ability to be used as dietary supplement as they contain many bioactive constituents (Wongsorn et al., 2011). Monascus pigments obtained as fermented products from rice and bread has been used as food colourants and as therapeutic agents (Fabre et al., 1993). Monascarubromine, a red pigment produced from Monascus strains is in high demand and widely used in meat, fish and ketchup to replace synthetic colours (Hamano and Kilikian, 2006). Colourants from Monascus sp. can also be added to fruit flavored yoghurt for enhancing colour of fruit (Juzlova et al. 1996). Monascus pigments by the name of 'AngKak' has been used in Asia as a natural food colouring matter of fish, Chinese cheese, red wine and sausages (Moore and Chiu, 2001). They are also used as nutrient supplements to reduce cholesterol level. In oriental countries, Monascus sp. are used to produce red rice used as natural food supplement to reduce blood cholesterol and used as food colorant in many products (Babitha et al., 2009).

Pigment extracted from R. glutinis DFR-PDY has been added as a food colourant in puffed products like popcorns, biscuit creams, cake icing and in ice-creams

(Latha and Jeevaratnam, 2010). The microbial pigments obtained from yeast are good precursors of vitamin- A and used as food supplement for modifying the colour of fats, oil, cheese and drinks. Anthocyanin obtained as pigment from microbial source is used as an indicator to evaluate quality of coloured foods (Boyles and Wrolstad, 1993). Arpink red from *P. oxalicum* used as red colourant in food industry in various food stuffs such as milk products, confectionary items and non-alcoholic drinks (Sardaryan et al., 2004).

Riboflavin from Clostridium acetobutylicum is used as a yellow food colourant in dressings, sherbets, beverages, ice-creams and in cereal based products (Chattopadhyay et al., 2008). Riboflavin produced by *Ashbya gossypii*, filamentous fungi *Candida* famata or bacterium *Bacillus subtilis* is used in baby foods, breakfast cereals, pastas, sauces, processed cheese, fruit drinks, vitamin-enriched milk products and some energy drinks (Malik, 2012).

Table 1: Applications of microbial pigments in food and feed

Microbial Pigment	Microbial Source	Applications
Astaxanthin	Xanthophyllomyces dendrorhous Phaffia rhodozyma Haematococcus pluvialis Halobacterium salinarion Agrobacterium aurantiacum	Powerful bioactive anti-oxidant Used for skin colouration in aquatic species Nutrient supplement in feed of farmed animals and humans
Arpink red	Penicillium oxalicum	Red colourant in meat products, drinks, ice-creams and confectionary items
Riboflavin	Ashbya gossypii Clostridium acetobutylicum Candida guilliermundii	Colourant in dressings, beverages, milk cereal products, baby foods and cheese
β-carotene	Blakeslea trispora Mucor circinelloides	Improves the appearance of confectionary items Act as Antioxidants
Monascorubrine, Monascine, Rubropunctamine	Monascus sp.	Food colourant of fish, cheese, wine and sausages. Used as substitute for nitrate for colouration of meat products Used as nutrient supplement in food Used as food preservative and flavour enhancing agent.
Canthaxanthin	Bradyrhizobium sp. Holobacterium sp.	Used as dark red colourant in food products, potent anti oxidants.
Zeaxanthin	Flavobacterium sp. Paracoccus xanthinifaciens	Additive in poultry feed, strengthen yellow colour of skin of animals, accentuate the colour of yolk of egg.

(*Source*: Dufosse, 2006; Joshi et al., 2003; Babitha et al., 2006)

β-carotene from *Blasklea trispora* is used as colouring matter for food stuffs and as antioxidant in pharmaceutical industry. β- carotene is one of the leading food colourant of world and used in range of food and beverage products to improve appearance of margarine, cheese, fruit juice, baked goods, canned goods and confectionery items (Dufosse, 2006). Moreover, it helps to mediate the harmful effects of free radicals implied in life threatening diseases. Lycopene from *Blasklea trispora* is used as a food ingredient and dietary supplement with proposed limit of 20 mg/day. Pycocyanin and pyorubrin produced by *P. aeruginosa* has been reported to be used as food colourants and are also antibacterial in nature (Saha et al., 2008).

Feed Industries

Red pigments produced from *R. gracis* are the mixture of carotenoids containing rhodoxanthin, β-carotene and rhodotorulin. They act as potential pharmaceutical ingredients, also as single cell protein for aqua cultured animals (Krinsky, 2001; Costa et al., 2005; Iriani et al., 2005). Pigments obtained from fungal cultures are also used as poultry and fish feeds (Joshi et al. 2003). β-carotene from B. trispora has been used as nutrient supplement for feed of many pets like dog, cat, fish and birds (Dufosse, 2009).

Pigments with wide colour variations are responsible for fetching higher price of aquatic animals in the market. As the fishes cannot synthesize their colouring pigments de novo, natural sources of pigments need to be incorporated into their diet (Gupta et al. 2007). Astaxanthin are added to feed of farm animals to impart orange-red colourant to increase consumer appeal in market and is considered an effective flesh pigmenter. Supplementation of astaxanthin in diet of rats and mice inhibits carcinogenesis of colon, oral cavity and urinary bladder (Dufosse, 2009). Astaxanthin from *P. rhodozyma* is used for flesh pigmentation of sea animals and used to enhance the immune response of fish and shrimps for maximum survival and growth. Zeaxanthin is used as additive in feed for poultry to intensify yellow colour of skin and yolk of egg of farm animals. This pigment has been also used for the feed of birds, swine and fish (Masetto et al., 2001).

Conclusions

Now-a-days the demand for pigments is multi-dimensional and the pigments are used in various fields of everyday life. Natural colourants from plants, animals or microbes being now preferred by consumers over traditionally used synthetic pigments. Natural pigments not only have the capacity to increase marketability of products but also possess biological, anti-oxidant and anti-cancerous activities. Micro-organisms due to their intrinsic properties are considered potential source of pigments with wide applications in food and feed industries. Moreover, the consumer preference for natural food additives is now adding to the increasing business of bio-pigments and various research efforts are being made worldwide in this direction.

Acknowledgements

Dr. Reeba Panesar acknowledges the financial support from Council for Scientific and Industrial Research (CSIR), New Delhi, India.

References

1. Babitha, S. Soccol, C.R. and Pandey, A. (2006). Jack fruit seed-a novel substrate for the production of Monascus pigment through solid state fermentation. Food Technology and Biotechnology, 44:465-471.
2. Babitha, S. (2009). Microbial pigments. In: Biotechnology for Agro-Industrial Residues (Nigam, P.S. and Pandey, A. eds.), Springer Science+Business Media B.V., Van Godewijckstraat, Netherlands, pp. 147-163.
3. Boyles, M.J. and Wrolstad R.E. (1993). Anthocyanin composition of red raspberry juice: influences of cultivar, processing, and environmental factors. Journal of Food Science, 58:1135-1141.
4. Carvalho, J.C. Oishi, B.O. Pandey, A. and Soccol, C.R. (2005). Biopigments from *Monascus*: strains selection, citrinin production and colour stability. Brazilian Archives of Biology and Technology, 48:885-894.
5. Chattopadhyay, P. Chatterjee, S. and Sen, S.K. (2008). Biotechnological potential of natural food grade biocolorants. African Journal of Biotechnolology, 7: 2972-2985.
6. Costa, H.L. Martelli Da Silva, I.M. and Pomeroy, D. (2005). Production of β-carotene by *Rhodotorula* strain. Biotechnology Letters, 9:373-375.
7. Dufosse, L. (2006). Microbial production of food grade pigments. Food Technology Biotechnol. 44: 313-321.
8. Dufosse, L. (2009). Pigments, Microbial. Encyclopedia Microbiol, 4: 457-471.
9. Fabre, C.E. Santerre, A.L. Loret, M.O. Baberian, R. Pareilleux, A. Goma, G. and Blanc, P.J. (1993). Production and food applications of red pigments of Monascus ruber. Journal of Food Science, 50:1099-1102.
10. Gupta, S.K. Jha, A.K., Pal, A.K. and Venkateshwarlu, G. (2007). Use of natural carotenoids for pigmentation in fishes. Nat. Pro. Rad., 6: 46-49.
11. Gupta, C. Garg, A.P. Prakash, D. Goyal, S. and Gupta, S. (2011). Microbes as potential source of biocolours. Pharmacologyonline, 2:1309-1318.
12. Hamano, P.S. and Kilikian, B.B. (2006). Production of red pigments by Monascus ruber in culture media containg corn steep liquor. Brazalian Journal of Chemical Engineering, 23:443-449.
13. Iriani, R.M. Adilma, R.P. Scamparini, D.B. and Rodriguez, A. (2005). Selection and characterization of carotenoid-producing yeasts from Campinas region. Brazalian Journal of Microbiology, 40:2985-2991.
14. Joshi, V.K. Atrri, D. Bala, A. and Bhushan, S. (2003). Microbial pigments. Indian Journal of Biotechnollogy, 2:362-369.
15. Juzlova, P. Martinkova, L. and Kren, V. (1996). Secondary metabolites of the fungus Monascus: a review. Journal of Industrial Microbiology, 16: 163-170.

16. Krinsky, N.I. (2001). Carotenoid antioxidants. Nutrition journal, 17: 815-817.
17. Saha, S. Thavasi, R. and Jayalakshmi, S. (2008). Phenazine pigments from Pseudomonas aeruginosa and their application as antibacterial agent and food colourants. Research Journal of Microbiology, 3:122-128.
18. Spears, K. (1988). Developments in food colourings: the natural alternatives. Trends in Biotechnollogy, 6: 283-288.
19. Latha, B.V. and Jeevaratnam, K. (2010). Purification and characterization of the pigments from Rhodotorula glutinis DFR-PDY isolated from natural source. Global Journal of Biotechnology and Biochemistry, 5:166-174.
20. Malik, K., Tokkas, J. and Goyal, S. (2012). Microbial pigments: a review. International Journal of Microbiology Research and Technology, 1:361-365.
21. Malisorn, C. and Suntornsuk, W. (2008). Optimization of â-carotene production by Rhodotorula glutinis DM28 in fermented radish brine. Bioresource Technology, 99: 2281-2287.
22. Masetto, A. Flores-Cotera, L.B., Diaz, C. Langley, E. and Sanchez, S. (2001). Application of a complete factorial design for the production of zeaxanthin by Flavobacterium sp. Jouran of Bioscience and Bioengineering, 9:55-58.
23. Moore, D. and Chiu, S.W. (2001). Fungal products as food. In: Bio-Exploitation of Filamentous Fungi (S.B. Pointing and K. D. Hyde, eds.), Fungal Diversity Press, Hong Kong, pp. 223-251.
24. Pattanagul, P. Pinthong, R. Phianmongkhol, A. and Leksawasdi, N. (2007). Review of Angkak production (Monascus purpureus). Chiang Mai Journa of Science, 34: 319-328.
25. Sardaryan E. Zihlova H. Strnad R. and Cermakova Z. (2004). Arpink red-meet a new natural red food colorant of microbial origin. In: Pigments in Food, More than Colours (L. Dufosse, ed.), Universite de Bretagne Occidentale Publ., Quimper, France, 207-208.
26. Vijayalakshmi, G Shobha, B. Vanajakshi, V. Divakar, S. and Manohar, B. (2001) Response surface methodology for optimization of growth parameters for the production of carotenoids by a mutant strain of Rhodotorula gracilis. European Food Research andTechnology, 213:234-239.
27. Wongsorn, H. Wongjewboot, I. and Kongruang, S. (2011). Solvent stability of ultrasonic mutants of Monascus purpureus pigments. Indian Journal of Biochemistry and Biophysics, 1:206-211.
28. Zheng, Y. Xin, Y. and Guo, Y. (2009). Study on the fingerprint profile of Monascus products with HPLC-FD. Food Chemistry, 113:705-711.

CHAPTER 6

Nutraceutical Characteristics of Minor Millet: A Review

Seema, Mandeep Singh Sibian, Romee Jan and C.S. Riar

Abstract

The aim is to review the potential health benefits of minor millet. There is in need to identify newer sources of nutraceuticals and other natural and nutritional materials with the desirable functional characteristics. They are an important staple food in India for people of low income groups. Minor millet is rich in several nutrients as well as non-nutrients such as phenols. Nutritionally, its importance is well recognised because of its high content of calcium (0.38%), dietary fiber (18%) and phenolic compounds (0.3–3%). Minor millets, one of the cereals are known for several health benefits including its polyphenol and dietary fiber contents. Health beneficial effects include anti-diabetic, anti-tumerogenic, atherosclerogenic effects, antioxidant and antimicrobial properties. The energy of minor millets is greater than other cereals and nearly equal to that of brown rice because the lipid content is generally higher (3 to 6%). This review deals with the nature of polyphenols and dietary fiber of minor millets and their role with respect to the health benefits associated with millet.

Introduction

According to world agriculture production millets are one of the most important drought-resistant crops and have the 6th rank in cereal crop. Major cereals fail to give substantial yields in marginal areas under agricultural conditions in that area they are grown mostly (Adekunle, 2012). As compared to major cereals millets have short growing season, productivity and resistance to pests under drought conditions (Devi et al., 2011).Therefore, millets are now receiving specific attention in the developing countries in terms of utilization as food and for manufacturing of bio-ethanol and biofilms (Li et al., 2008). Divergences exist be relevant to classification of family millet, with some references giving the family name Gramineae, and others classifying it in the family Poaceae.

Fig. 1: Pictures of commonly used Millet

Minor millets are a group of sedgy plants with short slender culm and small grains possessing remarkable ability to survive under severe drought conditions whereas major millets like pearl millet and sorghum are tall growing and fairly drought tolerant striking ability to survive under severe drought conditions. Their deep root system having efficient soil moisture extraction ability which makes them capable to grow under very low to severe moisture stress conditions, where most other cereal crops may not survive to produce grain. Minor millet have to bears on to their lesser importance in trade, but also on the smaller size of the grains. In India, minor millets are cultivated on 71.9 million acres, accounting for nearly 25% of the total acreage under cereal crops. According to FAO (2012) India was the top producer of minor millet with an annual production of 334500tons (43.85%) where as the world production of minor millets were 62712 metri tons. Pearl millet yield accomplished approximately 54% of the global production in 2004 as reported to (FAO, 2005).

Minor millets include Barnyard millet (Echinochloa spp.), Kodo millet (Paspalum scrobiculatum), little millet (Panicum sumatrense), Guinea millet (Brachiaria deflexa = Urochloa deflexa), Brown top millet (Urochloa ramosa = Brachiaria ramosa = Panicum ramosum) (ICRISAT, 2007; FAO, 2009). The major millet type in terms of world production is pearl millet (Pennisetum glaucum) which accounts for about 46% (Marathee, 1994) followed by foxtail, proso and finger millet. A number of other minor millets exist, namely kodo, little, Japanese barnyard, fonio, and teff millets. Table 1, presents common names of millets used in different countries and their taxonomy. Veena et al., (2004) reported that in the recent years, they are recognized as important substitutes for major cereal crops to cope up with the world food shortage and to meet the demands of increasing population of both developing and developed countries.

Table 1: Scientific and common names of major types of millets

Scientific name	Common name	Taxonomy (Tribe)	Major areas of production for grains
Pennisetum glaucum	Pearl, bajra, cattail, bulrush, candlestick, sanyo , munga, seno	Paniceae	India, Western and Central Africa,Eastern and Sout hern Africa
Elusine coracana	Finger, ragi, African, bird's foot, rapoko	Eragrostideae	India, Eastern and South ern Africa, Uganda, Tanza nia Sri Lanka, Mala ys ia
Setaria italic	Foxtail, Italian, Ger man, Hunga rian	Paniceae	China, India, Eastern Europe
Panicum milliaceum	Proso , common, hog, broom,	Paniceae	China, Russia, Kazhakastan, Ukraine,India, Japan
Panicum sumatrense	Little, blue panic, Heen meneri	Paniceae	India
Paspalum scrobiculatum	Kodo, varagu, basta rd, ditch,	Paniceae	India
Echinoc hola crus-galli	Barnyard, Japanese, sanwa,	Paniceae	India, Japan, China, Malaysia
Eragrostis tef	Teff, Abyssinian	Paniceae	Ethiopia, Eritrea, Au stralia

Source: Adapted from (ICRISAT, 2012; Serrna-Saldivar and Rooney, 1995; Tylor and Emma mbux, 2008).

Minor millets are arrogated to be future foods for better health and nutrition security, as compared to major cereals such as wheat and rice millets and are found to have high nutritive value. The protein content in millet is very close to wheat and provides about 11% protein by weight. It has also been reported that millet proteins are good sources of essential amino acids except lysine and threonine, but are relatively high in methionine. Millets are rich in B-complex vitamins especially niacin, B_6, folic acid and minerals such as calcium, iron, potassium, magnesium and zinc (Kalinova and Moudry, 2006). Millets are significantly rich in resistant starch, soluble and insoluble dietary fibers, minerals, and antioxidants. Black finger millet contains 8.71 mg/g dry weight fatty acid and 8.47 mg/g dry weight protein (Glew et al., 2008). Kodo millet and little millet were also reported to have 37% to 38% of dietary fiber, which is the highest among the cereals; and the fat has higher polyunsaturated fatty acids (Hegde and Chandra, 2005). Kalinova and Moudry (2006) reported the protein content of proso millet 11.6% richer in essential amino acids (leucine, isoleucine, and methionine).

Minor millets contain a range of substances which may have health promoting effects, these substances are often referred to as Phytochemicals or Plant bioactive substances (Goldberg, 2003). Nutraceuticals are natural bioactive compounds that have health promoting, disease preventing or medicinal properties. Bioactive compounds are extra nutritional elements that typically occur in small quantities in foods. These substances are beneficial to human health but are not essential for the human body (Kris-Etherton et al., 2002). Millets must also be accepted a functional food and nutraceuticals because they provide dietary fibers, proteins, energy, minerals, vitamins, and antioxidants required for human health. The majority of bioactive compounds of whole-millet grains are present in the bran/germ fraction of cereal-grains. Several potential health benefits such as preventing cancer and cardiovascular diseases, reducing tumor incidence, lowering blood pressure, risk of heart disease, cholesterol, and rate of fat absorption, delaying gastric emptying, and supplying gastrointestinal bulk were reported for millets (Gupta et al., 2012). This review discusses about the nutritional composition, nutraceuticals and functional properties of minor millet and their health benefits.

Nutritional Composition of Millets

Millets are significantly rich in resistant starch, soluble and insoluble dietary fibers, minerals, and antioxidants. It contains about 92.5% dry matter, 2.1% ash, 2.8% crude fiber, 7.8% crude fat, 13.6% crude protein, and 63.2% starch (Ali et al., 2003). Black finger millet contains 8.71 mg/g dry weight fatty acid and 8.47 g/g dry weight protein (Glew et al., 2008). Kodo millet and little millet were also reported to have 37% to 38% of dietary fiber, which is the highest among the cereals; and the fat has higher polyunsaturated fatty acids (Hegde and Chandra, 2005). Kalinova and Moudry (2006) reported the protein content of proso millet 11.6% richer in essential amino acids (leucine, isoleucine, and methionine). The average of nutrient composition of some millet grains and other grains is summarized in Table 2.

Table 2: Nutrient composition of major cereals and millets (per 100 g of edible portion at 12% moisture

Cereals	Energy (kcal)	CHO (g)	Protein (g)	Fat (g)	CF (g)	Calcium (mg)	Iron (mg)	Niacin (mg)
Rice (brown)	362	76.0	7.9	2.7	1.0	33.0	1.8	4.3
Pearl millet	363	67.0	11.8	4.8	2.3	42.0	11.0	2.8
Finger millet	336	72. 6	7.7	1.5	3.6	350	3.9	1.1
Foxtail millet	351	63.2	11.2	4.0	6.7	31.0	2.8	3.2
Proso millet	364	63.8	12.5	3.5	5.2	8.0	2.9	4.5
Little millet	329	60.9	9.7	5.2	7.6	17.0	9.3	3.2
Kodo millet	353	66.6	9.8	3.6	5.2	35.0	1.7	2.0

Abbreviations: CHO, carbohydrates; CF, crude fibre. (Source: data adapted from reference FAO, 1995).

Millets Nutraceuticals and their Functional Properties

Nutraceuticals are natural bioactive, chemical compounds that have health promoting, disease preventing or medicinal properties. Millets have been recommended as nutraceuticals or functional food due to presence of essential dietry components in valuable amounts such as dietary fibers, proteins, minerals, vitamins, and antioxidants necessary for human health. Millets being a rich source of various phytochemicals including tannins, phenolic acids, flavonoids, anthocyanins, and phytosterols and policosanols, which have potential of significant influence on human health. Millet having high antioxidant activity in vitro as compared to other cereals or fruits. These millets may represent health benefits similar to that associated with fruits. Phenolic acids and flavonoids are found in different parts of the grain the content and composition vary depending on the type of millet grain (Shirley, 1998). Chandrasekara and Shahidi (2011a) reported the content of a number of flavonoids, namely flavan3-ols (monomers, and dimmers), flavonols and their glycosides, flavones and flavononol, in millet phenolic extracts mainly in the soluble fraction (Table 3).

Table 3: Major phenolic acids and flavonoids in millet grains.

Millet	HBAS	HCAS	FLOS	References
Pearl	p-Hydroxybenzoic, protocatechuic, vanillic	Caffeic, p-coumaric, cinnamic,sinapic, trans-ferulic	Apig enin, myricetin	Chandrasekara andShahidi (2011c), N'Dri et al. (2012)
Finger	Gallic,p-hy droxybenzoic, protocatechuic, syring	Caffeic, p-coumaric, cinnamic,sinapic, transferulic	Catechin, galloca techin, epicatec hin, epigalloca-techin, taxifolin, vitexin,tricinic, gentisic, vanillic	Rao and Muralikrishna (2001, 2002, 2004) Chethan et al.
Foxtail	Gallic, p-hy droxybenzoic, protocatechuic, syringic, gentisic, vanillic	Caffeic, p-coumaric, sinapic trans-ferulic	Catechin, quercetin, apigenin, kempherol	Chandrasekara and Shahidi (2011c)
Proso	Gallic,p-Hydroxy benzoic, protocatechuic, syringic, gentisic, vanillic	Chlorogenic, caffeic, p-coumaric, sinapic, ferulic	Kempherol, apig enin, myricetin	Chandrasekara and Shahidi (2011c)
Kodo	Gallic,p-hy droxybenzoic, protocatechuic, syringic, vanillic	Chlorogenic, caffeic,p-coumaric, sinapic,trans-ferulic	Kempherol, apig enin, vitexin, isov itexin, luteolin, quercetin	Chandrasekara and Shahidi (2011c)
Little	Gallic, protocatechuic, syring ic, gentisic, vanillic	Caffeic, p-coumaric, sinapic,trans -ferulic	Apigenin	Chandrasekara and Shahidi (2011c)
Barnyard	Na		Luteolin, tricin, N-(p-coumaro yl) serotonin	Watanabe (1999)

Abbreviations: na, data not available; HABS, hydroxybenzoic acids and their derivatives; HACS, hydroxycinnamic acids and their derivatives; FLOS, flavonoids

Antioxidant contents and activities

Antioxidant compounds are gaining importance due to their main roles as lipid stabilizers and as suppressors of excessive oxidation that causes cancer and ageing (Namikii, 1990). Their stable radical intermediates prevent the oxidation of various food ingredients, particularly fatty acids and oils (Maillard et al., 1996). Phenolic acids and their derivatives, flavonoids and tannins present in millets seed coat are of multifunctional and can act as reducing agents (free radical terminators), metal chelators, and singlet oxygen quenchers (Sripriya et al., 1996). The potency of phenolic compounds to act as antioxidants arise from their ability to donate hydrogen atoms via hydroxyl groups on benzene rings to electron deficient free radicals and in turn form a resonance-stabilized and less reactive phenoxyl radical. Studies were carried out on the natural antioxidants in edible flours of small millets. Total antioxidant capacity of finger, little, foxtail and proso millets were found to be higher and their total carotenoids content varied from 78–366 mg/100 g in the millet varieties.

Millet for diabetics

Diabetes mellitus is a chronic metabolic disorder characterized by hyperglycemia with alterations in carbohydrate, protein, and lipid metabolism. Recent reports indicate that hyperglycemia could induce non-enzymatic glycosylation of various proteins, resulting in the development of chronic complications in diabetes (Lebovitz, 2001). Therefore, control of postprandial blood glucose surge is critical for treatment of diabetes and for reducing chronic vascular complications (Baron, 1998) which can be controlled by intake of high complex carbohydrate and high fiber diet. For example, consumption of finger-millet based diets resulted in significantly lower plasma glucose levels that might have been due to the higher fiber content of finger millet than rice and wheat. Gopalan (1981) reported that regular consumption of finger millet is known to reduce the risk of diabetes mellitus and gastrointestinal tract disorders, and these properties were attributed to its high polyphenols and dietary fiber contents (Chethan et al., 2008b). In another study, phenolic compounds from the millets seed coat showed strong inhibition towards α-glucosidase and pancreatic amylase (Shobana, 2009). The beneficial effect of phenolics is due to partial inhibition of amylase and α-glucosidase during enzymatic hydrolysis of complex carbohydrates and delays the absorption of glucose, which ultimately controls the postprandial blood glucose levels (Shobana et al., 2009). Therefore, millet grains have the potentials to be useful in preventing diabetes and for the treatment of diabetics.

Millet and cardiovascular disease

Most of the world countries face high and increasing rates of cardiovascular disease. It has been demonstrated that rats fed with a diet of native and treated starch from barnyard millet had the lowest blood glucose, serum cholesterol, and triglycerides compared with rice and other minor millets (Kumari and Thayumanavan 1997). Also, the feeding of proso millet protein improved plasma levels of adiponectin, high-density lipoprotein (HDL) cholesterol in genetically obese type-2 diabetic mice under high-fat feeding conditions (Park et al., 2008). Finger millet and proso millet may prevent cardiovascular disease by reducing plasma triglycerides in hyperlipidemic rats (Lee et al., 2010). In addition, phenolic extracts from kodo, finger, proso, foxtail, little, and pearl millets were evaluated for their inhibitory effects on lipid peroxidation in vitro copper-mediated human LDL cholesterol oxidation and several food model systems, namely, cooked comminuted pork, stripped corn oil, and a linoleic acid emulsion. At a final concentration of 0.05 mg/mL, millet extracts inhibited LDL cholesterol oxidation by 1% to 41%. Millets exhibited effective inhibition of lipid oxidation in food systems.

Millet against cancers

Millet grains based on literature values are known to be rich in phenolic acids, tannins, and phytate that act as "antinutrients" (Thompson, 1993). However, it has been established that these antinutrients reduce the risk for colon and breast cancer in animals (Graf and Eaton, 1990). It has also been reported that populations consuming sorghum and millets have lower incidences of esophageal cancer than those consuming wheat or maize (Van Rensburg, 1981). Furthermore, a recent study has demonstrated that millet phenolics may be effective in the prevention of cancer initiation and progressionin vitro (Chandrasekara and Shahidi, 2011c).

Celiac disease

The overall growing demand for novel, tasty, and "healthy", foods, together with the increasing number of people suffering from celiac disease, has given birth to a new market consisting of cereal products made from grains other than wheat and rye. In this challenging market, oat, sorghum, and millet have gained a special position (Angioloni and Collar, 2012a). Celiac disease is an immune-mediated enteropathy triggered by the ingestion of gluten in genetically susceptible individuals. It is one of the most common lifelong disorders worldwide. In the past, celiac disease was considered a rare disorder, mostly affecting children of European origin. In the developed countries, there is a growing demand for gluten-free foods and beverages from people with celiac disease and other intolerances to wheat, barley, or rye. However, millets are gluten-free; they have considerable potential in foods and beverages that can be suitable for individuals suffering from celiac disease (Chandrasekara and Shahidi, 2011b). Therefore, millet grains and their fractions have the potential to be useful for producing food products for celiac people.

Antimicrobial activity

Baranowski et al. (1980) have reported that phenolics have been implicated for minimising the intensity of several diseases and also to inhibit the in vitro growth of an assortment of fungal genera. Millet grain extracts were found to have antimicrobial activity. In one study, seed protein extracts of pearl millet, sorghum, Japanese barnyard millet, foxtail millet, samai millet, and proso millet were evaluated in vitro for their ability to inhibit the growth of *Rhizoctonia solani, Macrophomina phaseolina*, and *Fusarium oxysporum*. In another study, phenolic acids from finger millet milled fractions (whole flour, seed coat, 3%, 5%, and 7%) were isolated. The seed coat extract of millet showed higher antimicrobial activity against *Bacillus cereus* and *Aspergillus flavus* than whole flour extract. The results indicated that potential exists to utilize finger millet seed coat as an alternative natural antioxidant and food preservative (Viswanath et al., 2009). Therefore, extracts of phenolic acids and other bioactive components have the potential to be used as natural alternatives in food preservation and for therapeutic purposes.

Millet and aging

The chemical reaction between the aldehyde group of reducing sugars and the amino group of proteins, termed as non enzymatic glycosylation, is a major factor responsible for the complications of diabetes and aging (Monnier, 1990). Millet grains are rich in antioxidants and phenolics; however, it has been established that phytates, phenols, and tannins can contribute to antioxidant activity important in health, aging, and metabolic syndrome (Bravo, 1998). It has also been found that methanolic extracts from finger millet and kodo millet inhibited glycation and cross-linking of collagen (Hegde et al., 2002). Therefore, there is potential usefulness of millets in the protection against aging.

Conclusion

Based on the results of studies carried out, we can examine that millet grains contain many health-promoting components such as dietry fiber, minerals, vitamins including phenolic compounds, these bioactive components better than other major grains and fruits in term of their anti-diebetic and antioxidants properties. Millets consumption help in reduction of certain types of cancer in humans compared to other cereals due to their high concentration of phytochemicals.

References

1. Adekunle, A. A. (2012). Agricultural innovation in sub-saharan Africa: experiences from multiple stakeholder approaches. Forum for Agricultural Research in Africa, Ghana. ISBN 978-9988-8373-2-4.
2. Ali, M.A.M. Tinay, A.H. and Abdalla, A.H. (2003). Effect of fermentation on the in vitro protein digestibility of pearl millet. Food Chemistry, 80(1):51–4.
3. Angioloni, A. and Collar, C. (2012). Suitability of oat, millet and sorghum in bread making. Food Bioprocess Technol DOI: 10.1007/s11947-012-0786-9.
4. Baranowski, J.D. Davidson, P.M. Nagel, C.W. and Brannen, R.L. (1980). Inhibition of Saccharomyces cerevisiaeby naturally occurring hydroxyl cinnamates. Journal of Food Science, 45:592–594.
5. Bravo, L. (1998). Polyphenols: chemistry, dietary sources, metabolism and nutritional significance. Nutrition Review, 56:317–333.
6. Chandrasekara, A. and Shahidi, F. (2011b). Antioxidant phenolics of millet control lipid peroxidation in human ldl cholesterol and food systems. Journal of American Oil Chemcal Society. DOI: 10.1007/s11746-011-1918-5.
7. Chandrasekara, A. and Shahidi, F. (2011c). Antiproliferative potential and DNA scission inhibitory activity of phenolics from whole millet grains. Jouranal of Functional Foods 3:159–70.
8. Chethan, S. Sreerama, Y.N. and Malleshi, N.G. (2008b). Mode of inhibition of finger millet malt amylases by the millet phenolics. Food Chemistry,111:187–191.
9. Devi, P.B., Vijayabharathi, R., Sathyabama, S., Malleshi, N.G., and Priyadarisini, V.B. (2011). Health benefits of finger millet (Eleusine coracana L.) polyphenols and dietary fiber. A Review. Journal of of Food Science and Technology 16 (5):123-156.
10. ElShazali, A.M. Nahid, A. A. Salma, H.A. and Elfadil, E.B. (2011). Effect of radiation process on antinutrients, protein digestibility and sensory quality of pearl millet flour during processing and storage. International Food Reearchs Journal, 18(4):1401–1407.
11. Graf, E. and Eaton, J.W. (1990). Antioxidant functions of phytic acid. Free Radic Biol Med 8(1):61–9.
12. Gupta, N. Srivastava, A.K. and Pandey, V.N. (2012). Biodiversity and nutraceuticals quality of some Indian millet. Proceedings of the National Academy of Sciences, India Section B: Biological Sci [DOI: 10.1007/s40011-012-0035-z].
13. Hegde, P.S. Chandrakasan, G. and Chandraa, T.S. (2002). Inhibition of collagen glycation and crosslinking in vitro by methanolic extracts of Finger millet (Eleusine coracana) and Kodo millet (Paspalum scrobiculatum). Journal of Nutrition and Biochemistry, 13:517–21.
14. Hegde, P.S. Rajasekaran, N.S and Chandra, T.S. (2005). Effects of the antioxidant properties of millet species on oxidative stress and glycemic status in alloxan-induced rats. Nutrition Research, 25:1109–20.
15. Kalinova, J. and Moudry, J. (2006). Content and quality of protein in proso millet (Panicum miliaceum L.) varieties. Plant Foods Human Nutrition, 61:45–9.
16. Kim, J.S., Hyun, T.K. and Kim, M.J. (2011). The inhibitory effects of ethanol extracts from sorghum, foxtail millet and proso millet on á-glucosidase and a-amylase activities. Food Chemistry, 124:1647–51.
17. Kris-Etherton, P. M. Hecker, K. D. Bonanome, A. Coval, S.M. and Binkoski, A.E. (2002). Bioactive compounds in foods: their role in the prevention of cardiovascular disease and cancer. Amrican Journal of Medicine, 13(8):234-56.
18. Kumari, S.K. and Thayumanavan, B. (1997). Comparative study of resistant starch from minor millets on intestinal responses, blood glucose, serum cholesterol and triglycerides in rats. Journal of Science and Food Agriculture, 75:296–302.

19. Lebovitz, H.E. (2001). Effect of the postprandial state on non traditional risk factors. Amrican Journal of Cardiology, 88:204–205.
20. Lee, S.H. Chung, I.M. Cha, Y.S., Parka, Y. (2010). Millet consumption decreased serum concentration of triglyceride and C-reactive protein but not oxidative status in hyperlipidemic rats. Nutrition Research, 30:290–6.
21. Li, J., Chen, Z. Guan, X. Liu, J. Zhang, M. and Xu, B. (2008). Optimization of germination conditions to enhance hydroxyl radical inhibition by water soluble protein from stress millet. Journal of Cereal Science, 48: 619–24.
22. Maillard, M.N. Soum, M.H. Boivia, P. and Berset, C. (1996) Antioxidant activity of barley and malt: relationship with phenolic content. LWT Food Scienca and Technology, 3:238–244.
23. Monnier, V.M. (1990). Nonenzymatic glycosylation, the Maillard reaction and the aging process. Journal of Gerontology, 45:105–11.
24. Namikii, M. (1990) Antioxidants/antimutagens in food. Crittcal Review: Food Science and Nutrition, 29:273–300.
25. Parameswaran, K. and Sadasivam, S. 1994. Changes in the carbohydrates and nitrogenous components during germination of proso millet (Panicum miliaceum). Plant Foods Human Nutrition, 45:97–102.
26. Park, K.O., Ito, Y. Nagasawa, T. Choi, M.R. and Nishizawa, N. (2008). Effects of dietary korean proso-millet protein on plasma adiponectin, HDL cholesterol, insulin levels and gene expression in obese type 2 diabetic mice. Bioscience Biotechnology and Biochemistry. 72(11):2918–25.
27. Serrna -Saldivar, S., and Roone y, L. W. (1995). Structure and chemistry of sorghum and millets. In D. A. V. Dend y (Ed.), Sorghum and millets: Chemistry and Technology (pp. 69–124).
28. Shobana, S. Sreerama, Y.N. and Malleshi, N.G (2009). Composition and enzyme inhibitory properties of finger millet (Eleusine coracana L.) seed coat phenolics: mode of inhibition of α-glucosidase and pancreatic amylase. Food Chemistry, 115:1268–1273.
29. Sireesha, Y. Kasetti, R.B, Swapna, S.A.S. and Apparao, C. (2011). Antihyperglycemic and hypolipidemic activities of Setaria italica seeds in STZ diabetic rats. Journal of Pathophysiology, 18(2):159–64.
30. Sripriya, G., Chandrasekharan, K., Murty, V.S. and Chandra, T.S. (1996). ESR spectroscopic studies on free radical quenching action of finger millet (Eleusine coracana). Food Chemistry, 57(4):537–540.
31. Thompson, L.V. (1993). Potential health benefits and problems associated with antinutrients in foods. Food Research International, 26:131–49.
32. Tovey, F.I. (1994). Diet and duodenal ulcer. Journal iof Gastroenterol Hepatology, 9:177–185.
33. Ugare, R. Chimmad, B. Naik, R. Bharati, P. and Itagi, S. (2011). Glycemic index and significance of barnyard millet (Echinochloa frumentacae) in type II diabetics. Journal of Food Science and Technology, DOI: 10.1007/s13197-011-0516-8.
34. Van Rensbur, S.J. (1981). Epidemiological and dietary evidence for a specific nutritional predisposition to esophageal cancer. Journal of Natural Cancer Institute, 67(2):243–51.
35. Veena, B. Chimmad, B.V. Naik, R.K. and Shantakumar, G (2004). Development of barnyard millet based traditional foods. Journal of Agricultural Science, 17 (3): 522-527.
36. Viswanath, V. Urooj, A. and Malleshi, N.G. (2009). Evaluation of antioxidant and antimicrobial properties of finger millet polyphenols (Eleusine coracana). Food Chemistry, 114:340–346.

CHAPTER 7

Effects of Processing on Bioactive Compounds During Development of Food Formulations: A Review

Mandeep Singh and Charanjit Singh Riar

Abstract

Micronutrients and bioactive compounds must be part of the population's diet in sufficient amounts to meet the body's needs. The concentrations of micronutrients and bioactive compounds in plant foods at harvest are known to vary with such factors as the variety or cultivar, stage of maturity, geographic or climatic effects, etc. Most subclasses of polyphenols e.g., phenolics acids and flavonoids, the effects of processing on the contents were found negligible in comparison to the differences between varieties of plants but carotenoids are among the food components most affected by environmental and processing conditions. Thermal processing has long been known to cause significant losses of vitamins, carotenoids and anthocyanins. The heat treatment viz. blanching, pasteurization, sterilization etc. result in losses of thermally labile nutrients and bioactive compounds, but the inactivation of oxidative and other degradative enzymes (e.g., lipoxygenase for carotenoids, polyphenol oxidase for flavonoids, ascorbic acid oxidase for vitamin C). Nutrients in dehydrated products are more likely to undergo degradation during storage because of the greater surface area exposed to oxygen and light. Blanched products generally resist decomposition better than unblanched foods during storage. Compared to losses during processing, subsequent storage often has a small but significant effect on vitamin and bioactive compound contents. Reaction rates are relatively slow at ambient or reduced temperature. The processing method chosen, and storage condition applied finalise the retention of these compounds which decreases with longer processing time, higher processing temperatures and storage of the food.

Keywords: Micronutrients, bioactive compounds, phenolics acids, flavonoids.

Effects of Processing Levels on Bioactive Compounds and Micronutrients

Plant-based foods contain significant amounts of bioactive compounds, which provide desirable health benefits beyond basic nutrition. Epidemiological evidence suggests that consumption of a diet rich in vegetables and fruits has positive implications for human health. In the last decades, special attention has been paid towards edible plants, especially those that are rich in secondary metabolites. Highly important to human health, micronutrients and bioactive compounds must be part of the population's diet in sufficient amounts to meet the body's needs. They are found in foods at very low levels and their retention from the farm to the table should be maximized. Numerous studies have reported variable, often substantial losses of these food components, by physical removal or chemical reactions, during postharvest handling, industrial processing, storage and domestic preparation. Losses are inevitable to some extent, but investigation and better understanding of the possible causes, including the influencing factors, can help to find measures that can be taken to diminish such losses. Among unit operations, we distinguish different categories: (i) the thermal processes such as pasteurization, baking, cooling, freezing, (ii) the non-thermal processes such as high pressure, pulsed electric fields, filtration, (iii) the mechanical processes such as peeling, cutting or mixing and (iv) the domestic processes that is to say processes by means of preparation of the convenience foods at consumers home.

Effects of Processing and Storage

Discussions of processing and storage effects often focus on industrial processing and storage. However, losses during home preparation/cooking and storage are considerable, and may surpass those incurred in the food industry. Blanching, a mild heat treatment (accomplished with hot water, steam, hot air or microwave) is an important step in the processing of fruits and vegetables. It is done primarily to inactivate potentially harmful enzymes, reduce microbial loads and decrease interstitial gases prior to further processing. Blanching in hot water can provoke large losses of water-soluble vitamins by leaching. The heat treatment in blanching may result in some losses of thermally labile nutrients and bioactive compounds, but the inactivation of oxidative and other degradative enzymes (e.g., lipoxygenase for carotenoids, polyphenol oxidase for flavonoids, ascorbic acid oxidase for vitamin C) will prevent further and greater losses by enzyme-catalyzed degradation during slow processing (as in drying) and storage. Considerable losses of carotenoids and flavonoids have been observed when fruits and vegetables are processed into juice because substantial amounts of the bioactive compounds are left in the discarded skin and pulp (press cake), these losses often surpassing those of heat treatment. Thermal processing has long been known to cause significant losses of vitamins, carotenoids and anthocyanins. High temperature accelerates reactions, which would occur more slowly at ambient temperature. Whatever the processing method had chosen the retention of these compounds decreases with longer processing time, higher processing temperatures and cutting or maceration of the food.

On the other hand, as shown with carotenoids mechanical disruption of the food matrix or heat treatment increases bioavailability, presumably by softening or breaking cell walls and membranes, thereby facilitating the release of the health-promoting compounds from the food matrices as the first step in the digestion of food. Processing conditions should therefore be optimized so that bioavailability is increased, but without provoking significant degradation. Bioavailability refers to that portion of the nutrient or bioactive compound consumed that is released from the food matrix, absorbed and used by the body. In home preparation, retention of vitamins and bioactive compounds decreases in the following order: microwaving > steaming > boiling. For the fat-soluble carotenoids, boiling is less harmful than sautéing. Deep-frying, prolonged cooking, combination of several preparation methods all result in substantial losses. Loss of thiamin, riboflavin and Nicotinic acid during processing of tinned flesh food is as shown in Table-1. Whatever the cooking or processing method used, retention decreases with longer cooking time, higher cooking temperature and peeling/cutting of the food. Freezing (especially quick freezing) and frozen storage generally preserve nutrients and bioactive compounds, but slow thawing may be detrimental.

Table 1: Loss of thiamin, riboflavin and Nicotinic acid during processing of tinned flesh food.

S. No	Food item	Food form	Percentage loss of vitamin B forms		
			Thiamin	Riboflavin	Nicotinc acid
1.	Beef	Chopped	79	6	0
		Corned	56	12	0
		Roast	75	0	0
2.	Veel	Diced	79	0	0
	Pork	Chopped	55	0	0
		Bacon	59	29	16
		Sausage	55	0	52
3.	Lamb	Strained	84	0	13
4.	Poultry	Light meat	67		
		Dark meat	77		
5.	Fish		75		

Traditional sun-drying, although the cheapest and most accessible means of food preservation in poor countries, causes considerable destruction of nutrients and bioactive compounds. Drying in a solar dryer even of simple and inexpensive design can appreciably reduce losses. Protecting the food from direct sunlight also has a positive effect. Compared to losses during processing, subsequent storage often has a small but significant effect on vitamin and bioactive compound contents. Reaction rates are relatively slow at ambient or reduced temperature. Exclusion of oxygen (e.g., through vacuum or hot filling, oxygen impermeable packaging or inert atmosphere), protection from light, and storage at low temperature protect these compounds from decomposition. Nutrients in dehydrated products are more likely to undergo degradation during storage because of the greater surface area exposed to oxygen and light. Blanched products generally resist decomposition better than un-blanched foods during storage. In the last two decades, non-thermal processing technologies (e.g., high pressure processing, high-intensity pulsed electric field processing) have been introduced. These technologies have been reported to inactivate microorganisms and enzymes, without provoking adverse thermal effects on sensory and nutritional properties. Enzymatic degradation may be a problem in minimally processed foods. Loss of cellular integrity and of cell compartmentalization in fresh-cut fruits and vegetables, due to the cutting operations, destroys the natural protection of health-promoting substances and exposes them to oxidative and hydrolytic degradative enzymes. Modified atmosphere packaging and low temperature storage, used to extend the shelf-life of minimally processed foods, have also been shown to preserve micronutrients and bioactive compounds.

Effect of Processing on the Fruit Associated Bioactive Components

Plant produces various kinds of responses to stresses like microbiological attack, pest infestation and other physiological conditions and generates various chemical metabolites. These can help plant to increase resistance against these conditions. Phenolic compounds, which are mainly synthesized from phenylalanine produced by the shikimic acid pathway, are the most widely distributed in the plant kingdom. Plant tissues may contain up to several grams per kilogram of polyphenols. External stimuli such as microbial infections, ultraviolet radiation, low temperature, water and nutrition stresses induce their synthesis. Flavonoids (Fig. 1) are considered one of them. The most described property of flavonoids is their capacity to protect the organism against free radicals and oxygenated reactive species (ORS) produced during the metabolism of oxygen. The free radicals act also on the mediators of the inflammatory diseases, and accelerate the tissue damage. Moreover, cells lesions lead to an increase in the production of the ORS which induces the consumption and the depletion of the endogenous chelating agents. Several flavonoids reduce the cellular lesions related to ischaemia, by interfering with the activity of nitric oxide synthase. The nitric oxide is produced by various types of cells such as the endothelium cells and the macrophages. The nitric oxide is produced through the constitutive activity of nitric oxide synthase.

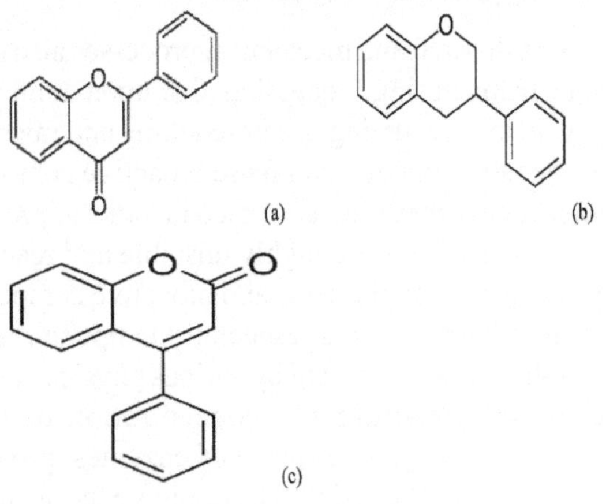

Fig. 1: Different forms of flavonoids: (a) Flavon; (b) Isoflavon; (c) phenylcoumarin

The xanthine dehydrogenase and the xanthine oxydase are implied in the metabolism of the xanthine into uric acid. The xanthine deshydrogenase is the configuration available under the normal physiological conditions, but it changes into xanthine oxydase during cells reperfusion (reoxygenation) and reacts with molecular oxygen to release the superoxide radical. Biological activities of flavonoids (anti-cancer, anti-inflammatory, and antioxidant) depend on the presence of several functional groups on the backbone of these compounds. Flavonoids occurs in food are as shown in Table 2.

Table 2: Functional components in Flavonoids from different sources

S. No	Flavonoids	Form
1.	Flavanons	Naringenin, Hesperidine
2.	Flavons	Luteolin, Apigenin
3.	Flavonols	Kaempferol, Quercetin
4.	Flavonols-3-ols	Catechin, Epicatechin
5	Anthocyanins	Petunidin 3-Oglucoside, Malvidin 3-Oglucoside

Different heating methods (drying, microwaving, heating by an autoclave, roasting, water immersion, pasteurization, pressured-steam heating, blanching) were used. Thermal processes have a large influence in flavonoid availability in foods which depends on their magnitude and duration. some authors studied the effects of thermal processes on model solutions of phenolic compounds; these studies are led especially on flavonoids. The data indicated that flavonoids in aqueous solutions show different sensitivity to heat treatment depending on their structures. However, whatever their structure a significant degradation is observed for temperature above 100°C. For rutin, a higher stability compared to its aglycon form (quercitin) is observed.

Apart from that, non thermal and mechanical processes also effect the flavonoid content in various fruits during processing. Common domestic processes such as boiling, frying, baking, sautéing, steam-cooking and microwaving were also reported to have negative impact over these bioactive components. Among the flavonoids, anthocyanins are the most studied in terms of processing and storage effects, probably because they are highly unstable and reactive, and they play the dual role of being natural pigments and bioactive compounds. They impart red, blue and purple hues to fruits, especially temperate fruits, Anthocyanin stability has been shown to be affected by various physical and chemical factors, such as the chemical structure and concentration of the anthocyanins, temperature, pH, light, oxygen, presence of enzymes, proteins, metallic ions and other flavonoids and phenolics. These pigments readily degrade during processing and storage of foods, resulting in dramatic impact on color and their health-promoting properties. Suggested pathways for anthocyanin degradation involve cleavage (loss of color) or polymerization (browning).

In alkaline or acid medium or in the presence of the enzyme b-glucosidase, anthocyanin is hydrolyzed, removing the sugar moiety and releasing the anthocyanidin, which is more easily degraded than the anthocyanin glycoside. Degradation commence with the opening of the middle ring followed by cleavage at this midpoint of the molecule, forming colorless products. With its reactivity and solubility in water, folates can suffer large losses during food processing and preparation. Aside from considerable leaching to water used in washing, blanching, canning or cooking, oxidative degradation can occur. Unfortunately, studies on processing effects on folates are still very limited. The naturally occurring form, tetrahydrofolic acid, is extremely susceptible to oxidative degradation while the synthetic form used in food fortification, folic acid, is very stable. Oxidative cleavage of tetrahydrofolic acid, and to a lesser extent, folic acid yields nutritionally inactive products.

Preliminary Operation and Effects

For most subclasses of polyphenols (e.g., phenolic acids, flavonoids), the effects of storage and processing on the contents were found negligible in comparison to the differences between varieties of plants. The concentrations of micronutrients and bioactive compounds in plant foods at harvest are known to vary with such factors as the variety or cultivar, stage of maturity, geographic or climatic effects, agricultural practices, soil composition. In animal foods, the composition varies with the breed and age of the animal, the composition of the feed and breeding practices. Postharvest compositional changes also occur, depending on the conditions (e.g., temperature, exposure to light) and duration of transport and storage. Carotenoids are among the food components most affected by environmental conditions. Sunlight and higher ambient temperature promotes the biosynthesis of carotenoids, thereby increasing the carotenoid levels during maturation, but accelerates oxidative degradation once the cellular integrity is destroyed (e.g., by bruising, cutting, chopping, maceration). Carotenoids and anthocyanin levels continue to rise after harvest, provided the fruit or root is kept intact, allowing enzymes to continue to catalyze biosynthesis.

Vitamins, carotenoids, flavonoids and fiber are more concentrated in the peel than in the pulp. Thus, the simple trimming or peeling of fruits and vegetables can discard appreciable amounts and significantly reduce the levels of these substances in the portions utilized. Alkaline treatments to facilitate peeling can also cause losses, although relatively small, of labile vitamins such as folate, ascorbic acid and thiamin at the surface of the product. Vitamins and bioactive compounds are naturally protected in plant tissues. Cutting, chopping, shredding and pulping of fruits and vegetables destroy this protection, increase exposure to oxygen and release enzymes that catalyze their degradation. Enzymatic degradation may be a more serious problem than thermal decomposition in many foods. Thus, thermal processing should be carried out immediately after peeling and cutting operations. Exposure of cut or bruised tissues of food products to water causes loss of water-soluble vitamins, such as ascorbic acid (vitamin C) and folate, and anthocyanin. This occurs by leaching during washing and exposure to water during cooking/processing. Grinding and fractionation during milling of cereal grains to remove the bran and the embryo cause major losses of minerals and vitamins, which are concentrated in the embryo and bran. Unlike vitamins and bioactive compounds, minerals are not destroyed by exposure to heat, light, oxidizing agents and extreme pH. They can be removed from foods by leaching or physical separation. A major cause of mineral loss is milling of cereals, which leaves the mineral poor endosperm.

Considering the research findings on the effects of processing and storage of foods on micronutrients and bioactive compounds, the following measures are recommended to maximize retention of these food components.

- For domestic purpose it is advised to bring varieties rich in micronutrient and bioactive components,
- Storage at refrigerated temperature will keep raw fruits and vegetables intact for prolonged time,
- Avoid cutting of fruits and vegetables into very small pieces and consume or cook them immediately after peeling, cutting, chopping or pulping.
- For commercial processing along with aforesaid precautions, minimize processing time and temperature, high-temperature, short time processing is a good alternative,
- Exclude oxygen through vacuum or hot-filling, oxygen-permeable packing or inert atmosphere etc.

References

1. Amarowicz, R. Carle, R. Dongowski, G. Durazzo, A. Galensa, R. Kammerer, D. Maiani, G. and Piskula, M.K. (2009). Influence of postharvest processing and storage on the content of phenolic acids and flavonoids in foods. Journal of Molecular Nutrition and Food Research, 53:S151-S183.
2. Boon, C. S. McClements, D.J. Weiss, J. and Decker, E.A. (2010). Factors influencing the chemical stability of carotenoids in foods. Critical Review in Food Science and Nutrition, 50:515-532.
3. Burger, I.H. and Walter C.L. (1973). The effect of processing on nutritive value of flesh foods. Proceedings of the nutrition society. Proc. Nutrition Society, 32 (1):1-8.
4. Castañeda-Ovando, A. Pacheco-Hernández, M.L. Páez-Hernández, M.E. Rodriguez, J.A. Galan-Vidal, C.A. (2009). Chemical studies of anthocyanins: A review. Food Chemistry, 113:859-871.

5. Damodaran, S. Parkin, K.L. and Fennema, O.R. (Eds.) (2008). Fenema's Food Chemistry, 4th ed., CRC Press, Boca Raton.
6. Lee, S.K. and Kader, A.A. (2000). Pre-harvest and postharvest factors influencing vitamin C content of horticultural crops. Postharvest Biology and Technology, 20:207-220.
7. Patras, A. Brunton, N.P. O'Donnell, C. and Tiwari, B.K. (2010) Effect of thermal processing on anthocyanion stability in foods; mechanisms and kinetics of degradation. Trends in Food Science and Technology, 21:3-11.
8. Rodriguez-Amaya, D.B. (1997). Carotenoids and Food Preparation: The Retention of Provitamin Carotenoids in Prepared, Processed, and Stored foods. US AID-OMNI, Washington DC.
9. Rodriguez-Amaya, D.B. (1999). Changes in carotenoids during processing and storage of foods. Arch. Latinoamer. Nutrition, 49:38S-47S.
10. Rodriguez-Amaya, D.B. (2002). Effects of processing and storage on food carotenoids. Sight Life Newsletter, 3 (Special Issue): 25-35.
11. Ruiz-Rodriguez, A. Marín, F.R. Ocaña A. and Soler-Rivas, C. (2008). Effect of domestic processing on bioactive compounds. Phytochemical Review, 7:345-384.
12. Sadilova, E. Carle, R. and Stintzing, F.C. (2007). Thermal degradation of anthocyanins and its impact on color and in vitro antioxidant capacity. Molecular Nutrition and Food Research, 51:1461-1471.
13. Wrolstad, R.E. Durst, R.W. and Lee, J. (2005). Tracking color and pigment changes in anthocyanin products. Trends in Food Science and Technology. 16: 423-428.
14. Xianquan, S. Shi, J. Kakuda, Y. and Yueming, J. (2005). Stability of lycopene during processing and storage. Journal of Medical and Food, 8:413-422.

Unit - II: Functional Foods and Nutraceuticals Their Health Benefits

CHAPTER 8

Aloe Vera: Bioactive Compounds Therapeutic Properties and Food Applications

Bhupendar Singh Khatkar

Abstract

Health benefits of aloe vera include its application in wound healing, treating burns, protection against skin damage, lung cancer, intestinal problems, increasing high density lipoprotein (HDL), reducing low density lipoprotein (LDL), reducing blood sugar in diabetics, fighting acquired immune deficiency syndrome (AIDS), allergies and improving immune system. Phytochemistry of aloe vera gel has revealed the presence of more than 200 bioactive compounds. Aloe vera gel is extracted from its leaves and appropriate processing techniques are needed for stabilization and preparation of the food products. Aloe vera gel products have become available in our country for internal consumption as a tonic and diversification in its products is on progress. The use of aloe vera has gained popularity because of herbal movement initiated by naturopaths, yog gurus, alternative medicine promoters and holistic healers. A recent market analysis report indicates that Americans have spent almost 40 billion dollars on functional foods, drinks and supplements for the improvement of their appearance as well as to provide energy and added nutrition to handle health issues such as hypercholesterolemia and diabetes.

Introduction

Aloe vera has been used for its medicinal value since rigvedic times. There are at least four main species of aloe which have therapeutic properties viz. Aloe barbadensis Mill Syn. Aloe vera (L), Aloe perryi Baker, Aloe ferox and Aloe arboresnens. Among these species Aloe vera is considered medicinally the most potent. The use of aloe vera has gained popularity because of herbal movement initiated by naturopaths, yog gurus, alternative medicine promoters and holistic healers. A recent market analysis report indicates that Americans have spent almost 40 billion dollars on functional foods, drinks and supplements for the improvement of their appearance as well as to provide energy and added nutrition to handle health issues such as hypercholesterolemia and diabetes. Aloe vera products are among the popular choices for these applications. Today, the aloe vera industry is flourishing and the gel is used in many food products such as fresh gel, juice and other formulations for health, medicinal and cosmetic purpose. As awareness about therapeutic effects of aloe vera is increasing among people, demand for aloe vera gel products is also gaining momentum. In our country a booming economy has resulted in overall increase in disposable income of population. Added to this unhealthy eating habits coupled with sedentary lifestyle have led to increase in incidence of diet related health issues. On the other hand, there is a growing awareness among public that health can be improved significantly by avoiding junk foods and eating foods that offer special health benefits beyond nutrition (Ahlawat and Khatkar, 2011).

Bioactive Compounds and their Therapeutic Properties

Aloe vera leaf yields two liquids; yellow exudates from the pericyclic tubules surrounding the vascular bundles located below the epidermis and a colourless mucilaginous gel from parenchyma. The major and active constituents of yellow exudates are hydroxyanthracene derivatives such as the anthraquinone glycosides aloin A and aloin B, collectively known as barbaloin (Saccu et al., 2001). The yellow exudates in dried form is known as 'bitter aloe', Barbados aloe or curacao aloes finds pharmaceutical applications. The aloe vera gel contains quite different compounds from the yellow exudates. The gel consists primarily of water (> 98%) and carbohydrates which constitute major fraction of remaining solid portion (Ahlawat et al., 2013). The total solid content of aloe vera gel is 0.66% and soluble solids are 0.56% with some seasonal fluctuation. On dry matter basis, aloe gel consists of polysaccharides (55%), sugars (17%), minerals (16%), proteins (7%), lipids (4%) and phenolic compounds (1%).

There are as many as 200 different types of bioactive compounds in aloe vera (Davis, 1997). Analysis of the aloe vera gel by Meadows (1980) indicated that at least four different partially acetylated glucomannans (acemannan) were thought to be responsible for producing the thick, stringy mucilage-like properties that are characteristic of the 'raw' aloe vera gel. Acemannan accelerates wound healing, modulate immune function and demonstrate antineoplastic and antiviral effects. The aloe vera gel contains many vitamins namely vitamins A, C, E, vitamin B_1 (thiamine), niacin, vitamin B_2 (riboflavin), choline and folic acid. Some authors also suggested the presence of vitamins B_{12} (cyanocobalamin) in trace amounts which is normally available in animal source. About 20 amino acids and 7 essential amino acids required by human body are also present in aloe vera gel.

Major bioactive compounds occurring in aloe vera pulp and exudates are summarized in Table 1. A novel anti-inflammatory compounds, C-glycosyl chromones, has been isolated from aloe gel. Saponins are the soapy substances which are considered to effect general cleansing and antiseptic properties to aloe gel. The sterols include comperterol, β-sitosterol and lupeol. Salicylic acid is an aspirin like compound possessing pain relieving properties. Aloe vera gel has been evaluated for antioxidant potential and the study showed significant presence of antioxidant in aloe extracts. A three years old plant extract exhibits the strongest free radical scavenging activity of 72.19%, which is significantly higher than that of Butyl hydroxyl toluene (BHT) having 70.52% and α-tocopherol with 65.65%.

It is suggested that growth stage in aloe plant plays a vital role in the composition and antioxidant activity. Aloe vera juice also has antibacterial properties against Gram- positive bacteria. A study showed that aloe vera gel increases the bioavailability of vitamin C and vitamin E in human beings. When vitamin C is ingested with aloe vera gel its bioavailability increases 3 times and that of vitamin E is 3.7 times. The mechanism of action of the aloe products to improve the bioavailability of the vitamins was explained to be a possible protection effect against the degradation of the vitamins in the intestinal tract as well as binding of the polysaccharides to the vitamins and thereby slowing down the absorption rate.

The bioactive compounds of aloe vera have been credited with a number of therapeutic effects (Table 2). Aloe vera is a well-known anti-inflammatory and wound healer, accelerating the rapid growth of epithelial tissue. Aloe vera is effective in treating wounds and reducing inflammation by the action of mannose-6-phosphate (a major sugar in aloe gel). Aloe vera is known to have a marked effect in the treatment of scar tissue and the prevention of scar formation following injury to the skin. This is because aloe vera stimulates cell production through the activity of the amino acids, which are the basis for new cell formation, and also, due to the ability of its enzymes, promotes regeneration at the deepest layers of the skin. Aloe vera has also been used in the treatment of burns. It is suggested that lectin may be responsible for the therapeutic effect of the gel on burns.

Aloe vera can be used successfully in the general treatment of skin ulcers, including mouth ulcers, cold sores (herpes simplex), and leg ulcers. This is possibly due to the anti-virucidal effect of the aloe vera gel at concentrations of about 80%. It is reported that the use of aloe vera gel internally taken to treat peptic ulcers and disturbances of the gastrointestinal tract. In the field of dentistry, aloe vera has been used to treat a variety of dental conditions and has been found to relieve pain and accelerate healing after periodontal flap surgery (Payne, 1970).

Table 1: Summary of the bioactive compounds of aloe vera gel and exudates

Class (es)	Compounds
Anthraquinones/anthrones	Aloe-emodin, aloetic-acid, anthranol, aloin A and B (or collectively known as barbaloin), isobarbaloin, emodin, ester of cinnamic acid
Carbohydrates	Pure mannan, acetylated mannan, acetylated glucomannan, glucogalactomannan, galactan, galactogalacturan, arabino galactan, galactoglucoarabinomannan, pectic substance, xylan, cellulose
Chromones	8-C-glucosyl-(2'-O-cinnamoyl)-7-O-methylaloediol A, 8-C-glucosyl-(S)-aloesol, 8-C-glucosyl-7-O-methyl-(S)-aloesol, 8-C-glucosyl-7-O-methylaloediol, 8-C-glucosyl-noreugenin, isoaloeresin D, isorabaichromone, neoaloesin A
Enzymes	Alkaline phosphatase, amylase, carboxypeptidase, catalase, cyclooxidase, cyclooxygenase, lipase, oxidase, phosphoenol, pyruvate carboxylase, superoxide dismutase
Minerals	Calcium, chlorine, chromium, copper, iron, magnesium, manganese, potassium, phosphorous, sodium, zinc
Lipids and miscellaneous organic compounds	Arachidonic acid, γ-linolenic acid, steroids (campestrol, cholesterol, β- sitosterol), triglicerides, triterpenoid, gibberillin, lignins, potassium sorbate, salicylic acid, uric acid
Amino acids	Alanine, arginine, aspartic acid, glutamic acid, glycine, histidine, hydroxyproline, isoleucine, leucine, lysine, methionine, phenylalanine, proline, threonine, tyrosine, valine
Proteins	Lectins, lectin-like substance
Saccharides	Mannose, glucose, L-rhamnose, aldopentose
Vitamins	B_1, B_2, B_6, C, β-carotene, choline, folic acid, α-tocopherol

Sources: Dagne et al., (2000)

Table 2: Health benefits of bioactive compounds of aloe vera

Chemical component	Health benefits
Acamannan	Accelerate wound healing, modulate immune system, Antineoplastic and antiviral effects
Alprogen	Anti-allergicC-glycosyl chromone
Anti-inflammatory	Bradykininase
Anti-inflammatory Magnisium lactate	Anti-allergic
Salicylic acid	Analgesic, anti-inflammatory

Sources: Shelton (1991)

The use of aloe vera gel has also been described in veterinary medicines. The gel extract has been used in the treatment of a number of external conditions in many animals (Northway, 1975). These conditions include ringworm, allergies, abscesses, fungal infections, various types of inflammation, pain, and itching. Recent studies indicate that aloe vera can be used in the treatment of HIV-AIDS. This is attributed to the anti-viral and immuno-modulating properties of acemannan (a group of mucopolysaccharides found in Aloe vera), which have direct effects on the cells of the immune system, activating and stimulating macrophages, monocytes, antibodies, and T-cells. The acemannan acts as a bridge between foreign proteins, such as virus particles and macrophages, thus facilitating phagocytosis. This is a key component in boosting cell-mediated immunity, which is deficient in HIV infection. Aloe vera has also found its use in the treatment of cancer and has been demonstrated to have a positive effect on the inhibition of tumour growth. The aloe vera gel has found an extensive application in the cosmetic and toiletry industries. This is principally due to its valuable moisturizing emollient effect. The cosmetic formulations available include a large range of moisturizing creams, cleansers, shampoos, and soaps. Aloe extracts have been incorporated into shaving creams and lotions to promote the healing of cuts. Aloe vera's application in the pharmaceutical industry is highly significant. It is used in the manufacture of medical products, such as burn treatments, ointments, and medicated creams and lotions for topical applications to fight various skin disorders.

Food Applications of Aloe Vera Gel

The industrial processing of aloe vera gel and its products invariably involve unit operations such agitation, filtration, pasteurization, evaporation and drying. Technical and economical feasibility of these industrial processes depends greatly on knowledge of physico-chemical properties and rheological behavior of aloe vera gel. The knowledge of rheological properties of aloe vera gel can be exploited for its functionality in food systems, including textural attributes and mouth feel. As a matter of fact, the polysaccharides found in aloe gel are not stable, especially under conditions such as heat, the presence of acid and enzymatic activities. Processing may cause irreversible modifications to the bioactive polysaccharides components, affecting their original structure which may diminish the proposed physiological and therapeutical properties of these constituents. Thus, preservation of bioactivity and physiological effectively of aloe vera gel constituents is of great concern. Seeing the popularity of aloe vera gel as food supplements there is great potential to process aloe vera gel to its products like juice, concentrate and powder to diversify its food uses (Table-3). Incorporation of aloe vera gel in already popular food products to impart therapeutic effects is a promising option in food processing sector.

Table 3: Food applications of aloe vera gel products

Aloe vera products	Food applications
Gel fillet	Candies, bar, munch, chewing gum, instant aloe vera tea granules, aloe vera gum for sore or bleeding gums, candy type aloe vitamins, aloe vera fruit smoothies
Juice	Ready to serve drink, health drink, soft drink, laxative drink, sherbet, sports drink (with electrolytes), diet drink with soluble fiber, hangover drink with B- vitamins, amino acids and acetaminophen, healthy vegetable juice mix, yoghurts, aloe vera mix for whiskey or other alcohol, white bread with aloe vera, and cucumber juice with aloe vera
Concentrate	Squash, jam, jellies, aloe vera concentrate can also be mixed with tea, water or juice
Powder	Yoghurt, curd, 'lassi', ice-cream, and aloe vera 'laddu'

Sources: Eshun and He (2004), Hamman (2008).

Conclusion

Aloe vera leaf yields two liquids viz. yellow exudates a colourless gel. The major bioactive compound of yellow exudates is aloin which is credited with pharmaceutical applications. Some researchers have reported acemannan as the main active substance present in the aloe vera gel which is reported to accelerate wound healing, modulate immune function and demonstrates antineoplastic and antiviral effects. Recent studies indicate that aloe vera gel can also be used in the treatment of HIV-AIDS. The aloe vera gel is presently used in many food products such as juice, jam, yoghurt, ice-cream and other formulations for health, medicinal and cosmetic purposes. The aloe vera gel has found an extensive application in the cosmetic and toiletry industries.

References

1. Ahlawat, K.S. and Khatkar, B. S. (2011). Processing, food applications and safety of aloe vera products: a review. Journal of Food Science and Technology, 48 (5):525-533.
2. Ahlawat, K.S. Khatkar, B.S. Sushil and Gulia, N. (2013). Effect of processing and storage on sensory properties and acemannan content of aloe vera gel. Annals of Agri-bio Research, 18 (1):70-73.
3. Davis, R.H. (1997). Aloe vera- A scientific approach. Vantage Press Inc. New York, USA, 290-306.
4. Dagne, E., Bisrat, D. Viljoen, A. and Van Wyk, B.E. (2000). Chemistry of Aloe species. Current Organic Chemistry, 4:1055-1078.
5. Eshun, K. and He, Q. (2004). Aloe vera: A valuable ingredient for the food, pharmaceutical and cosmetic industries: A review. Critical Review in Food Science and Nutrition, 44:91–96.
6. Hamman, J.H. (2008). Composition and application of aloe vera leaf gel. Molecules, 13:1599-1616.
7. Meadows, T.P. (1980). Aloe as a humectant in new skin preparation. Cosmetics and Toiletries, 95(11):51-56.
8. Northway, R.B. (1975). Experimental use of Aloe vera extract in clinical practice. Veterinary Medicine/Small Animal Clinician. 70:89.
9. Payne, J.M. (1970). Tissue response to aloe vera gel following periodontal surgery. Thesis submitted to Faculty of Baylor University in partial fulfillment of the requirements for the Degree of Master of Science.
10. Saccu, D. Bogoni, P. and Procida, G. (2001). Aloe exudate: characterization by reversed phase HPLC and headspace GC–MS. Journal of Agriculture and Food Chemistry, 49(10):4526–4530.
11. Shelton, M. (1991). Aloe vera, its chemical and therapeutic properties. International journal of Dermatology, 30:679-683.

CHAPTER 9

Phytochemicals and Health Potentials

Savita Sharma and Swati Kapoor

Abstract

Upcoming era of functional foods has led food scientists and food processors to deeply understand the role of phytochemicals in promoting human health. Phytochemicals are the secondary metabolites produced by plants and are commonly found in our daily food. Natural food products seem to be good substitute for pharmaceutical supplements to improve health status of people. Therefore, consumer's interest is growing in the food items naturally containing phytochemicals, each having there own functional and health aspects. Most discussed phytochemicals in the present paper are carotenoids such as lycopene, beta-carotene, lutein and, polyphenols such as phenolic acids, flavonoids, anthocyanins and isoflavones. Major dietary source for carotenoids include orange and yellow fruits and vegetables as well as green leafy vegetables whereas polyphenols are found abundantly in green tea, wine, berries, cocoa, broccoli. Carotenoids can also be used as colorants in foods during processing. The role of polyphenols as an antioxidant is well characterized because of there ability to scavenge free radicals or bind pro-oxidant metal ions by means of their hydroxyl group. Carotenoids also play an important role in protection against photooxidative processes by acting as singlet molecular oxygen as well as peroxyl radical scavengers, thus acting as potent antioxidant. Apart from antioxidative properties multitude of health benefits are associated with carotenoids and polyphenols such as reduced risk of age-related macular degeneration, certain cancers such as prostate, lung and colon, prevention of cardiovascular disease and enhancing immunity towards infection. Hence, continuous developments are being made to combine phytochemicals to avoid improper combinations that may nullify the effect of supplements and cause lower absorbance.

Keywords: Phytochemicals, carotenoids, polyphenols, antioxidant, photosensitive diseases

Introduction

In recent years, concern has risen among the consumers regarding the inappropriate dietary habits as a major cause of poor health and has resulted in increase health care costs. These concerns have made the food scientists to search for alternatives of supplement pills which has led to the discovery of functional foods or super foods that are rich in phytochemicals and form a part of daily diet. The term *Phytochemicals* mean "plant chemicals" or non-nutrient biologically active chemical compounds found in plants. These are the secondary metabolites secreted by plants and responsible for much of the disease prevention. The most extensively studied phytochemicals are carotenoids, polyphenols, phytoestrogens, polyunsaturated fatty acids, tocopherols and conjugated linoleic acid. A phytochemical is mostly found in the coloring agent of fruits and vegetables, so eating brighter colored varieties may have benefits (Tyagi et al., 2010). Carotenoids are mainly responsible for the red, yellow and orange color of fruits and vegetables. In addition to their function as a natural pigments and precursor of pro-vitamin A (certain carotenoids) they can be used as food additives for coloring (European Parliament and Council Directive, 1994). Polyphenols contribute to the healthful properties of plants and are responsible for organoleptic qualities such as coloring of leaves and fruits and physiological processes which include anti-bacterial, anti-viral, anti-inflammatory and anti-proliferative properties. Polyphenols are the most abundant antioxidants in the diet known to have a probable role in the prevention of various diseases associated with oxidative stress, such as cancer and cardiovascular and neurodegenerative diseases. This paper mainly deals with the role of carotenoids and polyphenols in promoting human health and there major dietary source.

Classification of Phytochemicals

Based on their chemical structure, phytochemicals are classified into the ten categories (Arts and Hollman, 2005) The major classes of phytochemicals are- i) Carotenoids such as beta-carotene, α-carotene, lutein, lycopene; ii) Phenolics which include phenolic acids (caffeic acid, ferulic acid), flavonoids (Flavones, anthocyanidins, isoflavones), tannins, stilbenes and coumarins; iii) Alkaloids; iv) Nitrogen containing compounds and v) Organosulphur compounds namely isothiocyanates, allicin (Fig.1) (Liu, 2004).

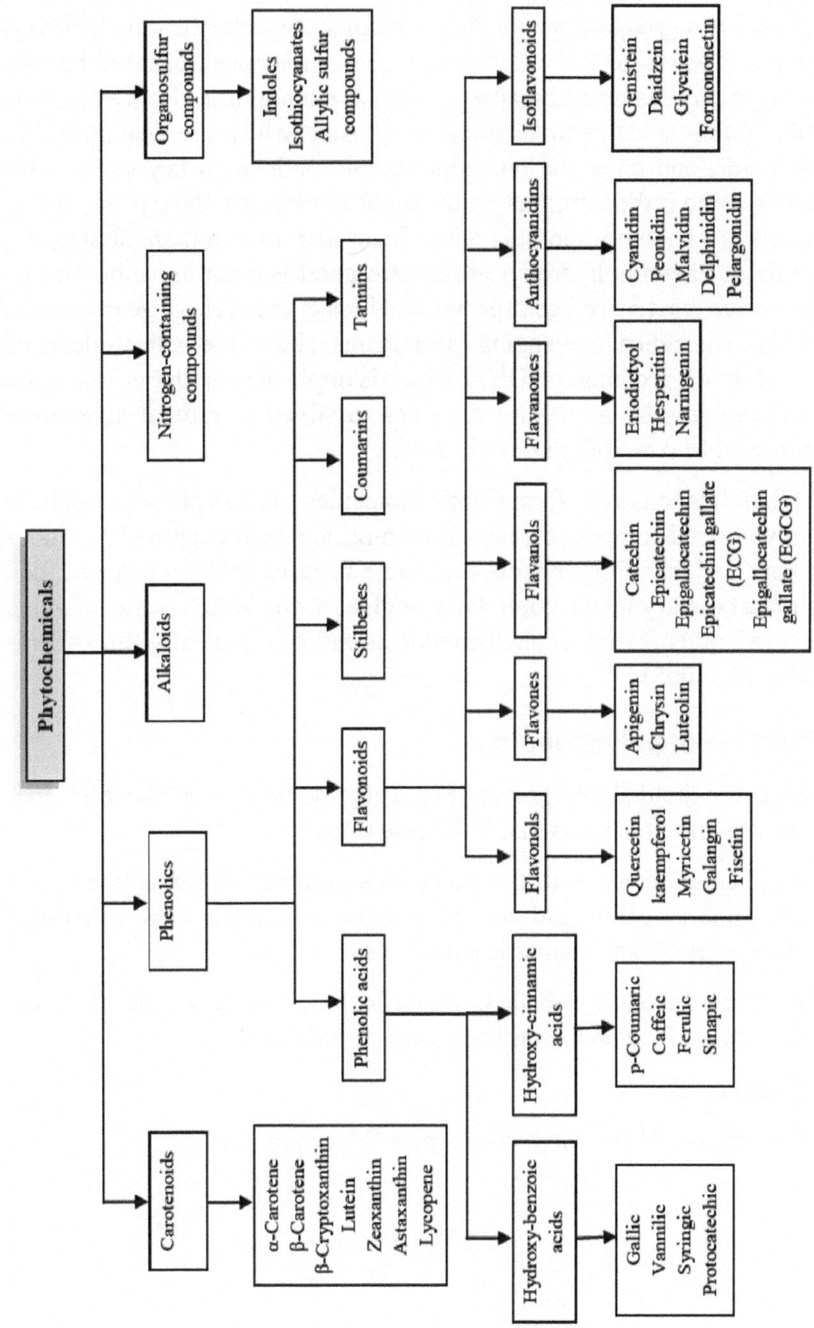

Fig.1: Classification of dietary phytochemicals (Liu, 2004)

Polyphenols

Polyphenols are a group of chemicals containing multiple phenolic hydroxyl groups and is found in many fruits, vegetables, and other plants, such as berries, walnuts, olives, tea leaves and grapes. Most polyphenols arise from a common origin: the amino acids phenylalanine or tyrosine which are deaminated to cinnamic acids, and enter the phenylpropanoid pathway a key step in this biosynthetic route is the introduction of one or more (Castellano et al., 2012). Polyphenols are known to possess antioxidant activities and are abundantly found in food. Phenolic hydroxyl groups are good H-donating antioxidants, which scavenge reactive oxygen species and breaks the cycle of generation of new radicals. The radical scavenging antioxidants inhibit free-radical-mediated oxidation of lipids, proteins and DNA, which is implicated in illness. Phenolics act as antioxidants by inhibiting enzymes involved in radical generation (Fukumoto and Mazza 2000; Dewick, 2002).

Natural phenolics can range from simple molecules, such as phenolic acids, to highly polymerized compounds, such as tannins, and their occurrence in foods is extremely variable. Polyphenols contribute to taste and/or a color in food and their total dietary intake could be as high as 1 g/d, which is much higher than that of all other classes of phytochemicals and known dietary antioxidants (Scalbert et al., 2005).

Classification of polyphenols

Polyphenols are divided into groups (Fig. 2) by the number of phenolic rings and of the structural elements that link these rings:

1. The phenolic acids with the subclasses derived from hydroxybenzoic acids such as gallic acid and from hydroxycinnamic acid, containing caffeic, ferulic, and coumaric acid
2. The large flavonoid subclass, which includes the flavonols, flavones, isoflavones, flavanones, anthocyanidins, and flavanols
3. The stilbenes
4. The lignans and the polymeric lignins (Han et al., 2007).

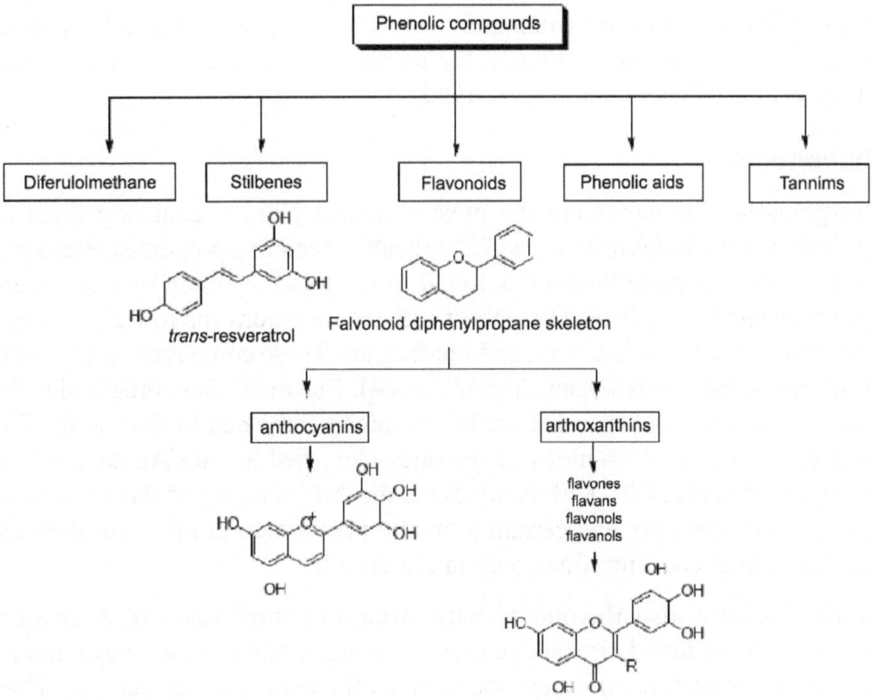

Fig. 2: Classification of dietary polyphenols (Han et al., 2007)

Phenolic acids

There are two classes of phenolic acids which can be distinguished: derivatives of benzoic acid and derivatives of cinnamic acid. Most of the plants are low in hydroxybenzoic acids e.g: hydrolyzable tannins such as gallotannins in mangoes and ellagitannins in red fruit such as strawberries, raspberries and blackberries. Hydroxycinnamic acids are found more extensively in plants and consist chiefly of *p*-coumaric, caffeic, ferulic, and sinapic acids. These acids are rarely found in the free form, except in processed food that has undergone freezing, sterilization or fermentation. Highest content of cinnamic acid is found in blueberries, kiwis, plums, cherries, apples etc (Manach et al., 2004).

Caffeic acid, both free and esterified, is generally the most abundant phenolic acid and represents between 75% and 100% of the total hydroxycinnamic acid content of most fruit. Ferulic acid is the most abundant phenolic acid found in cereal grains, which constitute its main dietary source. The ferulic acid content of wheat grain is 0.8–2 g/kg dry wt, which may represent up to 90% of total polyphenols (Sosulski et al., 1982; Lempereur et al., 1997). The ferulic acid content of different wheat flours is thus directly related to levels of sieving, and bran is the main source of polyphenols (Hatcher and Kruger, 1997). Rice and oat flours contain approximately the same quantity of phenolic acids as wheat flour (63 mg/kg), although the content in maize flour is about three times as high (Shahidi and Naczk, 1995).

Flavonoids

Polyphenolic flavonoids are the most abundant plant phenolics present as secondary metabolites in plants and have health promoting properties. Flavonols are the most common flavonoids found in food and the main flavonoids are quercetin and kaempferol. The richest sources are onions (up to 1.2 g/kg fresh wt), curly kale, leeks, broccoli, and blueberries. These compounds are present in glycosylated forms (Manach et al., 2004). Flavones also come under the category of flavonoids but thet are less common compared to flavonoids. The only important edible sources of flavones identified to date are parsley and celery (Hertog et al 1992). Flavonones are the third category of flavonoids and are found in tomatoes and certain aromatic plants such as mint, but they are present in high concentrations only in citrus fruit.

Isoflavones are also flavonoids with structural similarities to estrogens. Isoflavones are found almost exclusively in leguminous plants. Soya and its processed products are the main source of isoflavones in the human diet. They contain 3 main molecules: genistein, daidzein, and glycitein. Soybeans contain between 580 and 3800 mg isoflavones/kg fresh wt, and soymilk contains between 30 and 175 mg/L (Cassidy et al., 2000).

Anthocyanins

Anthocyanins are pigments dissolved in the vacuolar sap of the epidermal tissues of flowers and fruit, to which they impart a pink, red, blue, or purple color (Mazza and Maniati, 1993). According to pH they exist in different chemical forms either colored or uncolored. In the human diet, anthocyanins are found in red wine, certain varieties of cereals, and certain leafy and root vegetables (aubergines, cabbage, beans, onions, radishes), but they are most abundant in fruit.

Anthocyanins have variable structure and are associated with sugars including glucose, rhamnose, xylose, galactose, arabinose, and fructose. Currently, there are 19 naturally occurring anthocyanidins. Some of the commonly found anthocyanins are peonidin, malvidin, cyanidin, delphinidin, pelargonidin, and petunidin (Prior, 2003). Anthocyanins with the highest antioxidant activity are delphinidin, cyanidin and cyanidin-3-glucoside (Fukumoto and Mazza, 2000) and they are known to have more antioxidant activity then vitamin E and vitamin C (Duthie, 2007).

Anthocyanin content in food is generally proportional to color intensity and reach values up to 2–4 g/kg fresh wt in blackcurrants or blackberries. Anthocyanins are found mainly in the skin, except for certain types of red fruit, in which they also occur in the flesh (cherries and strawberries). The health effects of polyphenols depend on both their respective intakes and their bioavailability, which can vary greatly. However, anthocyanins are generally known to have low bioavailability and structure of anthocyanin determines the degree of bioavailability (McGhie et al., 2003).

Table-1 enlists some of the dietary sources of polyphenols that are present either naturally or in processed form. They are often used in food industry as natural colorants in foods and beverages (Manach et al., 2004) to replace synthetic red dyes. However, enocyanin and lees (sediment of the grape juice tanks) preparations are the only anthocyanin sources approved by FDA (in the U.S.) to be used for human food and anthocyanins from grape extract has been used in beverages and soft drinks @ approximately 3 kg of a 1% anthocyanin extract added to 1000 L of beverage to impart deep red colour (Bridle and Timberlake, 1996; Francis, 1989). Anthocyanins present in blue and purple corn are being used for the production of naturally colored blue tortillas (Sugiyama Chemical Institute, 1977). Fermented red rice is being sold as a dietary supplement and marketed as Cholestin to help reduce cholesterol levels (Pharmanex Inc.). The instability of anthocyanin pigment however limits its use in variety of food products as it is pH and temperature sensitive.

Health Benefits of Phenolic Compounds

- Dietary anthocyanins may play a role in health promotion and protection from cardiovascular disease and cancer by acting as dietary antioxidants, reducing inflammation and lipid oxidation.
- The potential benefits to human health of flavonoids include antiviral, antitoxic, anti-fungal, antibacterial, anti-allergic, anti-inflammatory and antioxidant activities.
- Caffeic acid protects against immunoregulation diseases, asthma and allergic reactions.
- Caffeic acid and several of its esters might have some activity against colon cancer and caffeic acid derivatives like dicaffeoylquinic and dicaffeolytartaric acids act as potent and selective inhibitors of HIV-1 integrase, being currently considered for their potential antiviral therapy.

Table 1: Dietary sources of polyphenols (Han *et al* 2007)

Class and subclass	Dietary polyphenol	Foods or beverages
FlavonoidsAnthocyanidins	Cyanidin 3-galactoside Cyanidin 3-glucoside Cyanidin 3-xyloside MalvidinDelphinidin Pelargonidin	**Fruits:** blackberries,black currant,blueberries, black grape, elderberries, strawberries, cherries, plums, cranberry, pomegranate juice, raspberry **Others:** red wine
AnthoxanthinsFlavonols	Myricetin Fisetin Quercetin Kaempferol Isorhamnetin	**Vegetables:** capers, celery, chives, onions, red onions, dock leaves, fennel, hot peppers, cherry tomatoes, spinach, sweet potato leaves, lettuce, celery, broccoli, Hartwort leaves, kaleCereal: buckwheat, beans(green/yellow) **Fruits:** apples, apricots, grapes, plums, bilberries, blackberries, blueberries, cranberries, olive elderberries, currants, cherries, black currant juice, apple juice, ginkgo bilobaSpices and herbs: dill weed **Others:** red wine, tea (green, black), tea (black beverage), cocoa powder, turnip (green), endive, leek
Flavanones	Naringenin Eriodictyol Hesperetin	**Citrus fruits and juices:** lemon, lemon juice, lime juice, orange, orange juice, grapefruit, tangerine juiceSpices and herbs: peppermint
Flavones	ApigeninLuteolin	**Fruits:** celery, olivesVegetables: hot peppers, celery hearts, fresh parsleySpices and herbs: oregano, rosemary, dry parsley, thyme
Isoflavones(Flavans)	Genistein Daidzein Equol	**Fruits:** grape seed/skinOthers: soybean, soy nuts, soy flour/bread, tofu, miso, soy milk, tofu yogurt, soy cheese/sauce/hot dog
Phenolic acids Hydroxycinnamic acids	Caffeic acid Chlorogenic acid Ferulic acid P-coumaric acid Sinapic acid	**Fruits:** bluberry, cranberry, pear, cherry(sweet), apple, orange, grapefruit, cherry juice, apple juice, lemon, peach,Vegetables: potato, lettuce, spinachOthers: coffee beans, tea, coffee, cider

Carotenoids

Among the phytochemicals, carotenoids have been the most studied. Carotenoids are tetraterpenoids responsible for the yellow, orange and red color of many fruits, vegetables, a few roots, egg yolk, fish like salmon and trout, crustaceans (Rodriguez et al. 2006). The principal carotenoids investigated mostly in terms of human health are: β-carotene, α-carotene, β-cryptoxanthin, lutein and lycopene. These are also the carotenoids most commonly found in foods (Rodriguez-Amaya, 1993, 1999). The three major dietary carotenoids that have provitamin A activity are α-carotene, β-carotene, and β-cryptoxanthin.

Carotenoids have extensive applications as anti-oxidants in dietary supplements, and as colors in foods and beverages as well as pigments in poultry and fish. The carotenoids used as food ingredients include astaxanthin, beta-apo-carotenal, canthaxanthin, beta-carotene, lutein, zeaxanthin and lycopene. Food carotenoids are tetraterpenoids built from isoprenoid units and a distinctive characteristic is an extensive conjugated double-bond system (Rodriguez–Amaya, 2001). Lycopene is the principle carotenoid primarily responsible for the red color of fruits and vegetables such as tomato, watermelon, red-fleshed papaya and guava, red or pink grapefruit and red carrots etc. and it has no provitamin A activity. It is the most potent antioxidant *in vitro* of all the carotenoids found in appreciable amounts in humans (Di Mascio et al., 1989). β-Cryptoxanthin is the main pigment of many orange-fleshed fruits, such as peach, nectarine, orange-fleshed papaya, persimmon etc. Lutein is the predominant carotenoid in leaves, green vegetables, and yellow flowers.

Food sources of carotenoids

There are about twenty carotenoids in the human diet. Their presence in leafy vegetables is masked by the green of chlorophyll. Carrots are the major source of β-carotene, although spinach, broccoli and watercress also contain substantial amounts. The predominant dietary source of lycopene is tomatoes. Good dietary sources of lutein include peas, sprouts, greens, broccoli, spinach and peppers. Mangoes, apricots and oranges are sources of cryptoxanthin and also contain some β carotene (Table 2).

Carotenoids and health benefits

Carotenoids particularly beta-carotene plays an important role in pro-vitamin A synthesis that is involve in vision, cell differentiation, synthesis of glycoprotein, mucus secretions from epithelial tissues, reproduction, overall growth and development of bones (Wolf, 1980). On a worldwide basis, about 60% of dietary vitamin A is estimated to come from pro-vitamin A (Simpson

1983). The daily pro-vitamin A intake recommended by the FAO is 250 to 400 retinol equivalents (RE) for children, 575 to 725 RE for adolescent and 750 RE for adults. Apart from this carotenoids have been linked with the enhancement of the immune system and decreased risk of degenerative disease such as cancer, cardiovascular disease, age related muscular degeneration and cataract formation (Mathews-Roth 1991; Bendich and Olson 1989). Carotenoids have also been identified as a potential inhibitor of Alzheimer's disease (Zaman et al., 1992). Fig.3 summarizes some of the common health benefits possessed by carotenoids.

Table 2: Dietary sources of carotenoids

Carotenoid	Food source
Lycopene	Tomatoes
β-cryptoxanthin	Citrus fruit
β-carotene	Carrot, mango, spinach, broccoli, watercress
Lutein	Green leafy veg.
Zeaxanthin	Green leafy veg., maize, orange peppers
α-carotene	Peppers, carrots
Capsanthin	Red peppers
Astaxanthin	Salmon, lobster
Canthaxanthin	Mushrooms

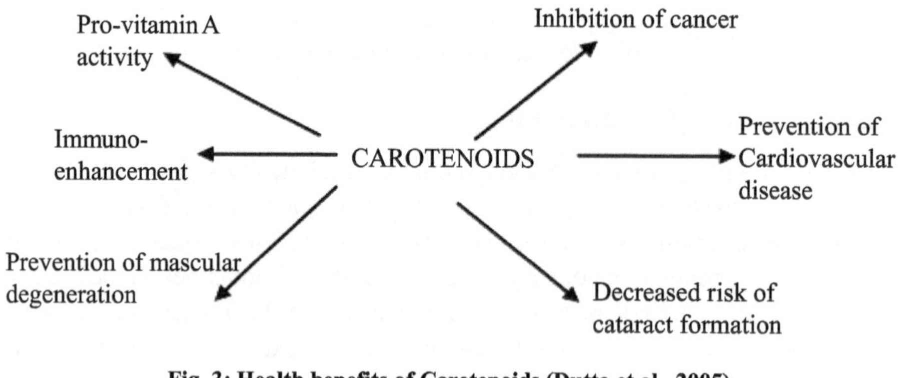

Fig. 3: Health benefits of Carotenoids (Dutta et al., 2005)

Antioxidant Property of Carotenoids

The electron rich conjugated double bond structure is primarily responsible for the excellent ability of β-carotene to physically quench singlet oxygen without degradation and for the chemical reactivity of β-carotene with free radicals and for its instability toward oxidation (Krinsky 1994). The maximum protection to quench singlet oxygen is given by those carotenoids having nine or more double bonds (Foote et al., 1970). The unique structure of lycopene molecule (13 double bonds) makes it a potent antioxidant twice as potent as β-carotene and ten times that of β-tocopherol in terms of its singlet oxygen quenching ability (Rao et al., 2003).

α- and β- carotene

α- carotene is one of the most commonly and widespread carotenoid found in foods. α- carotene is found as a minor component, accompanying β- carotene in carotenoid rich foods. Substantial amounts of α -carotene are found in carrot and some varieties of squash and pumpkin (Arima and Rodriguez-Amaya 1988, 1990). Structure of β- carotene is illustrated in Fig.4. α- carotene and β- carotene both have vitamin A activity but β-carotene has shown to be an effective source of vitamin A in dietary supplements and conventional foods. β-carotene is converted to vitamin A in the form of retinyl ester through cleavage of β-carotene molecule and majority of conversion takes place in intestinal mucosa (Gronowska- Senger and Wolf, 1970).

β-carotene is known to act as an immune modulator, quench singlet oxygen, and reduce peroxyl radicals at a low partial oxygen pressure (Wang and Russell, 1999). α-carotene inhibits the growth of neuroblastoma cells within 18 hours (Murakoshi et al., 1989) and in animal models it is reported to reduce the occurrence of liver, lung and skin cancers.

Bioavailability of β-carotene is usually influenced by matrix structure of fruits and vegetables and in relation to this absorption of β-carotene and other carotenoids is 5-30% compared to synthetic supplements due to presence of protein and fibre matrix around carotenoids which must be broken down to make it bioavailable (Castenmiller and West, 1998).

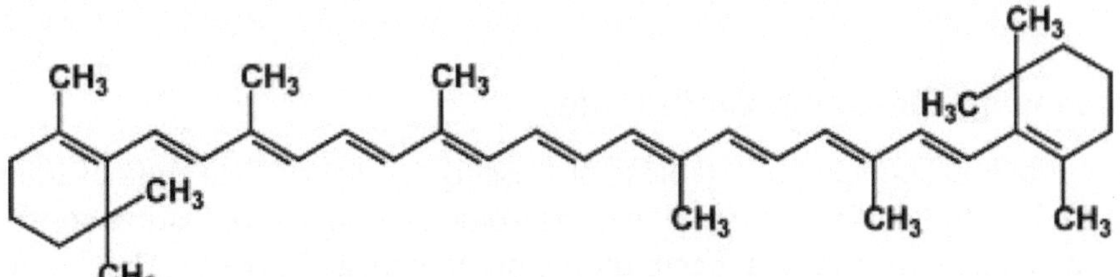

Fig. 4: β-carotene structure

Lycopene

Lycopene is a principal pigment responsible for coloration of many red fleshed fruits and vegetables. Lycopene is a vibrant tetrapenic carotenoid with molecular formula of $C_{40}H_{56}$ (Fig. 5) and contains 11 conjugated and 2 unconjugated double bonds (Fish et al., 2002). It undergoes *cis-trans* isomerisation upon interaction with light, temperature and chemicals (Ollanketo et al., 2001). Lycopene is better absorbed in body when accompanied with fats because of its lipophillic nature (Rao and Agarwal, 1999).

Among various carotenoids present in food, focus of the food scientists has shifted mainly towards lycopene because of the robust health benefits provided by it. An inverse relationship has been found between dietary lycopene with oxidative stress and positive impact on bone integrity (Mackinnon et al., 2011). Consumption of functional juice rich in lycopene led to a decline in total cholesterol (Jacob et al., 2008) and reduced the symptoms of hypercholesterolemia (Basuny et al., 2009). High intake of lycopene lowered the thickness of the internal layer of the blood vessels thus reducing the risk of myocardial infarction (Zhang and Hamauzu, 2004). Lycopene has also been found to reduce the risk of diabetes mellitus. An experiment was conducted by Jian et al., (2008) to support the statement by orally supplementing lycopene to normal rats and concluded that high dose of lycopene reduced glucose values significantly compared to control. Lycopene based functional drinks was reported to reduce malignant transformation of oxidized cholesterol in diabetic state (Naz et al., 2014). Lycopene is reported to be effective against immunodeficiency diseases like HIV and cerebral damage microangiopathy in Austrian stroke (Kun et al., 2006).

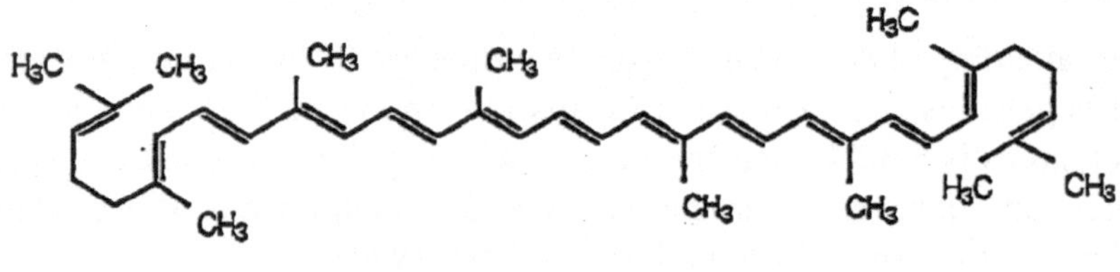

Fig. 5: Structure of Lycopene

Processing Effects on Carotenoids

Carotenoids are comparably stable to processing conditions than anthocyanins due to their lipophillic nature. However, isomerization and oxidation are the two main phenomenons that occur during processing and storage. The factors that induce oxidation are oxygen, metal traces, enzymes, prooxidants, exposure to light and severity of treatment (Rodriguez-Amaya, 2001). During heating of foods, *cis-trans* isomerization takes place altering the properties of carotenoids in processed foods e.g. processing of guava juice led to increase in *cis-* lycopene content (Padula and Rodriguez-Amaya, 1987) and insignificant decrease in trans-lycopene content. Processed tomato products such as tomato ketchup, puree and sauce have also been found to be high in *cis* isomers of lycopene.

According to Chasse et al., (2001) cis-isomers of lycopene were shown to be more stable, having higher antioxidant potential compared to the all-trans lycopene.

Processing of mango puree however, decreased β- carotene content by 13% and among the packaging material β- carotene degraded faster in bottled mango puree compared to canned mango puree (Rodriguez-Amaya, 2001). Canning of carrot juice that is one of the potential sources of β-carotene, resulted in high loss of carotenoids (Chen et al 1995) however, canning of sweet potatoes and tomatoes increased the percentage (39% and 33% respectively) of total *cis* isomers of provitamin A carotenoids (Lessin et al., 1997). Studies on the effect of canning and freezing process on the carotenoid content of carrot juice indicated that the canned juice contained significantly lower amounts of α- and β-carotene and total carotenoids than' home made' and frozen juice (Jun, 1987). Among the established novel technologies, effect of high-intensity pulsed electric fields on stabilization of carrot juice was studied and it was found that carrot juices had higher α-carotene and lower vitamin C contents than the untreated juices immediately after processing (Teixeira et al., 2009).

Conventional air drying is the most frequently used dehydration operation in food industry. Arya et al., (1982) reported that freezing before conventional hot air drying considerably improved the stability of carotenoids and lipids in dehydrated carrots. Carotenoids were more sensitive to high drying temperature than drying time (Mohamed and Hussain 1994). Excellent retention (96–98 %) of carotenoids in freeze dried carrots has been noticed (Rodriguez-Amaya, 1997).

Polyunsaturated Fatty Acids

Polyunsaturated fatty acids (PUFA's) are also known as Essential fatty acids and are known to play an important role in human health promotion. They are basically carboxylic acid containing 20-22 carbon chains having various double bonds. The first double bond when found between the third and the fourth carbon atom from the ω- carbon; these are called ω-3 fatty acids and if the first double bond is between the sixth and seventh carbon atom, then they are called ω-6 fatty acids (Rustan and Drevon, 2005). These cannot be synthesized in human body so have to be taken through diet. Out of all the polyunsaturated fatty acids, omega-3 and omega-6 fatty acids have received considerable attention since they contribute to the maintenance of whole hormone balance (El-Badry et al., 2007). The three major types of omega-3 fatty acids are alpha linolenic acid, which is the basic omega-3 fatty acid, eicosapentaenoic acid (EPA) and docosahexaenoic acid (DHA) and the primary omega-6 fatty acid is linoleic acid. Essential fatty acid health benefits include reduction in cardiovascular disease (Kinney 2006; Sijtsma and de Swaaf, 2004), lowering blood pressure, reduction in growth of artherosclerotic plaques (Balk et al., 2006), antibiotic in nature (El-Badry et al., 2007), mainatainance of central nervous system, enhancement of cognitive development (Mazza et al., 2007; Beltz et al., 2007) . The ratio between omega-3 and omega-6 fatty acid needs to be balanced for to maintain overall heath in humans. The recommended ratio of omega-6 to omega-3 has been set to 1-2 to 1 but due to consumption oh highly processed food products the ration has been disturbed (10-20 to 1) (Simopoulos, 2004) leading to high omega-6 content causing inflammation and cancers (Simopoulos 2003, 2008).

Dietary source

Good sources of omega 3 PUFA's are dark fleshed fishes such as herrings, mackerels, sardines and salmons. Alpha linolenic acid is also present in green leafy vegetables, flaxseed, rapeseed oils, soyabean and canola oils (El-Badry et al., 2007). Omega-6 PUFA's are contained in most vegetable oils, cereals, eggs, animal fat, whole grain bread, sunflower and corn oils (Kinney 2006; El-Badry et al., 2007). The recommended daily mount of EPA and DHA for adults is at least 220 mg.

Glucosinolates

Glucosinolates are a group of phytochemicals that contain nitrogen and sulphur. They are present in high concentration in cruciferous vegetables such as broccoli, cauliflower, kale, brussels sprouts, cabbage, raddish, turnips and oilseeds (Stahl et al., 2002). Glucosinolates are known to reduce the risk of different types of cancer, namely lung, breast, liver, esophagus, stomach, small intestine and colon (Hecht, 2004; Conaway, 2002). They are identified as potent cancer prevention due to there ability to enhance the activity of detoxification enzymes such as quinine reductase, glutathione-S-transferase (Krzyzanowska et al., 2010).

Allicin and Diallyl sulphide

Allicin is one of the phytnutrients that belongs to the group of thiosulfinate compounds. Thiosulfinates are volatile, unstable compounds which transform into other types of organosulphur compounds. Allicin is considered to be the main biologically active antimicrobial phytochemical produced in garlic extracts. It is found in high concentration in raw garlic and onion and is known to possess antimicrobial, antioxidant, antimutagenic and prebiotic properties (Corzo-Martinez et al., 2007). Allicin helps in preventing cardiovascular diseases, is anti-diabetic and prevents atherosclerosis.

Phytosterols

Phytosterols are triterpenoids which are structurally similar and functionally analogous to the animal sterol, cholesterol. Among all the phytosterols that are found β-sitosterol, stigmasterol and campesterol are the most abundant and are predominantly supplied by vegetable oils (Piironen et al., 2000) rice bran oil contains ferulic acid esters of phytosterols, commonly known as oryzanol. Cereals, nuts and vegetables are also sources of sterols. These compounds are known to inhibit intestinal absorption of cholesterol, for instance β-sitosterol has been used as a supplement and as a drug for lowering serum cholesterol levels in hypercholesterolemic individuals (Hicks and Moreau, 2001). US FDA in September 2000, has allowed a health claim for reducing the risk of coronary heart disease for foods (spreads and salad dressings) containing phytosterol and phytostanol esters (Jones and Raeini-Sarjaz, 2001).

Conclusion

In recent years increase in the number of degenerative diseases has been a driving force for the study of new phytochemicals to fight age related conditions and prevent obesity, cancer, cardiovascular diseases, arthritis and high blood pressure. The relationship between food and health is complex. By replacing foods with supplements, beneficial food components or important interactions between food components may be lost. Therefore, consumption of a balanced diet, high in a variety of fruits, vegetables, and whole grain foods, is associated with optimal health. A recommended intake for phytochemicals does not currently exist. Today, many health authorities such as the American Cancer Society and the American Heart Association recommend consuming a diet high in fruits and vegetables to ensure that an individual ingests an adequate amount of phytochemical compounds.

References

1. Arima, H. K. and Rodriguez-Amaya, D. B. (1988). Carotenoid composition and vitamin A value of commercial Brazilian squashes and pumpkins. J. Micronutr. Anal. 4:177-191.
2. Arima H K and Rodriguez-Amaya D B (1990) Carotenoid composition and vitamin A value of a squash and a pumpkin from Northeastern Brasil. Arch. Latinoam Nutr. 40: 284-292.
3. Arts, I. C. and Hollman, P. C. (2005) Polyphemons and disease risk in epidemiologic studies. Am. J. Clin. Nutr. 81: 317-325.
4. Arya, S. S. Netesan, V. Premaavalli, K. S. and Vijayraghavan, P. K. (1982). Effect of prefreezing on the stability of carotenoid in blanched air dried carrots. J. Food Technol. 17:109-113.
5. Balk, E. M. Lichtenstein, A. H.Chung, M. Kupelnick, B. Chew, P. and Lau, J. (2006). Effects of omega-3 fatty acids on serum markers of cardiovascular disease risk: A systematic review. Atherosclerosis, 189:19–30.

6. Basuny, A. M. Gaafar, A.M. and Arafat, S. M. (2009). Tomato lycopene is a natural antioxidant and can alleviate hypercholesterolemia. Afr. J. Biotechnol. 8:6627-6633.
7. Beltz, B. S. Tlusty, M. F. Benton, J. L. and Sandeman, D. C. (2007). Omega-3 fatty acids upregulate adult neurogenesis. Neuroscience Letters 415:154-158.
8. Bendich, A. and Olson, J. A. (1989). Biological actions of carotenoids. FASEB J. 3:1927-1932.
9. Bridle, P. and Timberlake, C. F. (1996). Anthocyanins as natural food colour-selected aspects. Food Chem. 58(1-2):103–109.
10. Cassidy, A. Hansley, B. and Lamuela-Raventos, R.M. (2000). Isoflavones, lignans and stilbenes-origins, metabolism and potential importance to human health. J. Sci. Food Agric. 80: 1044–1062.
11. Castenmiller, J. J. and West, C. E. (1998). Bioavailability and bioconversion of carotenoids. Annu. Rev Nutr. 18:19-38.
12. Castlellano, G. Tena, J. and Torrens, F. (2012). Classification of Phenolic Compounds by Chemical Structural Indicators and Its Relation to Antioxidant Properties of *Posidonia Oceanica* (L.) Delile. MATCH Commun. Math Comput. Chem. 67:231-250.
13. Chasse, G. A. Mak, M. L. Deretey, E.Farkas, I. Torday, L. L. Papp, J.G. Sarma, D. S. R. Agarwal, A. Chakravarthi, S.Agarwal, A. and Rao, A. V. (2001). An ab initio computational study on selected lycopene isomers. J. Mol. Struct. 571:27-37.
14. Chen, B. H. Peng, H. Y. and Chen, H. E. (1995). Changes of carotenoids, color, and vitamin A contents during processing of carrot juice. J. Agric. Food Chem. 43:1912-1918.
15. Conaway, C. C. Yang, Y. M. and Chung, F. L. (2002). Isothiocyanates as Cancer Chemopreventive Agents: Their Biological Activities and Metabolism in Rodents and Humans. Curr Drug Metab 3(3):233-255.
16. Corzo-Martinez, M. Corzo, N. and Villameil, M. (2007). Biological properties of onion and garlic. Trends Food Sci Technol 18:609-625.
17. Dewick, P. M. (2002). Medicinal Natural Products: A Biosynthetic Approach, Wiley, New York.
18. Duthie, S. (2007). Berry phytochemicals, genomic stability and cancer: evidence for chemoprevention at several stages in the carcinogenic process. Molecular Nutr. Food Res. 51:665-674.
19. Dutta, D. Chaudhuri, U. R. and Chakraborty, R. (2005). Structure, health benefits, antioxidant property and processing and storage of carotenoids. Afr. J.Biotechnol. 4(13):1510-1520.
20. El-Badry, A.M. Graf, R. and Clavien, P.A. (2007). Omega3- Omega 6: what is right for the liver? J Hepatol 47:718-725.
21. European Parliament and Council Directive (1994). On colors for use in food stuffs. Official J European Communities 237: 13-29.
22. Fish W W, Perkins-Veazie P and Collins J K (2002) A quantitative assay for lycopene that utilizes reduced volumes of organic solvents. J. Food Compost. Anal. 15:309-17.
23. Foote, C. S. Chang, Y. C. and Denny, R. W. (1970). Chemistry of singlet oxygen X. Carotenoid quenching parallels biological protection. J. Am. Chem. Soc. 92:5216-5218.
24. Francis, F. J. (1989). Food colorants: anthocyanins. Crit. Rev. Food Sci. Nutr. 28(4):273-314.
25. Fukumoto, L. R. and Mazza, G. (2000). Assessing antioxidant and prooxidant activities of phenolic compounds. J. Agric. Food Chem. 48:3597-3604.
26. Gronowska-Senger, A. and Wolf, G. (1970). Effect of dietary protein on the enzyme from rat and human intestine which converts beta-carotene to retinal. J. Nutr. 100:300-308.

27. Han, X. Shen, T. and Lou, H. (2007). Dietary Polyphenols and Their Biological Significance. Int. J. Mol. Sci. 8: 950-988.
28. Hatcher, D. W. and Kruger, J. E. (1997). Simple phenolic acids in flours prepared from Canadian wheat: relationship to ash content, color, and polyphenol oxidase activity. Cereal Chem. 74:337–343.
29. Hecht, S. S. (2004). Chemoprevention by Isothiocyanates. In (Kelloff, G J. Hawk, E. T. and Sigman, C. C. Eds), Promising Cancer Chemopreventive Agents, Volume 1: Cancer Chemopreventive Agents. New Jersey: Humana Press.
30. Hertog, M. G L. Hollman, P. C. H. and Katan, M. B. (1992). Content of potentially anticarcinogenic flavonoids of 28 vegetables and 9 fruits commonly consumed in the Netherlands. J. Agric. Food Chem. 40: 2379–2383.
31. Hicks, K. B. and Moreau, R. A. (2001). Phytosterols and phytostanols: Functional food cholesterol busters. Food Technol 55:63-67.
32. Jacob, K. Periago, M.J. Bohm, V. and Berruezo. GR. (2008). Influence of lycopene and vitamin C from juice on biomarkers of oxidative stress and inflammation. Brit. J. Nutr. 99:137-46.
33. Jian, W. Chiou, M.H. Chen, Y.T. Lin, C.N. Wub. M.C. Du, C.J. (2008). Twenty-eight-day oral toxicity study of lycopene from recombinant Escherichia coli in rats. Regul. Toxicol. Pharmacol, 52:163-8.
34. Jones, P. J. and Raeini-Sarjaz, M. (2001). Plant sterols and derivatives: The current spread of results. Nutr Rev 59: 21-24.
35. Jun, H. (1987). Bioavailbility of carotene in carrot juice. Dissert. Abstr. Int. B47 8:3300-301.
36. Kinney, A.J. (2006). Metabolic engineering in plants for human health and nutrition. Curr Opin Biotechnol 17:130-138.
37. Krinsky, N. I. (1994). The biological properties of carotenoids. Pure Appl. Chem. 66:1003-1010.
38. Krzyzanowska, J. Czubacka, A. and Oleszek, W. (2010). Dietary phytochemicals and human health. In Bio Farms for Nutraceuticals: Functional Food and Safety Control *by* Biosensors (Giardi, M. T. Rea, G and Berra, B. eds). Landes Bioscience and Springer Science+Business Media, New York, USA. Pp 74-98.
39. Kun, Y. Ssonko-Lule, U. Xiao-Lin, D. (2006). Lycopene: Its properties and relationship to human health. Food Rev. Int. 22:309-33.
40. Lempereur, I. Rouau, X. and Abecassis, J. (1997). Genetic and agronomic variation in arabinoxylan and ferulic acid contents of durum wheat (*Triticum durum* L.) grain and its milling fractions. J. Cereal Sci. 25:103–110.
41. Lessin W J, Catigani GL, Schwartz SJ (1997) Quantification of *cis-trans* isomers of provitamin A carotenoids in fresh and processed fruits and vegetables. J. Agric. Food Chem. 45:3728-3732.
42. Liu, R. H. (2004). Potential synergy of phytochemicals in cancer prevention: Mechanism of action. J. Nutr. 134:3479S–3485S.
43. Mackinnon E S, Rao AV and Rao LG (2011) Dietary restriction of lycopene for a period of one month resulted in significantly increased biomarkers of oxidative stress and bone resorption in postmenopausal women. J. Nutr. Health Aging 15:133-8.
44. Manach, C. Scalbert, A. Morand, C. Remesy, C. and Jimenez, L. (2004). Polyphenols: food sources and bioavailability. Am. J. Clin. Nutr. 79:727–47.
45. Mathews-Roth, M.M. (1991). Recent progress in the medical applications of carotenoids. Pure Appl. Chem., 63:147-156.
46. Mazza, G and Maniati, E. (1993). Anthocyanins in fruits, vegetables, and grains. Boca Raton, FL: CRC Press.

47. Mazza, M. Pomponi, M. Janiri, L. Bria, P. and Mazza, S. (2007). Omega-3 fatty acids and antioxidants in neurological and psychiatric diseases: An overview. Progress in Neuro-Psychopharmacology and Biological Psychiatry, 31:12–26.
48. McGhie, T. Ainge, G. Barnett, L. Cooney, J. and Jensen, D. (2003). Anthocyanin glucoside from berry fruit are absorbed and excreted unmetabolized by both human and rats. J. Agric. Food Chem. 51:4539-4548.
49. Mohamed, S. and Hussain, R. (1994). Effect of low temperature blanching, cysteine-HCl, N-acetyl- L- cysteine, sodium metabisulphite and drying temperatures on the firmness and nutrient content of dried carrots. J. Food Process. Pres. 18:343-48.
50. Murakoshi, M. Takayasu, J. Kimura, O. Kohmura, E. Nishino, H. Iwashima, A.Okuzumi, J. Sakai, T. Sugimoto, T. Imanishi, J. (1989). Inhibitory effects of alpha-carotene on proliferation of the human neuroblastoma cell line GOTO. J. Nat. Canc. Inst. 81:1649-52.
51. Naz, A. Butt, M. S. Sultan, M. T. Qayyum, M. M. N. and Niaz, R.S. (2014). Watermelon lycopene and allied health claims. EXCLI Journal 13:650-666.
52. Ollanketo, M. Hartonen, K. Riekkola, M.L. Holm, Y. Hiltunen, R. (2001). Supercritical carbon dioxide extraction of lycopene in tomato skins. Euro. Food Res. Technol. 212: 561-5.
53. Padula, M. Rodriguez-Amaya, D. B. (1987). Changes in individual carotenoids and vitamin C on processing and storage of guava juice. Acta Alimentaria 16:209-216.
54. Pharmanex [homepage on the internet]. Provo, UT: Pharmanex, Inc.; [cited 2009 Oct 15]. Available from: http://www.pharmanex. com/corp/product/solutions/cholestin.shtml
55. Piironen, V. Lindsay, D. G. Miettinen, T. A. Toivo, J. and Lampi, A. M. (2000). Plant sterols: biosynthesis, biological function and their importance to human nutrition. J. Sci. Food Agric. 80: 939-966.
56. Prior, R. L. (2003). Fruits and vegetables in the prevention of cellular oxidative damage.Am. J. Clin. Nutr. 78 (suppl):570S-578S.
57. Rao, A.V. and Agarwal, S. (1999). Role of lycopene as antioxidant carotenoid in the prevention of chronic diseases: a review. Nutr. Res.19:305-23.
58. Rao, L. G. Guns, E. and Rao, A. V. (2003). Lycopene: Its role in human health and disease. AGROFood industry hi-tech, July/August: 25-30.
59. Rodriguez, E. B. Flavier, M. E. Rodriguez-Amaya, D. B. and Amaya-Farfán, J. (2006). Phytochemicals and functional foods. Current situation and prospect for developing countries. Segurança Alimentar e Nutricional, Campinas 13(1): 1-22.
60. Rodriguez-Amaya, D. B. (1993). Nature and distribution of carotenoids in foods. In: Charalambous G (editor). Shelf-life studies of foods and beverages. Chemical, biological, physical and nutritional aspects. Amsterdam: Elsevier Science Publishers; Pp: 547-89.
61. Rodriguez-Amaya, D. B. (1997). Carotenoids and food preparation: the retention of provitamin A carotenoids in prepared, processed and stored foods. OMNI project, US Agency for International development office of Health and Nutrition, John Snow Inc.
62. Rodriguez-Amaya, D. B. (1999). Latin American food sources of carotenoids. Arch Latinoamer Nutr 49:74S-84S.
63. Rodriguez-Amaya, D.B. (2001). Nature of carotenoids in food. In A guide to carotenoid analysis in food (Rodriguez-Amaya DB ed.) OMNI reaserch, ILSI Press, Washington DC.
64. Scalbert, A. Johnson, I. T. and Saltmarsh, M. (2005). Polyphenols: antioxidants and beyond. Am. J. Clin. Nutr. 81(suppl): 215S–7S.
65. Shahidi, F. and Naczk, M. (1995) Food phenolics, sources, chemistry, effects, applications. Lancaster, PA: Technomic Publishing Co Inc.

66. Sijtsma, L. and de Swaaf, M. E. (2004). Biotechnological production and application of omega-3 polyunsaturated fatty acid docosahexaenoic acid. Appl Microbiol Biotechnol, 64:146-153.
67. Simopoulos, A. P. (2004). Omega-6/omega-3 essential fatty acid ratio and chronic disease. Food Rev. Int. 20: 77-90.
68. Simopoulos, A. P. (2003). Omega-3 fatty acids and cancer. Indoor Built Environ. 12: 405-412.
69. Simopoulos, A. P. (2008). The omega-6/omega-3 fatty acid ratio, genetic variation and cardiovascular disease. Asia Pac. J. Clin. Nutr. 17: 131-134.
70. Simpson, KL, Katayama, T and Chichester, C O (1981). In: J C Bauernfeind (ed) carotenoids as colorants and Vitamin A precursors, Academic Press, New York. Pp: 463.
71. Sosulski, F. Krygier, K. and Hogge, L. (1982). Free, esterified, and insoluble-bound phenolic acids. Composition of phenolic acids in cereal and potato flours. J. Agric. Food Chem. 30:337–340.
72. Stahl, W. van den Berg, H. Arthur, J. Bast A. Dainty, J. Faulks, R.M. Gartner, C. Haenen, G. Hollman, P. and Holst, B. (2002). Bioavailability and metabolism. Molecular aspects of medicine 23:39-100.
73. Sugiyama Chemical Institute (1977). Anthocyanin food colouring agent from purple corn. Japanese Patent 77130824.
74. Teixeira, L. J. Q. Serrano, I. O. Soliva-Fortuny, R. Mota-Ramos, A. and Martin-Belloso, O. (2009). Comparative study on antioxidant properties of carrot juice stabilised by high-intensity pulsed electric fields or heat treatments. J. Sci. Food Agric. 89:2636-42.
75. Tyagi, S. Singh, J. Sharma, A. and Aggarwal, G. (2010). Phytochemicals as candidate therapeutics: an overview. Int. J. Pharma. Sci. Rev. Res. 3(1): 53-55.
76. Wang, X.D. and Russell, R.M. (1999). Procarcinogenic and anticarcinogenic effects of beta-carotene. Nutr. Rev. 57:263-272.
77. Zaman, Z. Roche, S. Fielden, P. Frost, P.G. Nerilla, D. C. Cayley, A.C.D. (1992). Plasma concentration of vitamin a and E and carotenoids in Alzheimer's disease. Age and Aging 21:91-96.
78. Zhang, D. and Hamauzu, Y. (2004). Phenolic compounds and their antioxidant properties in different tissues of carrots (Daucus carota L.). J. Food Agric. Environ. 2:95-100.

CHAPTER 10

Role of Fibers Spices and Herbs as Nutraceuticals and Functional Food Ingredients in Regulating Human Health

A.B. Rodge

Abstract

Dietary fibers mostly include non-starch polysaccharides such as celluloses, hemicelluloses, gums and pectins, lignin, resistant dextrins and resistant starches. Based on their water solubility, dietary fibers may be divided into two forms insoluble dietary fiber which includes celluloses, some hemicelluloses and lignins which is fermented to a limited extent in the colon and soluble dietary fiber which includes β-glucans, pectins, gums, mucilages and hemicelluloses that are fermented in the colon. Resistant starch generally considered to be represented by the total amount of starch and the product of starch degradation that are not digested in the small intestine and pass into large intestine, and regarded as a component of dietary fiber. Herbal nutraceutical is used as a powerful instrument in maintaining health and to act against nutritionally induced acute and chronic diseases, thereby promoting optimal health, and quality of life. Dietary spices in their minute quantities has an immense influence on the human health by their antioxidative, chemopreventive, antimutagenic, anti-inflammatory, immune modulatory effects on cells and a wide range of beneficial effects on human health. Nutraceutical is the hybrid of 'nutrition' and 'pharmaceutical'. Nutraceuticals, in broad, are food or part of food playing a significant role in modifying and maintaining normal physiological function that maintains healthy human beings. Functional foods are including whole foods and fortified, enriched, or enhanced foods or dietary components that may reduce the risk of chronic disease and provide a health and physiological benefit beyond the traditional nutrients it contains.

Introduction

The principal reasons for the growth of the nutraceutical market worldwide are the current population and the health trends. The food products used as nutraceuticals can be categorized as dietary fiber, prebiotics, probiotics, polyunsaturated fatty acids, antioxidants and other different types of herbal/ natural foods. These nutraceuticals help in combating some of the major health problems of the century such as obesity, cardiovascular diseases, cancer, osteoporosis, arthritis, diabetes; cholesterol etc. Dietary fibers mostly include non-starch polysaccharides such as celluloses, hemicelluloses, gums and pectins, lignin, resistant dextrins and resistant starches. Based on their water solubility, dietary fibers may be divided into two forms insoluble dietary fiber which includes celluloses, some hemicelluloses and lignins which are fermented to a limited extend in the colon and soluble dietary fiber which includes β-glucans, pectins, gums, mucilages and hemicelluloses that are fermented in the colon. Resistant starch generally considered to be represented by the total amount of starch and the product of starch degradation that are not digested in the small intestine and pass into large intestine, and regarded as a component of dietary fiber.

Spices, the predominant flavoring, coloring and aromatic agents in foods and beverages, are now gaining importance for their diversified uses. The nutritional, anti-oxidant, anti-microbial and medicinal properties of spices have far-reaching implications. In the present scenario, the anti-diabetic, anti-hyper-cholesterolemic, anti-carcinogenic, anti-inflammatory effects of spices have paramount importance, as the key health issues of mankind nowadays are diabetes, cardio-vascular diseases, arthritis and cancer. Spices or their active principles could be used as possible ameliorative or preventive agents for these health disorders. Extensive studies on animal models carried out indicate that spices could be consumed at higher dietary levels without any adverse effects on growth, organ weights, food efficiency ratio and blood constituents. Dietary spices in their minute quantities has an immense influence on the human health by their antioxidative, chemopreventive, antimutagenic, anti-inflammatory, immune modulatory effects on cells and a wide range of beneficial effects on human health. Herbal nutraceutical is used as a powerful instrument in maintaining health and to act against nutritionally induced acute and chronic diseases, thereby promoting optimal health, longevity, and quality of life.

Dietary Fiber

Dietary fiber is the food material, more precisely the plant material that is not hydrolyzed by enzymes secreted by the digestive tract, but digested by microflora in the gut. Dietary fibers mostly include non-starch polysaccharides (NSP) such as celluloses, hemicelluloses, gums and pectins, lignin, resistant dextrins and resistant starches. Therapeutic areas covered by nutraceutical products in soluble fiber include fruits, oats, barley and beans. Chemically dietary fiber means carbohydrate polymers with a degree of polymerization not lower than three, which are neither digested nor absorbed in the small intestine. Based on their water solubility, dietary fibers may be divided into two forms Insoluble dietary fiber (IDF), which includes celluloses, some hemicelluloses and lignins which is fermented to a limited extend in the colon. Soluble dietary fiber (SDF), which includes β-glucans, pectins, gums, mucilages and hemicelluloses that are fermented in the colon. The IDF and SDF compounds are collectively known as non-starch polysaccharides (NSP).The soluble components of dietary fiber by virtue of their bulking and viscosity producing capabilities, retards the gastric emptying of the stomach. This affects the rate of digestion and the uptake of nutrients and creates a feeling of satiety. Soluble fiber has been shown to lower selectively serum LDL cholesterol and to improve glucose tolerance. They also enhance insulin receptor binding and improve glycaemic response. In colon, dietary fiber increases faecal bulking due to increased water retention, increased transit time and increased faecal bacterial mass caused by soluble fiber fermentation. The fiber also promotes the growth of Bifidobacteria in the gut (especially fructooligosaccharides).

Sources of soluble fiber: Oatmeal, oat cereal, lentils, apples, oranges, pears, oat bran, strawberries, nuts, flaxseeds, beans, dried peas, blueberries, psyllium, cucumbers, celery, and carrots.

Sources of insoluble fiber: whole wheat, whole grains, wheat bran, corn bran, seeds, nuts, barley, couscous, brown rice, bulgur, zucchini, celery, broccoli, cabbage, onions, tomatoes, carrots, cucumbers, green beans, dark leafy vegetables, raisins, grapes, fruit, and root vegetable skins.

Guar Gum as Dietary Fiber

Guar gum, one of the most promising dietary fibers, is a gel-forming galactomannan obtained by grinding the endosperm portion of *Cyamoposis tetragonolobus* L., a leguminous plant grown for centuries mainly in India and Pakistan to produce the seed for human and animal food (Goldstein et al., 1973). Since 1953, the seeds of the guar plant have been processed into guar gum, and widely used in modern food industry as thickener and emulsion stabilizer for both food and industrial purposes in all over the world with increasing amounts. Guar gum is structurally comprises long, straight chains of α-D-mannopyranosyl units linked together by β-D-(1-4)-glycosidic linkage. The hexose linked along this chain is α-D-galactopyranose and these side groups appear on both side of the mannose chain at about every other mannose link in the main chain (Stephen, 1983). Therefore, the ratio of mannose to galactose in galactomannan of guar gum has been known approximately 2:1 (Robinson et al., 1982), Englyst, 1988). Guar crop has a great industrial importance because the seed are major source of gum (galactomannan, 35.50%), which has a several industrial applications. Guar also has crude protein (34.22%) and total carbohydrate (43.42%) shown in Table 1 (Gadhe et al., 2007).

Table 1: Chemical composition of guar seed

Composition (%)	Guar Seed
Moisture	9.51
Ash	5.62
Crude lipid	7.23
Crude protein (N x 6.25)	34.22
Total carbohydrate	43.42
Gum	35.50

Although a single molecular weight cannot be given due to difficulty for determination, it is estimated to be in the range of 200,000 to 300,000 daltons (Glicksman, 1969). Galactomannan is presents in various natural sources such as coffee beans, soy beans, alfalfa seeds, pineapple, sugar beets, and locust beans (Dea and Morrison, 1975). There are certainly benefits when guar gums are included in human diets. However, guar gum is extremely viscous which results in the liquid products with high viscosity when it is added to enteral formulas at a physiologically effective concentration, thereby its utilization may be restricted for dietary fiber supplementation in actual food products, especially for liquid products. High viscosity of guar gum also decreases the protein efficacy (Poksay and Schneeman, 1983) and lipid utilization. Simons et al., (1982), by interfering with the digestion and absorption of nutrient (Ikegami et al., 1990), when it dissolved into water, results in a slower gastric emptying (Shah et al., 1986).

Rodge et al., (2006) studied the utilization of guar gum obtained from guar seed in a dairy product as a stabilizer. There was pronounced effect of guar gum on the size of the ice crystals formation, during the storage of ice cream. Rodge et al., (2009) used Guar gum to overcome the problem of cloudiness in Ginger ready to serve (RTS) beverages galactomannan was exploited as an additive to stabilize the Ginger RTS beverage for longer period.

Rodge et al., (2011) studied the effect of guar gum on rheological properties of dough, physico-chemical properties of bread, sensory characteristics of finished product and white bread. At all levels of incorporation, there were increases water absorption from 61.2 to 64.5 percent, lowered the dough development time from 6.20 to 5.60 min, increased gluten development increased dough stability increased break down time from 37.3 to 40.5 min, higher the missing tolerance index from 52.2 to 54.6 BTU and increased the elasticity i.e. 72.3 to 74.6 BU of the dough. The external characteristics like colour of crust, symmetry evenness, character of crust and aroma was improved linearly with increased the concentration of guar gum. Rodge et al., (2012) studied ready to serve beverage by blending aonla, ginger and lemon with incorporation of guar gum at various level i.e. 0.25%, 0.50% and 1.00%. Incorporation of guar gum (0.5 %) found to be over all acceptable in respect of colour, flavor

Resistant Starch

Food grain legumes are rich and less expensive source of dietary proteins, Legumes often fall below the popular grains and fruits and vegetables on the list of foods that people consume to meet the recommended intake of daily dietary fiber. Researchers are hoping to motivate greater consumption of legumes. Many researchers reported that legumes contain substantially higher levels of resistant starch than do cereal grains, flours and grain-based food products. Resistant starch (RS) content in raw whole and decorticated legumes as described in Table 2 (Katyal et al., 2005).

Table 2: Resistant starch (RS) content in raw whole and decorticated legumes

Legume	RS Content (g/100g)	
	Whole	Decorticated
Green gram	17.9	21.5
Kabuli Bengal gram	18.4	19.6
Desi Bengal gram	9.0	—
Red gram	10.5	14.2
Lentil	19.6	23.3
Cowpea	15.6	—
Field pea	12.6	—

Resistant starches (RS) have drawn broad interest worldwide for both their functional properties and potential health benefits. Initial clinical studies demonstrated that resistant starches have properties similar to dietary fiber and promising physiological benefits in humans, which may result in prevention of various diseases. There are four types of resistant starches as based on the mechanism that contributes to their indigestibility are $RS_{1,}$ $RS_{2,}$ RS_3 and RS_4.

Legumes come under RS_1 starches. RS_1 starch resists digestion because it is protected from digestive enzymes by other components normally present in a matrix of typical starch sources. Among the best characterized forms of RS are those derived from legumes. Resistant starches provide many health benefits such as improving bowel health, blood lipid profile, increased feeling of satiety and micronutrient absorption (magnesium and calcium) in the colon. These factors may affect the risk of developing diseases such as colorectal cancer, cardiovascular diseases, osteoporosis and inflammatory bowel diseases.

Potential Health Benefits of Spices

Most of the spice components are terpenes and other constituents of essential oils. They have been found to be effective in different forms. For instance, about 50 g of onion and 5–6 cloves of garlic in their raw form are adequate for lowering of cholesterol in human body. Recent studies on the lipid profile and blood pressure of moderately hypercholesterolemic subjects showed better beneficial effects of dietary supplement with aged garlic extract relative to the fresh ones. Co-administration of garlic with fish oil had a better beneficial effect on serum lipid and lipoprotein concentrations by providing a combined lowering of total cholesterol, LDL cholesterol and triglyceride concentration. Spices and herbs are in most cases harmless, when used as food, but may exhibit toxicity, when used as medicine, because of their relative higher dose administered, or rather due to the possibilities of their interactions with other pharmaceutical medications. Again reports say that excessive consumption of garlic (4 ml/kg for raw garlic juice or of 100 mg/kg for garlic oil) lead to adverse effects on health such as anemia, weight loss, heart, liver, kidney toxicity and other dermatological problems. High doses of onion (500 mg/kg) as well causes lung and tissue damage in rats. Fenugreek seeds (25–50 g), garlic (5–6 cloves), onion (50 g) and turmeric powder (1 pinch) in the daily diet of diabetics prevent and manage long-term complications of diabetes. Regular intake of curcuminoids at about 0.5 g reduces blood lipid peroxide level upto about 33% due to their antioxidant activity.

Herbal Health Benefits

Herbal nutraceutical is used as a powerful instrument in maintaining health and to act against nutritionally induced acute and chronic diseases, thereby promoting optimal health, longevity, and quality of life.

- Avoid the side effect.
- May increase the health beneficial effect.

- May have naturally dietary supplement, so do not have unpleasant side effect.
- May increase the health value, our diet and improve medical condition of human.
- May easily be available and economically affordable.

Phytonutrients basically are plant nutrients with particular biological activities in supporting human health.

Nutraceuticals

Nutraceuticals differ from dietary supplements in the following aspects: nutraceuticals must not only supplement the diet but should also aid in the prevention and treatment of disease and disorder, (Roberts and Rosenberg, 2006) nutraceuticals are used as conventional foods or as sole items of a meal or diet. Dietary components play beneficial roles beyond basic nutrition, leading to the development of the functional food concept and nutraceuticals. A functional food for one consumer can act as a nutraceutical for another consumer. Examples of nutraceuticals include fortified dairy products (e.g., milk) and citrus fruits (e.g., orange juice). Several naturally derived food substances have been studied in cancer therapies. Vitamin-E, selenium, vitamin-D, green tea, soy, and lycopene are examples of nutraceuticals widely studied in human health (Zeisel, 1999).

Polyunsaturated fatty acids (PUFAs) (which include the omega-3 and omega-6 fatty acids) and phytochemicals also play an important role as healthy dietary bioactive compounds. A balanced PUFA composition of food influences diverse aspects of immunity and metabolism. Moreover, interactions between PUFAs and components of the gut microbiota may also influence their biological roles. Phytochemicals (bioactive non-nutrient plant compounds), have raised interest in human nutrition because of their potential effects as antioxidants, antiestrogenics, anti-inflammatory, immunomodulatory, and anticarcinogenics. Gut microbiota can, for example, transform and influence the bioavailability and effects of polyphenols. Phytochemicals and their metabolic products may also inhibit pathogenic bacteria while stimulate the growth of beneficial bacteria, exerting prebiotic-like effects. Interactions between functional food components, such as prebiotics, probiotics, phytochemicals, and intestinal microbiota, have consequences on human health (Laparra, and Sanz, 2010).

Functional Foods

Functional foods are similar in appearance to conventional foods; the former being consumed as part of the normal diet. In contrast to conventional foods, functional foods, however, have demonstrated physiological benefits and can reduce the risk of chronic disease beyond basic nutritional functions, including maintenance of gut health. When food is being cooked or prepared using "scientific intelligence" with or without knowledge of how or why it is being used, the food is called "functional food". Thus, functional food provides the body with the required amount of vitamins, fats, proteins, carbohydrates, etc., needed for its healthy survival.

Prospects for the Functional Food Industry

Production of functional foods is being recognized as the number one global food industry as changing trends in population demography, consumer affluence, increased education, life expectancy and improved healthcare give rise to a rapidly emerging diet and health conscious consumer clientele. Increasing health consciousness has been one of the most important stimulating factors for rapid global growth of the nutraceutical and functional food industry.

References

1. Dea, I.C.M. Morrison, A. (1975). In: Chemistry and interactions of seed galactomannans, in Advances in Carbohydrate Chemistry and Biochemistry, vol. 3. Tipson R.S., Hórton D., editors. Academic Press; New York. pp. 241–312.
2. Englyst, H.N. Cummings, J.H. (1988). Improved method for measurement of dietary fiber as non-starch polysaccharides in plant foods. J. Assoc. Off. Anal. Chem. 71:808–814.
3. Gadhe, K.S. Ghatge, P.U. Rodge, A.B. (2007). Comparative studies on viscosity profile and functional attributes of carboxy-methylated gum of guar and sesbania. J.of Arid Legume 4(2):145:148.
4. Glicksman, M. (1969). Gum technology in the Food Industry. Academic Press; New York:. p. 590.
5. Goldstein, A.M., Alter, E.N. Seaman, J.K. (1973). In: Guar gum, in Industrial Gums, 2nd ed. Whistler, R.L., BeMiller, J.N., editors. Academic Press; New York. pp. 303–321.
6. Ikegami, S. Tsuchihashi, F. Harada, H. Tsuchihashi, N. Nishide, E. Innami, S. (1990). Effect of viscous indigestible polysaccharides on pancreatic-biliary secretion and digestive organs in rats. J. Nutr. 120:353–360.
7. Katyal, D., Padmini, S. G, Udipi, S.A. (2005). Resistant starch in selected raw amd processed legumes. J. Food Sci. and Technol. 42(6):506-510.
8. Laparra, J.M. Sanz, Y. (2010). Interactions of gut microbiota with functional food components and nutraceuticals. Pharmacol. Res., 61:219-225.
9. Poksay, K.S., Schneeman, B.O. (1983) Pancreatic and intestinal response to dietary guar gum in rats. J. Nutr. 113:1544–1549.
10. Roberts, S.B. Rosenberg, I. (2006). Nutrition and aging: changes in the regulation of energy metabolism with aging. Physiol. Rev., 86:651-667.
11. Robinson, G. Ross-Murphy, S.B. Morris, E.R. (1982). Viscosity-molecular weight relationships, intrinsic chains flexibility, and dynamic solution properties of guar galactomannan. Carbohyd. Res.107:17–32.

12. Rodge, A.B. Ghatge, P.U. Wankhede, D.B. and Ghule, D.R., (2006). Utilization of Guar gum as a stabilizer in ice-cream. Journal of Arid Legumes, 3 (2): 41-43.
13. Rodge, A. B. and Deshmukh, S.S (2009). An assessment of Guar gum as stabilizer in ginger RTS bevrages. Legumes in dry areas (Eds. D.Kumar, A. Henry and K.P.R. Vittal) published Indian Arid Legumes Society and Scientific publishers (India): 521-24.
14. Rodge, A.B. Sonkamble, S. M. Salve, R.V. and Hashmi S. I. (2011). Effect of Hydrocolloid (Guar Gum) incorporation on the quality characteristics of bread, J Food Process Technol, 3:1.
15. Rodge, A.B. Machewad, G.M. Mantri, G.D. Chappalwar, V.M. and Ghatge, P.U. (2012). Studies on incorporation of guar gum in aonla, ginger and lemon ready-to-serve drink.Beverage Food World, 39:58-59.
16. Shah, N. Mahoney, R.R. Pellett, P.L. (1986). Effect of guar gum, lignin and pectin on proteolytic enzyme levels in the gastrointestinal tract of the rat: a time-based study. J. Nutr.116:786–794.
17. Simons, L.A. Gayst, S., Balasubramaniam, S. Ruys. J. (1982). Long-term treatment of hypercholesterolaemia with a new palatable formulation of guar gum. Atherosclerosis; 45:101–108.
18. Stephen, A. M. (1983). In: Other plant polysaccharides, *in The polysaccharides*. Aspinall G.O., editor. Academic Press; New York:. pp. 97–193.
19. Zeisel, S.H. (1999). Regulation of "Nutraceuticals." Science,285:185-186.

CHAPTER 11

Dietary Supplements, Functional and Multi-Functional Foods, Health Benefits of Different Common Nutrients and Methods to Enhance Active Components in Foods

Amarjeet Kaur, Poonam A. Sachdev and Priyanka Sharma

Abstract

The concept of functional foods has received significant media attention and, as a result, interest has focused on the need for research into the non- nutrient constituents of food plants. This has enabled market penetration of functional foods containing increased amounts of bioactive food compounds or physiologically active nutrients and non-nutrients within food that impart proven health benefits. Functional foods can take many forms. Some may be conventional foods with bioactive components that can now be identified and linked to positive health outcomes. Some may be fortified or enhanced foods, specifically created to reduce disease risk for a certain group of people. Numerous observational studies have shown that omega-3 fatty acids enriched diets are associated with a reduction of cardiovascular mortality, heart attack and sudden death. In several studies, a higher fish intake has been associated with decreased incidence of coronary artery disease (CAD) and cardiovascular mortality. A trial revealed that fish oil supplementation of 130 mg/kg of body weight each day decreased the number of stiff joints, duration of morning stiffness, pain and global arthritis activity. Many supplements contain active ingredients that have strong biological effects in the body. The bioavailability of nutrients within a given food is governed by the composition and microstructure of the food, commonly referred to as the food matrix. More specifically, numerous studies investigating the development of functional foods have shown that the interactions between nutrients and non-nutrients within the food matrix can be synergistic, additive, or neutralizing in nature. The preventive and curative effects of the functional foods on health concludes it to be integral components of established public health programs to reduce the risk of specific diseases.

Introduction

The desire for health and pain-free living has created its own arena in today's world. Based on market research one third of the consumers involves food selection based on the knowledge of healthy attributes ascribed to specific nutrients within food. The concept of functional foods has received significant media attention and, as a result, interest has focused on the need for research into the non- nutrient constituents of food plants. This has enabled market penetration of functional foods containing increased amounts of bioactive food compounds or physiologically active nutrients and non-nutrients within food that impart proven health benefits. Factors driving the functional food market include rising health care costs and the growing trend to self-medicate to keep costs lower, the increasing age of the population, the obesity epidemic, and the high levels of lifestyle diseases afflicting millions of Americans. In general, functional foods have the potential to minimize health care costs while improving health and wellness, and giving consumers greater control over their health by providing a convenient form of health-enhancing ingredients. As a result of scientific advances and consumer trends, the US functional food market continues to grow. For example, based on primary and secondary market research from more than 300 food companies, the global functional food and drink market is expected to reach $130 billion by the year 2015.

Functional Foods

Functional foods may be defined as foods and food components that provide a health benefit beyond basic nutrition. "A food that beneficially affects one or more target functions in the body, beyond adequate nutritional effects, in a way that is relevant to either an improved state of health and well-being and/or reduction of risk of disease. It is part of a normal food pattern. It is not a pill, a capsule or any form of dietary supplement" (European Commission). Examples may include conventional foods; fortified, enriched or enhanced foods; and dietary supplements. These substances provide essential nutrients often beyond quantities necessary for normal maintenance, growth, and development, and/ or other biologically active components that impart health benefits or desirable physiological effects.

Functional foods can take many forms. Some may be conventional foods with bioactive components that can now be identified and linked to positive health outcomes. Some may be fortified or enhanced foods, specifically created to reduce disease risk for a certain group of people. Consumers can already select from a wide spectrum of foods that contain functional components either inherently (e.g., soy protein, cranberries) or via fortification (e.g., folate-fortified foods).

Functional foods fall into three general categories:

- Conventional foods containing natural bioactive food compounds e.g., antioxidant vitamins in orange juice, isoflavones in soy-based foods, and prebiotics and probiotics in yogurt.
- Modified foods containing bioactive food compounds through enrichment or fortification, such as n-3 fatty acids in margarine spreads and eggs.
- Food ingredients that are synthesized, such as indigestible carbohydrates, which provide prebiotic benefits like oligosaccharides or resistant starch.

Multifunctional foods

The foods may provide more than one functional health benefits due to the presence of one or more bioactive components. Tomatoes contain valuable phytochemicals, including carotenoids and polyphenols. For instance, carotenoids, such as the red pigmented lycopene, beta-carotene, a pro-vitamin A compound; phytoene, and phytofluene are all found in abundance in raw tomatoes and tomato products. Flavonol content of tomatoes is also high, with up to 98% of the total flavonols contained in the skin as the conjugated forms of quercetin and kaempferol. These tomato phytochemicals are thought to contribute to the reduced risk of human ailments such as cardiovascular disease (CVD) and prostate cancer.

Many berries are suitable to eat raw after rinsing and most types of berries vary from 50 to 100 calories per serving when eaten raw. Berries are loaded with vitamin C, potassium, and fiber. All berries with strong red and blue colors have phytochemicals that can potentially reduce cancer rates and other chronic diseases. Oxygen radical absorptive capacity (ORAC), is a way to measure the antioxidant capacity of fruits and vegetables. Berries are considered excellent source of antioxidant levels of any fresh fruits.

Numerous observational studies have shown that omega-3 fatty acids enriched diets are associated with a reduction of cardiovascular mortality, heart attack and sudden death. In several studies, a higher fish intake has been associated with decreased incidence of coronary artery disease (CAD) and cardiovascular mortality. A trial revealed that fish oil supplementation of 130 mg/kg of body weight each day decreased the number of stiff joints, duration of morning stiffness, pain and global arthritis activity.

Other foods like beans, cocoa, soy, grains, dairy and dairy products, tea, nuts and fruits and vegetables have shown several health benefits.

Dietary Supplements

The Dietary Supplement Health and Education Act (DSHEA) defines dietary supplements as food products that (a) are intended to be ingested in the form of a tablet, capsule, powder, soft gel, gel cap, or liquid droplet (or, if not intended for ingestion in such a form, that are not represented to be useful either as a conventional food or as a sole item of a meal or the diet) and (b) provide a vitamin, mineral, herb or other botanical, amino acid, or other "dietary substance" (including a concentrate, metabolite constituent, extract, or combination of any of the above). It generally is not permitted to make a health claim in labeling for a food (including a dietary supplement) unless the claim meets the FDA approval.

Supplements assure an adequate dietary intake of essential nutrients and reduce the risk of diseases. However, supplements should not replace the variety of foods that are important to a healthful diet. Unlike drugs, supplements are not intended to treat, diagnose, prevent, or cure diseases. Dietary supplements are categorized as nutrient supplement ingredients, botanical (plant-derived) supplement ingredients, and other dietary substances.

Functional foods are not considered dietary supplements because they are whole foods and part of the diet.

Many supplements contain active ingredients that have strong biological effects in the body. Consumption of excess of dietary supplements may lead to ill effects on health e.g intake of too much of some supplements, such as vitamin A, vitamin D, and iron may cause vitaminosis.

Methods to Enhance Functionality of Food

Foods that naturally provide a bioactive substance may be enhanced to increase the level present in the food (e.g., eggs with increased levels of omega-3 fatty acids). Alternately, foods that do not naturally contain a substance can be fortified to provide consumers with a broader selection of food sources for a particular component and its health benefit (e.g., calcium-fortified orange juice). However, currently the knowledge base on bioactive ingredients and food structure is very limited. The bioavailability of nutrients within a given food is governed by the composition and microstructure of the food, commonly referred to as the food matrix. The understanding of the following considerations is crucial for the development of novel products with added health beneficial value, and they could aid in predicting the absorption rate, metabolism and bioavailability of bioactive compounds within the human organism:

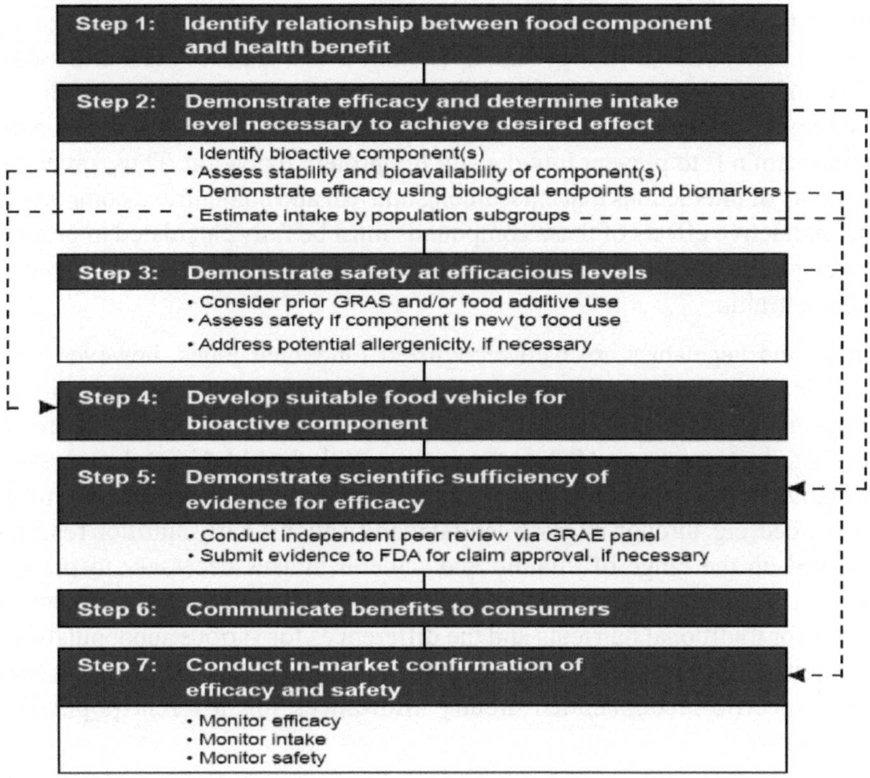

Process for Bringing Functional Foods to Market

i) Original compounds might be present in a form which is not directly available in the human digestive system.

ii) The food matrix has a significant effect on the release and availability of bioactive ingredients

iii) Compounds might need an additional carrier substance to aid solubility.

iv) Released ingredients might not be fully absorbed.

v) Functional responses to bioactive compounds may vary throughout the population according to their genetic makeup.

The complexity of food matrix determines both food texture and also the release of functional components. It has been suggested that quantitative structure function relationships (QSFR) can help the rational design and efficient production of such functional food system.

More specifically, numerous studies investigating the development of functional foods have shown that the interactions between nutrients and non-nutrients within the food matrix can be synergistic, additive, or neutralizing in nature. For example, vitamin C regenerates vitamin E and enhances the antioxidant effect of carotenoid compounds. Likewise, in vitro studies have shown that flavonoids, a subcategory of phenolic phytochemicals, behave synergistically with vitamin E to prevent low-density lipoprotein oxidation. Thus, given the diversity of interactions that exist among nutrient and non-nutrient compounds, the interactive effects of these compounds must be fully elucidated in order to develop functional foods with the greatest potential to synergistically impact human health.

Fruits and vegetables are well recognized functional foods, however, their beneficial ingredients also can be extracted, purified and used as dietary supplements and consumed in concentrated form, or after addition to a different food product exert an added dietary value. With the aid of novel processing technologies (such as microencapsulation), these effects could be minimized or avoided, e.g. through microstructural modifications. Early nutrition research focused on the range of vitamin and mineral intakes necessary to prevent deficiencies. However, now, researchers are investigating the optimum intake levels for traditional nutrients and the differences for various subpopulations. Understanding the role of nutrients at the molecular level will result in even more specific recommended dietary allowances for different population subgroups.

Similar research is needed to identify the role of other bioactive food components, an area of research that is still in its infancy. Only recently, several government agencies have begun developing a standard definition for "bioactive" food components. Research has proven that food and isolated food components reduce can the risk of disease, from the effect of vitamin A from eggs on blindness to the effect of zinc from high-protein foods on the immune system. Some examples of foods that may be considered functional foods include calcium-fortified orange juice, phytosterol/stanol-fortified spreads and juices, folate-enriched foods, soluble oat fiber, cranberry, and soy. In addition, new technologies will provide opportunities to produce bioactive food components from nontraditional sources e.g. transgenic plant oils enriched with very long chain polyunsaturated fatty acids have been developed. Other research has produced stearidonic acid (a precursor for eicosapentaenoic acid) in canola seeds to provide another source of omega-3 fatty acids in the diet.

Foods may be developed to promote the expression of specific metabolites, reducing or preventing common diseases that afflict consumers with a specific genotype. Consumers might select functional foods and tailor their diets to meet changing health goals and different requirements at different ages. Future benefits might include functional foods for increased energy, mental alertness, and better sleep. Food has traditionally been viewed as a means of providing normal growth and development. Regulatory policies were established to replace nutrients lost during processing and, in some cases, to prevent nutrient deficiencies in the population. Federal policies have generally required that other diseases be treated and managed through the use of drugs. A new self-care paradigm recognizes that foods can provide health benefits that can co-exist with traditional medical approaches to disease treatment.

Conclusion

The preventive and curative effects of the functional foods on health concludes it to be integral components of established public health programs to reduce the risk of specific diseases. Incorporation of dietary supplements into food products and research on functional foods has changed the scenario of present food market. The current Food and Drug Administration approach to drugs, foods, additives, and supplements is based on the concept of intended use. This approach integrates the concept research, safety, and regulation of foods.

References

1. Aarts, T. (1998). Nutrition industry overview. Nutritional Business Journal, 3:1–5.
2. Adams, K. C. Campbell, J. K. Zaripheh, S. Jeffery, E. H. and Erdman J. W. (2005). The Tomato As a Functional Food. Symposium: Relative Bioactivity of Functional Foods and Related Dietary Supplements. American Society for Nutritional Sciences.. 1226- 1230.
3. Adolfsson, O. Meydani, S. and Russell, R. (2004). Yogurt and gut function. Am J Clin Nutr.; 80: 245-56.
4. Ariefdjohan, M. and Savaiano, D. (2005). Chocolate and cardiovascular health: is it too good to be true? Nutrition Reviews. 63(12): 427-30.
5. Dietary Supplement Health and Education Act of 1994. Public Law. 1994, 103-417.
6. Guarner, F. Perdigon, G. Corthier, G. Salminen, S. Koletzko, B. and Morelli, L. (2005). Should yoghurt cultures be considered probiotic? Brazalian Journal of Nutrition, 93: 783-786.
7. Hannum, S. (2004). Potential impact of strawberries on human health: a review of the science. A Critical Review in Food Science and Nutrition, 44:1-17.
8. Heber, D. (2004). Vegetables, fruits and phytoestrogens in the prevention of diseases. Journa of Postgraduate and Medicine, 50(2): 145-149.

9. Keen, C. Holt, R. Oteiza, P. Fraga, C. and Schmitz, H. (2005). Cocoa antioxidants and cardiovascular health. Amrican Journal of Clinical Nutrition, 81(suppl): 298S-303S.
10. Marriott, B. M. (2000). Functional foods: an ecologic perspective. American Journal of Clinical Nutrition. 71(suppl):1728S–34S.
11. Minorsky, P. (2002). Lycopene and the prevention of prostate cancer. Plant Physiology. 130: 1077-1078.
12. National Institutes of Health. Office of Dietary Supplements. Version current 27 March 2000. World Wide Web: http://www. dietary-supplements.info.nih.gov.
13. Position of the Academy of Nutrition and Dietetics: Functional Foods. Position Paper. Journal of The Academy Of Nutrition And Dietetics. 2013. 113 (8): 1096-1103.
14. Schieber, A. (2012). Functional foods and nutraceuticals [editorial]. Food Research International. 46(2):437.
15. Sloan, A. E. (2011).Top 10 food trends. Food Technology. 65(4): 24-41.
16. Thompson, A. K. and Moughan, P. J. (2008). Innovation in the foods industry: Functional foods. Innov Manage Policy Pract. 10(1): 61-73.
17. Ubbink, J. and Krüger, J. (2006). Physical Approaches for the Delivery of Active Ingredients in Foods. Trends in Food Science and Technology, 17, 244-254.
18. Wang, L, and Bohn, T. Health-promoting food ingredients and functional food processing. Health-Promoting Food Ingredients and Functional Food Processing, Nutrition, Well-Being and Health, Dr. Jaouad Bouayed (Ed.), ISBN: 978-953-51-0125-3, http://www.intechopen.com/books/nutrition-well-being-and-health/health-promoting-food-ingredientsdevelopment- and-processing.
19 Wildman, R.E.C. (2007). Handbook of Nutraceuticals and Functional Foods, Second Edition, CRC Press, ISBN: 978-0-8493-6409-9.

CHAPTER 12

Probiotics, Prebiotics, Synbiotics Functionality and Health Benefits

Shilpa Vij, R.K Malik and Jagrani Minj

Abstract

The gut microbiota, microbe population living in the human large intestine is generally associated with various metabolic functions and it influences health and well-being. Taking into account the major role gut microbiota play in the normal functioning of the body and the different functions it accomplishes, experts nowadays consider it as an "organ". However, it is an "acquired" organ, as babies are born sterile; that is, intestine colonisation starts right after birth and evolves as we grow. Each of us has a unique microbiota, as the fingerprints, it always fullfils the same physiological functions and they have a direct impact on our health. Gut microbiota consists of a complex of microorganism species that live in the digestive tracts of animals and is the largest reservoir of micro-organisms mutual to humans. Gut microorganisms benefit the host by gleaning the energy from the fermentation of undigested carbohydrates and the subsequent absorption of short-chain fatty acids. The most important of these fatty acids are butyrates, metabolised by the colonic epithelium; propionates by the liver; and acetates by the muscle tissue. The human body carries about 100 trillion microorganisms in its intestines, a number ten times greater than the total number of human cells in the body Guarner and Malagelada, 2003). Gut microbiota, particularly the probiotics, have beneficial effects on gastroenteritis resistance, blood lipids, antitumor properties, lactose tolerance and gastrointestinal immunity. Probiotics have role in reducing the risk of rotavirus-induced diarrhea and colon cancer. The survivability of any probiotic bacteria till a longer time in the gut can be enhanced by the supplementation of prebiotics. Prebiotics are non-

digestible food ingredients that benefit the host by selectively stimulating the growth or activity of one or a limited number of bacteria in the colon. There are so many prebiotics which stimulate the growth and survivability of probiotic bacteria such as fructo-oligosaccharide (FOS), inulin, galacto-oligosaccharide (GOS), polydextrose (PDX), acacia gum (GUM), lactulose (LOS), lactitol etc. Many research studies have been done on above mentioned prebiotics. Mostly inulin-type fructans have been used as functional food ingredients. The prebiotics are known for reduction of disease risk such as constipation relief, suppression of diarrhea and reduction of the risks of osteoporosis, atherosclerotic cardiovascular disease associated with dyslipidemia and insulin resistance, obesity and possibly type-2 diabetes. Probiotics are beneficial microbes that confer a realistic health benefit on the host, which in combination with prebiotics, (indigestible dietary fibre/carbohydrate), also confer a health benefit on the host via products resulting from anaerobic fermentation. The combination of probiotics and prebiotics in a synbiotic might improve the survival of the bacteria crossing the upper part of the gastrointestinal tract, thereby enhancing their effects in the large bowel and immune-modulatory ability.

Probiotics

According to Joint FAO/WHO working group "Live microorganism which when administered in adequate amounts confer health benefit to the host". A preparation or product containing viable, defined micro-organisms in sufficient numbers, can alter the intestinal microflora and exert beneficial health effects on the host. The probiotics are most frequently used dietary method to influence the gut flora composition. Metchnikoff hypothesised that longevity in Bulgarian peasants was associated with their elevated intake of 'soured milks', i.e. dairy based drinks containing live bacteria. Lactic acid bacteria (LAB) and bifidobacteria are the most common types of microbes used as probiotics; but certain yeasts and bacilli may also be used. Probiotics are commonly consumed in fermented foods with specially added active live cultures; such as in yogurt, soy yogurt or as dietary supplements.

A recent definition of probiotics was given as 'a live microbial feed supplement that is beneficial to health'. Many species of microorganisms have been used as probiotics. They consist not only of lactic acid bacteria (e.g. lactobacilli, streptococci, enterococci, lactococci, and bifidobacteria) but also *Bacillus* spp., *E. coli* and fungi/yeasts such as *Saccharomyces* spp. and *Aspergillus* spp. The most common probiotics belong to the genera *Lactobacillus* (e.g. *L. casei, L. acidophilus, L. rhamnosus, L. johnsonii, L. reuteri*) and *Bifidobacterium* (e.g. *B. bifidum, B. longum, and B. breve*). To be effective, probiotics must be capable of being prepared in a viable manner and on a large scale (e.g. for industrial purposes), whilst during use and under storage the probiotic should remain viable and stable, be able to survive in the intestinal ecosystem and the host should gain beneficially from harbouring the probiotic. The strains used should be generally regarded as safe.

Probiotics are marketed as functional foods, whereby they are ingested for their purported positive advantages in the digestive tract and/or systemic areas like the liver, vagina or bloodstream. For probiotics to exert beneficial properties, they must have a high viability in the product and have robust survival properties in the gut, which is their first point of contact. Moreover, they should not adversely affect immune up-regulation, produce toxins, disrupt colonocyte function or have the ability to transfer antibiotic resistance to the normal gut microflora. Food vehicles include live yoghurts, fermented dairy drinks, freeze-dried supplements (capsules, pills, liquid suspensions and sprays), and cheese and fruit juices. Both single and multiple strain products are available. Probiotic have many well-established health effects.

1. Prevention and/or reduction of duration and complaints of rotavirus-induced or antibiotic-associated diarrhea as well as alleviation of complaints due to lactose intolerance.
2. Reduction of the concentration of cancer-promoting enzymes and/or putrefactive (bacterial) metabolites in the gut.
3. Prevention and alleviation of unspecific and irregular complaints of the gastrointestinal tracts in healthy people.
4. Beneficial effects on microbial aberrancies, inflammation and other complaints in connection with: inflammatory diseases of the gastrointestinal tract, *Helicobacter pylori* infection or bacterial overgrowth.
5. Normalization of passing stool and stool consistency in subjects suffering from constipation or an irritable colon.
6. Prevention or alleviation of allergies and atopic diseases in infants.
7. Prevention of respiratory tract infections (common cold, influenza) and other infectious diseases as well as treatment of urogenital infections.

Other effects are: prevention of cancer, hypocholesterolemic effect, improvement of the mouth flora and caries prevention or therapy of ischemic heart diseases or amelioration of autoimmune diseases (e.g. arthritis).

Table 1: Some of the commercially available probiotic strains

Probiotic bacteria name	*manufacturer*
L. rhamnosus R0011	Institut Rosell (Montreal, Canada)
L. rhamnosus GR-1	Urex Biotech (London, Ontario, Canada)
L. rhamnosus 271	Probi AB (Lund, Sweden)
L. rhamnosus GG	Valio Dairy (Helsinki, Finland)
L. rhamnosus HN001 (DR20)	Fonterra (Wellington, New Zealand)
L. rhamnosus LB21	Essum AB (Umea, Sweden)

Mechanism of action of probiotics

The mechanisms of probiotic can be seen in the picture. The mechanism of action of any probiotic bacteria in the gut can be different at different level and it varies from one to another strain specific probiotic bacteria. Some of the main mechanisms are mentioned here:

1. Colonization resistance against harmful pathogenic bacteria

This is one of the most important characteristics of any probiotic bacteria. This property of any probiotic bacteria is due to the competitive exclusion by the production of various antimicrobial components such as bacteriocins, organic acids, hydrogen peroxide (H_2O_2), and antibacterial peptides like lactoferrin, immunoglobulin etc.

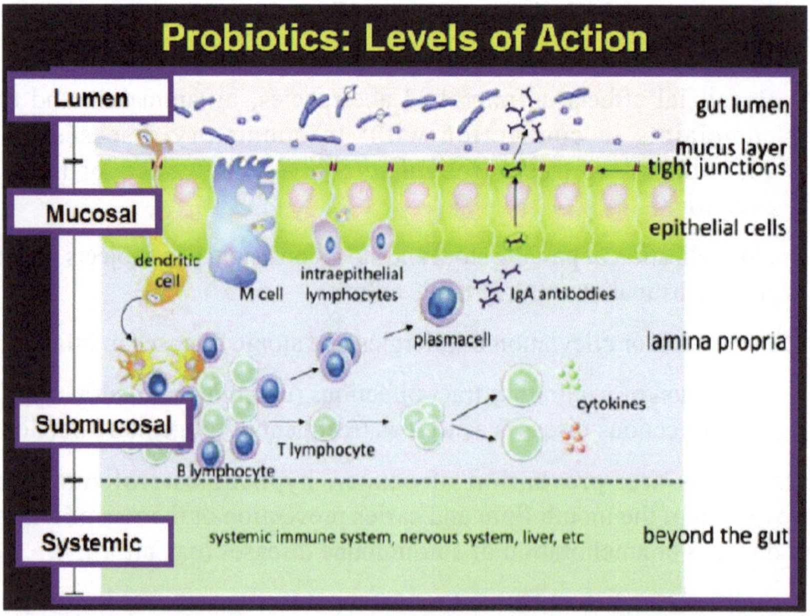

2. Maintenance of barrier functions

After colonization probiotic bacteria act as functional component in many ways such as it helps in the maintenance of gut barrier functions as well as epithelial cell. A Probiotic bacterium provides protection against attack of inflammatory agents and on the other side it promotes the anti-inflammatory agents who are very helpful in the gut barrier maintenance and function. Probiotic bacteria reduce the macromolecular permeability and bacterial translocation as well as it also maintains the tight junction proteins ZO-1 and claudin-1.

3. Modulation of immune system

Immunomodulatory action is one of the most studied properties of probiotic bacteria. Regular consumption of probiotics in the diet enhances the innate and adaptive immunity by the elevation of IgA, IgM and IgG. It has also been studied that probiotic bacteria increases the production of anti-inflammatory cytokines like IL-10 and TGF-β. Probiotic bacteria have the ability to modulate the signal transductions which are responsible for the development of pro-inflammatory cytokines such as NF-κβ, IFN-Ò and MAPK.

4. Enhancement of beneficial bacteria in the gut

It has been studied and found that probiotic bacteria increase the number of beneficial bacteria in the gut by suppressing the growth and multiplication of harmful and pathogenic bacteria.

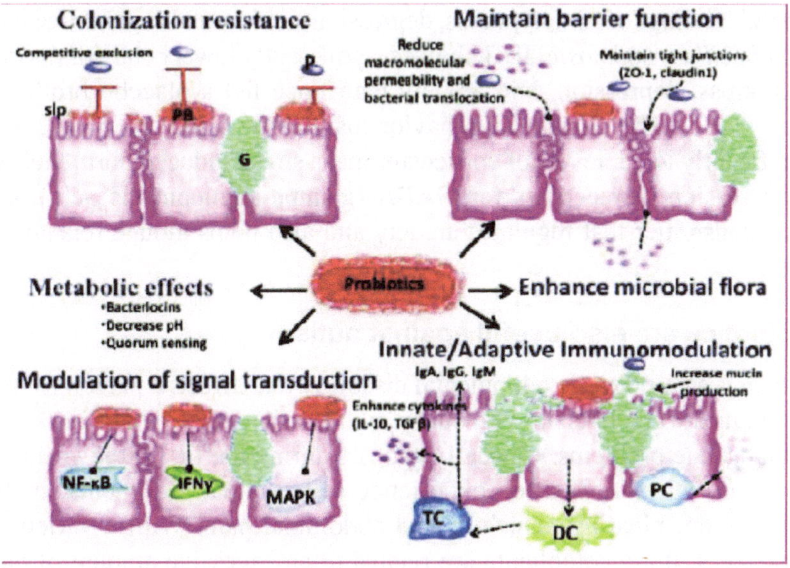

Source: Sherman NCP 2009 , Wallace TC 2011

Functional Properties of Probiotics

Probiotic bacteria promote the endogenous host defense mechanisms. In addition to the effects of probiotics on non immunologic gut defense, which is characterized by stabilization of the gut microflora, probiotic bacteria enhance humoral immune responses and thereby promote the intestine's immunologic barrier. Moreover, probiotic bacteria stimulate nonspecific host resistance to microbial pathogens and thereby aid in immune elimination, and to modulate the host's immune responses to potentially harmful antigens with a potential to down-regulate hypersensitivity reactions. Probiotic bacteria have several immunomodulatory effects: adjuvant-like properties and anti-inflammatory properties. Moreover, both quantitative and qualitative differences in immune exclusion, immune elimination, and immune regulation exist among probiotic bacteria. Distinct regulatory effects associated with probiotic consumption detected in healthy subjects and patients with inflammatory diseases (Isolauri et al., 2014). The intake of these bacteria as live cultures (probiotics) alters the expression of genes integral to neuro-development and complex behaviours in rodents. For instance, the oral administration of Bifidobacteria to rats elevated hippocampal brain-derived neurotrophic factor (BDNF) (Bercik et al., 2011; O'Sullivan et al., 2011), which may underlie some antidepressant actions (Kerman, 2012).

Probiotics are able to affect brain functioning and mood

Probiotics consumption at least for one month can significantly alleviate people's feelings of anxiety, stress, depression, anxiety and anger. Mice fed with *Lactobacillus rhamnosus* JB-1 showed significantly fewer behaviors associated with stress, depression, and anxiety than mice fed a placebo broth (Bravo et al., 2011). The change in behavior manifested in several ways. Such as significantly lower levels of corticosterone, a stress-induced hormone, change in expression of receptors for GABA (gamma-aminobutyric acid), a brain neurotransmitter that regulates anxiety and also helps induce relaxation and sleep.

Probiotics are also useful against autism

Autism is a complex developmental disability that affects a person's ability to communicate and interact with others, with a wide range of behavioral, social, and language problems. It usually appears during the first three years of life. Children with autism often experience severe dietary ailments, including constipation, bloating, diarrhea, and abdominal pain. With a variety of GI imbalances, these complaints are related to an excessive amount of harmful yeast and bacteria in the gut. Children experiencing trouble with their digestion may benefit from the addition of probiotics to their diet. Probiotics restore proper balance of good and bad bacteria, while improving absorption of nutrients. Improvements in digestion may also contribute to improved behaviors, as your child becomes more comfortable. A medical study on probiotics for autism showed very promising results in reducing the symptoms of autism and improving their behavior when given probiotics (Critchfield et al., 2011). A company named Custom Probiotics, Inc. provides D-Lactate Free Probiotic powder formulation containing *L. salivarius, L. rhamnosus, B. bifidum* and *B. infantis* that can be specifically used for children with Autism. D-Lactate is a by-product of probiotic fermentation that may produce neurological problems.

Probiotics are also effective against obesity

High energy intake and sedentary lifestyle are the main reasons for the worldwide obesity epidemic and related disorders such as type 2 diabetes, fatty liver disease and cardiovascular diseases. Recent evidence indicates a role for the gut microbiota in promoting obesity.

Probiotics available in worldwide markets

There are many food product companies manufacturing probiotic products which are now available in the market. These probiotic products are: (Table 2).

Prebiotics

A prebiotic is a non-digestible food ingredient that beneficially affects the host by selectively stimulating the growth and/or activity of one or a limited number of bacteria in the colon thus improving host health. The prebiotic approach advocates the administration of non-viable entities. Dietary carbohydrates such as fibres are candidate prebiotics, but most promise has been realised with non-digestible oligosaccharides, because of their selective metabolism. In particular, the ingestion of fructo-oligosaccharides (FOS) stimulates bifidobacteria in the lower gut. As prebiotics exploit non-viable food ingredients, their applicability in diets is wide ranging. A further approach is synbiotics, where probiotics and prebiotics are combined. Today, only bifidogenic, non-digestible oligosaccharides (particularly inulin, its hydrolysis product oligofructose, and (trans) galactooligosaccharides), fulfill all the criteria for prebiotic classification.

Table 2: Probiotics available in worldwide markets

Probiotics	Trade name	Bacteria present
Activa yogurt	Dannon Inc.	*Bifidobacterium animalis* DN-173010 aka Bifidus *Streptococcus thermophilusLactobacillus bulgaricus*
Adult Formula CP-1		*L. acidophilus*, *L. rhamnosus*, *L. plantarum*, *B. lactis* and *B. bifidum*
Align capsules		*Bifidobacterium infantis* 35624
Attune nutrition bars	Attune Foods	Koshar *Lactobacillus acidophilus* NCFM, *L. casei* Lc-11 and *Bifidobacterium lactis* HN019
Bio-K+ cultured milk based probiotic		*L. acidophilus* CL1285 and *L. casei* LBC804
Bio-K+ probiotic capsules (regular and Rx extra strength)	Bio-K+ International Inc	*L. acidophilus* CL1285 and *L. casei* LBC804
Culturelle capsules		*L. rhamnosus* GG
Florastor capsules (250 mg)		*Saccharomyces boulardii*
OWP probiotics		*B. longum*, *B. breves*, *B. infantis*, *L. plantarum*, *L. rhamnosus*, *L. acidophilus*
Philips Colon Health capsules		*Lactobacillus gasseri* KS-13, *Bifidobacterium bifidum* G9-1 and *Bifidobacterium longum* MM-2
Probulin		*B. lactis*, *B. breve*, *B. longum*, *L. acidophilus* DDS-1, *L.rhamnosus*, *L. casei*, *L. planatarum*)
VSL#3 packets	Sigma-Tau . Pharmaceuticals, Inc	*B. breve*, *B. infantis*, *B. longum*, *L. acidophilus*, *L. bulgaricus*, *L. casei*, *L. plantarum* and *Streptococcus thermophilus*
DanActive cultured milk (100 ml bottes)	Actimel	*S. thermophilus* and *L. bulgaricusL. casei* DN-114 001
Gerber Good Start Protect Plus powdered infant milk formula		*B. lactis* Bb-12
Good Belly fruit drink		*L. plantarum* 299v
Kefir Drinks		10-12 different probiotics
StonyField Farm yogurt (also makes other probiotic products like Oikos and Soy yogurt)	Stonyfield Farms, a subsidiary of Dannon Inc.	*L. rhamnosus* HN001, *B. lactis*, *L. acidophilus* and *L. casei* in addition to *S. thermophilus* and *L. bulgaricus*
Ultimate Probiotic Formula		*B. lactis*, *B. longum*, *L. plantarum*, *L. acidophilus*, *L. casei*, Kyo-Dophilus blend, *L. salivarius*, *L. rhamnosus, L. bulgaricus*, *L. sporogenes*
Yakult cultured milk	Yakult Honsha Co.	*L. casei* Shirota
Yo-Plus yogurt	Boulogne-Billancourt, France	*B. animalis* subsp l*acis* Bb-12 in addition to *S. thermophilus* abd *L. bulgaricus*

Yakult
DanActive
DANONE
L.CASEI DEFENSIS
Weekly Pack 7+1
Format hebdo
LaYogurt
Lowfat Probiotic
VANILLA
Yeo Valley
organic
FAT FREE
natural
Low-Fat
vaalia
Apricot Mango Peach
ROUNDY'S
PRObiotic
nonfat yogurt
Strawberry
Probiotic Formula
Dietary Supplement
For promoting the growth of beneficial intestinal bacteria*
Keep Refrigerated
HEALTHY CHOICE NATURALS
BALANCED PROBIOTIC
GI SUPPORT FORMULA
Dietary Supplement
phd
probiotic
Florax DS
Natural Flavor
ALL NATURAL
Ready-to-Drink
Probiotic
PROBIOTIC

Source of Prebiotics

Some examples of prebiotics are Polysaccharides – Inulin, lactulose, oligosaccharides: fructooligosaccharides (FOS), galactooligosaccharides (GOS), xylo-oligosachharides, isomaltooligosaccharides, soybean oligosaccharides, and other non-digestible oligosaccharides. Acacia Gums (Gum Arabic) are considered the richest natural source; other traditional dietary sources of prebiotics include beans, inulin sources (such as Jerusalem artichoke, jicama, and chicory root, garlic, onion and banana), raw oats, unrefined wheat, and unrefined barley. Some of the oligosaccharides occur naturally in breast milk and play an important role in the development of a healthy immune system in infants.

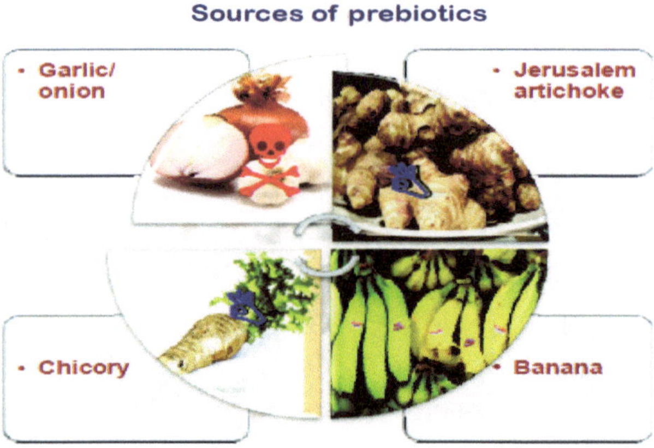

Mechanism of Action of Prebiotics

Prebiotics are dietary fibres with a well-established positive impact on the intestinal microflora. Other health effects of prebiotics (prevention of diarrhoea, modulation of the metabolism of the intestinal flora, cancer prevention, positive effects on lipid metabolism, stimulation of mineral adsorption and immunomodulatory properties) are indirect, i.e. mediated by the intestinal microflora, and therefore less-well proven. Infant formulas are prepared which is more breasts milk-like by the addition of fructo- and (primarily) galactooligosaccharides. A prebiotic dose of 5 grams/day should be sufficient to elicit a positive effect upon the gut microbiota (in some exceptional cases this may be nearer to 8g/d). A possible side effect of prebiotic intake is intestinal discomfort from gas production. However, bifidobacteria and lactobacilli cannot produce gas as part of their metabolic process. Therefore, at a rational dose of up to 20g/d, gas distension should not occur. If gas is being generated, then the carbohydrate is not acting as an authentic prebiotic. This is perhaps because dosage is too high and the prebiotic effect is being compromised i.e.bacteria other than the target organisms are becoming involved in the fermentation. Prebiotic-mediated proliferation of gut microbiota in rats, like probiotics, increases brain BDNF expression, possibly through the involvement of gut hormones. The effect of GOS on components of central NMDAR signalling was greater than FOS, and may reflect the proliferative potency of GOS on microbiota. The data of this study, therefore, provide a sound basis to further investigate the utility of prebiotics in the maintenance of brain health and adjunctive treatment of neuropsychiatric disorders (Savignac et al., 2013).

The mechanism of action of prebiotic is almost similar as probiotics. The main site of action is the colon. Thus, prebiotic should also resist the effects of various digestive enzymes and acidic conditions of the gut. Once the prebiotic can phase through all these conditions, it helps in the stimulation of growth and the various metabolic activities of live probiotic bacteria by fermentation itself. The target bacterial genera for the prebiotic actions are bifidobacteria and lactobacillus species. These bacterial species are able to take as a diet of any prebiotics and then produces various metabolic end products via the fermentation process. These fermented end products have a direct effect on health through the stimulation of host immune system as well as functions as a binding of receptor sites. The fermentation end products can be organic acids or biologically active peptides. These fermented products have a specific role in the direct antagonistic action against the harmful disease causing bacteria.

There are many health benefits of prebiotics some of which are mentioned here:

- Inflammatory Bowel Diseases
- Antihypertensive effects
- Diabetes
- Hypocholesterolemic effect

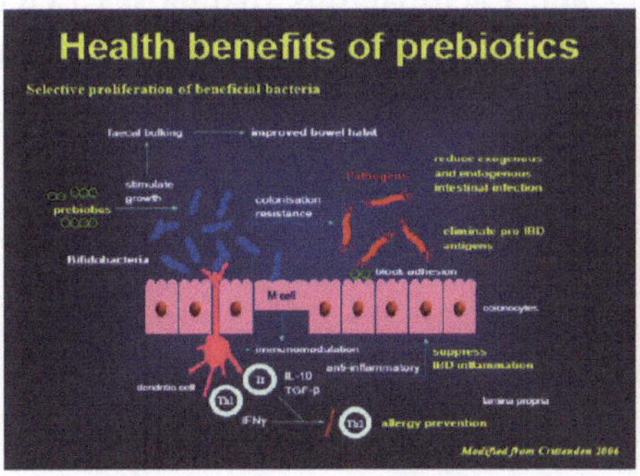

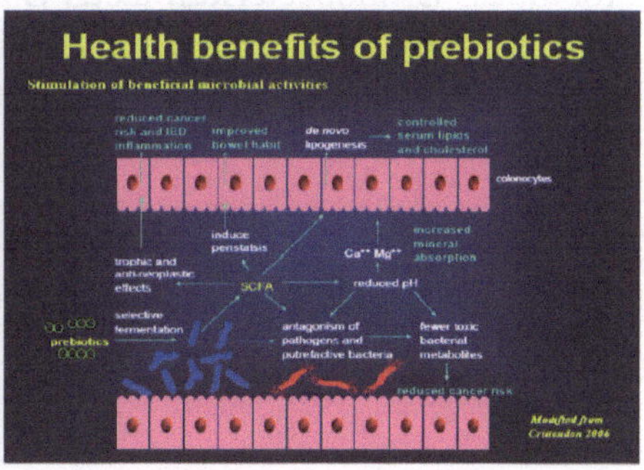

- Immunomodulatory effects
- Hepatic Encephalopathy

Mineral absorption and bone mineral density

It has been seen that prebiotics are very helpful in the elevation of mineral absorption and bone mineral density. Mostly studied prebiotics with regard to mineral absorption and mineral density are galactooligosaccharides, inulin and oligofructose. Most of the studies have been performed in rats to see the effect of prebiotics on bone development. It was studied that prebiotics stimulated the absorption of minerals like iron, calcium, magnesium and zinc in the experiment (Scholz-Ahrens, et al., 2001 and 2002; Coudray et al., 2003; and Cashman. 2003). The mechanism of action can be releasing of bone modulating factors such as the bone preserving effects of phytoestrogens (Ohta et al., 2002; Mathey et al., 2004), the improvement of isoflavone bioavailability (Rasmussen et al., 2003), increasing the luminal bacterial beta-glycoside activity, an enzyme that is responsible for hydrolyzing the glycosidic bond of isoflavone conjugates then and converts daidzein to equol which is an isoform having higher bone preserving potentials (Breinholt et al., 2000).

Prevention of colon cancer

Colon cancer is the most frequent type of cancer specially in western countries. Probiotics along with prebiotics are very useful in the reduction of colon cancer. The main mechanism is retarding of carcinogens by production of functional metabolic products in the colon. Because consumption of prebiotics is the main stimulants of probiotic bacteria thus prebiotics can be a preventive source of initiation of cancer in the colon. The mechanism of anticarcinogenic effects can be detoxification of genotoxins which are found in the gut. Prebiotic intake also increases the concentration of short chain fatty acids such as butyrate that is also a protective agent and it lowers the risk of development of cancer by inhibiting the genotoxic activity of nitrosamides and hydrogen peroxide. Probiotics Bifidobacterium lactis Bb12 and Lactobacillus rhamnosus LGG along with prebiotic inulin enriched with oligofructose have shown protection against azoxymethane (AOM)-insduced colon cancer (Table-3)

Table 3 : Effects of prebiotics as anti-cancerous in human studies

Prebiotic	Subjects and study design	Dose (g/day)	Increase in bifidobacteria	Markers of carcinogenesis	Reference
FOS	20, placebo-controlled, 12 days	125	1.2 log	No change in bile acids, neutral sterols, nitroreductase, azoreductase, or β-glucuronidase	Bouhnik et al., 1996
FOS	12, controlled diet, control and treatment periods, 26 days	4	0.5 log	Significant decrease in â-glucuronidase (75%) and glycocholic acid hydroxylase (90%)	Buddington et al., 1996
Resistant starch (RS)	12, controlled diet, high and low RS periods, 28 days	55.2±3.5 and 7.7±0.3	Not investigated	Significant decreases in neutral sterols (30%), 4-cholesten -3-one (36%), total bile acids (30%), secondary bile acids (32%) and â-glucosidase activity (26%) in high RS phase compared to low RS phase.	Hylla et al., 1998
GOS	40, placebo -controlled, 21 days	7.5 or 15	No significant increase	No significant changes in SCFA, bile acids, ammonia or skatoles	Alles et al., 1999

Immunological effects of prebiotics

It has been studied that prebiotic inulin and oligofructans can directly modulate the activity of the immune system (Bailey 2009; Meyer 2008; Seifert and Watz 2007). The effects of prebiotic mixture have been studied for their immune response in the healthy elderly people and they found that there are no changes in the serum proteins, albumins, secretory IaA and immunoglobulins. There is no effect of prebiotics on interferon-gamma and interleukin -4 secretions.

Systemic effects on blood lipids

Fermentation of any prebiotics alters the composition of probiotic bacteria in the gut and this alteration also influences the systemic blood lipid metabolism. It was studied that dietary fructans (oligosaccharides) exhibits the serum or hepatic lipid lowering properties. The prebiotic also responsible for suppression of hepatic triglyceride and very low density lipoproteins (VLDL) synthesis thus it reduced the triglycerides and cholesterol level in the blood. The mechanism can be the decrease in the systemic levels of blood lipids by the inhibition of hepatic cholesterol synthesis and redistribution of cholesterol from plasma to the liver. It has also been seen that prebiotics like inulin can have the ability to induce the inhibition of triglycerdes synthesis (Delzenne et al., 1999). Prebiotics indirectly interfere with the cholesterol absorption from the gut by deconjugation of bile salts which is the precursor of cholesterol synthesis by promoting of the growth of probiotic bacteria in the gut. The oligofructose and inulin are mostly known for their lipid lowering effects for serum total cholesterol and low density lipoprotein (LDL) cholesterol.

Anti-adhesive activities against pathogens and toxins

Prebiotics along with probiotics exert the anti adhesive activities against pathogenic bacteria. Prebiotics provides competitive advantage to the native bacteria in the gut as a source of nutrition. Adherence of human enteropathogenic *E. coli* in the human cell lines HEp-2 and Caco-2 have been studied in the presence of prebiotic galactooligosaccharides and found that prebiotic have the ability to reduce the adherence of *E. coli* (Shoaf *et al.*, 2006).

Health promoting effects in infant

One of the most important properties of prebiotics is the promotion of health beneficial effects in the infants. Research studies have shown that prebiotics galactooligosaccharides, lactulose, fructooligosaccharide, polydextrose and inulin are widely studied for their prebiotic potentials as food grade products. These prebiotics are generally fermented by the probiotic bacteria in the colon. Mainly bifidobacteria are significantly increases in the colon and this was observed by analyzing the stool samples of infants. Possible health effects of prebiotics in the gastrointestinal tracts are given in figure:

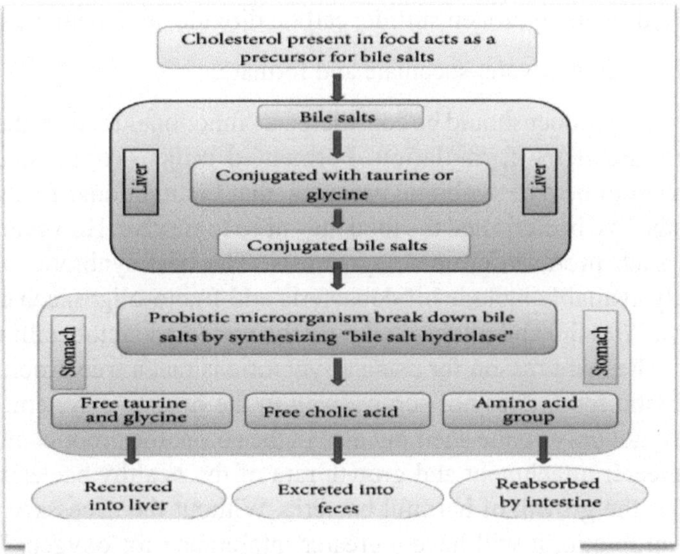

Source: Gopal et al.,(1996)

Synbiotic

A mixture of a Pre- and probiotics that helps to improve survival and implantation of live microbes in the GI tract by selectively stimulating the growth and/or activating the metabolism of one or a limited number of health-promoting bacteria. Thus according to the definition, "synbiotic is a supplement that contains both prebiotic and probiotic that work together to improve the "friendly flora" of the human intestine".

Effectiveness of a synbiotic depends directly on the various probiotic bacteria's ability to survive digestive juices as well as the alkaline environment of the duodenum. Another factor to consider is whether or not the bacteria are able to adhere to the intestinal wall to colonize the colon. As a result, prebiotics are necessary to assist the probiotic since prebiotic sugars (mainly oligosaccharides and disaccharides) are not digested. This enables prebiotics to travel through the digestive tract to help feed the bacteria in the intestines. These sugars are then fermented by bifidobacteria, lactobacilli and others in the colon to produce some beneficial by-products that include:

- Short-chain fatty acids acetate, propionate and butyrate (an important respiratory fuel for colon cells). These by-products are mainly absorbed from the colon and transported to the liver where they are extensively metabolized. Those that are not metabolized in the liver are circulated to various other tissues for use.

- Hydrogen, hydrogen sulfide, carbon dioxide, and methane gases.
- Lactate, pyruvate, succinate and formate.

A synbiotic product should be considered a "functional food" rather than some obscure chemistry formulation. Fermented milks (yogurt and kefir) are considered to be true synbiotic products, that is, functional foods, since they supply the live bacteria and the food they need to survive. However, evidentally not all such products promote symbiosis. The best synbiotic combinations currently available include bifidobacteria and fructo-oligosaccharides (FOS), *Lactobacillus* GG and inulins and bifidobacteria and lactobacilli with FOS or inulins. The main reason for using a synbiotic is that a true probiotic, without its prebiotic food, does not survive well in the digestive system. To enhance viability, not only on the shelf but also in the colon, the product must allow for much greater attachment and growth rate of the healthy bacteria in order to minimize the growth of harmful bacteria. Without the necessary food source for the probiotic, it will have a greater intolerance for oxygen, low pH, and temperature. In addition, the probiotic will have to compete against other bacteria that will take over if its specific food source is not available. Therefore, a "symbiotic" product (probiotic + prebiotic = synbiotic) makes for a better choice.

Functionality and Health Benefits of Synbiotic

How effective a synbiotic is depends on how the probiotic bacteria survive digestive stomach juices, as well as the alkaline environment of the duodenum. Effectiveness also depends on whether or not the bacteria are able to adhere to the intestinal wall to colonize the colon. As a result, prebiotics are necessary to assist the probiotic since prebiotic sugars (mainly oligosaccharides and disaccharides) are not digested. This enables prebiotics to travel through the digestive tract to help feed the bacteria in the intestines. These sugars are then fermented by bifidobacteria, lactobacilli, and others in the colon to produce some beneficial by-products that include short-chain fatty acids acetate, propionate and butyrate. These by-products are either used by the colon or absorbed and transported to the liver where they are extensively metabolized. Those that are not metabolized in the liver are circulated to various other tissues for use.

Potential health benefits of Pre and Probiotics are:

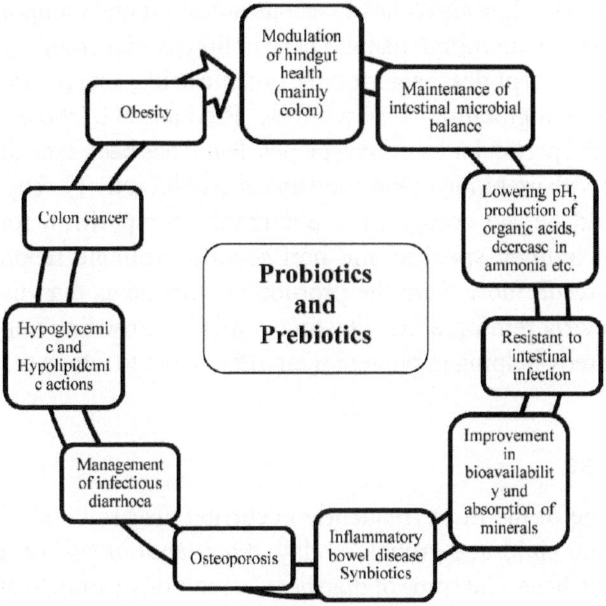

Healing and regulating the Intestinal Flora

Medical treatment like antibiotic therapy, chemotherapy or radiation therapy destructs and alters the gut microbiota composition thus it has significant adverse effects in metabolic process (Digestion and absorption). Because synbiotic products are made up of fibre materials and live probiotic bacteria so it immensely exhibits better results to heal the injurious bacteria and also enhance the amount of bacteria by supporting the growth and development host bacteria. Synbiotics play an important role in restoration of natural balance. A good synbiotic must have the ability to suppress some putrefactive processes in the stomach and intestine and thus helps to prevent the occurrence of a number of serious diseases: food allergies, ulcerous colitis, constipation, diarrhoea, cancers, gastrointestinal infections, IBS, IBD etc. Synbiotics containing one to ten billion active cells or CFU, should be taken several times a week, if not daily. In cases, where antibiotics have been taken, it is often suggested that synbiotics be taken several times a day, preferably at least 30 minutes before meals. Several different avenues are being explored for pre/probiotics (synbiotics). These are largely mediated by affecting an increase in beneficial bacteria within the gut flora.

Antimicrobial

Fermented dairy/food products have various antimicrobial components which are produced by the microorganisms. These antimicrobial components include biologically active peptides, bacteriocins, organic acids, short chain fatty acids, lactoferrin, immunoglobulins, lysozyme etc. Production of these components depends upon the particular bacterial species. But it has been seen that probiotic bacteria along with prebiotics have good antimicrobial activity. The mechanism of action could be the competitive exclusion, competition for nutrients, stimulation of antigen specific and non specific immune response. These qualities may result more from the probiotic components than the prebiotic. Beneficial bacteria reinforce the intestinal walls by crowding out pathogenic organisms, thereby helping to prevent their attachment to where they can cause disease.

Anti-Diarrhoea

Diarrhoea is one of the most frequently occurring disease conditions in India especially in the child age groups. Today, many antibiotics are used for the treatment of diarrhoea. But some of diarrhoeal conditions are not treated properly due to antibiotic nature of harmful pathogenic bacteria. In these situations one of the alternative ways to treat the diarrhoea is routinely consumption of probiotics. Benefits of probiotics are achieved by the crowding out of harmful bacteria known to cause diarrhoea. Since synbiotic products also strengthen the intestinal wall, this would help prevent diarrhoea also. The well-known use of functional dairy foods is for diarrhoeal diseases prevention and management of acute viral and bacterial diarrhoea as well as the control of antibiotic associated diarrhoea. A number of specific strains, including *Lactobacillus rhamnosus* GG, *Lactobacillus reuteri, Saccomyces boulardii, Bifidobacteria* spp., and others present in functional dairy products have been shown to have significant benefit from diarrhoea (Hilton et al., 1977; Isolauri et al., 1991; Duffy et al., 1994; Gionchetti et al., 2000a, 2000b; Gorbach, 2000; Guandalini et al., 2000; Levy, 2000; Danone 2001; Kyne, and Kelly, 2001; Marteau et al., 2001; Benchimol, and Mack, 2004), travellers' diarrhoea and diarrhoea disease in young children caused by rotaviruses.These functional products can prevent or ameliorate diarrhoea through their effects on the immune system. Moreover, they might prevent infection because probiotics present in them compete with pathogenic viruses or bacteria for binding sites on epithelial cells.

Protection from gastroenteritis

Gastroenteritis is also called the infectious diarrhoea. This mainly affects the stomach and the small intestine. In this condition patients have the symptoms of vomiting, diarrhoea, cramping and abdominal pains and finally dehydration occurs. Many studies have been done for the treatment of gastroenteritis through the application of functional foods like probiotics, prebiotics and or synbiotic. The most compelling evidence for the success of probiotics and prebiotics probably lies in their ability to improve resistance to pathogens. Lactic acid excreated by Lactobacillus bacteria is known for their inhibitory properties towards harmful gastroenteritis causing pathogens. There are a number of potential mechanisms for probiotic micro-organisms to reduce intestinal infections. Firstly, fermented metabolic end products such as acids excreted by these micro-organisms may lower the gut pH to levels below which pathogens are able to affectively compete. Also, many lactobacilli and bifidobacteria are able to excrete natural antibiotics which can have a broad spectrum of activity. Moreover, there is competition for nutrients and colonisation sites. This inhibitory effect also has relevance for more chronic diseases thought to have an involvement of pathogens.

Constipation

It is a chronic disease estimated to affect about 16% of the worldwide general population and to be 2–3 times more prevalent and symptomatic in women than men. Constipation frequency appears to augment with increasing age, particularly after 65 years and large amounts of healthcare resources are expended on its diagnosis and treatment. In addition, available therapies are unsatisfactory in one-third of patients. Constipation can be a consequence of intestinal dysbiosis, with an increase of potentially pathogenic microorganisms and a decrease of potentially beneficial microorganisms. These alterations may affect large bowel motility and secretory functions by changing the metabolic environment of the colon and the amount of available physiologically active substances. The ingestion of soluble fibres with prebiotic effects such as inulin and fructooligosaccharides (FOS), stimulates the growth of beneficial bifidobacteria and lactobacilli in the colon. Oral intake, at adequate concentrations, of specific strains of lactobacilli, bifidobacteria and other living commensal microorganisms, generically defined as probiotic, may be associated with health benefits, including improvements in bowel movements, permeability and microbial profile, improved function of intestinal immune barrier and prevention of colon cancer. The combination of probiotic strains and prebiotic fibers may provide synergistic effects after which they are named: synbiotics. The interaction between probiotics and prebiotics *in vivo* may improve the survival of probiotics. Synbiotic intake has been shown to modify microbiota composition and restore intestinal microbial balance, which may have positive effects on gastrointestinal functions. 3 weeks of treatment with a synbiotic containing FOS as a prebiotic and *L.paracasei* (Lpc-37), *L.rhamnosus* (HN001), *L. acidophilus* (NCFM) and *B. lactis* (HN019) as probiotics facilitated evacuation in constipated women by increasing stool frequency, improving stool consistency and shape, and decreasing constipation intensity. Further studies, in particular large, randomized clinical trials, are needed to confirm these results and to define the clinical role of synbiotic administration in constipated patients.

Antiallergenic

This quality may be helpful in cases such as food allergies since the probiotic bacteria help to reinforce the barrier function of the intestinal wall and thereby preventing the absorption of some antigens. It has been suggested that gut flora modulation may down-regulate gut inflammation and hypersensitivity that would otherwise lead to atopic eczema.

Treating Liver-related Brain Dysfunction

Synbiotics and fermentable fibre not only have a beneficial effect on the liver but on a brain dysfunction that affects many who have liver disease too. This dysfunction results in changes in behavior, intelligence, consciousness, and neuromuscular function called *Hepatic Encephalopathy.* Hepatic encephalopathy is a dreaded liver disease. Minimal encephalopathy is a condition of chronic liver disease with no clinical symptoms of brain dysfunction. Use of probiotics for minimal hepatic encephalopathy (MHE) has been rationalized based on various modes of action like decreasing bacterial urease activity, decreasing intestinal permeability, decreasing inflammation, decreasing uptake of other toxins, and other modes of action. Probiotics reduced concentrations of many bacteria, particularly gram-negative bacteria which produce urease. They also improve intestinal permeability in experimental human models. When patients with minimal hepatic encephalopathy were given *Bifidobacterium longum* with fructooligosaccharide for 9 weeks, their cognitive functions were seen to improve. Endotoxemia causes inflammation leading to cirrhosis of the liver. When fecal flora of cirrhosis patients was analysed, there was a substantial reduction in the levels of *Bifidobacteria*. Minimal hepatic encephalopathy (MHE) is a complication of cirrhosis during which accumulation of neurotoxic substances in the bloodstream produces neurological manifestations.

Anticarcinogenic

Anticarcinogenic qualities are not well understood, but what is speculated is that the sugars that are fermented by the bacteria form into substances that may inhibit the growth of carcinogenic cells. Prebiotic sugars help to increase calcium and magnesium in the colon which, in turn, assist in controlling the rate of cell turnover and the formation of insoluble bile or fatty acid salts, which can have damaging effects. Probiotic bacteria have the ability to bind to and inactivate some carcinogens. This may directly inhibit the growth of some tumours, as well as any bacteria that may convert matter into carcinogens. Probiotic bacteria are able to bound to and inactivate some carcinogens. They may directly inhibit the growth of some tumours, as well as any bacteria that convert matter into carcinogens. As early as 1995, in a controlled, double blind study, with 138 patients a *L. casei* Shirota preparation was shown to have a preventive effect on the recurrence rate of superficial bladder cancer after surgery. In different animal models (rats and mice) fed with inulin and/or oligofructose did reduce the genotoxicity of fecal water. It also decreased the number of chemically induced pre-cancerous lesions and stimulated defense functions. An increased level of IL-10 and of NK-cell activity was also observed. In the long term, the tumour incidence in the large intestine and in other organs (breast cancer in rats and mice, metastases in the lung was lowered by adding from 5 to 15% inulin or oligofructoseto the diet. This effect was even more pronounced when a combination of prebiotics and probiotics was given. Soybean oligosaccharide (SBO) and wheat bran oligosaccharide (WBO) were also fed in combination with bifidobacteria. SBO reduced the colonic mucosa proliferation compared with the control. The results suggested that the combination of bifidobacteria and oligofructose reduces colon cancer risk in carcinogen-treated rats. (Gallaher and Khil, 2014)

Osteoporosis prevention

Osteoporosis prevention may occur as a result of an improvement in mineral absorption and balance. This ability comes as a result of oligosaccharides (3-6 sugar chains) which bind such minerals as calcium and magnesium in the small intestine, but releasing them in the large intestine where they are better absorbed. Fatty acids, also formed from the fermentation process, will then assist with the absorption of these minerals.

Mineral absorption

A reduced pH in the bowel because of a lactic fermentation is thought to better sequester of calcium and perhaps magnesium which increases the Mineral bioavailability. The non digestible carbohydrates (dietary fibre) impair the small-intestinal absorption of minerals because of their binding or sequestering action. However, the minerals that are bound or sequestered and, consequently, not absorbed in the small intestine, do reach the colon, where they may be released from the carbohydrate matrix and absorbed. Moreover, a high concentration of short-chain carboxylic acids resulting from the colonic fermentation of the non-digestible carbohydrates facilitates the colonic absorption of minerals, particularly Ca^{2+} and Mg^{2+}. In addition, independent of any binding or sequestering of minerals, some non digestible carbohydrates (e.g., inulin-type fructans) may improve mineral absorption and balance because of an osmotic effect that transfers water into the large bowel, thus increasing the volume of fluid in which these minerals can dissolve. Furthermore, these carbohydrates, being extensively fermented, acidify the colonic content and consequently raise the concentration of ionized minerals, in particular Ca^{2+} and Mg^{2+}, a condition that favours passive diffusion.

More recently, *in vivo* human studies confirmed the positive effect of inulin and oligofructose on the absorption and balance of dietary calcium but not of iron, magnesium, or zinc. In the first published report, 9 men (mean age: 21.5 ± 2.5 y) who ingested 850 mg Ca/d and a dietary supplement of 40 g/d of inulin had a significant increase in the apparent absorption(±12%) and balance (100 mg/d) of calcium without any significant change in urinary excretion . In a second study, the calcium balance of 12 males between the age of 15 and 18 y who consumed 16.8 g oligofructose/d, as measured by the double stable isotope technique, increased to 11% ($P = 0.09$) with no significant effect on urinary excretion (Scholz-Ahrens, 2014).

Regulates the immune system

Since probiotic bacteria are known to have the ability to increase the levels of circulating immunoglobulin A (IgA). In addition, they enhance non-specific immune mechanisms, such as increasing phagocyte activity. Probiotic bacteria can modulate the activity of many cells of the immune system, including innate system NKs, DCs, macrophages, epithelial cells and granulocytes, as well as adaptive system Th1, Th2, Th17, Treg, Tc and B cells. Thus, probiotic bacteria have the potential to modulate any part of the immune system in the context of acute responses to intracellular or extracellular pathogens or chronic responses observed in dysregulated immune pathological conditions such as IBD, colorectal cancer and hypersensitivity. Probiotic bacterial strains can be generalised to exert immune-activation, -deviation or -regulation/suppression responses. Thus, selection of probiotic strains, and indeed combinations of probiotics, will be formulated upon careful consideration of the disease mechanisms and the desired immunomodulatory effect. Finally, the growing body of evidence for the suitability of prebiotics in immunomodulation and formulation with probiotic strains, as synbiotics, represents a realistic approach in disease modification. The probiotic species acting as a quick-fix immunomodulatory "plaster" and the prebiotic facilitating commensal organisms, modulating the dysbiosis in the gut microbiota and further modifying the immune system to our benefit—healing from within.

Hypocholesterolaemic and Cardioprotective Effects

Hypercholesterolemia, or elevated level of total cholesterol in the blood stream, is the result of high levels of low-density lipoprotein (LDL) as compared to high-density lipoprotein (HDL) cholesterol. Many *Lactobacilli*, being the natural inhabitants of the intestine, possess bile-salt hydrolase activity. This property has been used for developing probiotic formulations to combat hypercholesterolemia. Synbiotics have also been evaluated for their hypocholesterolemic effects. Twenty-four hypercholesterolemic male pigs fed with a synbiotic formulation of *L. acidophilus* ATCC 4962, fructooligosaccharides, mannitol, and inulin have given positive results at 8 weeks. Total plasma triacylglycerol, total cholesterol and LDL levels decreased which are the main causes of cardiovascular diseases. A synbiotic yogurt containing *L. acidophilus* 145, *B. longum* 913, and oligofructose in a randomized, placebo-controlled, and crossover study involving twenty-nine women. The HDL cholesterol increased. In yet another study, Schaafsma, et al., (1998) saw a significant decline in total cholesterol and LDL cholesterol in thirty volunteers who were fed synbiotic milk containing *L. acidophilus* and fructooligosaccharides. It was reported that consumption of probiotic and prebiotic significantly reduced the serum triglycerides level and also changed the total and LDL cholesterol in human subjects (Pereira and Gibson, 2002).

Reduces blood sugars in some individuals

Probiotic bacteria have the ability to normalize serum glucose levels by uptake of glucose via the highly activated receptor. The prebiotics may help by delaying gastric emptying and/or shorten the transit time through the small intestine. In addition, a substance released through fermentation (propionate) appears to make better use of the glucose molecule conversion.

Improved tolerance to lactose

It is thought that probiotics may help in this regard, through their β-galactosidase activity or lactase activity. The concentration of lactase enzyme is more in fermented milk and milk products as compare to normal milk. Inability to digest lactose because of inadequate levels of *lactase* (lactose-digesting enzyme) in such cases undigested lactose reaches to the large intestine and being fermented by the colonic microbes result in diarrhea, bloating, abdominal pain and flatulence. Probiotic microorganisms are able to produce lactase enzyme and digest lactose. *L. acidophilus* and *Bifidobacteria* fermented products improve digestion of lactose (Kiani et al., 2006). In another study it was reported that *Streptococcus thermophillus* and *L. delbruekii (bulgaricus)* improve lactose digestion and symptoms related to lactose intolerance.

Anti obesity and reduction of toxins

Probiotic bacteria along with prebiotic dietary fibres have potential role in the reduction of toxin levels and promotion of gut permeability by modulation of microbial balance. In humans, colorectal cancer is thought to have a bacterial origin, with around 10 different carcinogens described that have been attributed to microbial events. Dietary strategies that lead to a reduced accumulation of such products may be possible. Dietary fibres and resistant starches may be fermented in the large gut to increase faecal bulk and reduce the residence time of such materials in the gut. Moreover, probiotics and prebiotics may modify the activities of enzymes that are involved in carcinogenesis. The mechanism of action of probiotics can be the neutralization of carcinogens as well as the changes of highly carcinogenic compounds to low carcinogenic compounds.

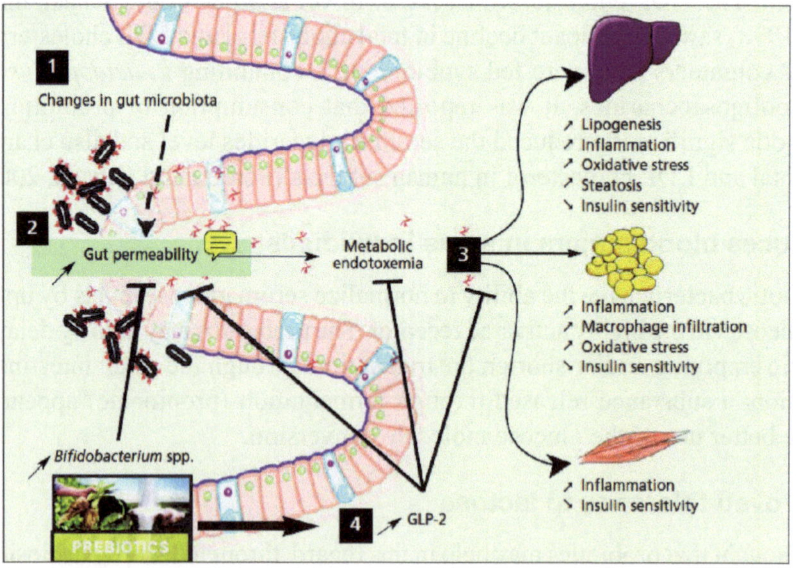

Source: Delzenne, and Cani, (2010)

Sedentary life style and diet are the main reason for obesity and changes of the gut microbiota associated with obesity. This leads to increase of metabolic endotoxemia, promotion of gut permeability and development of various metabolic disorders. From the figure we can see that the high fat diet changes the composition of gut microbiota thus leads to higher gut permeability and finally the higher level of plasma lipopolysaccharides or metabolic endotoxemia in the gut. And these metabolic endotoxemia induces the promotion of low-grade inflammation and metabolic disorders such as obesity, insulin resistance, diabetes mellitus, oxidative stress and steatosis. But the preventive measure can be taken by consumption of prebiotics and probiotics. Intake of prebiotics modulates the microbiota i. e. increasing the number of Bifidobacterium species. In addition it also increases the endogenous GLP-2 production and suppression of all the metabolic disorders.

Conclusions

Increasing consumer awareness, health consciousness and expenditure are the socio-economic factors responsible for the expanding world-wide interest in functional foods. The health concerns of most interest to consumers that could be addressed by food products include osteoporosis, cholesterol, heart disease, and general health. The main role of diet is to provide nutrients to meet host physiological requirements. As research behind diet and health has evolved, so has the concept of 'functional foods' become popular. Foods which are touted as being 'functional' are thought to exert certain positive properties over and above their normal nutritional value. The concept has now moved markedly towards gastrointestinal function, in particular the impact of gut bacteria. Possibly this is driven by the ubiquity of gastrointestinal disorders but also the fact that diet is an important controlling factor with regard to indigenous microbiota activities. The gut microflora contains pathogenic, benign and beneficial microbial species. A predominance of the former can lead towards gut upset which can be both acute (e.g. gastroenteritis) and chronic (e.g. inflammatory bowel disease). Functional foods directed towards the gut microbiota would serve to influence the composition of activities towards a more positive metabolism. The colon is the most heavily populated region of the gastrointestinal tract and, because of this resident microbiota, is one of the most metabolically active organs in the body. The concept of modulating activities directed towards improving gut microbial function has a long history, as diet can have a major effect on the gut microflora activities. The incidence of acute and chronic gut disorders continues to rise, with many diseases being untreatable. The functional food industry's perception of the importance of gut microbiology in human health and nutrition has led to a major increase in probiotic and prebiotic-based products.

Future Perspectives

As it has been widely studied that ingestion of synbiotics (Pro- and pre-biotics) are very important key factor in the regulation of human health and nutrition.

Prebiotics elevate the composition of lactobacilli and bifidobacteria in the colon and this increased number of bacteria automatically exhibits the functional properties. But further lot more research is needed to completely explore the applications of prebiotic and probiotics and understanding the mechanism of their action at molecular levels.

References

1. Alles, M. S. Hartemink, R. Meyboom, S. Harryvan, J. L. van Laere, K. M. J. and Nagengast, F. M. (1999). Effect of transgalactooligosaccharides on the composition of the human intestinal microflora and on putative risk markers for colon cancer. American Journal of Clinical Nutrition, 69:980-91.
2. Benchimol, E. I. and Mack, D. R. (2004). Probiotics in relapsing and chronic diarrhea. Journal of Pediatric Hematuiology Oncology, 26:515–517.
3. Bercik, P. Denou, E., Collins, J. Jackson, W. Lu, J. Jury, J. Deng, Y. Blennerhassett, P. Macri, J. McCoy, K. D. Verdu, E. F and Collins, S.M. (2011). The intestinal microbiota affect central levels of brain-derived neurotropic factor and behavior in mice. Gastroenterology, 141:599-609.
4. Bouhnik, Y. Flourie, B. Riottot, M. Bisetti, N. Gailing, M. F. and Guibert, A. (1996). Effects of fructo-oligosaccharides ingestion on faecal bifidobacteria and selected metabolic indexes of colon carcinogenesis in healthy humans. Nutrition Cancer, 26:21-29.
5. Bravo, J. A. Forsythe, P. Chew, M. V. Escaravage, E. Savignac, H. M. Dinan, T. G. Bienenstock, J and Cryan, J. F. (2011). Ingestion of Lactobacillus strain regulates emotional behavior and central GABA receptor expression in a mouse via the vagus nerve. Proc. Natl. Acad. Sci. USA 108 (38):16050–16055.
6. Breinholt, V. Hossaini, A. Svendsen, G. W. Brouwer, C and Nielsen, E. (2000). Estrogenic activity of flavonoids in mice. The importance of estrogen receptor distribution, metabolism and bioavailability. Food Chemical Toxiciology, 38:555–64.
7. Buddington, R. K. Williams, C. H. Chen, S. C and Witherly, S. A. (1996). Dietary supplementation of neosugar alters the fecal flora and decreases activities of some reductive enzymes in human subjects. American Journal of Clinical Nutrition, 63:709-16.
8. Cashman, K. (2003). Prebiotics and calcium bioavailability. Current Issues Interest in Microbiology, 4:21–32.
9. Coudray, C. Demigne, C and Rayssiguier, Y. (2003). Effects of dietary fibers on magnesium absorption in animals and humans. Journal of Nutrition, 133:1–4.
10. Critchfield, J. W. Hemert, S. V. Ash, M. Mulder, L. and Ashwood, P. (2011). The potential role of probiotics in the management of childhood autism spectrum disorders. Gastroenterology research and practice, Article ID 161358, doi:10.1155/2011/161358.
11. Danone (2001). Fermented foods and healthy digestive functions. France: Danone Publications, John LibraryEurotext.
12. Delzenne, N. M. and Cani, P. D. (2010). The role of intestinal microbiota in energy metabolism and metabolic disorders International Diabetes Monitor, Volume 22: Number 5.
13. Delzenne, N. M. and Kok, N. N. (1999). Biochemical basis of oligofructose-induced hypolipidaemia in animal models. Journal of Nutrition, 129:1467S-70S.
14. Duffy, L. C. Zielezny, M. A and Riepenhoff-Talty, M. (1994). Effectiveness of Bifdobacterium biudum in mediating the clinical course of murine rotavirus diarrhoea. Pediatr Research, 35:690–695.

15. Femia, A. P. Luceri, C. Dolara, P. Giannini, A. Biggeri, A. Salvadori, M. Clune, Y., Collins, K.J. Paglierani, M and Caderni, G (2002). Antitumorigenic activity of the prebiotic inulin enriched with oligofructose in combination with the probiotics Lactobacillus rhamnosus and Bifidobacterium lactis on azoxymethane-induced colon carcinogenesis in rats. Carcinogenesis, 23 (11):1953–1960.
16. Gallaher, D. D and Khil, J. (1999). The effect of synbiotics on colon carcinogenesis in rats. J. Nutr., 129: 1483S–1487S.
17. Gionchetti, P. Rizzello, F. Venturi, A and Campieri, M. (2000a). Probiotics in infective diarrhoea and inûammatory bowel diseases. Journal of Gastroenteriology Hepatology, 15: 489–493.
18. Gionchetti, P. Rizzello, F. Venturi, A and Campieri, M. (2000b). Oral bacteriotherapy as maintenance treatment in patients with chronic pouchitis: a double blind, pacebo controlled trials. Gasteroenterology, 119:305-309.
19. Gopal, A. Shah, N. P. and Roginski, H. (1996). Bile tolerance, taurocholate deconjugation, and cholesterol removal by *Lactobacillus acidophilus* and *Bifidobacterium* spp. Milchwissenschaft.51:619-623.
20. Gorbach, S. L. (2000). Probiotics and gastrointestinal health. American Journal of Gastroenteriology, 95 (1):S2-4.
21. Guandalini, S. (1998). Probiotics in the treatment of diarrhoeal disease in children. Gastroenterol Intest., 11:87–90.
22. Guarner, F. Malagelada, J (2003). Gut flora in health and disease. The Lancet, 361 (9356): 512–9
23. Hilton, E. Kolakawaki, P. Singer, C. and Smith, M. (1977). Efûcacy of Lactobacillus GG as a diarrheal preventive in travelers. Journal of Travel Medicine, 4: 41–43.
24. Hylla, S. Gostner, A. Dusel, G Anger, H. Bartram, H. P and Christl, S. U. (1998). Effects of resistant starch on the colon in healthy volunteers: possible implications for cancer prevention. American Journal of Clinical Nutrition, 67:136-42.
25. Isolauri, E. Juntunen, M. Rautanen, T. Sillanaukee, P. and Koivula, T. A. (1991). Human Lactobacillus strain (Lactobacillus casei sp strain GG) promotes recovery from acute diarrhea in children. Pediatrics, 88:90–7.
26. Isolauri, E. Yelda, S. Pasi, K. Heikki, A. and Salminen, S. (2001). Probiotics: effects on immunity. American Journal of Clinical Nutrition, 73(suppl):444S–50S.
27. Kerman, I. A. (2012). New insights into BDNF signalling: relevance to major depression and antidepressant action. American Journal of Psychiatryology, 169:1137–1140.
28. Kiani, L. (2006). Bugs in our guts – not all bacteria are bad: How probiotics keep us healthy. CSA Discovery Guides.
29. Kyne, L. and Kelly, C. P. (2001). Recurrent Clostridium difûcile diarrhoea. Gut., 49:152-153.
30. Levy, J. (2000). The effects of antibiotic use on gastrointestinal function. American Journal of Gastroenteriology, 95:S8–S10.
31. Marteau, P. R. de Vrese, M. Cellier, C. J and Schrezenmeir, J. (2001). Protection from gastrointestinal diseases with the use of probiotics. American Journal of Clinical Nutrition, 73 (suppl):430S–6S.
32. Mathey, J. Puel, C. Kati-Coulibaly, S. Bennetau-Pelissero, C. Davicco, M. J. Lebecque, P. Horcajada, M. N and Coxam, V. (2004). Fructooligosaccharides maximize bone-sparing effects of soy isoflavone-enriched diet in the ovariectomized rat. Calcif Tissue International, 75:169–79.
33. Meyer, D. (2008). Prebiotic dietary fibres and the immune system. Agro-Food Industry Hi-Techchnology 19:12-15.

34. Milck Bailey. (2009). The mucosal immune system: Recent developments and future directions in the pig. Developmental and Comparative Immunology, 33(3):375–383.
35. O'Sullivan, E. Barrett, E., Grenham, S. Fitzgerald, P. Stanton, C. Ross, R. P. Quigley, E. M. Cryan, J. F. Dinan, T. G. B. D. N. F. (2011). Expression in the hippocampus of maternally separated rats: does Bifidobacterium *breve* 6330 alter BDNF levels? Benefits of Microbial research, 2(3):199–207.
36. Ohta, A., Uehara, M., Sakai, K., Takasaki, M., Adlercreutz, H., Morohashi, T and Ishimi, Y. (2002). A combination of dietary fructooligosaccharides and isoflavone conjugates increases femoral bone mineral density and equol production in ovariectomized mice. Journal of Nutrition,132:2048–54.
37. Pereira, D.I.A. and Gibson, G.R. (2002). Effects of Consumption of Probiotics and Prebiotics on Serum Lipid Levels in Humans. Critical Reviews in Biochemistry and Molecular Biology, 37(4): 259–281.
38. Rasmussen, S. E and Breinholt, V. M. (2003). Non-nutritive bioactive food constituents of plants: bioavailability of flavonoids. International Journal of Vitamin and Nutritional Research, 73:101–11.
39. Savignac, H. M. Corona, G. Mills, H. Chen, Li. Spencer, J. P. E. Tzortzis, G and Burnet, P. W. J. (2013). Prebiotic feeding elevates central brain derived neurotrophic factor, N-methyl-D-aspartate receptor subunits and D-serine. Neurochemistry International, 63: 756–764.
40. Schaafsma, G. Meulig, W. J. A. van Dokkum, W. and Bouley, C. (1998). Effects of a milk product, fermented by Lactobacillus acidophilus and with fructo-oligosaccarides added, on blood lipids in male volunteers. European Journal of Clinical Nutrition, 52:436-440.
41. Scholz-Ahren, K.E. Ade, P. Marten, B. Weber, P. Timm, W. Ail, Y. Gluer, C.C. and Jurgen Schrezenmei, J. (2007). Prebiotics, probiotics, and synbiotics affect mineral absorption, bone mineral content, and bone structure. Journal of Nutrition, 137:838S–846S.
42. Scholz-Ahrens, K. E and Schrezenmeir, J. (2002). Inulin, oligofructose and mineral metabolism—experimental data and mechanism. British Journal of Nutrition, 87: Suppl 2:S179–86.
43. Scholz-Ahrens, K. E. Ade, P. Marten, B. Weber, P. Timm, W. Ail, Y. Gluer, C.C. and Schrezenmeir, J. (2007). Prebiotics, Probiotics, and Synbiotics Affect Mineral Absorption, Bone Mineral Content, and Bone Structure. Journal of Nutrition, 137: 838S–846S.
44. Scholz-Ahrens, K. E. Schaafsma, G. van den Heuvel, E.G.H.M. and Schrezenmeir, J. (2001). Effects of prebiotics on mineral metabolism. American Journal of Clinical Nutrition, 73: 2 Suppl:459s–64s.
45. Schoterman, H.C. and Timmermans, H.J.A.R. (2000). Galacto-oligosaccharides. In: Gibson, G.R. and Angus, F. editors. Prebiotics and probiotics. LFRA Ingredients Handbook. Leatherhead Food RA Publishing; p. 19-46.
46. Seifert, S. and Watz, B. (2007). Inulin and oligofructose: Review of experimental data on immune modulation. Journal of Nutrition, 137:2563S-2567S.

CHAPTER 13

Probiotics and Phytonutrients: Potential Candidates for the Immunobooster Nutraceutical Formulations

Bhatia Aruna

Abstract

Nutraceuticals have potential to be used as food supplement, preventive medicine and the growing evidence points in the direction that certain foods fight and or prevent against diseases. The development of next generation nutraceutical "super foods" or products consist of value-addition in the traditional natural diets.The word Nutraceuticals combine 'nutrition' and 'pharmaceuticals' to mean that they can be used as preventive drugs or food supplements. The increase in knowledge about the immune response unveils the fact that besides immunological disorders and infections the outcome of many other diseases like diabetes, hypercholesterolemia, rheumatoid arthritis, etc. can be modified by modulating Immune response. Hence, the efficient immunomodulators can be employed to prevent or cure the diseases as immunotherapeutic agents. The most acceptable alternative therapeutic agents are the edible sources. Hence the researchers shifted towards formation of nutraceuticals or foods acting as medicine, a concept which was given by Hippocrates long back.

Introduction

Probiotics the beneficial bacteria and the medicinal plants extracts or their phytochemicals are on the top of the list of these agents. Many reports reveal the beneficial effects of probiotics as gut microflora, inhibiting the growth of pathogens, and immunostimulating efficacy. Similarly, literature is adding up revealing the health benefits of the plant sources.The use of probiotics and of medicinal plants in nutraceuticals formulations has increased in past 25 years with the increase awareness about the harmful effects cost and nonavailability of chemical drugs and more consumer acceptance of the nutraceuticals. This resulted in the raise in their market and continuously people are looking for more formulations or candidates to make new safe therapeutic agents.

Work has been carried out since past twenty five years to find out the various plants' and bacteria' efficacy as immunoboosters and their applications as immunorestorers, antioxidants, anticancerous, antimicobials, anti-cholesterolemic and antidiabetics (1-13). The results showed that the extracts of plants in various solvents and from various parts show variability in their effect (Rani, 2009; Sheikh, 2013). The variation in the bioactivity depends upon the type of constituents which are solubilized or extracted in different solvents. The isolation of the constituents showed that oat seeds β-glucan, (Puneet 2008) glycosides from methanolic flowers extract of Nyctanthes harbour trisitis. Three out of four major components rutin, diadezine, quercitin and chlorogenic acid isolated from Prunus cereaseus (Cherry) (Sheikh, 2013), Woodfordia fruiticosa flower extract (under preparation), seeds of Ziziphus mauritiana (Mishra, 2010) were some of the best immunomodulators.

The results of the experiments with more than 30 strains of probiotics of Lactobacillus and Bifidobacterium species revealed that the probiotics are immunoboosters, anticholesterolemic (Ritu, 2008), and antioxidants, not only as cells but also the killed bacteria were immunoboosters. The bioactivity of probiotics could be raised by delivering them (i) as synbiotics with prebiotics (Rani, 2009) (ii) coencapsulation of the synbiotics to prevent them from the harsh environement (Bhatia et al., 2012a; Bhatia et al., 2012b, Bhatia et al., 2012c) (iii) by using the combination of the most bioactive probiotics (Randhawa, 2013) as all the probiotics donot have the same bioactivity and variation in the bioactivity varies in between the strains of the same species (Bhatia and Kaur, 2011) (iv) by delivering only probiotic bacterial DNA as DNA is a better immune enhancer than its whole cell (Randhawa, 2013) (v) delivering the bioactive materials in biosafe nanoparticles like chitosan (Bhatia et al., 2013).

In the present paper the results of the two major experiments are presented which included the i) ameiolaration of the bioactivity mainly the immunorestorative activity of the plant extract through nanodelivery ii) the isolated constituents of the plant extract are responsible for the bioactivity and the isolated compounds can be used for enhanced activity rather than the whole plant extract iii) DNA of the probiotics is better immune enhancer than the whole probiotic.

Materials and Methods

Strains of microorganisms

Different Strains of Lactobacillus and Bifidobacterium procured from National Dairy Research Institute (NDRI), Karnal, Haryana and some strains isolated in the lab were used in different set of experiments. The cultures so obtained were revived, grown and maintained on MRS broth for further use. A concentration of 109 cells/ 10 ^{6} /ml of Cells were used in the experimens (The concentrations for various experiments varied).

In vivo experiments

Swiss albino male mice (18-22 g) maintained on a standard laboratory diet (Kisan Feeds Ltd., Mumbai, India) and water ad libitum were employed in the study. The animals were divided into different group including the test groups and the positive and the negative controls house and were exposed to 12 hr cycles of light and darkness.

Experimental animal design: Animals were divided into many groups:

The groups included minimum test groups, control groups, postitive and negative and vehicle treated. e.g. to study the effect of chitosan nanoparticles encapsulation on immuneactivity and immunorestorative activity of plant extracts following groups were made

Group I: Control group (not subjected to any treatment i.e. kept only on diet)

Group II: Positive control (mice subjected to immune enhancer i.e Levamisole (2.5 mg kg^{-1}) with normal diet)

Group III: Antigen sensitized control (mice sensitized with Sheep Red Blood Cells (SRBC) and kept on normaldiet).

Group IV: Mice made immunosuppressed by hydrocortisone treatment

Group V: Immunosuppressed mice given nano encapsulated plant extract

Group VI: Immunosuppressed Mice dosed with unencapsulated plant extract

Group VII: Immunosuppressed animals given only nanoparticles

Wherever the test material was changed, correspondingly the control groups (positive and negative were changed). Similarly for the experiment "To see the effect of probiotic DNA on the immune response, two additional groups of animals, one given DNA of probiotics and other given only probiotic cells were also set up along with the control groups.

Immunization: Immunization was carried out by SRBC's, or BSA

The intraperitoneal injection with a single dose (100ìl/ml of 1 × 107 cells/ml) of sheep red blood cells (SRBC) was given to the animals to make them antigenically challenged.

Humoral Immune Response: It was assessed by direct haemagglkutination assay or ELISA

The results were expressed as mean ± S.E.M. log titer of individual animals.

Cell mediated immune response: CMI was assessed by Delayed Type Hypersensitivity assay reactions elicted in the mice by foot pad assay. Paw thickness was measured with vernier-caliper at 24h interval up to 72h.The difference in paw thickness was compared with control. Results are expressed as mean ± S.E.M. of footpad thickness up to 72h.

Nitroblue Tetrazolium Reduction assay

NBT test was employed to measure the macrophage function. This assay is based on reduction yellow coloured NBT dye results in the formation of coloured complex which can be phagocytosed by macrophages.which is measured spectro photometerically at 520 nm using dioxane as blank.

Inducible Nitric Oxide Synthase activity

Inducible nitric oxide synthase activity in splenocytes suspension was evaluated by using arginine as described by Bhatia et al.The color developed (indicating presence of citrulline) was measured spectrophotometrically at 540nm against MEM and Griess reagent as blank

Bactericidal activity

Bactericidal activity is one of the important parameter to measure the phagocytosis of splenocytes as described by procedure

Plant extracts

Preparation of plant extracts and the Isolation of the constituents of the plant extracts was carried out using the standard techniques of TLC, HPLC (Sheikh, 2013)

DNA Isolation of Probiotics

The isolation of the DNA from the probiotics as per the method of (Randhawa, 2013)

Results and Discussions

Experiment: Effect of nanoparticles encapsulation on immunorestorative activity of plant extract of *Zizyphus mauriatiana* are given in Figures 1, 2, below.

The Figure-2 reveals that the Cell mediated immune response and functions of the macrophages as depicted by Delayed type of hypersensitivity, Nitroblue tetrazolium reduction test and INOS activity and Phagocytosis were the best in extract Loaded nanoparticles .

DTH response

- DTH response of mice group treated with *Ziziphus mauritiana* leaf extract loaded chitosan nanoparticles is better as compared to all other groups.
- Not only this the response with loaded nanoparticles was two times better than that with plant extract (0.3mg/ml)given as such.

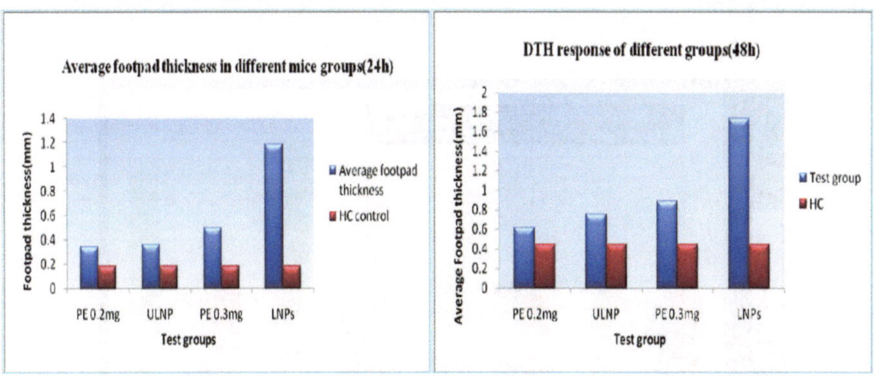

Fig. 1:

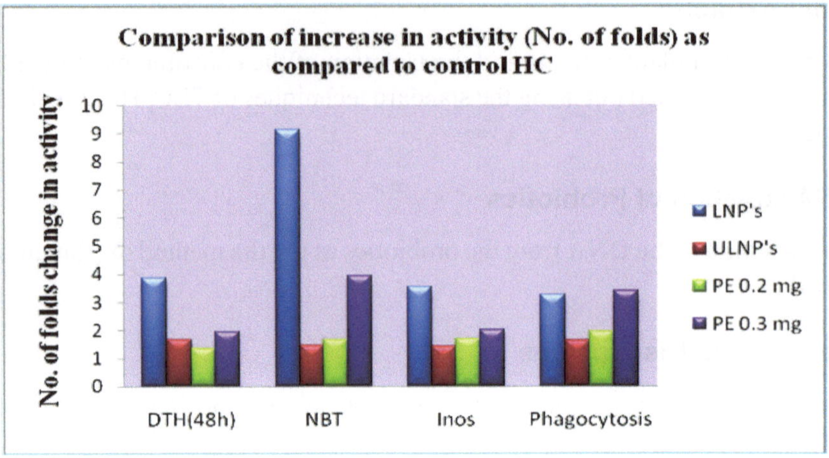

Fig. 2:

Resuts of Experiment: The immunomodulatory activity of the isolated compounds from the extract of Prunus cereasus varies. In the figures given below the results revealed that out of the four compounds, i.e. Rutin, chlorogeneic acid, Daidezin and quercetin isolated from the best immunomodulatory extract ie methanolic fruit extract of *P. cereasus*.

Diadezein showed the best immuneactivity in vitro and further the activity is equally good in in vivo experiments also. Hence diadezein can be employed as the best immunomodulator in the foods. The results are shown in the following Figures 3, 4 and 5.

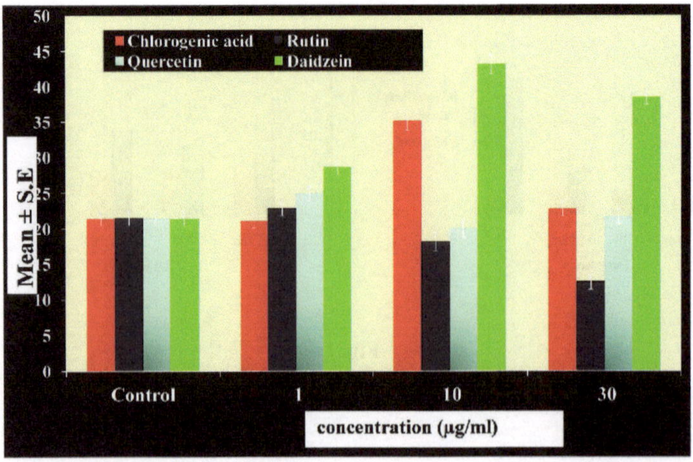

Fig. 3: Effect of isolated compounds on NBT reduction

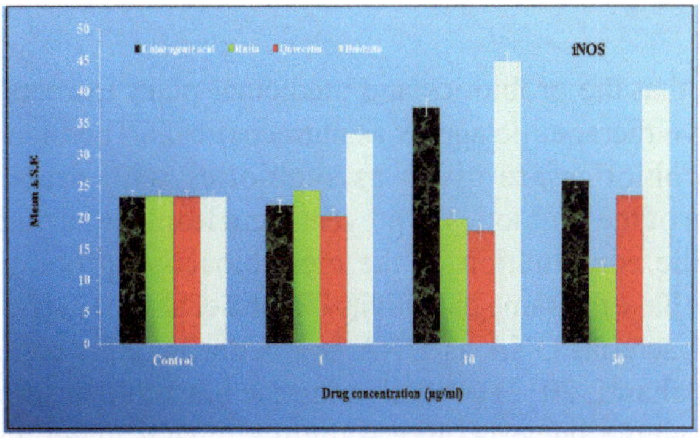

Fig. 4: Effect of isolated compounds on iNOS activity

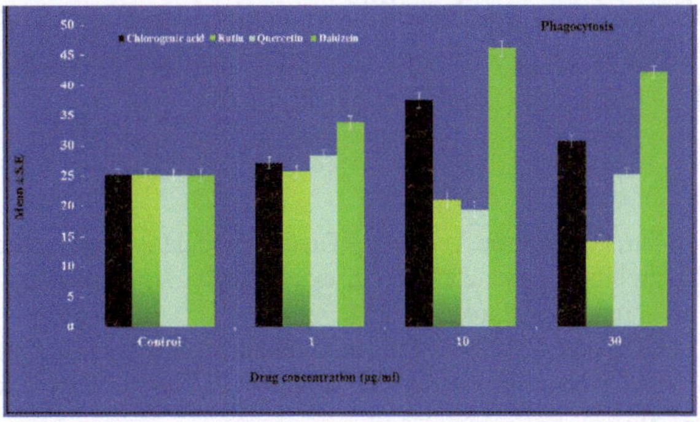

Fig. 5: Effect of isolated compounds on Bactericidal activity

Results of the experiment carried out to see the immunoefficacy of isolated DNA from the probiotics and its comparison with corresponding whole probiotic microorganisms is given below in the Figure-6. It is clear that the DNA was a better immuneenhancer than the whole cell. This shows that the incorporation of DNA than the whole cell will ameliorate the bioactivity in the formulation.

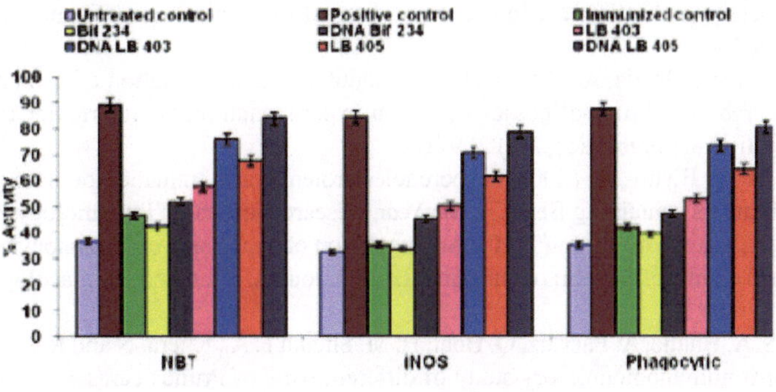

Fig. 6: Comparative evaluation of probiotics and their DNA for their effect on immune response in *in vivo* study. The results are expressed as mean ± S.E.M. (n=6).

Conclusion

It is concluded that the probiotics and medicinal plant extracts are potential edible alternative therapeutic agents as nutraceuticals. The bioactivity can be enhanced if probiotics are given as synbiotics with prebiotics, or (ii) coencapsulation of the synbiotics to prevent them from the harsh environement (iii) by using the combination of the most bioactive probiotics as all the probiotics donot have the same bioactivity (Bhatia et al., 2012) (iv) by delivering only probiotic bacterial DNA as DNA is a better immune enhancer than its whole cell (Randhawa, 2013) (v) delivering the bioactive materials in biosafe encapsulating (whole bacteria) like sodium alginate/ materials in chitosan nanoparticles (Bhatia et al., 2012). The bioactivity of any plant extract could be raised by delivering only selected immunomodulatory extract or one of its extracted content (Sheikh, 2013) or delivering the bioactive extract in biosafe nanoparticles like chitosan (Bhatia et al., 2013)'

References

1. Bhatia, A. Gulati, R. and Rani, U. (2007). Dietary modulation and restoration of immune response by prebiotic i.e. lactulose: an experimental study. International journal of probiotics and prebiotics, 2, (4):209-214.
2. Bhatia, A. and Dhillon, P. (2007). Puneet Modulation of hyperglycemia by oat extracts: an experimental study. International Journal of Biotechnology and Biochemistry, 3(2-3):213-220.
3. Bhatia, A. and Pawan, R. (2007). Therapeutic effect of Probiotics on immune response and hypercholesteremia: An experimental study. Research Journal of Biotech.nology, 2 (3):45-48.
4. Malik, F. Singh, J. Khajuria, A. Suri, K.A. Satti, N.K. Singh,S. Kaul, M.K. Bhatia, A. Kumar, A. and Qazi, G.N. (2007). A standardized root extract of Withania somnifera and its major constituent withanolide-A elicit humoral and cell-mediated immune responses by up regulation of Th1-dominant polarization in BALB/c mice. Life Science, 80(16): 1525-38.
5. Pawan, R. and Bhatia, A. (2007). Systemic immunomodulation and hypocholesteraemia by Dietary probiotics: a clinical study. Journal of Clinical and Diagnostic Research, 1(6):467-475.
6. Rani, U, and Bhatia, A. (2007). Dietary modulation and restoration of immune response by a prebiotic, Fructooligosacharides: an experimental study. International journal of probiotics and prebiotics, 2(4):209-214.
7. Dhillon, P. Bhatia, A. (2008) Hypercholesterolemic and Immunomodulatory effects of Oat Extracts containing Beta-Glucan Year. Research journal of Immunology. 1(1):29-35.
8. Mishra, T. and Bhatia, A. (2010). Augmentation of expression of immunocytes functions by seed extract of Ziziphus mauritiana (Lamk.). Journal of Ethanopharmacology, 127:345-349.
9. Ali, S. A. Bhatia, A. Parvaiz, Q. Bhat, H. M. Sheikh F. A., Khera, N and Refaz, A. (2013). *In vitro* immunomodulatory study of different parts of Prunus cerasus L. (sour cherry) Plant, Asian Journal of Plant Science and Research, 3(3):35-43.

10. Bhatia, A. Sharad, P. Dimple, C. and Mishra, T. (2013). Chitosan nanoparticles as carriers of immunorestoratory plant extract: synthesis, characterization and immunorestoratory efficacy. International Journal of Drug Delivery.
11. Randhawa, M.K. Bhatia A, Chugh C. and Jasrotia K (2011). Synergisitic hypocholesterolomic and immunomodulatory effect of two probiotic strains in Vivo. Journal of Clinical and and Diagnostic Research, 5:312-315.
12. Bhatia, A. and Kaur, P. (2011). Biotyping of various strains of Lactobacillus acidophilus based on their probiotic properties. The IUP Journal of Biotechnology, 1(3):42-49.
13. Randhawa, M. K., Bhatia A. Ali, S. A. and Balgir, P. P. (2012). *In Vivo* comparison of probiotic bacteria's with their DNA: As augmenter of immune efficacy. Advances in Applied Science Research, 3(2):826-832.
14. Kaur P. and Bhatia A. (2012). Comparison of the immunomodulatory properties of four prbiotic strains of lactobacillus: Prediction for in vivo efficacy. International Journal of Life Science and Biotechnology and Pharmacology Research, 1(2):2250-3137.
15. Bhatia, A. Kaur, G. Kaur, M and Singla R. (2012). Coencapsulation of Synbiotics for the evaluation of in vivo antidiabetic activity. Advances in Applied Science Research 3(5): 3020-3024.
16. Bhatia, A. Kaur, M. Randhawa M.K. Sharma A. and Singla R. (2012). The anti-cholesterolemic activity of encapsulated synbiotics. Journal of Microbiology and Biotechnology Research, 2(5):747-749.
17. Abid, S. Khajuria, A. Parvaiz, Q. Sidiq, T. Bhatia, A. Singh, S. Ahmad, S. Randhawa, K.M. Satti, N.K. and Dutt, P. (2012). Immunomodulatory studies of a bioactive fraction from the fruit of Prunus cerasus in BALB/c mice. International Journal of Immunopharmacol. 12(4):626-34.
18. Bhatia, A. Kaur, A, Randhawa, K.M. and Singla, R. (2012). Coencapsulation of Synbiotics as enhancer of Immunological activity: An in vivo study. Journal of Microbiology and Biotechnology Research, 2(4):606-611.
19. Bhatia, A. Rana, P. Sharma, A. Singla, R. and Randhawa, K.M. (2012). Preparation, Characterization and hypocholesterolemic effect of sodium alginate encapsulated lab isolate. Journal of Microbiology and Biotechnology Research, 2(5): 741-746. (ISSN: 2231-3168).
20. Ritu, P. (2008). Characterization of immunomodulatory potential of probiotics and its application as immunotherapeutic agent. A PhD thesis submitted to Punjabi University Patiala, (Under the supervision of Prof Dr Aruna Bhatia).
21. Mishra T. (2010). Characterization of immunomodulatory, anticarcinogenic and phytotherapeutic potential of Zizyphus mauriatiana (Lamk.) fruit. A PhD thesis submitted to Punjabi University Patiala, (supervision of Prof A. Bhatia).
22. Randhawa M. K. (2013). Immunomodulatory and biotherapeutic potential of Probiotic bacterial DNA. A PhD thesis submitted to Punjabi University Patiala, (Under the supervision of Prof Dr Aruna Bhatia).
23. Rani, U. (2009). Synbiotic effect of Streptococcus thermophilus and fructooligosacchrides on systemic immune response and its immunotherapeutic potential inswiss albino mice. A PhD thesis submitted to Punjabi University Patiala, (Under the supervision of Prof Dr Aruna Bhatia).
24. Sheikh A. A. (2013). Screening and isolation of an immunomodulatory compound from Prunus cerasus fruit and its biotherapeutic potential.A PhD thesis submitted to Punjabi University Patiala, (Under the supervision of Prof Dr Aruna Bhatia).
25. Puneet, D (2008). Characterization of immunomodulatory potential of oat bata- glucan and its application as immunotherapeutic agent. A PhD thesis submitted to Punjabi University Patiala, (Under the supervision of Prof Dr Aruna Bhatia).

CHAPTER 14

Effect of Storage Temperature on the Stability of Vitamin A and Iron in Re-Constituted Rice

Syed Zameer Hussain, Baljit Singh, H.R. Naik and A.H.Rather

Abstract

Re-constituted rice-premixes made by extrusion, has been successfully formulated for the fortification of market rice with vitamin A (retinyl palmitate and beta-carotene) and iron. The statistics of vitamin A deficiency (VAD) and Iron deficiency anaemia (IDA) are alarming in the developing world, particularly in India. In fact, India is second largest producer as well as consumer of rice in the world, so fortification of rice with vitamin A and iron seems to be a successful solution to reduce the VAD and IDA in India. In order to design the final formulation, eight formulations containing target micronutrient profile were tested for vitamin A (both retinyl palmitate and beta-carotene) and iron retention for a period of six months under ambient (24-26 ^{0}C) and elevated temperature (40 ^{0}C) conditions. Non-significant iron loss was observed during six months of storage time under both ambient and elevated conditions. The higher retention of retinyl palmitate and beta-carotene was observed in formulation 5 and 8 respectively. In F5, the retention of retinyl palmitate was 73% and 66% and in F8, retention of beta carotene was 70% and 63% under ambient and elevated temperature conditions respectively after six months of storage. Substituting the original, vegetable shortening with palm shortening resulted in increased retention of retinyl palmitate and beta carotene in formulation 5 and 8. Thus, increased antioxidant effect of natural antioxidants in palm shortenings contributed to stability of micronutrients.

Key words: Extrusion, Re-constituted rice, Fortification, Retinyl palmitate, β-carotene

Introduction

Micronutrient malnutrition is now focus of the world's nutrition community because of its largely covert deleterious effects on large portion of the population, including physical and mental deformities, high risk of infection, low work productivity and increased mortality in young children (FAO / WHO 2002). With one sixth of global population residing in India, one third of about two billion people suffering from vitamin and micronutrient deficiency are in India. The intake of daily diet is far from satisfactory and less than 50 percent Recommended Dietary Allowance (RDA) is consumed by over 70 percent of Indian population. The losses due to micronutrient deficiency costs India 1 per cent of its GDP. This amounts to a loss of Rs. 27,720 corer per annum in terms of productivity, illness, increased health care costs and death. Everyday 6,000 children below the age of five die in India. More than half of these deaths are caused by malnutrition - mainly the lack of vitamin A, Iron, Iodine, Zinc and Folic acid (Prakash, 2008).

In general, the emphasis to reduce the micronutrient deficiency has been on agronomic management and plant breeding (conventional breeding and transgenic selection) which are considered as time consuming and expensive and may not be feasible for economically disadvantaged developing countries (Fidler et al., 2003). In addition, genetically engineered crops may face the opposition of consumer and governments due to their concerns about genetically modified (GM) food. In the meantime, an alternative cost effective way to deliver the adequate quantity of vitamins and minerals through the staple food system is required for urgently addressing the prevailing deficiency problems. Short term strategy such as nutrient supplementation (Giving a large dose of the micronutrient as a medicinal supplement) have been effective in providing immediate relief in several countries, but there is concern, this approach is not sustainable in the long term. Consumption of large amount of vitamin A in the form of supplement (capsule) leads to increased levels of circulative retinoic acid, which can cause teratogenicity (Birth defects) in the fetus. However, vitamin A in food results in much lower increases in retinoic acid (Buss et al., 1994). In comparison to supplementation, food fortification has a less immediate effect on the target population; however, it covers a much wider range of recipients and has a more sustained impact. Education and dietary diversification is another sustainable and desirable strategy, yet it takes significantly longer to implement, requiring another option, such as food fortification to be carried out in the meantime (Allen et al., 2006). The wide population coverage and relative rapid micronutrient status improvement of food fortification strategies makes this a very cost effective strategy.

Rice is one of the world's most important cereals for human consumption. In the density populated areas of Asia, rice has always been the most important staple food. India is second largest producer as well as consumer of rice in the world, so fortification of rice with vitamin A and iron seems to be a successful solution to reduce the VAD and IDA in India.

Previously we have standardized the extrusion conditions for development of re-fabricated rice by studying the effect of extrusion conditions on pasting behaviour and microstructure of re-fabricated rice (Hussain and Baljit, 2013). It seems logical to study the effect of storage temperatures on the stability of vitamin A and iron in re-constituted rice during storage. Only a handful studies have been published so far on stability of micronutrients in re-constituted rice during storage (Yao et al., 2011; Lee et al., 2000). In order to ensure safe, efficient and effective clinical trials, the shelf life studies of micronutrient fortified re-constituted rice merit some thorough investigations particularly in tropical countries like India with extreme temperature conditions.

Materials and Methods

The paddy cultivar (PR-116) was dehulled and brown rice was whitened in a single-pass friction rice pearler (BS08 A, Satake, Japan). The small brokens d" 1/8th of actual kernel size were ground in a lab mill model 3303 (Perten, Sweden) to fineness that passes through 200µm sieve. Sodium alginate and calcium salts ($CaCl_2$ and $CaSO_4$) were used as structural and cross-linking agents in forming extruded grains with desired grain structure and integrity. The micronized ferric pyrophosphate was used as iron source. The target Iron concentration of the rice-premix was set at 0.4g/100g of rice.

The concentration in the final product, mixed at a ratio of 1:100 with natural rice grains would be 4 mgFe/100g. This amount would provide 8mg of additional iron to a child consuming 200g of rice per day. Assuming 5% absorption in iron-deficient subject, this fortification level would provide an additional 0.4 mg of absorbed Iron per day. The retinyl palmitate and beta-carotene fortification target level was set at 600IU/g and 2000IU/g respectively. The loss of retinyl palmitate and beta-carotene resulting in extrusion process was compensated in the final re-constituted rice-premixes. Other formulation ingredients, including water soluble antioxidants-sodium tripolyphosphate (STPP) and citric acid; fat-soluble antioxidants- butylated hydroxylanisole (BHA) and butylated hydroxyl toluene (BHT); and a whitening agent - titanium oxide (Tio_2) were added as described by Li, et al., (2009). All formulation ingredients were of food grade.

The moisture contents of the flours were determined by oven drying method (AACC 2000) before mixing. The blended materials were mixed for 30 min using a Hobart mixer (Hobart Corp., Troy, Ohio, U.S.A), and target feed moisture content of 36% was achieved by injection of water in the feeding zone of the extruder with a plunger type reciprocating water pump (DKM, Clextral, Firminy, France) equipped with a control for water addition. The re-constituted rice grains were collected on stainless steel, mesh trays. Trays were loaded into a Tray dryer (NSW-154, S-Narang, Scientific works, New Delhi, India) with a small fan and 220 volt Thermolyne © heating unit circulating air from bottom of the cabinet to the top. The micronutrient fortified re-constituted rice grains were pre-dried for 2-2.5 hours at 70°C. The partially dried grains were then stacked in trays and placed in conditioning chamber at 25°C and 40% relative humidity for 8 hours for final drying at 60-70°C to reach a moisture content of 12% (w.b.). The samples were sealed in plastic woven sacks and stored at ambient (24 -26°C) and elevated temperature conditions (40°C) for a period of six months.

Extruder and Processing conditions

Extrusion experiments were performed on a co-rotating intermeshing twin screw extruder model BCH (Clextral, Firminy, France) as shown in Figure 1. The barrel diameter and its length to diameter ratio (L/D) were 2.5 mm and 16:1 respectively. The extruder barrel is divided into four zones. Temperature of first, second and third zone was maintained at 20°C , 30°C and 40°C respectively, throughout the experiments. The screw configuration used in the entire experiment as shown in Figure 2 is given in table1.

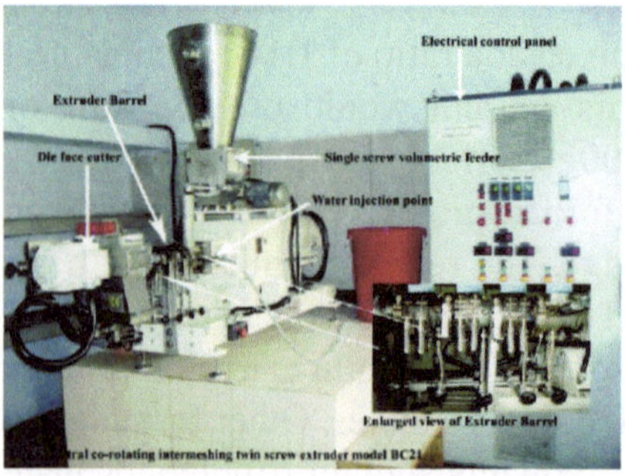

Fig. 1: Co-rotating intermeshing twin screw extruder

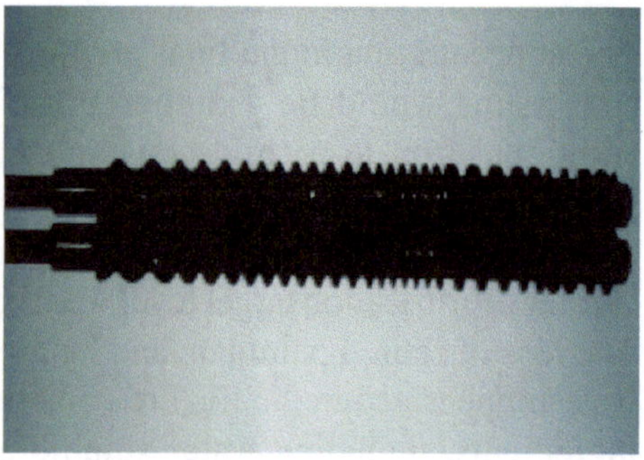

Fig. 2: Screw configuration

The extruder was equipped with a torque indicator, which showed percent of torque in proportion to the current drawn by the drive motor. The screw speed, moisture content and barrel temperature of fourth zone was maintained at 150 rpm, 36 percent and 89°C respectively (as pre-optimized conditions).

Table 1: Screw configuration in different sections of the extruder (From hopper to die)

Screw Section	1	2	3	4	5	6	7	8	9	10
Screw element	BAGUE	C2F	C2F	C2F	C2F	C2F	INOO	CIF	CIF	CIF
Length (mm)	20	50	50	50	50	50	5	50	50	25
Pitch (mm)	-	50	33.33	25	25	16.66	-	16.66	12.5	12.5

A specially designed cutter shown in figure III with four bladed knives (Length 415mm, width 315mm, height 360mm and weight 20 Kg) and pre-standardized die plate with single insert having one hole (8×1.5 mm dimensions) shown in Figure 4 and 5 made of stainless steel were used for shaping of the re-constituted rice grains.

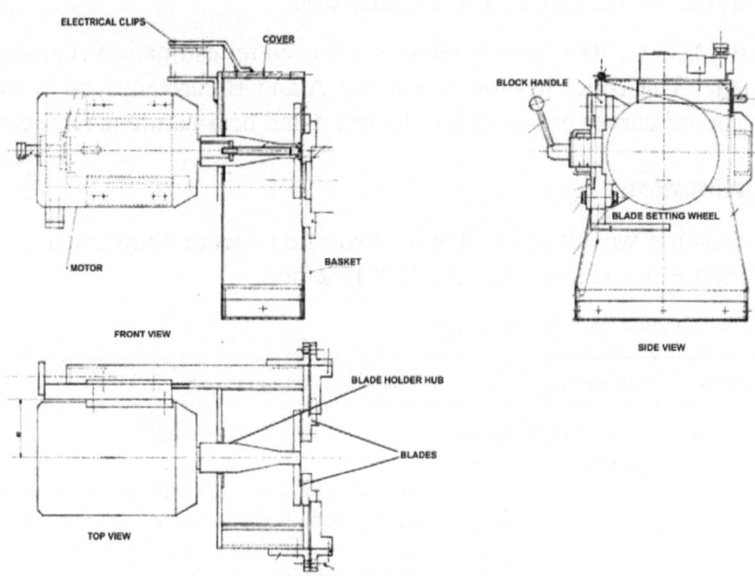

Fig. 3: Line diagram of cutter

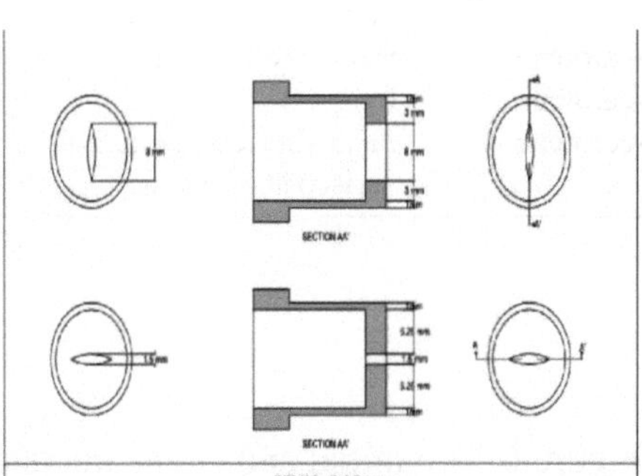

Fig. 4: Line diagram of dies Formulations

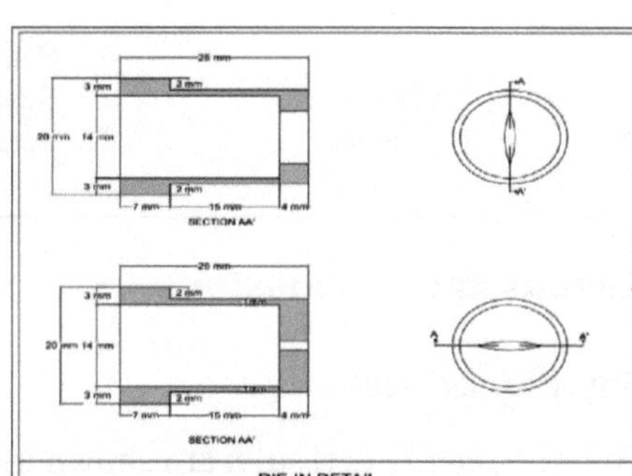

Fig. 5: Detailed dimensional view of die

In order to design the final formulation and to compensate the loss of micronutrients during storage, eight formulations (Table-2) containing the target micronutrient profile were tested for retention of iron, retinyl palmitate and beta-carotene.

In addition to sodium aliginate and calcium salts (used as structural and cross-linking agents in forming extruded grains with desired grain structure and integrity) water soluble anti-oxidants, (STPP and citric acid) fat soluble antioxidants (BHA and BHT), Tio_2 (whitening agent) and crisco shortenings were added. All formulation components were food grade and used as per permissible limits.

Estimation of Vitamin A and B-carotene

Standard (AOAC, 2000) procedure was followed for estimation vitamin A and β-carotene. The concentration of vitamin A and B-carotene was calculated from standard curves prepared for vitamin A and beta-carotene respectively.

Iron estimation

Iron estimation was done by atomic absorption spectrophotometer (Analyst 200, Perkin Elmer) using (AOAC, 2000) method.

Table 2: Rice premix formulation design

Formulations	Iron source	Vitamin A source	Other*
F - 1	Ferric pyrophosphate (Fe pp)	-	Fe only
F - 2	-	Retinyl palmitate	Retinyl palmitate only
F - 3	-	Retinyl palmitate	Retinyl palmitate with ascorbic acid
F - 4	Fepp	Retinyl palmitate	Original
F - 5	Fepp	Retinyl palmitate	Palm shortenings instead of crisco shortenings
F - 6	-	Beta-carotene	Beta-carotene only
F-7	Fepp	Beta-carotene	Original
F8	Fepp	Beta-carotene	Palm shortenings instead of crisco shortenings

Results and Discussions

Stability of iron

The iron retention results are shown in Figure 6. The initial and target amount of iron was 4 mg/g of rice premix, and after 6 months, under both temperature conditions almost, all the iron was retained in five designed formulations. The variability between the samples was less than the standard error of the mean values.

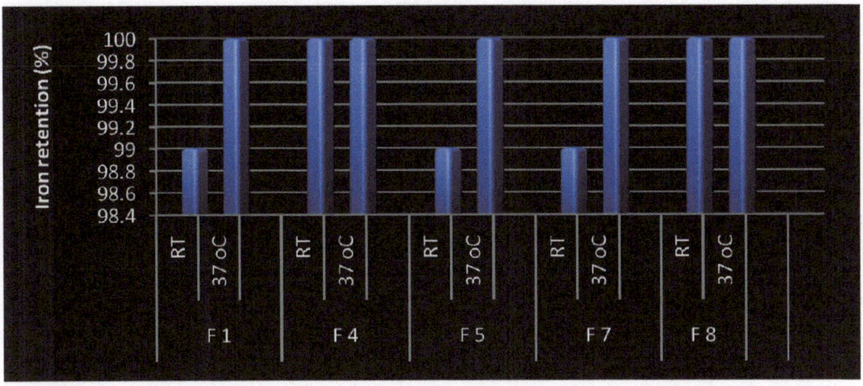

Fig. 6: Total iron retention in rice pre-mixes during storage

Retention of vitamin A/Beta - carotene

The vitamin A and Beta-carotene retention data results are shown in Figure 7 and 8 respectively. The initial load of retinyl palmitate and beta carotene was 600 IU/g and 1000 IU/g respectively. After extrusion process, the vitamin A and beta-carotene was assessed within 24 hours. The loss was attributed to processing and was compensated accordingly in the final formulations in order to ensure the iron and vitamin A content in rice-premixes as per the RDA of children between 5-8 years of age.

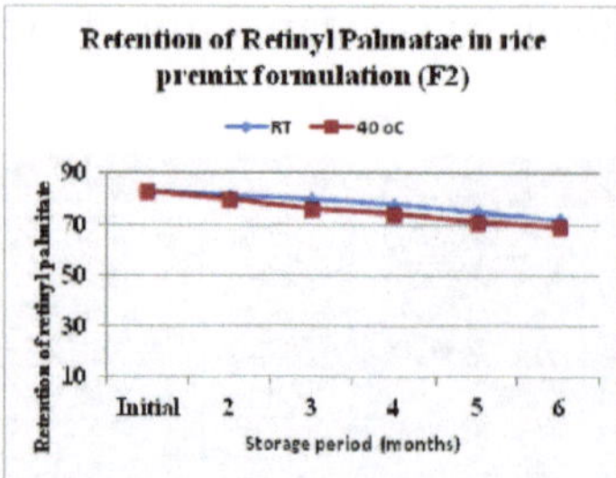

Retention of vitamin A in formulation - I

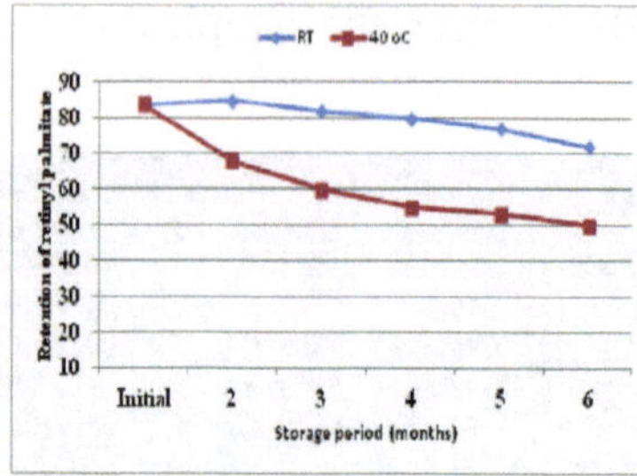

Retention of vitamin A in formulation - II

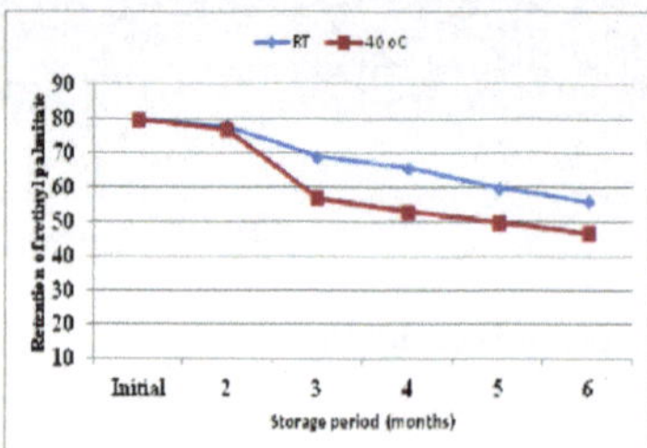

Retention of vitamin A in formulation -III

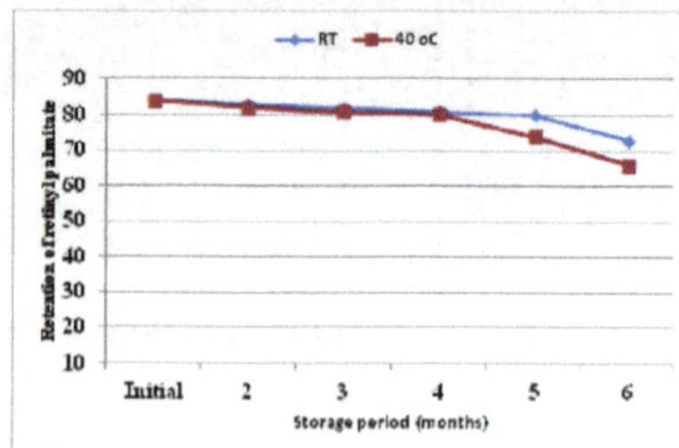

Retention of vitamin A in formulation - IV

Fig. 7: Retention of retinyl palmitate during storage

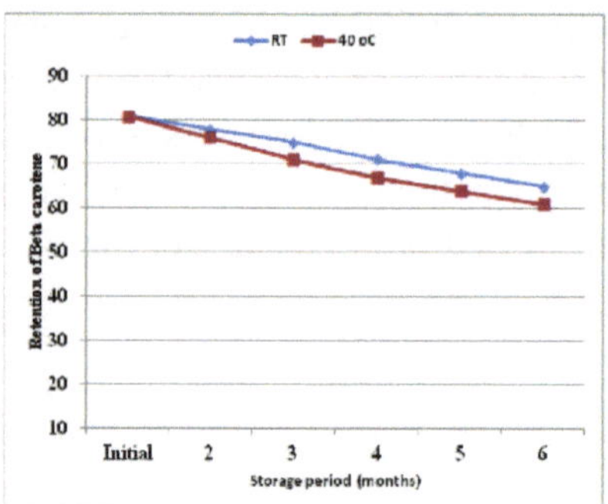

Retention of Beta Carotene in formulation -VI

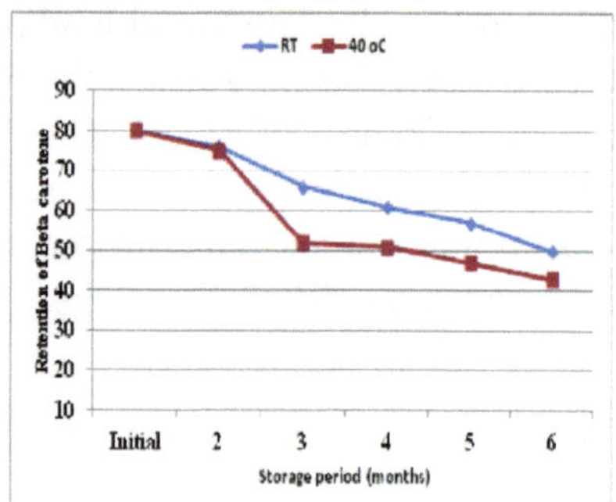

Retention of Beta Carotene in formulation - VII

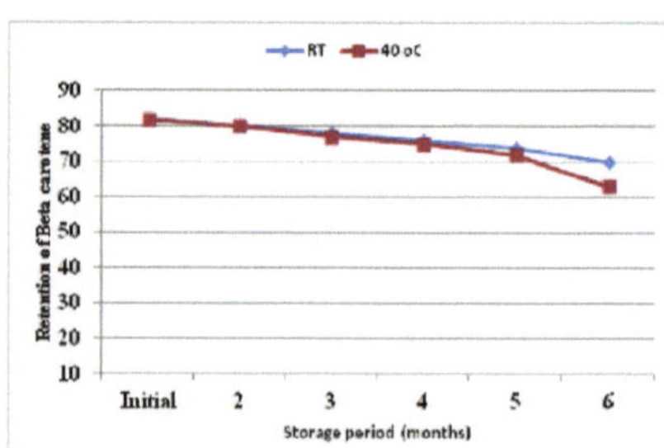

Retention of Beta Carotene in formulation - VIII

Fig. 8: Retention of beta-carotene during storage

Overall the Stability of retinyl palmitate was relatively high in formulation-II and III under ambient temperature conditions even after six months of storage. The only difference in formulation - II and III was that of incorporation of vitamin C as ascorbic acid in Formulation-III. Ascorbic acid was not added to the combined formulations due to dark coloured complex it forms with iron. The retinyl palmitate content remained almost same (> 70%) with or without ascorbic acid incorporation under ambient temperature condition. However, under elevated temperature condition (40°C) the retinyl palmitate content reduced drastically after two months of storage and by the end of six months, there was only 50% retention of retinyl palmitate in Formulation-III. Possibly the higher storage temperature resulted in abrupt loss of retinyl palmitate due to fast oxidation.

The stability of beta-carotene was relatively high (65%) after six months of storage in formulation -VI under ambient condition. A similar trend for retention of retinyl palmitate and beta-carotene was recorded in formulation IV and VII respectively. In Formulation - IV, the retinyl palmitate retention was 56% and 48% under ambient and elevated temperature conditions respectively. In a similar way, 50% and 43% beta-carotene retention was observed in formulation - VII under ambient and elevated temperature conditions respectively. Similar results were reported by Yun et al(1987) and Lee et al., (2000). Out of all the formulations tested, lowest retention of retinyl palmitate and beta-carotene was noticed in Formulation - IV and Formulation - VII respectively. The iron was incorporated as ferric pyrophosphate in both of these formulations (F- IV and F- VII) which might have degraded both retinyl palmitate and beta-carotene at faster rates.

The highest retention of retinyl palmitate and beta-carotene was observed in Formulation - V and formulation - VIII under both storage conditions. In Formulation - V, the retention of retinyl palmitate was 73% and 66% respectively under ambient and elevated temperature conditions, even after six months of storage. Likewise beta-carotene retention was assessed as 70% and 63% after six months of storage time in Formulation - VIII under ambient and elevated temperature conditions respectively. Although iron was incorporated as ferric pyrophosphate in both the formulations (F- V and F-VIII) as well. But, unlike F-IV and F-VII, the crisco-shortening, were replaced by palm shortening in such formulations. Palm shortening is a rich source of naturally occurring antioxidants like tocopherols and tocotrienols which donate their phenolic hydrogen to lipid free radicals (Kamal and Appelqvist, 1996). Simonne and Eitenimiller (1998), while conducting a study on retention of vitamin-E and added retinyl palmitate in selected vegetable oils during deep-fat frying and in fried breaded products also reported the better retention and stability of retinyl palmitate in presence tocopherols and tocotrienols in palm oil.

Conclusions

Based on the result of storage studies, the palm shortenings were incorporated for the production of rice-premixes fortified with retinyl palmitate, beta-carotene, iron and retinyl palmitate and iron and beta-carotene (Fig. 9). The rice-premixes were stored at ambient conditions (24-26°C) for six months of efficacy trial.

Iron fortified rice

Vitamin A fortified rice

Beta-carotene fortified rice

Beta-carotene and iron fortified rice

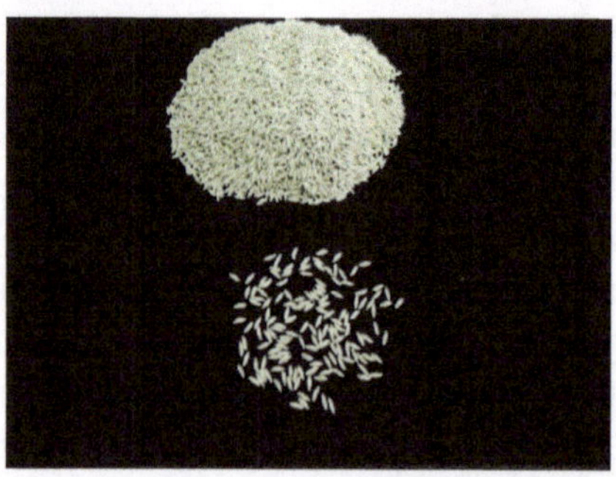

Vitamin A and iron fortified rice

Fig. 9: Micronutrient fortified rice-premixes

References

1. AACC (2000), Approved laboratory methods. American Association of Cereal Chemists, St. Paul, Mn.
2. Allen, L. de Benoist, D. Dary, O. and Hurrell, R. (2006). Guidelines on food fortification with micronutrients. Geneva, Switzerland: World Healthy Organization and Food and Agriculture Organization of the United Nations.
3. AOAC International (2000) Official Methods of Analysis. Gaithersburg, MD: Association of Official Analytical Chemists. p 16-17.
4. Buss, N. E. Tembe, E. A. Prendergast B D, Renwick, A. G and George, C.F. (1994) the teratogenic metabolites of Vitamin A in women following supplements and liver. Human and Experiment Toxicology, 13:33-43.
5. FAO and WHO (2002) Human vitamin and mineral requirements. Bangkok, Thailand.
6. Fiddler, M. C. Davidsson, L. Walczy, K. T. and Hurrell, R. F. (2003). Iron absorption from fish sauce and soy sauce fortified with sodium, Iron EDTA. Amrican Journal Of Clinical Nutrition, 78:274-278.
7. Kamal-Eldin, A. and Appelqvist, L. (1996). The chemistry and antioxidant properties of tocopherols and tocotrienols Springer Berlin / Heidelberg.
8. Lee, J. Hamer, M. and Eitenmiller, R. (2000). Stability of retinyl palmitate during cooking and storage in rice fortified with Ultra Rice™ fortification technology. Journal of Food Science, 65 (5):915-19.
9. Li, Y. Diosady, L. and Jankowski, S. (2009) Antioxidant system for the preservation of vitamin A in ultra rice. Food and Nutrition Bulletin, 30 (1):82-89.
10. Prakash, V. K. (2008). Micronutrient malnutrition in India- Let us say no to it now. Industrial Journal of Community Medicine, 33:9-10.
11. Simonne, A.H. and Eitenmiller, R.R. (1998). Retention of vitamin E and added retinyl palmitate in selected vegetable oil during deep-fat frying and in fried breaded products. Journal of Agriculture and Food Chemistry, 46(12):5273-5277.
12. Hussain, S. Z. and Singh, B (2013). Effect extrusion conditions on pasting behavior and micro-structure Rice: of Re-fabricated A response Surface Analysis. Cereal Chemistry, 90 (5):480-489.
13. Li, Y. O. Diosady, L. L. and Jankowski, S (2011). Folic acid stability in the presence of various formulation components including iron compounds in fortified extruded ultra rice over prolonged storage at 40^0 C and 60% Relative humidity, 46:379-385.
14. Yun, J. Shahidi, F. Rubin, L. and Diosady, L. (1987). Oxidative Stability and Flavour Acceptability of Nitrite-Free Meat-Curing Systems. Canadian Institute of Food Science and Technology Journal, 20(4):246-51.

CHAPTER 15

Antioxidant Properties of Wheat Based Breakfast Foods

Suresh Bhise and Amarjeet Kaur

Abstract

The cereal grains are grown around the world, the most common is wheat, which accounts for one third of the total grain crop worldwide. In addition to providing energy, whole grains contain several key nutrients, including fibers, lipids, B vitamins and minerals as well as other beneficial components such as phenolic acids, flavonoids, lignans and phytosterols. Several of these components function as antioxidants. While multiple studies have been conducted to analyze antioxidants in different grains. In wheat and other grains, naturally occurring phytonutrients with antioxidant activity include phenolic acids, flavonoids, coumarins, polyphenols, phytates, terpenes, carotenoids, tocopherols and tocotrienols. Potential roles of the naturally occurring antioxidants in grains include free radical scavenger, reducing agent, complexation of prooxidant metals, quenchers of singlet-oxygen molecules and antioxidant defense enzyme system activators for biological systems. Wheat grain contains several components that have potential antioxidant activity and thus may have a protective role in human health. Traditional wheat-based breakfast foods include hot and cold ready-to-eat (RTE) cereals, as well as a variety of yeast-leavened and chemically leavened bakery products. Increasing baking time and temperature increases antioxidant activity in the bread crust and lowers antioxidant activity in the bread crumb portion. Wheat and wheat based products, particularly partial and 100% whole grain products have been shown to provide a variety of beneficial components including fiber, vitamins, minerals and other phytonutrients with antioxidant activity. Scientific evidence indicates that consumption of three or more ounce equivalents of whole grain foods per day may reduce the risk of coronary heart disease, type 2 diabetes and certain cancers and help with weight management.

Introduction

The diet rich in whole grain foods may reduce the risk of coronary heart disease, type 2 diabetes and certain cancers and help with weight management. At least half of daily grain intake should be from whole grains or at least three ounce equivalent servings for a 2000 calorie diet. Increased consumption of whole grain foods is also recommended by the American Heart Association, the American Diabetes Association and the American Cancer Society. The FDA has approved the health claim for foods containing at least 51% whole grain by product weight and meeting other criteria: "Diets rich in whole grain foods and other plant foods and low in total fat, saturated fat and cholesterol may help reduce the risk of heart disease and some cancers" (FDA and CFSAN, 1999).

Grains Consumption

Cereal grains are the seeds (fruits) of grasses from the Pomaceae family. Wheat, corn, rice, oats, barley, rye, millet, sorghum, teff, wild rice, canary seed and Job's tears are cereal grains. Amaranth, buckwheat and quinoa are pseudocereal grains, which are not cereal grasses but are included in the grain category because of similarities in composition and in use. Legumes and oilseeds are not considered as grains. More than 70% of the total grain intake is wheat based grain at 117.5 g/day. Consumption of corn, rice and oats followed at 27, 17.9 and 3.05 g/day, respectively.

World grain production is given in Fig.1 from 1961 to 2012 all figures are in Million Metric Tonne. Barley and rye each are consumed at less than 0.5 g/day. Whole grains are the entire seed also known as a caryopsis or kernel and must contain all three parts of the seed the endosperm, bran, and germ along with the naturally occurring nutrients of the entire seed. Food manufacturers are encouraged to use the term "whole" or "whole grain" in the ingredient name to help consumers identify foods that are formulated with whole grain ingredients. Despite increased availability of foods made with whole grains and strong recommendations to increase whole grain consumption. Less than 10% of adult Americans meet the recommended whole grain consumption level.

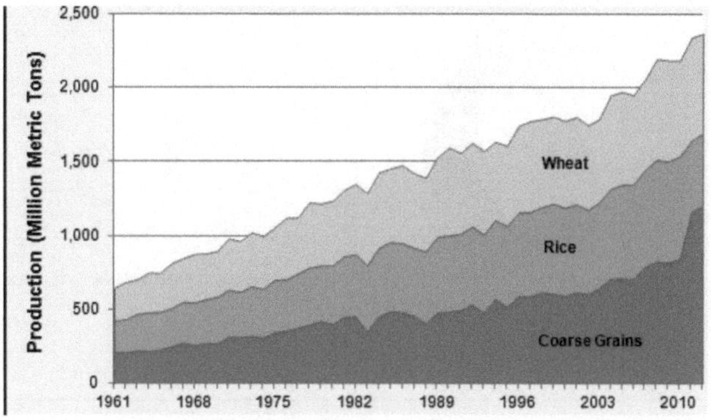

Fig. 1: World Grain Production from 1961-2012

Wheat Composition

The wheat endosperm is approximately 82% of the kernel weight and primarily consists of starch as well as protein. The endosperm functions as a source of energy used in the germination and the early growth of a new plant.

Wheat grain composition and nutritional value per 100g was given in Fig-2. The aleurone is the outermost cell layer of the endosperm; however, the composition is more similar to the outer bran layers. The aleurone cell walls are known to contain phenolic acids, in particular ferulic acid. The bran is about 15% of the kernel weight and is comprised of several cell layers. The bran contains approximately 16% protein, 65% carbohydrates (40% fiber), 6% ash and 4% lipid, as well as several vitamins, minerals and other phytonutrients, including ferulic acid. A 15 g serving of wheat bran provides 25% of the daily value of fiber. The germ is 2-3% of the kernel weight and is the embryonic plant. The germ consists of the embryonic axis (primary root and shoot) and scutellum (storage and transport organ). The germ contains approximately 25% protein, 52% carbohydrates (13% fiber), 5% ash and 10% lipid. The germ is also an important source of several vitamins, minerals and other phytonutrients, including apigenin and luteolin-based flavonoids. A 15 g serving of wheat germ provides 13% of the daily value of vitamin E. Whole grain wheat contributes several important nutrients to the diet including dietary fiber, several B vitamins (thiamin, riboflavin, niacin, and folic acid), vitamin E, iron, magnesium, selenium, manganese and chromium. Three whole grain ounce equivalents based on whole wheat (16 g per ounce-equivalent) provides needed nutrients including 23% of the daily value of dietary fiber, 15% of magnesium, 6% of potassium, 50% of selenium and 90% of manganese (USDA, 2007).

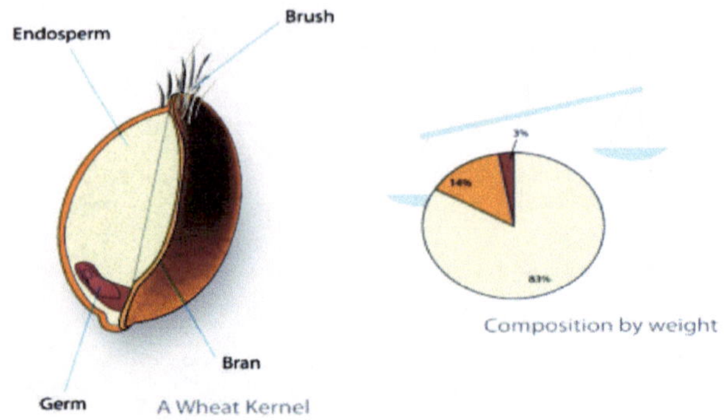

	Carb./g	Protein/g	Fat/g	Fiber/g	Iron (% daily req.)	Others
Bran	63	16	3	43	59	vitamin Bs
Endosperm	79	7	0	4	7	
Germ	52	23	10	14	35	vitamin Bs omega-3/6 lipids

Nutritional Value (per 100g)

Fig. 2: Wheat grain composition (per 100g)

Wheat milling fractions were analyzed for both hydrophilic and lipophilic antioxidant fractions and were reported as a collective total antioxidant capacity (mmole Trolox equivalents/100 g) (Table-1). Wheat germ had the highest average total antioxidant capacity, which was 2.4 times higher compared to coarse wheat bran (3500; 3200-3600) and more than 5.7 times higher than the refined wheat flour fraction (1450; 1000-1900).

Table 1: Total Antioxidant capacity of wheat milling fractions

Wheat Milling Fraction	μMTE/100g
Refined flour	1450
Coarse wheat bran	3500
Red do flour	4300
Shorts	4200
Wheat bran and germ flour	7400
Wheat germ	8400

Role of Antioxidants in Wheat and Other Cereal Grains

In wheat and other grains, naturally occurring phytonutrients with antioxidant activity include phenolic acids, flavonoids, coumarins, polyphenols, phytates, terpenes, carotenoids, tocopherols, and tocotrienols (Shahidi 2004). Potential roles of the naturally occurring antioxidants in grains include free radical scavenger, reducing agent, complexation of prooxidant metals, quenchers of singlet oxygen molecules and antioxidant defense enzyme system activators for biological systems (Andlauer and Furst 1998). Compared to phytate and phenolics, much less is known about the role and mechanisms of grain based polysaccharides and proteins as antioxidants in food systems or in human health.

Wheat Milling and Distribution of Antioxidants

In the traditional wheat milling process, the goal is to produce refined wheat flour by maximizing the separation of the starchy endosperm from the endosperm aleurone layer, bran and germ. Antioxidants are concentrated in the outer parts of the kernel, particularly the endosperm aleurone layer, pericarp, nucellar envelope and germ.

White bread and 100% whole grain bread contains approximately 1100 and 2000 TE per 100g (Fig-3). The combined bran and germ fractions from three wheat varieties, two of which were each grown in different locations, were found to contain 15-18 times more total phenolics, 52-70 times more ferulic acid and on average 4 times more lutein, 12 times more zeaxanthin and 2 times more cryptoxanthin as compared to the endosperm fractions. The carotenoid content of wheat appears to vary based on genetics with diploid and tetraploid wheat varieties, such as einkorn and durum, respectively, containing higher levels of lutein compared to common bread wheat varieties. Wheat milling fractions were analyzed for both hydrophilic and lipophilic antioxidant fractions and were reported as a collective total antioxidant capacity. Wheat germ had the highest average total antioxidant capacity, which was 2.4 times higher compared to coarse wheat bran and more than 5.7 times higher than the refined wheat flour fraction. The total antioxidant capacities of red dog flour, shorts, and wheat bran and germ flour were intermediate between the germ and coarse bran fractions, which is reflective of the mixed composition of each of these mill fractions.

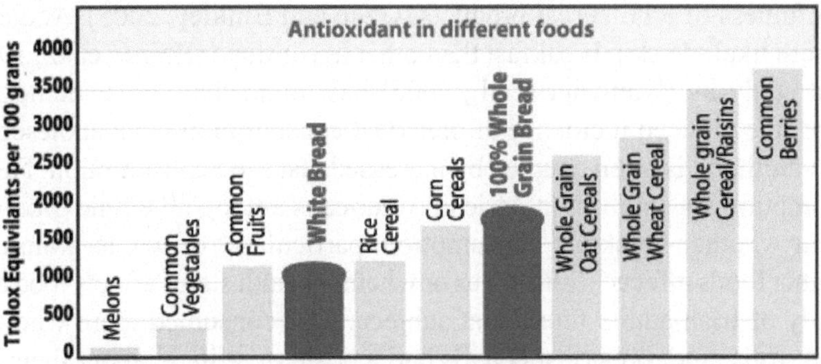

The average antioxidant activity is compared for different foods. Averages compared are from analysis of 3 melons, 20 vegetables, 12 fruit, 2 white breads, 12 fruit, 2 white breads, 1 rice cereal, 3 corn cereals, 2 whole grain breads, 3 whole grain oat cereals, 3 whole wheat cereals, 2 whole grain cereals with raisins and 5 berries.

Fig 3: Antioxidant in different foods

Wheat Based Breakfast Foods

Traditional wheat based breakfast foods include hot and cold ready to eat (RTE) cereals, as well as a variety of yeast leavened and chemically leavened bakery products. Popular yeast leavened products that may be eaten for the breakfast meal include pan and artisan style breads, buns, rolls, pastries, danishes, and croissants. Likewise, popular chemically leavened baked goods include quick breads, muffins, pancakes, waffles and tortillas. Wheat based foods used for the breakfast meal include refined, partial whole wheat and 100% whole wheat and multigrain forms. High bran content cereals are readily available.

Fig 4: Multigrain bread

Wheat bran and germ are available as individually packaged foods, which are generally used by consumers as breakfast cereals or as ingredients for home baking. Whole Wheat Bread incorporated with Oats, Bran and Flaxseed rich in antioxidant (Fig-4).

Breakfast Meal Consumption

Gender is not considered to be a significant factor influencing breakfast consumption. Reports on the effects of education and economic status on breakfast consumption and food choices have been inconsistent, possibly due to the availability of school breakfast programs and cost versus perceived healthfulness of RTE cereal products (Gaub and Binkley, 2005). Adolescents are more likely to skip breakfast than other age groups (Haines et al., 1965). A recent study by (Niemeier et al., 2006) has found that across demographic groups, the reduced incidence of breakfast consumption from adolescence to early adulthood is often coupled by increased fast food consumption. Fast food consumption was highest in African Americans among all ethnic groups. It is unclear whether breakfast consumption, particularly of whole grain cereals and other foods affects health status or whether health status affects food choice. A study of Irish adults found that subjects who consumed more whole meal breads, rolls, and, breakfast cereals had a positive attitude about their efforts to eat a healthy diet, while subjects who ate white breads and had a lower cereal intake had a negative attitude about their efforts to eat a healthy diet (Hearty et al., 2007).

Table 2: List of RTE breakfast cereal products available in market

Sr. No.	Name of products
1	Ready to Eat Cereals
2	General Mills Honey Nut Cheerios
3	Kelloggs Frosted Falkes
4	General Mill Cinnamon TST Cranch
5	Kellog Froot Loops
6	General Mills Lucky Charms
7	Kelloggs Apple Jacks
8	Kelloggs Corn Pops
9	Post Fruity Pebbles

Antioxidants in Breakfast Cereals

RTE breakfast cereals may increase macronutrient and micronutrient intake in individuals, as they are generally high in carbohydrates and many are high in fiber. Studies conducted on adolescent females found that breakfast consumption, particularly RTE cereals, increased micronutrient intake and reduced total fat and cholesterol intakes (Haines et al., 1965). Miller et al., (2000) determined the antioxidant content in a variety of fruits, vegetables, breakfast cereals and breads. The average antioxidant content of the RTE breakfast cereals was higher than that of the fruits and vegetables tested, except for berries. The antioxidant content of whole grain cereal with raisins was only slightly lower than that of berries. Among the grain based foods tested, whole grain cereals and breads were higher in antioxidant content than their refined grain counter parts. Also, bran based RTE cereals were higher in antioxidant content than 100% whole wheat cereals. Antioxidant activity measured in mmoles Products produced with wheat bran have higher levels of antioxidant activity, hydroxyl and hydrogen peroxide scavenging activity and lipid peroxidation inhibition than those produced from oat bran.

Additionally, RTE breakfast cereals appear to be a stable source of antioxidants that do not substantially diminish during normal storage conditions, with limited losses even under temperature abuse. When compared on an equal weight basis, spices are generally very high in antioxidant activity, followed by cocoa products and berries. There is a wide range of antioxidant activity among fruits and vegetables, although berries are usually among the highest of fruits and vegetables. The antioxidant activity range of whole grain and bran based RTE cereals is generally higher than many common vegetables and comparable to many fruits. Breakfast cereals are commonly fortified with iron, which may function as a prooxidant. The wheat bran based aqueous extract had the highest antioxidant activity, although not significantly different from the whole wheat based extract. The antioxidant capacities were greatly increased by subjecting the whole wheat and wheat bran based extracts to pH changes similar to the human gastrointestinal system, suggesting that antioxidant capacity in grain-based foods may be enhanced by digestion. The components found in whole wheat and other whole grains that may have antioxidant activity include phytate, polysaccharides, proteins, and phenolics. It is difficult to make direct comparisons between products analyzed in different studies since there is variability among product formulas and resulting compositions, as well as in the analytical methods used. A descriptor such as whole grain cereal or high fiber bran cereal does not reflect potential formulation differences among these products that may affect antioxidant activity.

Antioxidants and Bread Making

Bread making results in losses of vitamins and antioxidants (Wennermark and Jagerstad 1992). A study on French wheat bread making found that there were significant losses in tocopherols and tocotrienols during the dough formation and baking stages (Wennermark and Jagerstad, 1992). Leenhardt et al., (2006) measured carotenoids and vitamin E during kneading, fermentation and baking using bread, einkorn, and durum wheat. Vitamin E content decreased 30% during bread preparation with losses attributed to oxygenation during dough production and heat destruction during baking. There were minimal differences between wheat species. In bread wheat, the carotenoid content decreased 66% in whole and 40% in refined grain dough after kneading. By comparison, there was less than 10% loss in carotenoids in whole grain einkorn dough after kneading. Einkorn has a much higher carotenoid content and a lower lipoxygenase activity compared to bread wheat. Carotenoid losses during fermentation and baking were about 10% and 36-45%, respectively and did not differ appreciably between wheat species.

Choice of wheat species, as well as breeding common (bread) wheat varieties with higher carotenoid levels may serve as approaches to increase the overall phytonutrient content of breads. The losses reported in the study by Leenhardt et al. (2006) are higher than other baking studies reporting carotenoid losses of approximately 20% for bagels and cakes, 30% for cookies and total carotene losses of 4-15% in bread products and 18-23% in crackers (Park et al., 1997). Fermentation processes, such as those used in sourdough bread production, have variable effects on specific bioactive compounds. While limited data are available, flour that has undergone sourdough fermentation has higher folate and total phenolic content after extraction compared with unfermented flour (Katina et al., 2005). In contrast, the amount of alkylresorcinols (ARs) was diminished when wheat flour was made into sourdough wheat bread; however, it is important to note that complete extraction of ARs may be difficult due to complexation with starch in the food matrix (Ross et al., 2004). The dough fermentation stage of bread making may maintain more phytonutrients than other bread processing conditions such as kneading, which introduces oxygen.

Baking conditions may also alter antioxidant content. Increasing baking time and temperature increases antioxidant activity in the bread crust and lowers antioxidant activity in the bread crumb portion (Lindenmeier and Hofmann, 2004). Maillard reaction products have received significant attention as they may produce potential carcinogens; however, study of bakery product fractions also indicates that areas subject to the browning reaction have higher antioxidative potential in contrast to their corresponding inner counterparts (Lindenmeier and Hofmann, 2004). Halvorsen et al., (2006) found that the antioxidant content of whole wheat bread is more than 1.9 times higher compared to refined wheat bread. Additionally, the process of toasting bread was found to increase the antioxidant content 134-367% higher in bagels, French bread, white bread and whole wheat bread compared to their untoasted counterparts.

Conclusion

Wheat and wheat based products, particularly partial and 100% whole-grain product, have been shown to provide a variety of beneficial components including fiber, vitamins, minerals and other phytonutrients with antioxidant activity. Further work is Tocopherols in Wheat Bread Product needed to understand the effects of formulation, processing and storage on the quantity and bioavailability of the naturally occurring antioxidants in finished products. An additional critical factor is that the final product must have sensory properties that are acceptable to consumers. Recently, whole wheat and other wholegrain products have become more commonplace as research suggests that consumption of these products may have health beneficial effects. As reviewed in the Dietary Guidelines for Americans 2005 report, scientific evidence indicates that consumption of three or more ounce equivalents of whole grain foods per day may reduce the risk of coronary heart disease, type 2 diabetes and certain cancers and help with weight management.

References

1. Andlauer, W. and Furst, P. (1998). Antioxidant power of phytochemicals with special reference to cereals. Cereal Foods World, 43:356-360.
2. FDA, and CFSAN (1999). Health claim notification for whole grain foods. Docket No. 99P-2209.www.cfsan.fda.gov/dms/flgrains.html.U.S. Government Printing Office, Washington, DC.

3. Golub, A. and Binkley, J. (2005). Determinants of household choice of breakfast cereals: Healthy or unhealthy? American Agricultural Economics Association Annual Meeting, Providence Rhode Island.
4. Haines, P. S. Guilkey, D. K. and Popkin, B. M. (1996). Trends in breakfast consumption of US adults between 1965 and 1991. Journal of American Diet Assocciation, 96:464-470.
5. Halvorsen, B. L. Carlsen, M. H. Phillips, K. M. Bohn, S. K. Holte. K. Jacobs, D. R .J. and Blomhoff, R. (2006). Content of redox active compounds (ie, antioxidants) in foods consumed in the United States. American Journal of Clinical Nutrition, 84: 95-135.
6. Hearty, A. P. McCarthy, S. N. Kearney, J. M. and Gibney, M. J. (2007). Relationship between healthy eating and dietary behavior, lifestyle, and demographic factors in a representative sample of Irish adults. Appetite, 48: 1-11.
7. Katina, K. Arendt, E. Liukkonen, K. H. Autio, K. Flander, L. and Poutanen, K. (2005). Potential of sourdough breakfast cereals, fruits and vegetables. Journa of American Col Nutrition, 19: 312S-319S.
8. Leenhardt, F.Lyan, B. Rock, E. Boussard, A. Potus, J. Chanliaud, E. and Remesy, C. (2006). Wheat lipoxygenase activity induces greater loss of carotenoids than vitamin E during breadmaking. Journal of Agriculture Food Chemistry, 54: 1710-1715.
9. Lindenmeier, M. and Hofmann, T. (2004). Influence of baking conditions and precursor supplementation on the amounts of the antioxidant pronyl-L-lysine in bakery products. Journal of Agriculture Food Chemistry, 52:350-354.
10. Miller H E, Rigelhof F, Marquart L, Prakash A and Kanter M (2000) Antioxidant content of whole grain for healthier cereal products. Trend in Food Science and Technollogy, 16: 104-112.
11. Niemeier, H. M. Raynor, H. A. Lloyd-Richardson, E. E. Rogers, M. L. and Wing, R. R. (2006). Fast food consumption and breakfast skipping: Predictors of weight gain from adolescence to adulthood in a nationally representative sample. Journal of Adolsent Health, 39:842-849.
12. Park, H. Seib, P. A. Chung, O. K. and Seitz, L. M. (1997). Fortifying bread with each of three antioxidants. Cereal Chemistry, 74:202-206.
13. Ross, A. B. Kamal-Eldin, A. and Aman, P. (2004). Dietary Alkylresorcinols: Absorption, bioactives and possible use as biomarkers of whole-grain wheat and rye rich foods. Nutrition Review, 62: 81-95.
14. Shahidi, F. (2004). Functional foods: Their role in health promotion and disease prevention method, Journal of Food Science, 69:146-149.
15. USDA (2007). National Nutrient Database for Standard Reference, Release. http://www.ars.usda.gov/ main/site_main.htm?modecode=12-35-45-00.
16. Wennermark, B. and Jagerstad, M. (1992). Breadmaking and storage of various wheat fractions affect vitamin E. Journal of Food Science, 57:1205-1209.

CHAPTER 16

Significance of Amaranth Grains in Health and Nutrition

Arti Chauhan, D.C. Saxena and Sukhcharn Singh

Abstract

This review assesses the nutritional attributes of Amaranth grains and its utilization as food and as formulated foods. Amaranth is a pseudo-cereal and one of the most ancient cultural food plants with excellent source of superior quality protein (rich source of lysine) which is deficient in wheat and other cereals. Amino acid composition of amaranth protein is close to ideal and, as a result, has high food value. Amaranth grain is loaded with phytochemicals including phenolic acids, tannins, anthocyanins, and phytosterols. Amaranth is potentially a good dietary source of the pro-vitamin A carotenoid (β-carotene), antioxidants (nutraceuticals) and canthaxanthin (which has also been reported to be an antitumor agent) and lutein, which is reported to slow down the development of age-related eye diseases. There are also some anti-nutritional factors (Phytic acid, Tannin and Trypsin inhibitor etc.) that may be reduced by certain processing treatments. Several epidemiological studies show Amaranth grains are helpful in reducing several kinds of chronic diseases like cancers, cardiovascular diseases, type II diabetes and various gastrointestinal disorders. Grain amaranth has been tested and recognised by many authorities as a gluten-free foodstuff appropriate for incorporation into the diet for celiac disease patients. Amaranth grain or young leaves could completely or partially substitute common ingredients, so the crop could be potentially exploited as an ingredient in food formulation in form of flour, flakes, sprouts, grain undergoing fermentation, popping, malting, extrusion cooking, nixtimisation, and special compounds isolation. The interest in this crop is developing due to its several advantageous properties. The nutritionists find that the grain has higher nutritional value than conventional cereals.

Keywords: amaranth, nutritive value, health benefits, celiac disease, pseudocereal

Introduction

Amaranths, which comprise the genus Amaranthus, are widely distributed, short-lived herbs, occurring in temperate and tropical regions. The genus Amaranthus consists of approximately 60 species, several of which are cultivated as leaf vegetables, grains, or ornamental plants, while others are weeds (Kauffman and Weber, 1990). A. tricolor, A. dubius, A. lividus, A. creuntus, A. palmeri and A. hybridus are the main species grown as vegetable, while A. hypochondriacus, A. cruentus and A. caudatus are the main grain species (Teutonico and Knorr, 1985). In India, Amaranths are chiefly grown in Himalayas from Kashmir to Bhutan, and also in south Indian hills. Amaranth is a pseudocereal as well as alternative crop is plant species that are used traditionally for their food, fibre, fodder, oil or medicinal properties. These plants have an under-exploited potential to contribute to food security, nutrition, health, income generation and environmental services. Amaranth grain has some agricultural advantages and noted ability to grow successfully in adverse environmental conditions such as high irradiance, temperature and drought. Along this, Amaranth is one of the few plants whose leaves are eaten as a vegetable while the seeds are eaten in the same way as cereals and there is no distinct separation between the vegetable and grain type since the leaves of young grain type plants can be eaten as both human and animal food. Gluten is absent in amaranth grains so, it is beneficial for celiac disease patient (Alvarez-Jubete, 2010).

Gluten is a proteins present in wheat, barley, and rye which induced Celiac disease. Celiac disease (CD) is an autoimmune disease that is characterized by an aberrant response to dietary gluten in genetically susceptible persons that results in small intestinal injury and can be associated with diverse systemic consequences (Reilly and Green, 2014). Moreover, Amaranth is a highly nutritious pseudocereal that presents higher protein content than cereal species as well as many health benefits, such as decreasing plasma cholesterol levels, stimulating the immune system, antitumor activity, reducing blood glucose levels and improving conditions of hypertension and anemia (Caselato-souza and Amaya-farfan, 2012). Therefore, amaranth grain has great potential as a food that could improve nutritional status of daily diet with many health benefits.

Nutritive Value of Amaranth

Average chemical composition of amaranth grain is shown in Table 1. Amaranth grain has higher protein, lipid, crude fibre, ash, calcium, iron and vitamin c as compared with traditional cereal.

Table 1: Chemical composition of amaranth grain (on 100 g dwb)

Chemical composition	Amaranth	Maize	Wheat
Protein (g)	11-18	10	14
Lipid (g)	7-8.5	4 - 5	1.7-2.5
Carbohydrate (g)	65-70	80-82	75-77
Crude fibre(g)	6.5-9	3 - 4	2 – 3
Ash (g)	2.8-3.8	1.5-1.8	1.9-2.2
Calcium (mg)	160-212	5 - 7	33 – 35
Iron (mg)	7 - 15	2 - 3	2.5-3.3
Vitamin C (mg)	4 - 7	-	-

Sources: Bressani, 1993; Caselato-Sousa, 2012; Barba de la Rosa, 2009; Valcárcel-Yamani, 2012.

The crude protein content of grain amaranth ranges from 11 to 17.6 % dry matter (Bressani et al., 1987 a, and Bressani et al., 1987 b). This is higher than in most common grains except soybeans. The amino acid composition of amaranth protein compares well with the FAD/WHO protein standard for instance lysine in amaranth is 6.233g/100g while the standard 5.4g/100g.The lysine content is given as the main reason for the high protein quality of amaranth (Saunders et al., 1983; Teutonico and Knorr, 1985). Amaranth complete protein also contains significantly more sulphur amino acids than soya complete protein. The importance of Amaranth's protein is its essential amino acids which complement those of corn, rice, and wheat. However, Amaranth protein is low in the amino acid leucine, but leucine found in excess in most common grains like corn, while corn protein is low in both tryptophan and lysine, whereas amaranth has high levels of both. So, when amaranth flour is mixed with corn flour, the combination almost reaches the perfect 100 score. The nutritional value of amaranth protein is excellent. The amino acids composition of amaranth and wheat grain shown in Table 2. Data show the higher value of amaranth grain protein fraction, also, its balanced essential amino acid composition is very similar to the FAO/WHO guideline for human diet.

Table 2: Essential amino acids of amaranth and wheat grain (g per 100 g of protein)

Grain	Trp	Met/Cys	Thr	Isl	Val	Lys	Tyr/ Phe	Leu
Amaranth	0.9 – 1.8	4-4.9	3.3-4	2.7-4	3.9-4.7	5-6	6-8.5	4.2-6.3
Wheat	1.2	3.5	2.7	4.1	4.3	2.6	8.1	6.3
FAO/WHO	1	3.5	4	4	5	5.5	6	7

Sources: Amicarelli 2002b, pp. 94-101 and Grobelnik Mlakar, 2010, pp. 135-145

The range of protein efficiency ratio from 1.5 to 2.0 % (2.5% for casein) for cooked grain, and its total digestibility is about 90 percent. Amaranth protein has maximum biological value of 75, comes closer than any other grain protein to the perfect balance of essential amino acids, which theoretically would score 100 on the nutritionists' scale of protein quality based on amino acid composition. While, corn scores about 44, wheat 60, soybean 68, and cow's milk 72 (National Research Council, 1984).

Amaranth grain has higher lipid content as compared to most cereal grains. About 76 % of fatty acids are unsaturated; the share of linoleic fatty acid is 25 to 62 %, oleic acid 19 to 35 %, palmitic acid 12 to 25 %, stearic acid 2 to 8.6 % and linolenic fatty acid 0.3 to 2.2 %. The amaranth grain oil, saturated/ unsaturated fatty acid ratio ranged from 0.26 to 0.31%, while oil of vegetable, ornamental genotypes and weedy species are 0.33 to 0.36%. A large amount of squalene (2 to 7 % or even more than 8 % in refined oil) has been reported in amaranth seeds oil. The squalene is commercially obtained from the liver of whales, sharks, and some other marine species (from 33 to 64 % in oil) or is industrially synthesised and generally used and appreciated as a lubricant for fine electronic instruments, in cosmetics and medicine. Other alternative plant sources the olive, wheat germ and rice bran oils contain squalene in the 0.1 to 1.7 % range (Becker et al., 1981; Leon- Camacho et al., 2001; Berghofer and Schoenlechner 2002). High squalene content in oil allows its effective use as wound healing and regenerative substance in medicine and perfumery (Ofitserov, 2001).In addition, Amaranth seeds are the source of superior quality oil with some useful properties – bactericide, antitumuor and antiburn properties. In addition, Amaranth oil provides an excellent source for omega fatty acids.

The carbohydrates in amaranth grain consist primarily of starch made up of both glutinous and non-glutinous fractions. The unique aspect of amaranth grain starch is that the size of the starch granules (1 to 3 um) are much smaller than found in other cereal grains. Due to the unique size and composition of amaranth starch, it has been suggested that the starch may posses unique gelatinization and freeze/thaw characteristics which could be of benefit to the food industry (Becker et al., 1981; Lehman 1988). Several considerations for the use of amaranth starch in food preparation of custards, pastes, and salad dressing have been published in three papers (Singhal and Kulkarni 1990a, b, c). Moreover, the seed of grain amaranth is a rich source of iron (72-174 ppm), calcium (1300- 2850 ppm), natrium (160-480 ppm), magnesium (2300-3360 ppm) and zinc (36.2- 40 ppm) as well as vitamin riboflavin (0.19-0.23 mg/ 100g of flour) and ascorbic acid (4.5 mg/100 gm of flour), niacin (1.17-1.45 mg/100 g of flour), thiamine (0.07-0.1 mg/100 g of flour) and other microelements (Becker et al., 1981). Amaranth grains contain some antinutritional factors which can reduce the availability of nutrients. Light colour amaranth grains contains lower or comparable to the quantity of growth inhibitors in common grain (trypsin inhibitor, phytates, saponins, and tannins), or are heat labile as lectins (Souci et al., 1994; Berghofer and Schoenlecher; 2002).

Utilisation and Application of Amaranth Grains

Amaranth leaves and grains have been used in a wide variety of food. Fresh leaves of amaranth generally used as greens in salads or blanched, steamed, boiled, stir fried, or baked to taste. Cooked greens can be used as a side dish, in soups, as an ingredient in baby food, lasagne, pasta, pie, soufle, etc. Amaranth grain has a mild nutty flavour and they can be consumed in various ways. The whole seed can be used toasted or popped in breakfast cereals or in crunchy snacks. The industrial production of goods based on amaranth seeds and flour demonstrate that there are no remarkable changes in organoleptic characteristics but significant improvements in nutritive value are recorded (Sanz-Penella, 2012). Rolled or popped amaranth grains can be used in muesli and in granola bars. Amaranth grain can also be germinated for sprouts, malted for beer production, fermented or can serve as a starchy material in spirit production. Also, amaranth, like maize and buckwheat, can be popped through intense, short and dry heat without addition of fat. Ground grain can be used as a flour ingredient in different mixtures for pancakes, bread, muffins, dumplings, crackers, cookies, and pudding, etc. (Bejosano and Corke; 1998, Early 1990, Berghofer and Schoenlecher 2002 and Mlakar et al., 2009). In India amaranth grains popped to make a popular confection (laddoos) with honey or sugar syrup. The possible uses of amaranth grain can be seen in Table 3.

Table 3: Possible utilization of amaranth seeds

Process	Products
Cooking	Cooked seeds
Puffing and popping	Popped and expanded seeds
Milling	Whole meal flour
Milling and classification	Flour fraction, protein rich or starch rich flour fraction
Cooking and flaking	Flakes
Drum drying	Pregelatinized flours
Cooking extrusion	Cooked, expended or non expended product
Cooking extrusion and Milling	Pregelatinized flours
Cooking extrusion and flaking	Flakes
Germination (malting)	Sprouts and malts
Direct starch hydrolysis	Amaranth protein concentrates, glucose syrup
Starch isolation	Amaranth starch

Sources: Peter S. Belton, and John R.N. Taylor, 2002.

Health Benefits of Amaranth Grains

Hypocholesterolemic activity of amaranth

Kulakova et al. (2006) reported cholesterol-lowering effect in rats. They observed a decrease in blood cholesterol level but no increase in liver cholesterol. Whereas, an increase in liver sterol was found and it was thought to occur due to the presence of squalene in the amaranth oil. Such property of the oil was also confirmed by Martirosyan et al., (2007), while testing the properties of amaranth against coronary heart disease and hypertension, helping to decrease blood pressure. They reported that, when 18 – ml amaranth oil is used per day, its beneficial action was highest. Although most studies have been conducted with animals and/or hypercholesterolemic diets, Betim (2008) have found that if normolipidemic Wistar rats are chronically fed with normolipidemic diets in which 35% of the protein is replaced with the standard diet (AIN 93-G), they respond with a decrease in LDL cholesterol without going beyond the normal limits or adversely affecting the HDL cholesterol level. The authors conclude that moderate and chronic consumption, as a replacement of standard protein, may have a prophylactic effect. The cholesterol-lowering properties of amaranth grain and oil, reported, amaranth oil significantly lowered LDL cholesterol and improved HDL cholesterol, as well as lowering very low density lipoprotein cholesterol (VLDL cholesterol) by 21–50%. Amaranth grains can therefore be recommended as a functional food product for the prevention and treatment of cardiovascular diseases (Berger et al., 2003).

Antitumor effect of amaranth

Amaranth based diet is considered a risk factor for numerous chronic diseases, including cancer. Bennink and Om (1998) estimated that a period of 10 to 30 y, normal cells may be transformed in clinically detectable tumours, and the diet may promote or inhibit the development of the disease. Namely, Barrio and Anon (2010) studied the antitumor properties of the A. mantegazzianus isolated protein (MPI) and elucidated the possible mechanism of action. They studied on 4 different tumor cells: MC3T3E1, UMR106, Caco-2, and TC7. MPI showed an antiproliferative effect on the 4 cell lines with different potentials. MPI inhibits cell adhesion and induces apoptosis and necrosis in the malignant cell UMR 106, and the authors concluded antitumor potential in A. Mantegazzianus grains.

Antidiabetic activity of amaranth

There are some controversies on the action of amaranth on the blood glucose level. Although several investigators report that the eating of either the grain or the oil may protect against insulin deficiency, others declare that the amaranth grain has high glycemic index (GI) so it is a problem to diabetic patients. Conforti et al., (2005) detected antidiabetic activity in 2 varieties of A. caudatus seeds using an in vitro assay consisting of the inhibition of á-amylase, which acts in the digestion of starch and reduces the absorption of glucose. Kim et al., (2006a, b) examined the effect of amaranth oil and grain supplementation (A. esculentus L.) on blood glucose and the lipid profile of streptozotocin-induced diabetic rats. They divided rats in to four groups like normal control, diabetic control, diabetic group supplemented with amaranth grain, and diabetic group supplemented with amaranth oil for 3 week. They reported that a decrease was noticed in blood glucose and an increase in serum insulin levels.

Effect amaranth on Hypertension

The renin–angiotensin system is a biochemical system that helps control blood pressure. As angiotensinogen is released in response to a drop in blood pressure, sodium depletion or a reduction in plasma volume, the decapeptide angiotensin I is produced (Ondetti and Cushman, 1982), followed by the cleavage of the dipeptide His-Leu by the angiotensin-converting enzyme (ACE) to form angiotensin II, a potent vasoconstrictor (Li et al., 2004).Vecchi and Anon (2009), reported the occurrence of antihypertensive peptides in globulin 11S and is a most important constituents of the A. Hypochondriacus grain. A three-dimensional model of globulin 11S was built and the antihypertensive peptides were mapped through a computational method. The occurrence of 2 tetrapeptides, ALEP and VIKP, was experimentally observed through an in vitro evaluation of ACE inhibition, with values of 6.32 mM and 175 ìM, respectively.

Anti-Anemic effect of amaranth

Iron deficiency is the most common kind of anemia (Queiroz and Torres, 2000), and amaranth grain may be used as an interventional carrier to fight the deficiency. The anemia may be due to a number of circumstances, including chronic infections; hereditary blood problems; or the lack of one or more essential nutrients necessary for the formation of hemoglobin, such as iron, folic acid, B_{12}, B_6 and C vitamins, and proteins.

Prevention of Celiac Disease

Celiac disease (CD) is an autoimmune disease that is characterized by an aberrant response to dietary gluten in genetically susceptible individuals that results in small intestinal injury and can be associated with diverse systemic consequences (Reilly and Green, 2014).The only known treatment for celiac disease patients is strictly prohibit the consumption of dietary products that contain gluten. Therefore, the most sensible approach is to develop a variety of nutritious foods, such as those derived from amaranth grain. Amaranth and other pseudocereal are the best option to prepare a variety of nutritious product. Amaranth has been used for some time by patients with celiac disease because it does not cause allergic reactions for the intestinal mucosa Thompson (2001).

Conclusion

In this paper amaranth's nutritional and health beneficial properties have been described. Amaranth has an interesting characteristic to increase protein, starch, gluten-free products and squalene availability. In addition to its nutritional characteristics, amaranth plants have agronomic features identifying it as an alternative crop which might increase the use of marginal lands i.e. dry soil, high altitude and high temperature, where cereal and vegetable cannot be grown making it more environmental friendly. Amaranth could play an important role in Indian context because it produces high quality grain and leaves which can improve daily diet. Although, it has certain antinutritional factors which can be remove by using certain processing like germination, roasting, popping and extrusion etc.The amaranth grains is an alternative plant sources of squalene , which is commercially obtained from the liver of whales, sharks, and some other marine species, so it is indirect way to protect marine species. Amaranth grain could be used in celiac disease diet representing an alternative source of gluten-free products improving, in the same time the protein and mineral intake. Amaranth could be an interesting crop to meet out the increasing food demand by increasing population as well as to the rising good quality food products.

References

1. Alvarez-Jubetea, L. Arendtb, E.K. and Gallaghera, E. (2010). Nutritive value of pseudocereals and their increasing use as functional gluten-free ingredients, Trends in Food Science and Technology, 21, 106-113.
2. Amicarelli, V. Giungato P. Aveni, M. and Notarnicola, L., (2002b). La coltivazione di alcune specie di amaranto in Puglia. Nota II: I Lipidi e le proteine, In: University and enterprise. A partnership for training, research, employment and social development. Roma, 26-28 September, pp.94-101.
3. Barba de la Rosa, A.P. Fomsgaard, I.S. Larsen, B. Mortensen, A.G. Olvera-Martýnez, L. Silva-Sanchez, C. Mendoza-Herrera, A. Gonzalez-Castaneda J. and De Leon-Rodrýguez A. (2009). Amaranth (Amaranthus hypochondriacus) as an alternative crop for sustainable food production: Phenolic acids and flavonoids with potential impact on its nutraceutical quality, International Journal of Cereal Science, 49:117–121.
4. Barrio, D.A., and Anon, M.C. (2010). Potential antitumor properties of a protein isolate obtained from the seeds of Amaranthus mantegazzianus. European Journal of Nutrition, 49, 73–82.
5. Becker, R. Wheeler, E.L. Lorenz, K. Stafford, A.E. Grosjean, O.K. Betschart, A.A. and Saunders, R.M. (1981). A compositional study of Amaranth grain. Journal of Food Science, 46, 1175-80.
6. Bejosano, F.P. and Corke, H. (1998). Protein quality evaluation of Amaranthus wholemeal flours and protein concentrates. Journal of Science Food and Agriculture, 76:100-6.
7. Bennink, M.R. and Om, A.S. (1998). Inhibition of colon cancer (CC) by soy phytochemicals but not by soy protein. FASEB J, 12, 3806.
8. Berger, A. et al. (2003). Cholesterol-lowering properties of amaranth grain and oil in hamsters. International Journal for Vitamin and Nutrition Research, 73:39-47.
9. Berghofer, E. and Schoenlechner, R. (2002).Grain Amaranth. In: Belton P, Taylor J (eds), Pseudocereals and less common cereals, grain properties and utilization potential. Springer-Verlag, 219-60.
10. Betim, C.B. (2008). Metabolic Changes Beneficial to Rats Arising from the Consumption of Amaranth Master's Thesis. School of Food Engineering, University of Campinas Library. Available from: http://www.fea. unicamp.br/alimentarium/ver_documento. php?did=583.Accessed February 28, 2012.
11. Bressani, R. (1993). Amaranth, In: Macrae, R., Robinson, R.K., Sadler M.J., (Editors). Encyclopaedia of Food Science, Food Technology and Nutrition, 1:135-140.
12. Bressani, R. Elías, L.G. González, J.M. and Gómez-Brenes, R. (1987a). Arch Latinoam Nutrition, 37:364-377.
13. Bressani, R., González, J.M. Zúñiga, J. Breuner, M.and Elías, L.G. (1987b). Journal of Science Food and Agriculture, 38:347-356.
14. Caselato-Sousa, V.M. and Amaya-Farfán J. (2012). State of knowledge on amaranth grain: a comprehensive review, International Journal of Food Science, 4:93-104.
15. Caselato-Souza, V. M. and Amaya-Farfán, J. (2012). State of Knowledge on Amaranth Grain: A Comprehensive Review. Journal of Food Science, v. 77, n. 4, p. R93-R104, PMid: 22515252. http:// dx.doi.org/10.1111/j.1750-3841.2012.02645.x
16. Conforti, F. Giancarlo, A. Statti, A. Loizzo, M.R. Gianni, A. Sacchetti, B. Poli, F. and Menichini, F. (2005). In vitro antioxidant effect and inhibition of a alpha-amylase of two varieties of Amaranthus caudatus seeds. Biolchemical Pharmceutical Bulliten, 28:1098–102.
17. Early, D.K. (1990). Amaranth production in Mexico and Peru. In: Janick J, Simon JE (eds), Advances in new crops. Timber Press, Portland, OR, 140-2.

18. Grobelnik Mlakar, S. et al, (2009). Nutrition value and use of grain amaranth: potential future application in bread making. Agricultural, 7:5-15.
19. Grobelnik Mlakar, S. Turinek, M. Jakop, M. and Bavec, M. (2010). Grain amaranth as an alternative and perspective crop in temperate climate, International Journal for Geography 5:135-145.
20. Kauffman, C.S. and Weber, L.E. (1990). Grain amaranth. In: J. Janick and J.E. Simon (eds.), Advances in new crops. Timber Press, Portland, OR. 127-139.
21. Kim, H. K. Kim, M.J. and Shin, D.H. (2006b). Improvement of lipid profile by Amaranth (Amaranthus esculantus) supplementation in Streptozotocin-induced diabetic rats. Ann. Nutr. Metab, 50:277-281.
22. Kim, H.K. Kim, M.J. Cho, H.Y. Kim, E.K. and Shin, D.H. (2006a). Antioxidative and antidiabetic effects of amaranth (Amaranthus esculantus) in streptozotocin-induced diabetic rats. Cell Biochemical Functions, 24:195–9.
23. Kulakova, S.N. Pozdniakov, A.L. Korf I.I. Karagodina, Z.V. Medvedev, F.A. Viktorova, E.V. Gonor, K.V. Kamysheva, I.M. and Gadzhieva, Z.M. (2006). Amaranths oil: peculiarities of chemical composition and influence on lipid metabolism by rats. Vopr Pitan, 75:36–42.
24. Lehman, J. (1988). Carbohydrates of amaranth. Legacy 1:4-8 Amer. Amaranth Inst. Bricelyn, MN.
25. Leon-Camacho, M. Garcia-Gonzalez, D.L. and Aparicio, R. (2001). A detailed study of amaranth (Amaranthus cruentus L.) oil fatty profile. Eur. Food Res. Technol., 213:349-55.
26. Li, G.H. Le, G.W. Shi, Y.H. and Sundar, S. (2004). Angiotensin I-converting enzyme inhibitory peptides derived from food proteins and their physiological and pharmacological effects. Nutrition Research, 24:469–89.
27. Martirosyan, D.M. Miroshnichenko, L.A. Kulakova, S.N. Pogojeva, A.V. and Zoloedov, V. (2007). Amaranth oil application for coronary heart disease and hypertension. Lipids Health Disorders, 6:1–12.
28. Mlakar, G.S. Turinek, M. Jakop,M. Bavec,M. and Bavec, F. (2009). Nutrition value and use of grain amaranth: potential future application in bread making, Agricultura, 6,43-53.
29. National Research Council. (1984). Amaranth: Modern Prospects for an Ancient Crop.National Academy Press, Washington, D.C.
30. Ofitserov, N.E. (2001). Amaranth: Perspective raw material for food processing and pharmaceutical industry, chemistry and computational simulation. Butlerov communication, Vol.2. No. 5.
31. Ondetti, M.A. and Cushman D.W. (1982). Enzymes of the renin-angiotensin system and their inhibitors. Ann Rev Biochem, 51:283–308.
32. Peter S. Belton, and John R.N. Taylor (Eds.) (2002). Pseudocereals and less common cereals: grain properties and utilization potential, springer-Verlag Berlin, Heidelberg Newyork.
33. Queiroz, S.S. and Torres, M.A.A. (2000). Anemia ferropriva Na inf`ancia. Journal of Pediatratic, 76:298–304.
34. Reilly, N. and Green,P.R. (2014).Presentation of Celiac Disease in Children and Adults, in Celiac Disease, S.D. Rampertab and G.E. Mullin, Editors., Springer New York, 95-105.
35. Sanz-Penella, J.M. Wronkowska, M. Soral-Smietana, M. and Haros, M. (2012). Effect of whole amaranth flour on bread properties and nutritive value, In: LWT – Food Science and Technology, 50: 679-685.
36. Saunders R.M. and Becker R. (1983). In: Pomeranz Y. (ed.) Advances in cereal science and Technology, Vol. 6. St. Paul, Minn, USA: American Association of Cereal Chemistry.

37. Singhal, R.S. and P.R. Kulkarni. (1990a). Some properties of Amaranthus paniculatas (Rajgeera) starch pastes. Starch/Starke, 42:5-7.
38. Singhal, R.S. and P.R. Kulkarni. (1990b). Utilization of Amaranthus paniculatas (Rajgeera) starch in salad dressing. Starch/Starke, 42:52-53.
39. Singhal, R.S. and P.R. Kulkarni. (1990c). Studies on applicability of Amaranthus paniculatas (Rajgeeraa) starch for custard preparation. Starch/Starke, 42, 102-103.
40. Souci, S.W. Fachmann, W. and Kraut, H. (1994) Food composition and nutrition tables, 5th revised and completed edition. Medpharm, Stuttgart Medpharm GmbH, Scientific Publishers, Stuttgart and CRC Press, Boca Raton, FL.
41. Teutonico, R.A. and D. Knorr, D. (1985). Amaranth: composition, properties, andapplications of a rediscovered food crop. Food Technology, 39(4):49-61.
42. Thompson, T. (2001). Case problem: questions regarding the acceptability of buckwheat, amaranth, quinoa, and oats from a patient with celiac disease. Journal of Amrican Diet Association, 101: 586–7.
43. Valcárcel-Yamani, B. and Caetano da Silva Lannes, S. (2012). Applications of Quinoa (Chenopodium Quinoa Willd.) and Amaranth (Amaranthus Spp.) and Their Influence in the Nutritional Value of Cereal Based Foods, In: Food and Public Health, 2:265-275.
44. Vecchi, B. and Anon, M.C. (2009). ACE inhibitory tetrapeptides from Amaranthus hypochondriacus 11S globulin. Phytochemistry, 70:864–70.

CHAPTER 17

Effect of Frying Oils on Textural and Sensory Properties of Egg Slices

Sanju B. Dhull, Manju V. Nehra, Gurjant Singh

Abstract

Egg is a nutrient dense food, containing high quality proteins and a wide range of vitamins, minerals, trace elements, PUFA, MUFA etc. Whole egg slices were prepared by boiling in water technique for a predetermined time i.e. 15 min. The egg slices were prepared with spice extract at different levels of 2%, 5%, 7% and 10%, out of which 5% was found acceptable by sensory evaluation. The egg slices with 5% spice extract were then subjected to frying in different oils such as mustard oil, rice bran oil, sunflower oil and soya bean oil. Among all of the fried egg slices and control (without any spice extract and frying), on sensory evaluation, fried egg slices with mustard oil was most preferred. Texture analysis results shows that hardness was increased on frying and highest in fried egg slices with rice bran oil. Also the values for other TPA component viz. adhesiveness, cohesiveness, gumminess and chewiness were increased on frying as compared with control. Among the fried egg slices, highest values for adhesiveness, cohesiveness, gumminess was observed in egg slices with rice bran oil. However, chewiness of fried egg slices with sunflower oil was highest. It can be concluded from present study that the egg slices fried in mustard oil was organoleptically acceptable with good textural properties.

Introduction

The eggs of many species of birds have been in use as an item of food from prehistoric times. But chiken eggs are the most commonly eaten eggs. They supply all essential amino acids for humans (a source of 'complete protein') (FAO, 2010), and provide several vitamins and minerals, including retinol (USDA,2005) (vitamin A), riboflavin (vitamin B_2), folic acid (vitamin B_9), vitamin B_6, vitamin B_{12}, choline, iron, calcium, phosphorus and potassium. Although not as abundant as red meats, eggs are a source of CoQ10 depending on how they are prepared.

All of the egg's vitamins A, D, and E are in the egg yolk. The egg is one of the few foods to naturally contain vitamin D. A large egg yolk contains approximately 60 calories (250 kilojoules); the egg white contains about 15 calories (60 kilojoules). A large yolk contains more than two-thirds of the recommended daily intake of 300 mg of cholesterol (although one study indicates the human body may not absorb much cholesterol from egg. They have desirable unsaturated to saturated fatty acid ratio (2:1). People concerned about the amount of cholesterol in eggs, can use the egg white which is cholesterol -free. Growing children and person under special conditions require energy rich food products. Eggs are marketed and consumed mainly in fresh form. But demand for egg products and RTE egg product are increasing day by day.Of the 76.2 billion eggs consumed in 2009, 30 percent were in the form of egg products (eggs removed from their shells) (USDA, 2011). Based o the review of literature and with the aim of product sensory acceptability anhancement the persent study was planned keeping in mind the following objectives:

- Development of egg slices using various formulations and spice extract.
- Sensory analysis of the developed product.
- Frying of whole egg slices in different oils.
- Sensory and texture analysis of fried egg slices.

Methods

Preparation of spice extract

Weighted amounts of dry and wet spices and salt (Table-2) were ground and then boiled in a beaker with 100ml of water. The beaker was covered with petriplates to avoid the loss of volatile compounds during boiling the mixture. After some time, the concentrate was passed through muslin cloth to remove solids. It was then stored in covered glass container under refrigerated condition till further use.

Preparation of egg slices (Boiled and Fried)

Fresh hen eggs were procured from local market of Sirsa, Haryana. Preliminary quality evaluation was carried out prior to further use. Ten eggs were picked randomly and checked out for various quality parameters i.e. shell quality, weight and pH. Any unclean or cracked eggs were removed from the lot. Egg slices (boiled and fried) were prepared by the process shown in Figure-1.

Sensory Evaluation

Freshly prepared egg slices with spice extract and fried whole egg slices in different oils were analyzed by a panel of 15 semi-trained members. For each sample, participant was asked to score color and appearance, aroma, texture, taste and overall acceptability using a 9 point Hedonic scale.

Fresh egg

↓

Egg material (albumen,/whole egg/double yolk egg

↓

Blending ← Spice extract

(2%, 5%, 7%, 10%)

↓

Filling egg material in polythene tube and sealing

Cooking in Boiling water (15 min)

↓

Removal from polythene tubes and slicing → Some of the whole egg slices

↓ Frying (Mustard oil, rice bran oil,Sunflower oil, soya bean oil)

↓

Storage

(Room temp. refrig. temp., freezing temp.)

Fig.1: Flow diagram for preparation of Whey based Papaya-Carrot RTS beverage.

Texture analysis of egg slices

The TA- XT2 Warner Bratzler texture analyser available at SDS instrument lab in Delhi was used to measure the hardness final product formed. Each of the five samples (1-boiled standard and four fried egg slices) were evaluated instrumentally at the temperature recommended by Szczesniak et al., (1963) and Munoz et al. (1986) when developing the standard texture rating scales. The instrumental texture profile analysis (TPA) test described by Bourne (1978) was used. A two cycle compression was set for a 70% stain. A compression load cell (500 kg, INSTRON) interfaced with the series IX data acquisition software (INSTRON) was used to evaluate the stress applied to the samples during compression. The crosshead speed or loading rate was set to 1.0 mm/sec for the evaluation of all samples. Parameters corresponding to sensory attributes were obtained from the curves. The test was configured so that the product TPA parameters, hardness, cohesiveness, springiness and chewiness, were calculated at the time of the test by determining the load and displacement at predetermined points on the TPA curve. Hardness (h,) was the maximum load, expressed in kg, applied to the samples during the first compression. Cohesiveness (A,/A,) was the ratio of the area under the curve for the second compression (A,) to that under the curve for the first compression (A,). Springiness, expressed as dJd, was the ratio of the duration of contact with the sample during the first compression (d,) to that during the first compression (d,). Chewiness was the mathematical product of hardness, cohesiveness, and springiness (h, x A /A, x d, /d,).

Results and Discussion

Quality parameter of egg

The results for quality parameters (Table-1) shows that the weight of the egg ranges from 52.0 -53.7 gm. The pH also ranges (7.11-7.13) nearby neutral which shows that the egg were freshly laid (Stadelman and Cotterill, 1986).

Table 2: Quality Parameters of Eggs

Sample No.	Shell Quality	Weight (g)	pH
1.	+	52.0	7.11
2.	+	53.4	7.13
3.	±	53.0	7.12
4.	+	52.8	7.11
5.	+	52.2	7.12
6.	+	52.7	7.12
7.	+	53.1	7.13
8.	±	52.0	7.11
9.	+	53.7	7.13
10.	+	53.6	7.13

+ absolutely clean
± slightly unclean

Level of spice extract

The weight of ingredient used for preparation of spice extract (Table 2) acceptability of egg slices with different level of spice extract was estimated by trial and check method. The egg slices with spices levels (Table 3) at a level of 5% level were found to be acceptable organoleptically. The spices with 2% spices extract was found tasteless and less acceptable. Also the slices with spices at 7% and 10% level were found unsatisfactory as it was too spicy to eat. Therefore, the egg slices with 5% spice extract were selected for further trials.

Table 2: Estimated weight of ingredients used for preparation of Spice extract.

Ingredient	Weight (g)
Coriander	2.0
Red chilli	2.0
Cinnamon	4.0
Black pepper	2.0
Salt	4.0
Onion	4.0
Garlic	4.0
Green chilli	4.0

Table 3: Acceptability of egg slices with spice extract at different levels.

Products	2%	5%	7%	10%
Whole Egg Slices	LA	A	UA	UA
Egg albumen slices	LA	A	UA	UA
Double yolk slices	LA	A	UA	UA

A – Acceptable; LA - Less Acceptable; UA - Unacceptable

Texture Analysis of different egg slices

Texture plays an important role in the quality of a product also in food identification. Texture perception is dynamic process that usually takes place in the mouth, where the food is masticated. Despite the majority of textural responses occuring in mouth, humans use several senses to perceive texture, such as vision, touch and hearing (Wiikins et al., 1995). One of the important physical characteristic of egg slice is the texture. The texture will depend on the formulation and the cooking regims used, but a key factor is the component that forms the main matrix that links the different elements together. The texture analyzer can measure several textural properties such as hardness, chewiness, adhesiveness, gumminess, which was invented in texture profile analysis utilizing the universal testing machine which mimics chewing through the use of large destructive shears, and the hardness, springiness, and cohesive nature of the food are calculated (Van Hekken et al., 2004). The texture properties of egg slices are shown in Table 4. Hardness of egg slices was increased after the frying in oils. The hardness of boiled egg slices was found to be lowest (9900.60 g) and highest in fried egg slices 16770.52 (rice bran oil). This may be due to the decreasing in moisture content in fried egg slices. Adhesiveness is a negative force after the first compression cycle and is a measure of how sticky, tacky, or gooey the food are (Whistler, 1997). The adhesiveness of egg slices was also highest in fried egg slices in rice bran oil (-0.71g/sec), fried in mustard oil (-0.85 g/sec) and lowest in boiled egg slices (-296.70 g/sec). The springiness is calculated by Length 2/Length 1 (mm/mm) (Abdelghafor et al., 2011). Springiness of the boiled egg slices was (0.86 J/F) significantly (p d"0.05) increased on frying in sunflower oil (1.77 J/F), in soya bean oil (1.84 J/F), in mustard oil (1.75 J/F) and in rice bran oil (1.015 J/F). The cohesiveness of the egg slices was found highest in freid egg slices in rice bran oil (0.83) and lowest in boiled egg slices (0.62). The results showed that gumminess increased after the frying in oils, the gumminess of the egg slices was lowest in boiled egg slices (6145.63) and highest in freid egg slices in rice bran oil (13924.62). Chewiness of boiled egg slices was in the range of boiled egg slices (5304.45 P/N) to fried egg slices in soya bean oil (22699.47 P/N). Abdelghafor et al., (2011) reported that gumminess and chewiness values are highly dependent on hardness rather than on cohesiveness or springiness values. Resilience was how well a product "fights to regain its original position". It was calculated by Area 5 / Area 4 (Nmm/Nmm). Among the different egg slices samples, highest value (0.51) was observed in fried egg slices in rice bran oil and lowest (0.15) in boiled egg slices.

Table 4: Texture analysis of different egg slices

Parameters	A	B	C	D	E
Hardness(gm force)	9900.60	12780.00	16770.32	15232.15	15902.30
Adhesiveness(gm sec)	-296.70	-16.34	-0.71	-0.94	-0.85
Springiness (J/F)	0.86	1.77	1.01	1.84	1.75
Cohesiveness (I/G)	0.62	0.82	0.83	0.80	0.79
Gumminess (K/O)	6145.63	10512.26	13924.62	12300.36	12669.89
Chewiness (P/N)	5304.45	18659.48	14127.11	22699.47	22227.71
Resilience	0.15	0.42	0.51	0.42	0.39

A = Boiled egg slices; B = Egg slices fried in sunflower oil; C = Egg slices fried in rice bran oil ; D = Egg slices fried in soyabean oil; E = Egg slices fried in mustard oil.

The values are mean ± S.D. of three independent determinations. The values with different superscripts in a row differ significantly ($p \leq 0.05$).

Sensory evaluation of different egg slices

Boiled egg slices with different level of spice extracts and fried egg slices made from different type of oils (mustard oil, rice bran oil, sun flower oil, soya been oil) were evaluated for color, appearance, flavor, taste, texture and overall acceptability by sensory evaluation.

Fig 2 shows that colour and appearance of plain salty egg slices (8.4) was preferred over othe slices with different level of spice extract. However there was no significant difference in the score for texture and taste. Overall acceptability score was also good for spicy egg albumen slices (7.9), spicy whole egg slices (7.8) and spicy egg yolk slices (7.5), compared with standard plain salty egg slices (8.4).

Sensory evaluation score for fried egg slice (Fig. 3) showed that all the fried egg slices were preferred over control i.e. boiled egg slices. Among all of the fried egg slices, fried egg slices with mustard oil was most preferred having highest overall acceptability score (8.3).

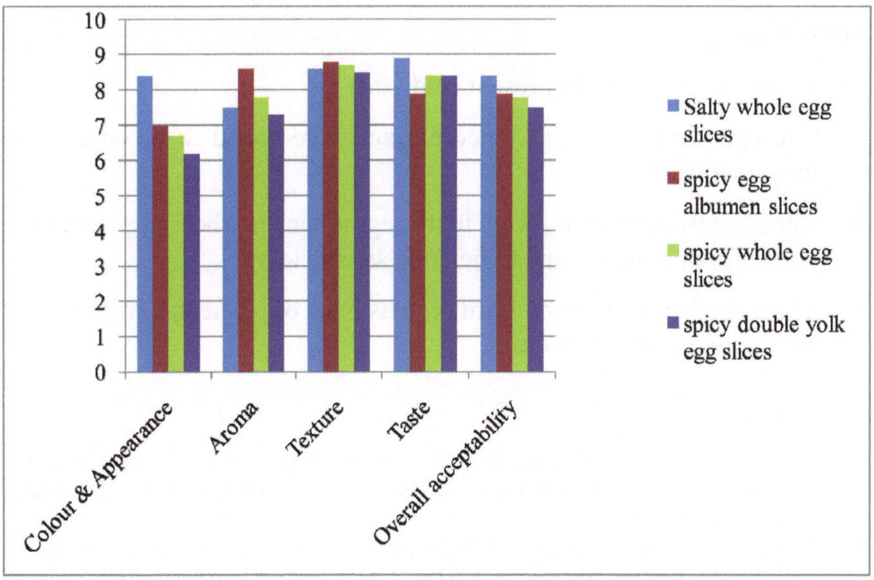

Fig. 2: Sensory evaluation of different type of egg slice fried in different oil.

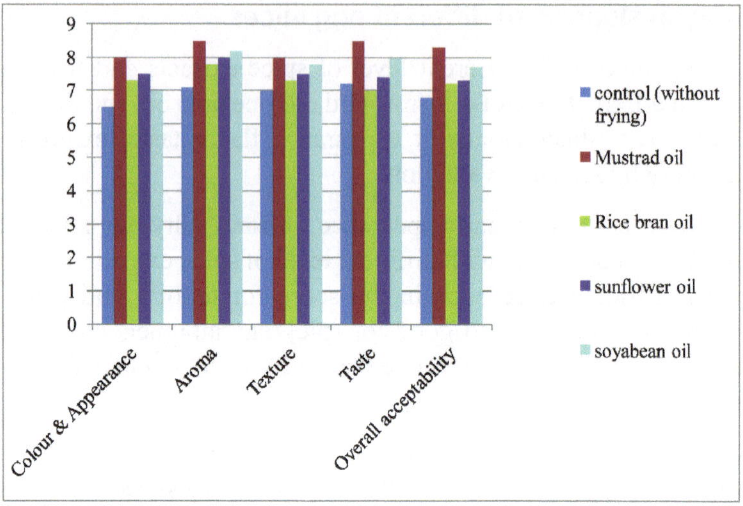

Fig. 3: Sensory Evaluation of different type of egg slice fried in different oil.

Conclusion

It can be concluded from the present study that:

- The egg slices with 5% spice extract were found to be acceptable organoleptically.
- The overall acceptability score for the egg albumen, whole egg and double yolk egg slices was found to be quite satisfactory.
- The egg slices fried in mustard oil was most organoleptically preferred with good textural properties.

References

1. Ahn, D.U. Sell, J.L. Jo, C. and Jeffrey, M., (1999). Effect of dietary conjugated linoleic acid on the quality characteristics of chicken eggs during refrigerated storage. Poultry Science, 78:922-928.
2. Alovarez, C. Mendez, J. and Garcia-Rebollar, P. (2004). Effects of conjugated linoleic acid addition on its deposition in eggs of laying hens, fed with no other fat source, Journal of Agriculture Ressearch, 2 (2):203-209.
3. American Egg Board (1974). A Scientist Speaks About Egg. American Egg Board, Chicago.
4. American Egg Board (1981). Egg cyclopedia, American Egg Board, Park ridge, Illinois.
5. Anonymous (1975). Glossary of terms relating to the sensory analysis of food, British Standard 5098, British standards Institution, London.
6. Campbell, L. Raikos, V. and Euston, S.R. (2004). Heat stability and Emulsifying ability of whole egg and egg yolk as related to heat treatment, Journal Aerosol Science.
7. Chang, H.C. and WU, L.C. (2008). Texture and quality properties of Chinese fresh egg noodles formulated with green seaweed powder, Journal of Food Science, 73(8)398-404.

8. Food Standards Agency (2002). McCance and Widdowson's, the composition of foods, sixth summary edition. Cambridge Royal society of chemistry.
9. Food Technology Corporation (2008) Why Measure Texture, Food Technology Corporation Virginia.
10. "Food and Agriculture Organization article on eggs". Fao.org. Archived from the original on 2004-03-07. Retrieved 2010-01-10.
11. Jack, J.W. (1964). new marketable poultry and egg products, prepacked hard cooked eggs, Cornell University.
12. Johnson, H.S. and Ridlen, S.F. (2009). Structure of the egg incubation and embryology, Illinois University.
13. Ng, W. (1971). Method of improving colour of cooked egg products, U.S. Patent No. 3, 598, 612.
14. Raikos, V. Campbell, L. and Euston, S.R. (2006). Rheology and texture of Hen's egg protein heat set gels as affected by pH and the addition of sugar and salt, Herriot watt university, Scotland, U.K.
15. Rose, E.L. and Bruce, J.H. (2006). Effects of a liquid egg product containing fish oil on selected cardiovascular disease risk factors; a randomized crossover trial, Food Res. Int. 39 (8), 910-916.
16. Stadelman, W.J. and Cotterill, O.J. (1986). Functional properties of egg in; egg science and technology, 3rd edition, AVI Publishing Co. 230-272.
17. Vitamin A, RAE Content of Selected Foods per Common Measure, sorted by nutrient content. United States Department of Agriculture (USDA). 2005. p. 4
18. Westly, R.D. Rousselle, J.R. Schwan, D.R. and Stadelman, W.J. (1982). Improvement in quality of scrambled egg products served from steam table display, Poultry Science, 61, 457-462.
19. Whistler, R.L. and BeMiller, J.N. (1997). Carbohydrate chemistry for food scientists chapter 5. American Ass. of Cereal Chemists St. Paul, MN.

Unit-III: Functional Foods and Nutraceuticals Their Development Techniques

CHAPTER 18

Concept and Techniques in Development of Functional Foods

R.K. Gupta and Monika Sharma

Abstract

Tremendous changes in lifestyle, eating habits and shifting rural habitations are causing an irreversible change that is leading to manifold multiplication of health problems and these health issues have attracted the attention of every segment of the society in the recent past. Today foods are not intended to only satisfy hunger and to provide necessary nutrients for humans but also to prevent nutrition-related diseases and improve physical and mental well-being of the consumers. Due to increasing control over nutrition and health, consumers are becoming more proactive towards their health. In this regard, functional foods play an outstanding role and they are gaining public acceptance and there is great scope for value-added as well as health promoting food products in developing countries. Various terms have used for functional foods throughout the world such as designer foods, medicinal foods, nutraceuticals, therapeutic foods, superfoods, foodiceuticals, and medifoods. People are interested in foods which boost immune system, reduce the risk of diseases and improve health. The increasing demand of such foods can be explained by the increasing cost of healthcare, the steady increase in life expectancy, and the desire of older people for improved quality of their later years. Typically, a food marketed as functional contains added, technologically developed ingredients with a specific health benefit. These range of products cannot be truly classified as 'food' and a new hybrid term between nutrients and pharmaceuticals, 'nutraceuticals', has been coined to designate them. The development of functional foods involves the application of various food processing techniques such as microencapsulation of functional ingredients for controlled release, nanotechnology for the development of nanostructured (or nanotextured) food ingredients and delivery systems for nutrients and supplements etc. Inclusion of probiotic bacteria and probiotic supplements in various food systems has also become very popular and this is becoming a very vital sector in functional food market. Systematic RandD efforts are, therefore, required to develop health promoting foods for all segments of the society at affordable cost. Prior to promoting the developed health foods, it is of utmost importance to conduct clinical trials to ascertain safety and efficacy, and also establish appropriate regulatory framework for their distribution to masses. All these issues have been discussed in this paper and recommendations were drawn to formulate future research.

Introduction

The basic human need and prerequisite for healthy life is nutritive food. Growth and development is dependent upon a proper diet which is essential from the very early stages of life. Food consumption, which largely depends on production and distribution, determines health and nutrition of the population. Importance of the health promoting foods is increasing steadily with the growing health concerns of people across the globe. Although there is no strict definition of "Functional Food", in general, it is used to describe foods that are considered to be beneficial for human health, besides providing basic nutrition. It is mostly used as an umbrella term covering natural foods, organic foods, whole foods, health foods, nutraceuticals and sometimes dietary supplements. Functional foods are also known as designer foods, medicinal foods, nutraceuticals, therapeutic foods, superfoods, foodiceuticals, and medifoods. In general health foods should have high nutritional value, high contents of functional and bioactive compounds and no anti-nutritional or harmful compounds. Health foods are basic food stuffs and not a nutraceutical or a drug. They are safe to consume without any prescription.

Production of various food items has increased many folds during the last five decades however; a substantial part of developing nations' population still faces challenges of ill health, malnutrition, chronic diseases etc. which is more acute among infants, children and women. In contemporary societies, however, many individuals do not consume a balanced diet and this has led to an increase in the incidence of serious metabolic imbalances. These in turn increase the risk of diseases such as obesity, type-2 diabetes, hypertension, food allergies and intolerances, and gastrointestinal and inflammatory disorders. Good nutrition is vital for health, optimal growth and development, and prevention of disease. Health issues and malnutrition have already taken an alarming state in various countries. Malnutrition is a broad term commonly used as an alternative to undernutrition but technically it also refers to over nutrition. People are malnourished if their diet does not provide adequate calories and protein for growth and maintenance or they are unable to fully utilize the food they eat due to illness (undernutrition). They are also malnourished if they consume too many calories (overnutrition). The nations which are in developmental transition face the dual burden of pre-transition diseases like under-nutrition and infectious diseases as well as post-transition, lifestyle-related degenerative diseases such as obesity, diabetes, hypertension, cardiovascular diseases (CVDs) and cancers. According to most recent estimate released in October 2010 by FAO, 925 million people are undernourished all over the world. Thus, there is an urgent need of development of functional food products for these low-income groups based upon the major staples.

Functional foods are defined as 'foods that contain some health-promoting component(s) beyond traditional nutrients' (Shah, 2001). They are also referred as "Health food" and sometimes nutritional supplements for which a specific claim of health benefits is made. Functional foods used synonymously to Health food encompasses; nutraceuticals, nutrient or food supplements, fortified or biofortified foods, probiotics etc. If we compare the health foods with medical drugs then it has following advantages:

- No prescription or supervision is required in case of health foods whereas drugs require medical supervision.
- These can be distributed through supermarkets, drugstores or retailers whereas drugs can be provided only through hospital, pharmacies etc.
- These can be consumed as per the desire whereas drugs can be consumed according to requirement only.
- These can be consumed for lifelong benefits whereas drugs have short term benefits.

Types of Functional Foods

As mentioned earlier that health food is a term that covers various types of foods. Functional food is one claiming to have a disease-preventing property beyond the basic nutritional function of supplying nutrients. Nutraceutical refers to a component of food thought to have a beneficial effect on human health. Designer Food is a processed food that is supplemented with food ingredients naturally rich in disease-preventing substances. Natural food implies to foods that are minimally processed and do not contain manufactured ingredients. Organic food are foods that are produced using methods that do not involve modern synthetic pesticides, chemical fertilizers etc., and do not contain genetically modified organisms. Whole foods are foods that are unprocessed and unrefined. Probiotics are live microbial food supplements containing beneficial bacteria supporting human health. Broadly, we can say that health food is an umbrella term used for functional foods, probiotics, natural foods, organic foods, whole foods, nutraceuticals, and designer foods.

Table 1: Functional foods and examples

Type of health food	Examples
Natural food	Peeled carrot , peeled apple, Orange juice
Organic food	Organically grown rice, vegetables or fruits
Whole food	Whole wheat flour bread, unpeeled carrot
Probiotics	Curd, sauerkraut
Designer food	Ca fortified flour, high protein biscuits
Functional foods	Fe fortified flour, high fibre biscuits, carotenoid rich carrot, high lycopene tomato

Functional Foods Trends and Market

From a market perspective, health foods are difficult to quantify because different definitions are used in different countries. If we view functional foods as foods that make specific health claims, then the market in the USA, Japan and Europe is estimated to be worth about 7 billion pounds. Considering health food marketing i.e. broader health-based/driven foods market including natural and organic foods, 'low and lite', weight management and vitamin- and mineral-fortified products as well as functional foods, the market is very large with a global estimate of around 95 billion pounds. The most popular foods are energy/ sports drinks, probiotic dairy products, heart health spreads and ready-to-eat cereals. Major functionality claims are for gut health (especially in Japan and Europe), heart health (especially in the USA and Europe), promoting natural defences, and boosting energy levels. The outlook for the future for health based / driven foods are bright. An overall growth rate of 10% per annum for the next five years is possible, significantly outperforming the overall foods and beverage market's growth of about 2% per annum.

The factors governing the success of a health food are:

- Awareness and education of the consumer bout the product
- Product meeting consumer needs and requirements
- Product do not disturbing the general dietary patters / habits of the consumer i.e. the product should fit exactly into the traditional food habits / cuisines.

- Powerful communication of health benefits to the consumer
- Uncompromised taste, flavour, colour etc.
- Optimal convenience of preparation, handling, serving etc.
- Adequate retail availability even in the remotest of the areas
- Proven and certified safety and efficacy
- Affordable price
- Favourable scientific opinion leaders
- Support by clear regulatory framework for making claims
- Providing equal opportunities to all companies.

The health foods we find in the market today are mostly based on general discoveries in nutritional science, and less on a deliberate research strategy to develop specific health foods. For the future, an approach that integrates insights into consumer needs and demands (market pull) and a structured scientific research process (science push) will give the largest chance of real innovations. The scientific research process into functional foods will be powered by technology and insights available from other disciplines, such as informatics, pharmacology, engineering, proteomics and genomics. Insights into genetic susceptibility arising from the human genome project and possibilities for individualised health monitoring using micro-engineering and sensor technology may alter dramatically the way in which we deal with food and health. For this, however, consumer acceptance is a first prerequisite. This is also true for the attainments of modern biotechnology that may also have a major impact on possibilities to optimise the health foods.

Developing countries can enjoy the benefits of the health food sector to expand options for producers and to promote growth in the sector through partnerships between various stakeholders i.e. research centers, private entrepreneurs, and indigenous communities. However, the success requires sufficient proof to establish the health claim and capacity to accurately market functional foods to consumers in high-end markets. Countries that are interested in this sector should also assess the opportunities at the national level because functional foods cover such a broad group of products and production systems that some can find demand in the domestic market, while others can be targeted for export. Identification of specific export markets, certification and other regulations, and consumer demand are product and/or ingredient specific and largely dictate the possibilities for development. The target in developing countries should be to link health with the nutritional security and develop need based products suiting to specific area and dietary pattern requirements. This would no doubt open new avenues for rural employment, health enhancement and industrial growth.

Techniques Involved in Development of Functional Foods

The development of functional foods involves the application of various food processing techniques such as microencapsulation of functional ingredients for controlled release, nanotechnology for the development of nanostructured (or nanotextured) food ingredients and delivery systems for nutrients and supplements etc. Inclusion of probiotic bacteria and probiotic supplements in various food systems has also become very popular and this is becoming a very vital sector in functional food market. Some of these techniques have been discussed below.

Nanotechnology and its Applications in Functional Foods

Nanotechnology is an exciting and rapidly emerging technology allowing us to work, manipulate and create tools, materials and structures at the molecular level, often atom by atom into functional structures having nanometer dimensions. "Nano" usually refers to a size scale between 1 and 1000 nanometers (nm). Nanotechnology is the design, fabrication and utilization of materials, structures, devices and systems through control of matter on the nanometer length scale and exploitation of novel phenomena and properties (physical, chemical, biological) at that length scale in at least one dimension. The nanoparticles have various functionalities which can be exploited for development of functional foods.

Unique Properties of nanoparticles

- Change in physical properties as they follow laws of quantum physics
- Surface behaviour dominates bulk material behaviour
- Metals become harder and ceramics become softer
- Stronger, more heat resistant, transparent
- Increased chemical resistance, reduced weight
- Different interactions with light/other radiation
- New electrical properties
- Novel biological properties

Nanotechnologies for functional foods

Convergence of nanotechnology with other technologies is also leading to further innovations that are expected to make a major impact on production, processing, storage, transportation, traceability, safety and security of food. For example, integration of biotechnology, nanotechnology and IT has opened up new opportunities for the development of nano-biosensors for the detection of pathogens and contaminants in food. Such integration of technologies, combined with understanding of taste receptors and flavour perception, is leading to the development of an "electronic tongue" for describing the taste attributes of food. Indeed, the manipulation of substances so close to the molecular level has blurred the boundaries between a numbers of traditional food science disciplines and opened up ways for the development of new food textures, tastes and sensations.

A major focus of current nanotechnology applications in food is the development of nanostructured (or nanotextured) food ingredients and delivery systems for nutrients and supplements. For this, a variety of processes are being utilized, including nano-emulsions, surfactant micelles, emulsion bilayers and reverse micelles (Weiss et al., 2006). A number of nano-micelle based carriers for nutraceuticals and nutritional supplements are currently available. The nano-micelle-based carrier system NovaSOL® from Aquanova® Germany for introduction of antioxidants and supplements in food and beverage products is one such system. The NovaSOL® technology, which is based on "product micelles" with a diameter of around 30 nm, has been applied to develop a range of solubilisates. Example applications include food additives, such as benzoic acid, citric acid, and ascorbic acid, dietary supplements and functional food ingredients, such as vitamins A and E, α-lipoic acid, soybean isoflavones, β-carotene, lutein, omega-3 fatty acids, and coenzyme Q10.

In addition to Novasol® from Aquanova® Germany, these include nano-structured supplements based on "Nano-Sized Self-assembled Liquid Structures (NSSL)" from NutraLease Ltd. Israel, and NanoCluster™ delivery system for food products from RBC Life Sciences® Inc. USA. However, the problem with most emulsions is that they are dependent on conditions - just as they spontaneously form under certain conditions, they can also spontaneously disassociate if the temperature or concentration changes too much - not very useful for a system which may need to remain stable through various areas of the human digestive tract. Stabilization of some materials also requires addition of significant amounts of surfactant, or emulsifier, which has to be food-grade, and may adversely affect the taste or texture of the food. Recently, self-assembled nanotubes have been developed from hydrolysed milk protein β-lactoglobulin, which can offer a new naturally derived carrier for nano-encapsulation of nutrients, supplements and pharmaceuticals (Graveland-Bikker and de Kruif 2006). The use of microencapsulated food additives is already well established. For example, microencapsulation has been used to mask the taste and odour of tuna fish oil added to bread for health benefits (e.g. “Tip Top-up” brand bread from George Weston Foods, Australia. The nano-selenium-enriched Nanotea from Shenzhen Become Industry and Trade Co., Ltd. China is another example. In some parts of China, selenium deficiency is the main cause of a number of ailments and enhanced uptake and bioavailability of selenium through Nanotea is claimed to provide a variety of health benefits. A health supplement Nutri-Nano™ CoQ-10 from Solgar, USA, based on Aquanova’s technology and claimed to increase absorption of the fat soluble CoQ-10, is currently being marketed in the UK.

Microencapsulation

Microencapsulation is a process of entrapping solid particles, liquid droplets or gases in thin polymeric coatings. The outer film (wall) protects the encapsulated material (core) and allows for its controlled release. Application of this technique is wide spread in several industrial fields, including the manufacture of foods, pharmaceuticals, cosmetics and pesticides. Common food ingredients subjected to microencapsulation include probiotic bacteria, acidulants, flavours, colours, sweeteners, minerals and vitamins. A number of coating materials are employed in production of microcapsules, such as gums, starch, and starch derivatives, cellulosic materials, lipids, proteins and inorganic materials. Microencapsulation can improve the convenience of food. The shell provides a barrier between reactive components (for instance, delaying the release of leavening agents for fluffier bread products or protecting oxygen-sensitive materials during processing and storage).

Microcapsules can help fragile and sensitive materials survive processing and packaging conditions and stabilize the shelf-life of the active ingredient. Karel (1994) described various methods of encapsulation of food ingredients. In addition, microencapsulation can simplify the food manufacturing process by converting liquids to solid powder, decreasing production costs by allowing batch processing using low-cost, powder-handling equipment. The reasons for microencapsulation are countless. In some cases, the core must be isolated from its surroundings, as in isolating vitamins from the deteriorating effects of oxygen, retarding evaporation of a volatile core, improving the handling properties of a sticky material, or isolating a reactive core from chemical attack. The other objective is the controlled release rate at which it leaves the microcapsule, as in the controlled release of drugs or pesticides.

Among various types of microencapsulation methods, liposome technology is often used to deliver fluid food products, as it reduces the ability of iron to react with food components. By encapsulating calcium lactate in lecithin liposomes, it was possible to fortify soymilk with levels of calcium equivalent to those found in cow's milk (Hirotsuka et. al., 1984), while preventing undesirable calcium-protein reactions. As with probiotics, the co-encapsulation of vitamins and minerals could be beneficial. Thus, the presence of calcium in a hydrophilic phase and vitamins A or D in a hydrophobic phase could promote calcium absorption in the GI tract. In one such example, encapsulation of ascorbic acid in a liposome together with vitamin E produced a synergistic antioxidant effect (Reineccius, 1995). In this case, the vitamin E is incorporated into the liposome wall, while the ascorbic acid is entrapped in the aqueous interior.

Omega-3 and omega-6 fatty acids are used for food fortification having considerable health benefits; however, major issues have been encountered with regards the taste and smell of these oils and their propensity to oxidize rapidly (Augustin and Sanguansri, 2003). It was demonstrated that the consumption of food enriched with microencapsulated fish oil obtained by emulsion spray-drying was as effective as the daily intake of fish oil gelatine capsules in meeting the dietary requirements of this omega-3 long-chain fatty acid (Wallace, et. al., 2000). Entrapment of retinoids in the lipid bilayers of multilamellar liposomes has been shown to improve their stability in the presence of oxygen. Incorporation of retinol into liposomes at a retinol: phospholipid ratio of 1:100 is up to 99% efficient, but liposomes offer little protection in acidic conditions or at temperatures at or above ambient. The process used to manufacture liposomes in this case used chloroform and methanol (Lee et al., 2002). Retinol shows greater affinity than retinyl palmitate for entrapment into liposomes (Singh and Das, 1998).

Probiotics

Due to their perceived health benefits, probiotic bacteria have been increasingly included in yoghurts and fermented milks during the past two decades. Most commonly they have been lactobacilli such as Lactobacillus acidophilus, and Bifidobacteria often referred to as 'bifidus' (Daly and Davis, 1998). Probiotics must have good technological properties so that it can be manufactured and incorporated into food products without losing viability and functionality or creating unpleasant flavours or textures; it must survive passage through the upper gastrointestinal (GI) tract and arrive alive at its site of action; and it must be able to function in the gut environment. The basic notion behind probiotics is pretty straightforward i.e. to restore the depleted ecology of the intestine with new, healthful bacteria.

Food substrate/diet is considered as one of the major factors in regulating colonization of micro-organisms in gastrointestinal tract. Food helps to buffer the bacteria through the stomach and may contain other functional ingredients that could interact with probiotics to alter their functionality. By increasing the amount of prebiotics in the diet, it is possible to increase and maintain healthy bacterial gut flora in the host (Sanders, 1998; Gibson et al., 2003). Better growth and survivability during food manufacturing and storage as well as in gastrointestinal tract, protection against acid, bile and gastrointestinal enzymes, adhesion to intestinal epithelium, antimicrobial properties and antibiotic resistance could be considered as factors that might be important in maintaining probiotic efficacy. Foods used for dissemination of probiotics are usually fermented foods even if probiotics also could be present in infant formula, fruit drinks, whey drinks and sweet milk. However, with microencapsulation technologies, the probiotics can become an important ingredient in the functional foods, expanding the probiotic application outside the pharmaceutical and supplement industries. Microencapsulation technologies have been developed and successfully applied using various matrices to protect the bacterial cells from the damage caused by the external environment (Del Piano et al., 2006). Probiotic products are usually standardised based on the presumption that culture viability is a reasonable measure of probiotic activity, thus the ability of the strain to attain high cell population is of primary importance. A concentration of approximately 10^7 cells/ ml at the time of consumption is considered functional (Shortt, 1999).

Gaps and Constraints in Developing Countries

Although, food products at subsidized prices are being routed to the masses through various government schemes in order to tackle the problem of malnutrition in the country, but still we are facing a number of challenges. In order to develop capabilities to meet future demands of functional foods, the following gaps need to be addressed in scientific perspective:

- Processes and technologies for production of health foods are either not available or are patented
- Indigenous equipments and pilot plants for production of health foods are presently not available
- Limited availability of health promoting foods in rural and remote areas of developing countries
- Majority of the presently available products for health promotion are diet supplements

- Newly developed varieties like vitamin A rich rice, high lycopene tomatoes have not been properly processed and utilized.
- A part of population suffers from gluten allergy and lactose intolerance
- Unavailability of low cost (economic) protein and calorie dense food products suitable to children.
- Lack of cost effective health foods
- Much of the health foods do not meet local/area specific requirements because of food habits.
- No defined regulatory framework for manufacturing, labelling, testing and distribution of health foods in various nations
- Clinical testing and safety issues of functional foods

Major Health Issues

The review of the situation of developing nations reveals that the major health problems presently prevailing in these countries may be broadly classified into three groups as follows:

a. Due to under nutrition
    i. Protein Energy Malnutrition (PEM)
    ii. Anaemia
    iii. Vitamin A deficiency
    iv. Calcium deficiency
    v. Folic acid deficiency
b. Due to over nutrition / lifestyle related reasons (mostly in urban population)
    i. Hypertension (High blood pressure)
    ii. Diabetes
    iii. Heart diseases
    iv. High cholesterol
    v. Obesity
c. Due to food allergies and miscellaneous reasons
    i. Allergy

ii. Lactose intolerance
iii. Gluten intolerance
iv. Chronic Lung diseases
v. Cancer/ tumour
vi. Asthma
vii. Skin infections

Thrust Areas for R and D

Research on nutrition and foods in some of the developing nations e.g. India has been largely confined to balanced diet and fortification. Further, these supplements are not available in rural areas and are also not a part of staple diet. Under these circumstances, there is a need to develop foods for health promotion from locally available raw materials, such as cereals, millets, pulses, oilseeds, fruits, vegetables, milk, meat, fish and their byproducts. Further, these health foods should be cheaper and acceptable in traditional staple diets. These foods when routed through public distribution system shall reach the masses and providing a long term solution to the prevailing problem of malnutrition and food related health issues. Systematic RandD efforts are, therefore, required to develop health promoting foods for all segments of the society. Some of these health foods need to be developed to overcome micronutrient deficiency for the target population, such as infants, children, women and elderly people. It is equally important to develop health foods for normal adult population as well. It is widely recognized that the long term solution to malnutrition must rest on the provision of adequate quantities of all nutritional components in the traditionally acceptable staple food itself. Prior to promoting the developed health foods, it is of utmost importance to conduct clinical trials to ascertain safety and efficacy, and also establish appropriate regulatory framework in the country. All these issues have been discussed in paper.

1. Integration of biological, nutritional science and engineering tools for cost effective production of
    - protein enriched traditional foods
    - micronutrient enriched existing foods
    - foods for specific groups (weaning, women, geriatric population)
    - foods for high risk groups (hypertension, diabetes, cardiac diseases, obesity etc.)

- foods for mitigation of allergies (lactose, gluten intolerance)

2. Utilization of locally available un-exploited commodities and by-products of food processing industries for developing nutraceutical based foods from
   - bioactive compounds of fruits and vegetable residues
   - bioactive compounds of food grains (millets etc.)
   - functional compounds from oilseed meals
   - bioactive compounds of fish, meat and dairy by-products
3. Establishment of pilot plants for
   - Fruits and vegetable based health foods
   - Extrusion based health foods
   - Traditional and fermented health foods
   - Meat based health foods
   - Fish/ marine animals based health foods
4. Regulatory framework for health foods
5. Outreach of developed health foods at affordable cost to the masses.

Development of specific health foods for various regions of the country would lead to their high acceptance in different demographic areas. Such cost effective health foods when routed through public distribution system or normal marketing channel in the country would have high impact on the health status of people. The cost effectiveness of the products would ensure their regular purchase and consumption by a common man of the country.

Safety and Regulation

India is one of the countries, where the market of health foods and supplements is growing enormously. Nutraceuticals and dietary supplements are sold in India under the name of Fast Moving Healthcare Goods (FMHG). Of course India has passed Food Safety and Standard Act in a year 2006 – a modern integrated food law to serve as a single reference point in relation to regulation of food products including nutraceutical dietary supplements and functional food. The food safety and standard act has needed to still make considerably substantive with infrastructure and appropriate stewardship for it to match with international standards of U.S. and Europe.

Conclusion

Research and development efforts in the area of functional foods are largely done in fragmented manner in various developing nations and to the laboratory scale only. Testing, clinical trials, safety issues, certification etc are usually lacking for these products. Further, the developed products and process may not be suitable for mass scale production to cater the need of large population. Our approach should be to identify the food habits of a particular area and to formulate nutritionally balanced food products keeping in view their food habits. So that developed food had higher acceptability in particular area and masses significantly contribute in the nutritional security of the masses. There are various unexplored promising technological packages for health foods with better nutritional and health attributes, which can give us a way of development of low cost health foods and combat the health related problems in an effective manner.

References

1. Augustin, M. A. and Sanguansri, L. (2003). Polyunsaturated fatty acids: delivery, innovation and incorporation in foods. Food Australia, 55:294-296.
2. Daly, C. and Davis, R. (1998). The biotechnology of lactic acid bacteria with emphasis on applications in food safety and human health. Agriculture Food Science, 7:219–250.
3. Del Piano, M. Morelli, L. Strozzi, G. P. Allesina, S. Barba, M. and Deidda, F. (2006). Probiotics: From research to consumer. Digestive and Liver Diseases, 38:248–255.
4. Gibson, G.R. Rastall, R.A. and Fuller, R. (2003). The Health Benefits of Probiotics and
5. Graveland-Bikker, J.F. and de Kruif, C.G. (2006) Unique milk protein-based nanotubes: food and nanotechnology meet. Trends in Food Science and Technology 175, pp.196-203.
6. Hirotsuka, M. Taniguchi, H. Narita, H. and Kito, M. (1984). Calcium fortification of soy milk with calcium-lecithin liposome system. Journal of Food Science, 49:1111-1112, 1127.
7. Ismail, A. (2006) India: The Land of Opportunity. Functional Foods and Nutraceuticals. Japan Development Institute (2006) Functional Food: Country Case Studies of China and India Draft internal report prepared for the World Bank.
8. Karel, M. (1990). Encapsulation and controlled release of food components. In: Biotechnology and Food Process Engineering Eds. G Schwartzberg and MA Rao, Marcel Dekker, NY, pp. 277-294.
9. Lee, S. C. Yuk, H. G. Lee, D. H. Lee, K. E. Hwang, Y. I. and Ludescher, R. D. (2002). Stabilization of retinol through incorporation into liposomes. Journal of Biochemistry and Molecular Biology, 35(4):358-363.
10. Pang, G. Xie, J. Chena, Q. and Hua, Z. (2012). How functional foods play critical roles in human health. Food Science and Human Wellness, 1:26–60.
11. Prebiotics. In R. Fuller an G. Perdigon (Eds.), Gut flora, nutrition, immunity and health (1st ed., pp. 52–76). Oxford: Blackwell Publishing.
12. Reineccius, G.A. (1995) Liposomes for controlled release in the food industries. In Encapsulation and Controlled Release of Food Ingredients. ACS Symposium Series, 590:113-131.

13. Sanders, M. A. (1998). Overview of functional foods: Emphasis on probiotic bacteria. International Dairy Journal, 8:341–347.
14. Shortt, C. (1999). The probiotic century: historical and current perspectives. Trends Food Science and Technology 10: 411– 417.
15. Singh, A. K. and Das, J. (1998). Liposome encapsulated vitamin A compounds exhibit greater stability and diminished toxicity. Biophysical Chemistry, 73(1e2),155-162.
16. Watson, J. (2006) Middle Class India Joins Global Organic Food Wave. Terra daily (UPI); Feb 28.
17. Weiss, J. Takhistov, P. and McClements, D.J. (2006). Functional Materials in Food Nanotechnology. Journal of Food Science 719:R107-R116.

CHAPTER 19

Role of Nanotechnology in Development of Functional Food

Jaspreet Kaur and Amarjeet Kaur

Abstract

Nanotechnology deals with understanding and control of matter at 1-100 nm dimensions. Nano-sized materials have been part of everyday life in the form of biological entities and processes. Many of the structural building elements in foods are colloids in nature and are built up as a result of self assembly of nano-sized molecules in particles or at interfaces. The most important food raw materials-proteins, starches and fats undergo structural changes at the nanometer and micrometer scales during normal food processing. Unique phenomena occurring at this length scale are being exploited for novel food applications. Functional foods are products, in food or drink form, which influence specific functions in the body and thereby offer benefits for health, well-being, or performance beyond their regular nutritional value. Nanotechnology promises to provide means of altering and manipulating food products to efficiently deliver nutrients like proteins and antioxidants for precisely targeted nutritional and health benefits. Nanotechnology applications for functional food ingredients and additives offer range of benefits to the consumer in terms of innovative, tasteful and healthy food products. Nanoencapsulation systems aid smart protection and delivery of bioactives so as to ensure their stability, taste, colour, safety and bioavailability. The advent of nanotechnology has opened up innovations in food packaging with improved stability and barrier properties. However, need exists to develop regulatory systems for managing any risks associated with functional nanofoods and the use of nanotechnologies in the food industry.

Keywords: Nanotechnology, Nanomaterials, Functional foods, Bioactives, Nanopackaging

Introduction

Public interest and consumer demand in nutraceuticals and functional foods is growing due to the current population and health trends. These are the principal reasons for the growth of the functional food market. Hence the focus of current food research is at identifying properties and potential applications of nutraceutical substances (Williams et al., 2006). There is need for maintaining sustainable food production and at the same time delivering high quality food products with an added functionality to prevent life-style related diseases such as cancer, obesity, diabetes, heart disease and stroke. Functional foods that contain bioactive components may provide desirable health benefits beyond basic nutrition. Nanotechnology is an upcoming technology that has the potential to revolutionize the food industry. The word "nano" means dwarf in Greek. Nanotechnology deals with the understanding and control of matter at dimensions of roughly 1-100 nanometers, where unique phenomena enable novel applications. The science, engineering and technology components of nanotechnology involve imaging, measuring, modeling, and manipulating matter at this length scale (Weiss et al., 2006). These can be applied to develop nanoscale materials, controlled delivery systems, and contaminant detection and to create nanodevices. For food, it would mean better texture, better flavour, contamination detection, pesticide reduction, green packaging and so on.

Nanotechnology offers tools for entrapping or immobilizing bio-functional compounds and delivering these to the desired site of action. However, these developments are still in their infancy. The underlying text emphasizes some of the current and potential applications of nanotechnology in the important field of functional food development.

Nanotechnology in Nature

The important structural elements in foods exist at the nano-scale. These include proteins, starches and fats which undergo structural changes during food processing. Starch granules expand when heated and hydrated releasing biopolymers that can be recrystallized into nano-sized structures (e.g., recrystallized amylose regions may be about 10–20 nm). Similarly, monoglycerides self-assemble at the nanoscale level into triglycerides, crystallites (10–100 nm), followed by arrangement into large clusters, then flocs and fat crystal networks. These fat crystal networks give foods spreadability, texture, and other similar properties (Aguilera, 2009).

Nanotechnology in food science and technology

Food production may benefit from nanotechnology in four major areas in: development of new functional materials, microscale and nanoscale processing, product development, and methods and instrumentation design for improved food safety and biosecurity (Weiss et al., 2006). Since foods are complex biological systems, they undergo a variety of postharvest and processing induced modifications that affect the biological and biochemical functionality of the system. A food product which has been cultivated, produced, processed or packaged using nanotechnology techniques or tools or to which manufactured nanomaterials have been added is called nanofood (Joseph and Morrison, 2006). The benefits of nanofood include health-promoting additives, longer shelf-lives or new flavor varieties. Examples include structured lipids replacing hydrogenated lipids as in olive oil where monoglyceride systems have been incorporated with a view to formulating shelf-stable and spreadable products.

Nanomaterials

Food and Drug Administration (FDA) defined nanomaterials as "particles with dimensions less than micrometer scale that exhibit unique properties" (Miller and Senjen, 2008). Systems with structural features on the nanoscale have physical, chemical and biological properties substantially different from their macroscopic counterparts. These material properties of foods at the nanoscale level strongly influence their bioavailability and nutritional value. Functional ingredients are rarely utilized directly in their pure form. Instead, they are often incorporated into some form of delivery system that protects the functional ingredient from degradation during processing and storage. Hence, they should be compatible with the components in the system (Weiss et al., 2006).

Nanoparticles

Nanoparticles used for health food applications can be broadly classified as follows:-

Metal and metal oxides: These include engineered nanoparticles (ENPs) of metals such as silver, iron, calcium, magnesium and non-metals like selenium and silicates. They are used in health foods and for food packaging. Nano silver has been used in health food supplements for natural antimicrobial action (Chaudhry and Groves, 2010). Antibacterial wheat flour contains 100-1000 ppm of silver nanaparticles of about 10 nm (Park, 2005). Similarly, it has been suggested that nano-titanium dioxide may be used for coating of confectionary products (Chaudhry and Groves, 2010).

Surface functionalized nanoparticles: These are second generation nanoparticles that add certain functionality to the matrix. Surface functionalized nanomaterials may bind to the polymer matrix to offer superior packaging attributes such as mechanical strength or act as barrier against movement of gases, volatile components or moisture (Chaudhry and Groves, 2010).

Biopolymeric nanoparticles: Particles in the nanometer-sized range can often be produced using food-grade biopolymers such as proteins or polysaccharides (Gupta and Gupta, 2005). Examples include polylatic acid (PLA) and polyethylene glycol (PEG) (Weiss et al., 2006).

Nanoencapsulation

Nanoencapsulation is incorporation of ingredients in small vesicles or walled material with nano (or submicron) sizes. Nanoencapsulation systems could be in form of association colloids such as surfactant micelles, vesicles, bilayers, reverse micelles, and liquid crystals. They encapsulate and deliver polar, nonpolar, and/or amphiphilic functional ingredients (Flanagan and Singh, 2006). Functional ingredients (for example, drugs, vitamins, antimicrobials, antioxidants, flavorings, colorants, and preservatives) are essential components of health-care products and foods. Besides serving as vehicles for carrying the functional ingredient to desired site of action, nanocapsules protect the functional ingredient from chemical/biological degradation during processing, storage and utilization. This is similar to microencapsulation of nutraceuticals. Nutraceutical carriers as microcapsules have been prepared by using spray drying, complexation and liposome entrapment procedures to obtain stable granules which can be used in food applications (Nasirullah et al., 2011). They also aid in controlled release of the functional ingredient, such as the release rate or the specific environmental conditions that trigger release (for example, pH, ionic strength, or temperature) (Weiss et al., 2006).

Nanofibers

These are fibers with diameters of 10-1000 nm that are developed using electrospinning process. Electrospun polymer fibers exhibit unusual functionalities with respect to their mechanical, electrical and thermal properties. Because of the large surface-to-volume ratio, nanofibers may have some novel applications within the food industry for producing materials with new or improved properties. Nanofibres can be used to strengthen food packaging material and as nanostructured and microstructured support structure for bacterial cultures (Weiss et al., 2006).

Nanotubes

Self-assembled nanotubes from hydrolyzed milk protein á-lactalbumin can act as potential raw carrier for nutrients and supplements. Carbon nanotubes can be used in food packaging with improved mechanical properties. It has been recently discovered that they exhibit powerful antimicrobial effects against E. coli (Lanzon et al., 2009).

Processes for nanomaterial production

Although "bottom-up" and "top down" approaches are used in nanomaterials production, the commercial scale production of nanomaterials primarily involves the "top-down" methods. This involves physical reduction to nanometre size range by employing processes such as grinding, milling, and lithography. "Bottom-up" approach is based upon growing larger structures from smaller particles such as atom by atom. Several techniques in this approach have been suggested by Sanguansri and Augustin (2006). Dry milling is the most common form of size reduction method.

Scope of nanotechnology in functional foods

Naturally occurring nanostructures in foods can be manipulated to improve the functional behavior of the food (Sekhon, 2010). Techniques of nanotechnology may help in improved delivery of micronutrients and bioactive food components through improved nutrient delivery viz. stability (i.e., against heat, pH and oxidation during food processing), better taste and color (i.e., avoiding unpleasant tastes or colors), safety, and increased bioavailability (Yada, 2010). This would enable ease in handling by transforming volatile liquids into powders, enhanced stability by isolating active ingredient from other ingredients, protection against oxidation, retention of volatile ingredients, taste masking, moisture/ph/heat triggered controlled release, consecutive delivery of multiple active ingredients, long lasting organoleptic perception and enhanced bioavailability and efficacy. Nanoencapsulated designer probiotic bacterial preparations may act as de novo vaccines, with the capability of modulating immune responses (Vidhyalakshmi et al., 2009). Some general applications of nanotechnology to functional (food) products are given in Table-1.

Lipophyllic bioactives

Lipophilic bioactives from natural sources, including phytosterols, antimicrobial, antioxidants, ω-3 fatty acids, flavours and numerous other components, are widely utilized as functional ingredients in food industry. However, most of these bioactives are almost insoluble in water and unstable at a specific environmental stimulus such as light, oxygen and temperature during manufacture, storage, transport and utilization. Nanoemulsions can encapsulate a diverse range of lipophilic compounds, including bioactive lipids, flavors, antimicrobials, antioxidants, and drugs (Chen et al., 2006) leading to the increase of their bioactivity, desirability, and palatability (McClements et al., 2007). Some bioactives along with their nanotechnique and functional benefit are given in Table 1.

Table 1: Applications of nanotechnology for functional foods

Product	Nanotechnology/Process used	Likely Benefits
Antibacterial wheat flour (Park, 2005)	Silver nanoparticles	Inhibit growth of many species of bacteria
Coated confectionary products (Chaudhry and Groves 2010)	Nano-titanium dioxide	Improved barrier properties
Edible antibacterial nano coating (Miller and Sejnon, 2008)	Silver nanoparticles, nanomagnesium oxide, nano copper oxide	Inhibit growth of many species of bacteria
Encapsulated β-Carotene	Nanoemulsions	Increased availability of the encapsulated component
Nutritional drink	Iron nanoparticles	Increased reactivity and bioavailability
Slim chocolate shake	Nanoclusters	Low-calorie
Poly(methyl methacrylate) nanoparticles (Kwon et al., 2002).	Microfluidization in combination with solvent evaporation.	Encapsulation of coenzyme Q_{10}
Structured olive lipids (Aguilera, 2009)	Processed nanostructures	Spreadable, Shelf-stable
Bioactives		
Carotenoids (Stringham and Hammond, 2005)	Oil-in-Water nanoemulsions	Contribute to red to yellow colours,antioxidative activity
Coenzyme Q_{10} (Thanatuksorn et al., 2009)	Molecular complexation, emulsification and liposomal systems	Essential for cell functions, antioxidant
Phytosterol (Cercaci et al., 2007)	Oil-in-Water nanoemulsions	Inhibit cholesterol absorption
ω-3 PUFA	Encapsulation	Reduce risks of cardiovascular diseases, immune-related and mental disorders

Nanopackaging

Innovative food packaging methods have ensured fresh-like foods with longer shelf-life. Nanotechnology may transform the nature of food packaging materials in future. Certain nanoscale innovations in food packaging would be in the forms of detection of pathogens, smart and active packaging, and barrier and mechanical properties. Technologies such as active packaging are designed to control various critical parameters responsible for the quality and shelf life of fresh foods (Singh et al. 2011). Through nanotechnology, it would be possible to incorporate "smart" nano materials which would respond to environmental conditions, repair themselves, alert a consumer to contamination and the presence of pathogens. Antimicrobial food packaging can be achieved through incorporation of nanomaterials such as nano silver, nano magnesium oxide, nano copper oxide, nano titanium dioxide and carbon nanotubes (Miller and Sejnon, 2008). Besides, nanolaminates and nanocomposites provide superior barrier properties.

Nanolaminates

A nanolaminate consists of two or more layers of material with nanometer dimensions that are physically or chemically bonded to each other. One of the most powerful methods of nanolaminate development is based on the layer-by-layer deposition technique. In this case, charged surfaces are coated with interfacial films consisting of multiple nanolayers of different materials. Alternatively, solutions containing the adsorbing substances could be sprayed onto the surface of the object (Cagri et al., 2004). Edible coatings and films are currently used on a wide variety of foods such as fruits, vegetables, meats, chocolate, candies, bakery products and French fries (Rhim, 2004). These coatings or films serve as moisture, lipid, and gas barriers as well as improve textural properties of foods or serve as carriers of functional agents such as colors, flavors, antioxidants, nutrients, and antimicrobials.

Nanocomposites

Bionanocomposites are hybrid nanostructured materials with improved mechanical, thermal and gas barrier properties. The use of bionanocomposites for food packaging not only protects the food and increases its shelf life but can also be considered a more environmentally friendly solution because it reduces the requirement to use plastics as packaging materials.

Polymer nanocomposites (PNCs) are created by dispersing an inert, nanoscale filler throughout a polymeric matrix. Polysaccharide/clay nanocomposites contain exfoliated clay layers having substantially reduced water-vapor permeability. They greatly improve the overall mechanical strength. These include nanocomposites from starch and amorphous poly beta-hydroxy octanoate from starch and tunicin whiskers (Mathew and Dufresne, 2002).

Safety Concerns and Regulations

Despite rapid developments in food nanotechnology, knowledge about the occurrence, fate and toxicity of nanoparticles is scarce. Once the engineered nanoparticles pass the gastro-intestinal epithelium and end up in the blood circulation, they can interact with various blood components (Cagri et al., 2004). Currently, there are no special regulations for the application or utilization of nanotechnology in foods. According to the FDA, a new substance is one that has been altered chemically and has not been marketed previously. In the case of nanoceuticals, the substances are not chemically altered; they are just reduced to a microscale or nanoscale. Recently, European Food Safety Authority published a draft scientific opinion on risk assessment of nanotechnology applications for food and feed (EFSA, 2011). This document guides the general public on matters of risk assessment in nanofoods. Ingredients in nanoparticles must undergo a full safety assessment by the relevant scientific advisory association before these are permitted to be used in the food products. Also, the ingredient lists of food products should disclose the fact that nanoparticle materials have been added. Further, release of nanoparticles and nanotubes into the environment must be avoided as far as possible. Hence, a balanced approach is required to prevent any perceived risk and provide maximum benefits from nano-functional foods (Cagri et al., 2004).

Conclusion

Development of novel functional foods is an important application of nanotechnology. Different types of lipid and protein based nanoencapsulation systems can be used to protect and deliver nutraceuticals. Nanostructured materials allow manipulating or self-assembling of active ingredients. Product safety concerns need to be assessed and proper regulations need to be in place before launching of functional nanofood into the market. Applications of nanotechnology in the area of functional foods are currently at an elementary stage but have enormous potential for the future.

References

1. Aguilera, J.M. (2009). Applications of nanosciences to nutrients and foods, in Pray L. and Yaktine A. (Eds) Nanotechnology in Food Products: Workshop Summary, National Academy of Sciences, Washington DC, USA, pp. 22-36.
2. Cagri, A. Ustunol, Z. and Ryser, E.T. (2004). Antimicrobial edible films and coatings, Journal of Food Properties, 67:833-848.
3. Cercaci, L. Rodriguez-Estrada, M.T. Lercker, G. and Decker, E.A. (2007). Phytosterol oxidation in oil-in-water emulsions and bulk oil, Food Chemistry, 102(1):161-167.
4. Chaudhry, Q. and Groves, K. (2010). Nanotechnology applications for food ingredients, additives and supplements, in Chaudhury, Q. Castle, L. and Watkins, R. (Eds.) Nanotechnologies in food, Royal Society of Chemistry, Cambridge, UK.
5. EFSA (2011). Public consultation on: Guidance on risk assessment concerning potential risks arising from applications of nanoscience and nanotechnologies to food and feed, available at www.efsa.europa.eu/en/consultations/call/scafl110114.htm (Accessed 14 June 2013).
6. Flanagan, J. and Singh, H. (2006). Microemulsions: a potential delivery system for bioactives in food, Criticla Review in Food Science and Nutrition, 46(3):221–237.

7. Joseph, T. and Morrison M. (2006). Nanotechnology in Agriculture Food: A Nanoforum report", available at www.nanoforum.org (accessed 25 December 2012).
8. Kwon, S.S. Nam, Y.S. Lee, J.S. Ku, B.S. Han, S.H. Lee, J.Y. and Chang, I.S. (2002). Preparation and characterisation of coenzyme Q10-loaded PMMA nanoparticles bay a new emulsification process based on microfluidisation, Colloids Surface A., 210:95-104.
9. Lanzon, N. Kahl, J. and Ploeger, A. (2009). Nanotechnology in the context of organic food processing" Wissenschaftstagung Ökologischer Landbau, Zürich, 2009, Feb 11–13, available at: http://orgprints.
10. org/14180/1/Lanzon_14180.pdf. (Accessed 12 April 2012).
11. Mathew, A.P. and Dufresne, A. (2002). Morphological investigation of nanocomposites from sorbitol plasticized starch and tunicin whiskers, Biomacromolecules, 3(3):609-617.
12. McClements, D.J. Decker, E.A. and Weiss, J. (2007). Emulsion-based delivery systems for lipophilic bioactive components, Journal of. Food Science, 72 (8)109-124.
13. Miller, G and Senjen, R. (2008). Out of the laboratory and on to our plates -Nanotechnology in food and agriculture", available at: http://www.foeeurope.org/activities /nanotechnology/ Documents/Nano_food_report. PDF (accessed 20 June 2013).
14. Nasirullah, K, P. and Shariff, R. (2011). Development of nutraceutical carriers for functional food applications, Nutrition andFood Science, 41(1):34-43.
15. Park, K.H. (2005). Preparation method of antibacterial wheat flour particles by using silver nanoparticles, Korean International Property Office (KIPO), Publication number/ date 1020050101529A/24/10/2005.
16. Rhim, J.W. (2004). Increase in water vapor barrier property of biopolymer-based edible films and coatings by compositing with lipid materials, Journal of Food Science and Biotechnology, 13:528- 535.
17. Sekhon, B.S. (2010). Food nanotechnology – an overview, Nanotechnology Science and Applications, 3:1-15.
18. Singh, H. (2011). "Nanotechnology and the development of novel dairy foods", in Souvenir: International Conference on Functional Dairy Foods, November, 16-19, 2011, Karnal, India.
19. Stringham, J.M. and Hammond, B.R. (2005). Dietary lutein and zeaxanthin: possible effects on visual function, Nutritional Review, Vol. 63 No. 2:59-64.
20. Thanatuksorn, P. Kawai, K. Hayakawa, M. Hayashi M. and Kajiwara K. (2009), "Improvement of the oral bioavailability of coenzyme Q_{10} by emulsification with fats and emulsifiers used in the food industry", Lebensm.-Wiss. Technol. Food Science and Technology, Vol. 42:285-390.
21. Vidhyalakshmi, R. Bhakyaraj, R. and Subhasree R.S. (2009). Encapsulation: The future of probiotics– A review, Advanced Biological Research, Vol. 3 No. 3-4:96-103.
22. Weiss, J. Takhistov P. and Mcclements D.J. (2006). Functional materials in food nanotechnology, J. Food Sci., Vol. 71 No. 9:107-116.
23. Williams, M., Pehu E. and Ragasa C. (2006). "Functional Foods: Opportunities and Challenges for Developing Countries", Agricultural and Rural Development Notes, World Bank.
24. Yada, R. (2010). Nanotechnology: a new frontier in foods, food packaging, and nutrient delivery, in Pray L. and Yaktine A. (Eds). Nanotechnology in Food Products: Workshop Summary, National Academy of Sciences. Washington DC, USA, pp. 8-19.

CHAPTER 20

PCR: Advanced Tool for Functional Foods

Aasima Rafiq, Savita Sharma and Baljit Singh

Abstract

Consumer now-a-days relies on manufacturing and processing standards, particularly to know what ingredients are present, due to dietary, nutritional requirements, or medical conditions (e.g., diabetes or allergies). Modified PCR and combination of PCR with other analytical tools have been successfully used for analysis of various parameters related to functional foods such as detection and quantification of beneficial as well as food borne pathogenic microorganisms, food allergens, raw material source, bioactive compounds, genetically modified foods. Multiplex PCR method has been found to be an efficient tool for simple, rapid, and reliable identification of Lactobacillus and Bifidobacterium in probiotic products. Food allergens like gluten, peanut, hazelnut etc. could be easily detected in foods using PCR, even if present in small quantities. BOX-PCR and PCR-DGGE analysis have been used for identification of bacterial diversity of a naturally fermented functional food (Enzyme). HPLC combined with real time –PCR were successfully used to quantify and analyse compounds of secondary metabolites like phenylpropanoid amides, which represent an important class of nutraceutical having strong antioxidant and chemotherapeutic effects in various transgenic tomato tissues. Detection of active compounds in natural plant products were successfully achieved using PCR techniques. In general PCR have gained more importance as it has high sensitivity, improved precision, wide dynamic range for quantification, genotype based and is quick method of detection and quality control of foods than traditional culture based methods.

Keywords: Functional foods, probiotic, antioxidant effects, bioactive compounds, genotype

Introduction

International Food Information Council defined functional food as a food which provides a health benefits beyond basic nutrition. Functional foods, according to generally accepted definition, is a food including whole foods and fortified, enriched, or enhanced foods or dietary components that may reduce the risk of chronic disease and provide a health and physiological benefit beyond the traditional nutrients it contains (Elisabeth, 1999). Food is considered as best, if it has maximum nutrient-density. Nutrient-density is a measure of the amount of nutrients a food contains in comparison with the number of calories it has.

Applied molecular microbiology is a fast-moving area. One of the branches of this discipline is involved in the development of molecular methods for the identification and monitoring of microorganisms in natural ecosystems. Methods in molecular microbiology have become a valid support to traditional techniques. During the last years, bacterial identification based on molecular methods, especially those including the sequencing of genes coding for ribosomal 16SrDNA, has become a very important tool in studying bacterial communities in environmental samples (Head et al., 1998)

Polymerase chain reaction (PCR) is a biochemical technique in molecular biology which involves amplification of a single or a few copies of a piece of DNA across several orders of magnitude, generating thousands to millions of copies of a particular DNA sequence. Simply states it is called "molecular photocopying,". It is a fast and inexpensive technique used to "amplify" - copy - small segments of DNA. Because significant amounts of a sample of DNA for molecular and genetic analyses, studies of isolated pieces of DNA are nearly impossible without PCR amplification.

PCR represented a revolution in biological techniques; it was first developed in 1983 by Kary Mullis (Saiki et al., 1985) and he was given Nobel Prize for his work on the use and development of PCR in 1993. It allows much easier characterization and comparisons of genetic material from different individuals and organisms.

Food safety and quality programs are of utmost importance while preparing functional foods in order to meet regulatory, customer, and certification agency standards. Priorities are Product Specification Verification, Preproduction APT testing, Good Manufacturing Practices (GMP), Sanitation, Allergens Programs and Micro-biological Testing.

Principle of PCR

The principle of PCR is based on thermal cycling, which exploits the thermodynamics of nucleic-acid interactions. The vast majority of PCR machines now use thermal cycling, i.e., alternately heating and cooling of the PCR samples following a defined series of temperature steps. These thermal cycling steps are necessary first to physically separate the two strands in a dsDNA double helix, in the high-temperature process known as DNA melting. At lower temperatures, each strand is then used as a template in dsDNA synthesis, aided by the enzyme DNA polymerase, for the synthesis of the new, complementary, DNA strands. A typical thermo cycler consists of metal thermal block with holes for the tubes containing PCR reaction mixture. Thermal cycler raises and lowers the temperature of block in preprogrammed steps.

Components and Reagents

A basic PCR set up requires several components and reagents (Joseph and David, 2001). These components include:

- **DNA template** that contain the DNA region (target) to be amplified.
- **Two primers** that are complementary to the 3' (three prime) ends of each of the sense and anti-sense strand of the DNA target commonly called forward and reverse primers.
- **Taq polymerase** or another DNA polymerase with a temperature optimum at around 70 °C. DNA polymerase was originally isolated from the bacterium *Thermus aquaticus*, by Thomas Brock in 1965 (Brock and Boylen, 1973; Chien et al., 1976). *T.aquaticus* is a bacterium that lives in hot springs and hydrothermal vents, and it has a DNA polymerase enzyme that can withstand the protein-denaturing conditions that are required during PCR (Chien *et al* 1976 and Saiki *et al* 1988).
- **Deoxynucleoside triphosphates** (dNTPs, sometimes called deoxynucleotid triphosphates nucleotides containing triphosphate groups), the building-blocks from which the DNA polymerase synthesizes a new DNA strand.
- **Buffer solution** providing a suitable chemical environment for optimum activity and stability of the DNA polymerase.
- **Monovalent and Divalent cations** (magnesium or manganese ions),the DNA polymerase requires a catalyst in the form of divalent cations, as either the magnesium (Mg^{2+}) or manganese (Mn^{2+}) cations. These cations also serve as a co-factor to help stabilises the two ssDNA strands. The

usual concentration of Mg^{2+} in the PCR reaction is approximately 2.5 μM. The right concentration of cations is critical, because at higher concentrations they can promote greater promiscuity of the Taq polymerase. Generally Mg^{2+} is used, but Mn^{2+} can be utilized for PCR-mediated DNA mutagenesis, as higher Mn^{2+} concentration increases the error rate during DNA synthesis (Pavlov et al., 2004). Potassium ions are used as monovalent cation.

- **Enzymes** helicase (untwists the two parallel DNA strands), topoisomerase (relieves the stress of this twisting), single strand binding or SSB proteins (binds to and stabilizes the unpaired DNA strands), primase.

Why PCR Based Detection?

- Isolation of Micro-organisms in pure culture from a complex mixture like clinical, environmental and food matrix is important for its detection.
- Conventional culture-dependent detection uses morphological and physiological properties (i.e phenotypic characterization) but micro-organisms can undergo morphological changes due to environmental stresses.
- The main advantage of microbial identification by genetic markers (PCR) is the relative stability of the genotype rather than the phenotype.

How it Works?

PCR is an iterative process, consisting of 3 elements; denaturation of the template by heat, annealing of the oligonucleotide primers to the single stranded target sequence(s), and extension of the annealed primers by a thermostable DNA polymerase. Sample preparation is primary and most important step for PCR.

Breaking of the cells using cell disruption method

↓

Removing membrane lipids by adding a detergent or surfactants

↓

Removing proteins by adding a protease

↓

Removing RNA by adding an RNase

↓

DNA purification and isolation

Fig. 1: Sample preparation for PCR analysis

To amplify a segment of DNA using PCR, the sample is heated to denature DNA, and separate into two pieces of single-stranded DNA. Next, an enzyme called "Taq polymerase" synthesizes - builds - two new strands of DNA, using the original strands as templates. This process results in the duplication of the original DNA, with each of the new molecules containing one old and one new strand of DNA. Then each of these strands can be used to create two new copies, and so on, and so on.

The cycle of denaturing and synthesizing new DNA is repeated as many as 30 or 40 times, leading to more than one billion exact copies of the original DNA segment.

The entire cycling process of PCR is automated and can be completed in just a few hours. It is directed by a machine called a thermocycler and finally detection is done using gel electrophoresis and ultra-violent visualisation or image analysis.

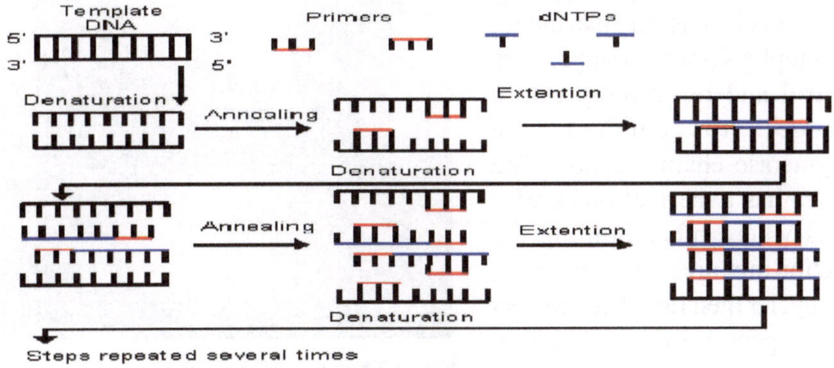

Fig. 2: General outline for Polymerase chain reaction (By courtesy of Anna Kaksonen)

Table 1: Modifications of PCR

Long PCR	Used to amplify DNA over the entire length up to 25kb of genomic DNA segments cloned.
Nested PCR	Involves two consecutive PCR reactions of 25 cycles. The first PCR uses primers external to the sequence of interest. The second PCR uses the product of the first PCR in conjunction with one or more nested primers to amplify the sequence within the region flanked by the initial set of primers.
Inverse PCR	Used to amplify DNA of unknown sequence that is adjacent to known DNA sequence.
Hot start PCR	Used to optimize the yield of the desired amplified product in PCR and simultaneously to suppress nonspecific amplification.

(Contd.)

Colony PCR	Screening of bacterial (E.Coli) or yeast clones for correct ligation or plasmid products.
Multiplex PCR	It is a variant of PCR which enabling simultaneous amplification of many targets of interest in one reaction by using more than one pair of primers. Multiplex PCR methods save considerable time and effort by decreasing the number of reactions required to assess the possible presence of target DNA in the food sample (Mafra et al., 2008).

Quantitative PCR or Real time PCR

Product amplification w.r.t time, which is compared with a standard DNA. Real-time PCR monitors the fluorescence emitted during the reaction as an indicator of amplicon production at each PCR cycle (in real time) as opposed to the endpoint detection.

PCR Machine

A thermal cycler (also called thermocycler) is an instrument that employs precise temperature control and rapid temperature changes to conduct the polymerase chain reaction. The device has a thermal block with holes where tubes holding the reaction mixtures can be inserted. The cycler then raises and lowers the temperature of the block in discrete, pre-programmed steps. These instruments are renowned for their reliability, accuracy, and user-friendly interfaces.

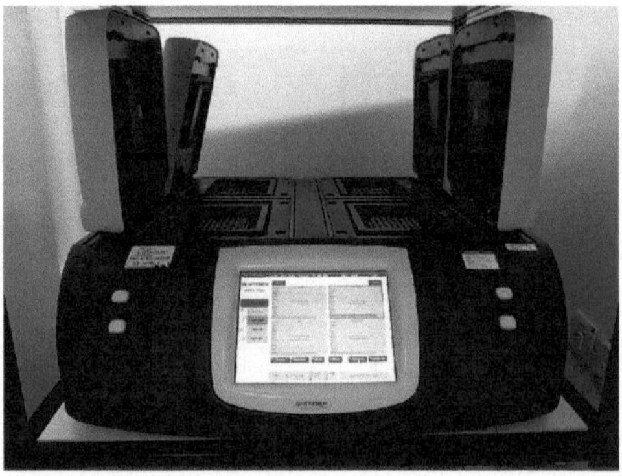

Fig. 3: Thermal cycler (G-STORM GS4 thermal cycler) [2]

Uses of Polymerase Chain Reaction

Once amplified, the DNA produced by PCR can be used in many different laboratory procedures. For example, most mapping techniques in the Human Genome Project (HGP) relied on PCR. PCR is also valuable in a number of laboratory and clinical techniques, including detection of bacteria or viruses, allergen detection, raw material detection, adultrations, DNA fingerprinting and diagnosis of genetic disorders.

PCR as A Tool for Functional Foods

Gluten free product

Gluten is a form of protein found in some grains. Gluten is defined as a fraction of wheat, rye, barley, or oat proteins which is insoluble in water and in 0.5 M NaCl. The individuals suffering from celiac sprue or gluten-sensitive enteropathy may have serious health problems after the consumption of even a small quantity of products containing gluten (Wieser and Koehler 2008). Celiac disease is a digestive and autoimmune disorder that results in damage to the lining of the small intestine when foods with gluten are eaten. The damage to the intestine makes it hard for the body to absorb nutrients, especially fat, calcium, iron, and folate. The only treatment in celiac disease is a lifelong avoidance of wheat, barley and rye (WBR) in the daily nutrition.

According to the Codex Alimentarius Commission of the Joint Food and Agricultural Organisation and the World Health Organisation of the United Nations, 100 ppm gliadin (10 mg gliadin/100g dry weight) is the maximum allowable limit in food labelled as 'gluten-free'. Various immunochemical methods like ELISA or lateral flow analysis where frequently used to detect the presence, or to determine the content, of gluten.

DNA analysis of the cereal by the polymerase chain reaction (PCR) method constitutes a suitable alternative method for detecting the presence of toxic protein (gluten) in a sample. It exceeds ELISA methods in sensitivity but its shortcoming includes presence of DNA in the sample and selective detection of its contamination with exactly one single kind of cereal. Other methods useful for determining gluten proteins are electrophoretic, chromatographic, or mass spectrometry based methods (Hulin *et al* 2008).

Naturally fermented functional food (Enzymes)

Functional microorganisms and health benefits represent a binomial with great potential for fermented functional foods. The health benefits of fermented functional foods are expressed either directly through the interactions of ingested live microorganisms with the host (probiotic effect) or indirectly as the result of the ingestion of microbial metabolites synthesized during fermentation (biogenic effect). PCR techniques were used to identify different type of microorganisms that grow during fermentation process and their behaviour with changing external conditions. Ampe et al., (1999) used PCR techniques to identify type of microorganisms, their possible biological role during fermentation of Mexican maize dough. Ben Omar and Ampe (2000) further investigated and identified the microbial species dominating at different steps of the process.

Linjiang et al., (2014) used culture-dependent Box elements PCR analysis and culture-independent Polymerase chain reaction denaturing gradient gel electrophoresis (PCR-DGGE) analysis simultaneously for study of bacterial diversity in a commercial available Enzyme product. BOX-PCR involve repetitive sequence elements in bacterial genomes to copy and identify the particular micro-organism where as PCR-DGGE is a form of electrophoresis which use chemical gradient to denature the sample as it moves across an acrylamide gel. It works by applying a small sample of DNA (or RNA) to an electrophoresis gel that contains a denaturing agent; the DNA spreads through the gel and can be analyzed for single components.

Bioactive compound detection

Bioactive compounds are extra nutritional constituents that typically occur in small quantities in foods. They are being intensively studied to evaluate their effects on health. A novel bioassay system based on RT-PCR has been established to search for biologically active compounds which can modulate the expression of genes encoding important regulatory proteins. Changes in target mRNA level in the treated cell cultures were quantitatively determined by competitive PCR using internal standard DNA. The compounds discovered using this bioassay will be useful tools to elucidate the mechanisms of cellular signaling guiding the expression of the target genes (Hakamatsuka and Tanaka, 1997).

PCR had been successfully used to detect the presence of bioactive compounds in foods like Phenylpropanoid amides which are plant-specific secondary metabolites, represent an important class of nutraceuticals, having strong antioxidant and chemotherapeutic effects. They have been studied by Kang et al., (2002), in a work using HPLC combined with real time polymerase chain reaction (RT-PCR), to quantify and analyse these compounds in transgenic tomato tissues.

Myxobacteria are considered to be a rich source of antibiotics that are rarely produced by other microorganisms. The presence of novel polyketide genes (PKS) in an organism is a good indicator of the production of novel bioactive compounds polyketide molecules. But the PKS genes in taxonomically diverse myxobacterial strains have been limited. In order to detect the presence of polyketide genes in myobacterial strains, PCR based techniques had been used (Hisayuki et al., 2008). Polyketides are structurally complex organic compounds that are often highly active biologically. Many pharmaceuticals are derived from or inspired by polyketides.

Another quantitative polymerase chain reaction (qPCR) assay was developed to detect bioactive compounds in various extracts that decreased *phosphoenolpyruvate carboxykinase* (Dmitry et al., 2007). In humans and other mammals, pyruvate carboxykinase (PCK) is a central enzyme in carbohydrate metabolism, helping to regulate the blood glucose level. The importance of , pyruvate carboxykinase to carbohydrate metabolism in humans is such that it has been suggested as a potential drug target in the treatment of non-insulin-dependent diabetes mellitus (Valera et al., 1994). Thus it can be seen that PCR can be used as a effective tool for detection of bioactive compound in natural foods.

Food pathogen detection

Bacteria are the source of many food poisoning cases, usually due to improper food handling. Some bacteria, in small amounts, are not harmful to most healthy adults because the human body is equipped to fight them off. However, the trouble begins when certain bacteria and other harmful pathogens multiply and spread, which can happen when food is mishandled or not cooked or cooled properly. Foods that are contaminated may not look, taste or smell any different from foods that are safe to eat. Symptoms of food poisoning can vary and develop as quickly as thirty minutes to a few hours or take longer (up to several days) after eating food that's been infected. As identified by the Centers for Disease Control and Prevention (CDC), eight known pathogens (bacteria, viruses and parasites that cause disease) account for the vast majority of foodborne illness, hospitalization and death in the United States: Salmonella, Clostridium perfringens, Campylobacter, Staphylococcus aureus, E. coli O157:H7, Listeria monocytogenes, Norovirus and Toxoplasma gondii

Enrica et al., (2009) developed a multiple PCR-based platform for the simultaneous detection of the widespread milk-associated pathogens *Salmonella* spp., *Listeria monocytogenes* and *Escherichia coli* O157. Salmonellosis is one of the leading causes of food borne bacterial enteritis in many countries (Tirado and Schmidt 2001; Centers for Disease Control and Prevention 2004a). Commercially available real-time PCR assay was used to detect *Salmonella* spp in different food commodities.

Allergen detection

PCR is efficient and reliable method for the detection of allergenic ingredients in functional foods like peanut allergen, halzelnut allergen etc. which is an important parameter for quality of these products. These allergies occur due to the ingestion of foods containing even very small quantities of these products and represent a severe threat to the health of sensitive individuals.

Peanut allergy causes an overreaction of the immune system and may lead to severe physical symptoms. Symptoms of peanut allergy are related to the action of immunoglobulin E (IgE) and other anaphylatoxin (complement peptides produced during clearance of pathogens by antibodies and phagocytic cells in an organism) (Hugli, 1986), which act to release histamine and other mediator substances from mast cells (degranulation). In addition to other effects, histamine induces vasodilation of arterioles and constriction of bronchioles in the lungs, also known as bronchospasm (constriction of the airways) and can lead to death.

Hazelnuts (Corylus avellana) are among the common tree nuts that lead to allergic reactions. They can be eaten as in-shell nuts, however, the majority is cracked, and the kernels are sold to candy makers, bakers and other food processors. Hazelnuts can be found in many food products including cookies, cakes, pastries, chocolates, confectionary products, ice cream, breakfast cereals, and bread. In addition, hazelnut oil may pose a threat to patients with hazelnut allergy, depending on the method of processing with oil undergone less processing at lower temperature tended to be more allergenic. Arlorio *et al* in 2007 developed two sensitive real-time PCR methods to quantify hazelnut in foods, one using the Sybr green dye and one based on hazelnut-specific Taqman probe.

Probiotic products

Food and Agriculture Organization (FAO) of the United Nations not the FDA defined "probiotics" as live microorganisms which, when administered in adequate amounts, confer a health benefit on the host. Most probiotic products are either foods or dietary supplements. A few probiotics are marketed as medical foods. Although fermented dairy products such as yogurt and kefir are typically associated with delivery of "beneficial cultures", the types of foods claiming to deliver probiotics has expanded to include granola and candy bars, frozen yogurt, cereal, juice and cookies. Products are required by law to be labeled in a truthful and not misleading fashion, safe for their intended use, and be manufactured under Good Manufacturing Practices. However, it is not practical for regulatory authorities to verify compliance with these laws for all products.

In food products, the probiotics used are primarily *Lactobacillus, Bifidobacterium* and *Streptococcus thermophilus*, and more recently *Bacillus*. Among the LAB, *Lactobacillus acidophilus, Lactobacillus rhamnosus, Bifidobacterium longum,* and *Bifidobacterium bifidum* are frequently used for probiotic products. In order to monitor this LAB in commercial probiotic products, a multiplex PCR method was developed and it has successfully detected four different LAB with high specificity in functional foods (Sul et al., 2007).

Yogurt is a primary food vehicle for probiotics. It is produced by fermentation of milk by two starter cultures, *Lactobacillus bulgaricus* and *Streptococcus thermophilus*. Lactic acid bacteria (LAB) are beneficial for the gastrointestinal tract and reinforce immunity in human health. Polymerase Chain Reaction-Denaturing Gradient Gel Electrophoresis (PCR-DGGE) technique had been used in identifying the microorganisms present in commercial probiotic yoghurts and lyophilised products was evaluated. Multiple competitive PCR-DGGE has also been used as a tool for quantifying and profiling defined mixed cultures of lactic acid bacteria during production of probiotics from complex polysaccharides (Pintado et al., 2003).

Genetically modified foods

The term GM (genetically modified foods) or GMOs (genetically-modified organisms) is most commonly used to refer to crop plants created for human or animal consumption using the latest molecular biology techniques. These plants have been modified in the laboratory to enhance desired traits such as increased resistance to herbicides or improved nutritional content. The enhancement of desired traits has traditionally been undertaken through breeding, but conventional plant breeding methods can be very time consuming and are often not very accurate. Genetic engineering, on the other hand, can create plants with the exact desired trait very rapidly and with great accuracy. But there are some criticisms against use of genetically modified foods. Environmental activists, religious organizations, public interest groups, professional associations and other scientists and government officials have all raised concerns about GM foods, and criticized agribusiness for pursuing profit without concern for potential hazards, and the government for failing to exercise adequate regulatory oversight.

Real-time PCR is currently considered the most precise and straight forward methodology for the quantification of specific nucleic acid sequences and, at the moment, the most widespread applications for genetically modified organisms (GMO) quantification (Weighardt 2003). Real-time PCR and Multiplex PCR techniques have achieved success in detection of GM foods like detection of soybean using specific RR soybean gene and GM organisms.

Partridge and Murphy, (2004) in UK tested 25 samples of food products containing unrefined soybean ingredients, ten tested positively for the presence of GM material, and eight out of ten GM-positive samples were either labelled as "GM free" and/or were labeled as "organic", both of which imply the absence of GM ingredients.

Conclusion

There is no doubt that the functional food industry is growing steadily worldwide. However, successful commercialization of new products is proving to be challenging especially due to the need of meeting quality standards. Quality control for functional foods is very important from technical and health point of view any sort of negligence can lead to even death of person. Polymerase chain reaction is a fast and inexpensive technique used to amplify small segments of DNA. Conventional culture-dependent detection uses morphological and physiological properties (i.e phenotypic characterization) but micro-organisms can undergo morphological changes due to environmental stresses. The main advantage of microbial identification by genetic markers (PCR) is the relative stability of the genotype rather than the phenotype. PCR has been successfully used for detection of food- borne pathogens such as Salmonella, Campylobacter, Bacillus cereus, pathogenic Escherichia coli (EHEC), food allergens, probiotic cultures, raw material source, genetically modified foods etc. Earlier PCR was used for qualitative detection, but now its being used for identification and quantification of micro-organisms.

References

1. Ampe, F. Sirvent, A. and Zakhia, N. (2001). Dynamics of the microbial community responsible for traditional sour cassava starch fermentation studied by denaturing gradient gel electrophoresis and quantitative rRNA hybridization. International Journal of Food Microbiology, 65:45–54.
2. Arlorio, M. Cereti, E. Coïsson, J. D. Travaglia, F. and Martelli, A. (2007). Detection ofhazelnut (Corylus spp.) in processed foods using real-time PCR. Food Control 18: 140–48.
3. Dmitry, G. Sithes, L. Yanxin, W. Debora, E. Slavko, K. David, R. Alexander, P. Zhong, W. William, T. C. and Ilya, R. (2007). Polyphenolic compounds from Artemisia dracunculus L. inhibit PEPCK gene expression and gluconeogenesis in an H4IIE hepatoma cell line. American Journal of Physiology Endocrinology Metab 293:E1503–E1510.
4. Elisabeth, S. (1999). What you need to know about health benefits of functional foods. Iowa State University. Adapted from "Focus on Functional Foods". Melinda Hemmelgarn, University of Missouri.
5. Enrica, O. Giulia, A. Giorgio, B. and Mauro, M. (2009). A new platform for Real-Time PCR detection of Salmonella spp., Listeria monocytogenes and Escherichia coli O157 in milk. Food Microbiology, 26: 615–622.
6. Hakamatsuka, T. and Tanaka, N. (1997). Screening for bioactive compounds targeting the cellular signal transduction pathway using an RT-PCR-based bioassay system, 20(4):464-466.
7. Head, I. M. Saunders, J. R. and Pickup, R. W. (1998). Microbial evolution,diversity, and ecology: a decade of ribosomal analysis of uncultivated microorganisms. Microbiology Ecology, 35:1–21.
8. Hisayuki, K. Ryosuke, F. Takashi, I. Daisuke, N. Koei, O. Daisuke, S. Makoto, O. and Shigeaki, H. (2008). PCR Detection ofType I Polyketide Synthase Genes in Myxobacteria. Applied and Environmental Microbiology, 74(17):5571-5574.

9. Hugli, T.E. (1986). Biochemistry and biology of anaphylatoxins. Complement 3 (3): 111–27. PMID 3542363.
10. Hulin, P. Dostalek, P. and Hochel, I. (2008). Metody stanoveni lepkovych bilkovin v potravinach. Chemicke listy, 102:327–337.
11. Kang, K. Lee, K.Sohn, S. O. Park, S. Lee, S. Kim, S, Y. Kim, Y. S. and Back, K. (2009). Ectopic expression of serotonin N-hydroxycinnamoyltransferase and differential production of phenylpropanoid amides in transgenic tomato tissues, Science Horticulture Amsterdam, 504-510.
12. Linjiang, Z. Hongtao, X. Yanping, Z. Guoliang, F. Paul, Q. W. and Yin, L. (2014). BOX-PCR and PCR-DGGE analysis for bacterial diversity of a naturally fermented functional food (Enzyme®). Food Bioscience, 5:115–122.
13. Partridge, M. and Murphy, D. J. (2004). Detection of genetically modified soya in a range of organic and health food products. British Food Journal, 106:166–180.
14. Pintado, J. Guyot, J.P. and Ampe, F. (2003). Multiple competitive PCR-DGGE as a tool for quantifying and profiling defined mixed cultures of lactic acid bacteria during production of probiotics from complex polysaccharides. Journal of Applied Microbiology, 95(5):921-33.
15. Saiki, R. K. Scharf, S. Faloona, F. Mullis, K. B.Horn, G. T. Erlich, H. A. and Arnheim, N. (1985). Enzymatic amplification of beta-globin genomic sequences and restriction site analysis for diagnosis of sickle cell anemia. Science, 230(4732): 1350–54.
16. Sul, S.Y.Kim, H.J. Kim, T.W. and Kim, H.Y. (2007). Rapid identification of Lactobacillus and Bifidobacterium in probiotic products using multiplex PCR. Journal of Microbiol Biotechnology, 17(3):490-5.
17. Tirado, C. Schmidt, K. (2001). WHO surveillance programmes for control of foodborne infections and intoxications: preliminary results and trends across greater Europe. World Health Organization. Journal of Infections 43:80–4.
18. Valera, A. Pujol, A. Pelegrin, M. and Bosch, F. (1994). Transgenic mice overexpressing phosphoenolpyruvate carboxykinase develop non-insulin-dependent diabetes mellitus. PNAS 91(19): 9151-54.
19. Weighardt, F. (2003). Quantitative PCR for the detection of GMOs. In: M. Querci, M. Jermini, G. Van den Eede (Eds.). The analysis of food samples for the presence of genetically modified organisms (European C European Commission DG-JRC Special publication No. I.03.114 (10): 1–20.
20. Wieser, H. and Koehler, P. (2008). The biochemical basis of celiatic disease. Cereal Chemistry, 85:1–13.
21. http://wiki.biomine.skelleftea.se/wiki/index.php/PCR
22. http://en.wikipedia.org/wiki/Thermal_cycler

[illegible] [illegible] and [illegible] (2006). [illegible] [illegible] [illegible] and [illegible] [illegible] [illegible] 2006: [illegible]
[illegible] [illegible] [illegible] a [illegible] [illegible] [illegible] [illegible]. Food Microbiol. [illegible]
20. [illegible] [illegible] [illegible]

CHAPTER 21

Novel Technology for Development of Low Calorie Functional Beverages from Fruits and Vegetables Using Non-Nutritive Sweeteners

Rakesh Sharma and V.K. Joshi

Abstract

Sedentary lifestyle coupled with excess calorie intake is partially responsible for increased incidence of lifestyle related diseases like coronary heart disease, obesity, diabetes, cancer and hypertension. Non-nutritive sweeteners can offer consumers a way to enjoy the sweet taste with little or no energy. Further, functional beverages are the drinks which are altered in such a way to provide specific health benefits and disease preventing properties beyond general nutrition. In the present paper, authors have summarized various research works on development of different low calorie functional fruit and vegetable products by replacing sugar with non-nutritive sweeteners like saccharin, cyclamate, sorbitol and stevioside at different levels of sweetness (equivalent used to sucrose). The cyclamate sweetened spiced plum squash at 50 per cent sweetness proportion gave higher taste scores in comparison to saccharin sweetened beverage. In case of sorbitol sweetened beverages, with the increase in the proportion of sorbitol sweetness, the organoleptic score for colour, body and flavour increased compared to saccharin and cyclamate sweetened products. Whereas, the stevia sweetened beverages indicated highest sensory acceptability at 75% stevioside sweetness level. Further, sorbitol sweetened health products were found to have similar calculated energy value as that of sucrose sweetened products (100%), but sorbitol is absorbed slowly and does not raise the insulin levels. The development efforts had successfully reduced calorie value of plum squash up to 42% with cyclamate at 50% sweetness level. The calorie value of the sorbitol and stevioside sweetened products has been reduced by 70% and 60% per serving at 100% and 75% sweetness level, respectively. The developed products were found to have strong antimicrobial activity against E. coli as well as high antioxidant potential. Overall, it was concluded that, the developed products had better taste, palatability and storage stability beside reduced energy value, hence their availability in the market will definitely benefit the at risk group i.e. diabetics, obese and health conscious people.

Keywords: Functional foods, low calorie fruit products, beverages, non-nutritive sweeteners

Introduction

Presently, the tenet "Let food be thy medicine and medicine be thy food", quoted by Hippocrates about 2,500 years ago is receiving overwhelming interest. The transition from a traditional to modern lifestyle, consumption of diets rich in fat and calories combined with a high level of mental stress has been the major cause of lifestyle related diseases like coronary heart disease, obesity, diabetes, cancer and hypertension (Moczara et al., 2012). Whereas, the use of foods for promoting health and relieving symptoms is as old as the practice of phyto-medicine for treating or preventing various types of illness have been documented in Hippocratic and Vedic texts and the canons of Traditional Chinese Medicine (Grover et al., 2002). Further, fruit and vegetables are indeed important components in both modern and traditional healing modalities throughout the world. Their consumption has strongly been associated with reduced risk of cardiovascular diseases, cancer, diabetes, Alzheimer disease, cataracts etc. (Temple, 2000). In this context, fruit based beverages are gaining popularity among the masses, but these are generally found to be concentrated source of sugar which provide quick burst of energy and large amount of calories. Unfortunately, excess calorie intake is partially responsible for hypertension, cardiovascular diseases, increased incidence of diabetes mellitus and obesity (WHO, 2010; Moczara et al., 2012). According to recent reports, chronic diseases have been the leading cause of mortality in the world (Sharma et al., 2013a), representing over 60% of all deaths, out of which 30% death is comprised of cardiovascular diseases followed by obesity and diabetes together (Fig-1). Further, sweet taste is one of the fundamental sensations for which people continue to demonstrate their desire. Therefore, availability of fruit beverages which are sweet in taste but with reduced calorie value can play an important role in health promotion and prevention of such diseases. Functional beverages are the drinks which are altered in such a way so to as provide specific health benefits and disease preventing properties beyond general nutrition (Shikha and Bajwa, 2011; Din et al., 2011; Sharma et al., 2013b). One of the alternatives for development of such functional beverages seems to replace bulk caloric sweeteners with high intensity non-nutritive sweeteners (Verdi and Hood, 1993; Din et al., 2011; Sharma et al., 2013aandb). The panorama of RandD work conducted on use of non-nutritive sweeteners for development of reduced calorie fruit based beverages is reported in this article.

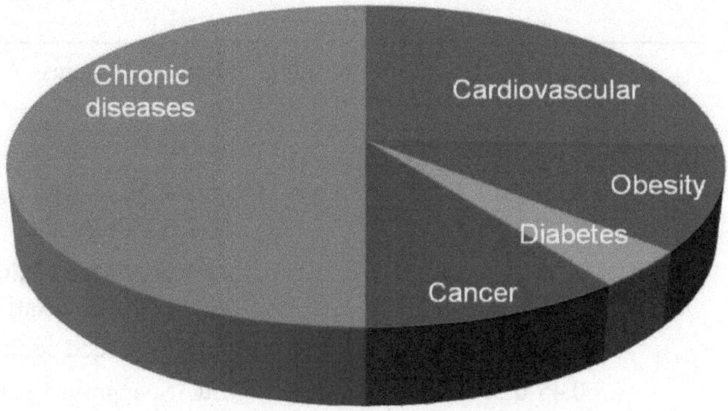

Fig. 1: Global scenario of chronic diseases

Non-Nutritive Sweeteners

Non-nutritive sweeteners are those compounds which sweeten the products with very small volume and do not provide significant energy on consumption. They have no or less than 2 percent of the calorie value of sucrose per equivalent unit of the sweetening capacity. They have a sweet taste but are effectively non-caloric (Pandey and Nigam, 1987; Bright, 1999). They are also referred as high intensity sweeteners. These high intensity sweeteners can offer consumers a way to enjoy the taste of a sweetener with little or no energy and glycaemia response. Six non-nutritive sweeteners viz. Saccharin, Aspartame, Acesulfame potassium, Sucralose, Neotame and Stevia are currently approved by the FDA for use in the U.S. and with the addition of stevia to this list, the group can no longer are called as artificial sweeteners. The relative sweetness as well as regulatory status of various alternative sweeteners is summarized in Table-1. Saccharin is commercially available and used as a substitute for sucrose in soft drinks and processed foods (Grenby, 1991). The joint committee of FAO/ WHO has fixed the acceptable daily intake of saccharin as 2.5 mg/kg of body weight. Aspartame is a nutritive sweetener containing 4 Kcal per gram, however, very little quantity is needed to impart the same sweetness as sugar, resulting in minimal calories added to foods. It is completely breaks down upon digestion into small amounts of methanol and the amino acids aspartic acid and phenylalanine. These components are then absorbed into the blood and used by the body in exactly the same ways as when they come from other foods and beverages (Homler, 1984). No accumulation of aspartame or its components occurs in the body over time. The FDA has established an ADI for aspartame of 50 mg/kg body weight.

Table 1: Relative sweetness and regulatory status of some potential approved alternative sweeteners

Sweeteners	Relative sweetness (Sucrose=1)	K cal/ g	Regulatory status
Less Sweet than sucrose (Polyols or novel sugars)			
Sorbitol	0.6	2.6	GRAS, label must warn about its laxative effect
Mannitol	0.7	1.6	Approved food additive, label must warn about its laxative effect
Xylitol	0.8	2.4	Approved food additive
Isomalt	0.45-0.55	2.0	GRAS, approved as bulking agent
Sweeter than sucrose			
Cyclamate	30	0	GRAS, approved in many countries
Acesulfame-K	130	0	Approved as general purpose sweetener
Aspartame	180	4.0	Approved as general purpose sweetener
Saccharine	500-600	0	Approved as a sweetener for beverages and as a tabletop sweetener
Sucralose	600	0	Approved as general purpose sweetener
Neotam	8000	0	Approved as general purpose sweetener
Stevioside (Stevia)	250-300	0	Legal in US and has received GRAS approval

Source: Pandey and Nigam, 1987; BeMiller, 1992; Barwal 1999; Dias, 1999; Sharma et al., 2013

Sucralose, a high intensity sweetener is the only non-caloric sweetener prepared from sucrose. It is sugar-like in taste and do not give any objectionable aftertaste. Further, food products prepared using sugar alcohols or polyols, also known as novel sugars (like sorbitol etc.) have been regarded as sugar-free because they are absorbed more slowly from the digestive tract and do not raise the post- parandial blood sugar and insulin levels (Barwal et al. 2002). It is generally recognized as safe (GRAS) for use in special dietary foods (Table-1). It has been reported that polyols do not undergo millard reaction and hence browning is minimal on heating (Dias, 1999). Stevioside is a low-calorie sweetener derived from Stevia rebaudiana, a sweet herb (Singh and Rao 2005). It is about 300 times sweeter than sucrose and is highly soluble in water, sweet in taste, and synergistic with other sweeteners. The dried leaves of stevia have been reported 18-45 times sweeter than sugar and have zero calories, and zero cholesterol (Savita et al., 2004; Das and Dang, 2005).

Application of non nutritive sweeteners for development of low calorie beverages

Various works have suggested the use of low calorie non-nutritive sweeteners for the development of different food products (Pandey and Nigam, 1987; BeMiller, 1992; Barwal and Kalia, 1997; Barwal 1999; Sharma et al., 2003; Umbelino et al., 2005; Thakur, 2010; Shikha and Bajwa, 2011; Sharma et al., 2013b). However, very less work has so far been reported on development of fruit and vegetable based low calorie beverages. Saccharin has been used as a substitute for sucrose in soft drinks; processed foods like jam, jellies juices, chewing gums and carbonated drinks (Polasa, 1995). Different artificial non-nutritive sweeteners like saccharin, cyclamate and sorbitol have been used for the preparation of fruit beverages by various researchers (Barwal et al., 2002; Sharma et al., 2003; Barwal et al., 2005). The process for preparation of low calorie beverages using non-nutritive sweetener is shown in Fig-2 (Sharma et al., 2013a). In a study conducted by Sharma et al., (2003), it was found that total soluble solids (TSS) decreased continuously (Table-2) as the per cent share of saccharin and cyclamate increased, due to the fact that these sweeteners do not add to the TSS (Pandey and Nigam, 1987). Whereas, in another study, highest TSS of 42.50 and 15.50 was recorded in 100 % sorbitol sweetened beverages (at par with 100% sucrose sweetened) prepared from plum and bitter gourd, respectively (Barwal et al., 2002). Further, it was reported that with the increase in proportion of sorbitol, there was a significant fall in total sugars due to the fact that sorbitol does not contribute to total sugars during analysis (BeMiller, 1992). Sharma et al., (2003) has reported that cyclamate sweetened beverages at 50 per cent sweetness proportion obtained higher taste scores in comparison to saccharin sweetened beverages. Whereas, Barwal et al., (2002) found that with the increase in proportion of sorbitol sweetness, the organoleptic score increased compared to saccharin and cyclamate sweetened products. Sharma et al., (2003) had successfully reduced calorie value of dietetic plum squash up to 15% and 40 % per serving with cyclamate and saccharin (25% sweetness level) and cyclamate at 50% sweetness level, respectively (Fig. 3). In another study, sorbitol sweetened beverages were found to have similar energy value (Fig. 4) as that of sucrose sweetened products (100%), but sorbitol is reported to be absorbed more slowly from the digestive tract and does not raise the post- parandial blood sugar and insulin levels (Dias, 1999). The calorie value of the sorbitol sweetened products has been reduced by 70% per serving at 100% sweetness level without compromising sensory quality (Sharma et al., 2013a).

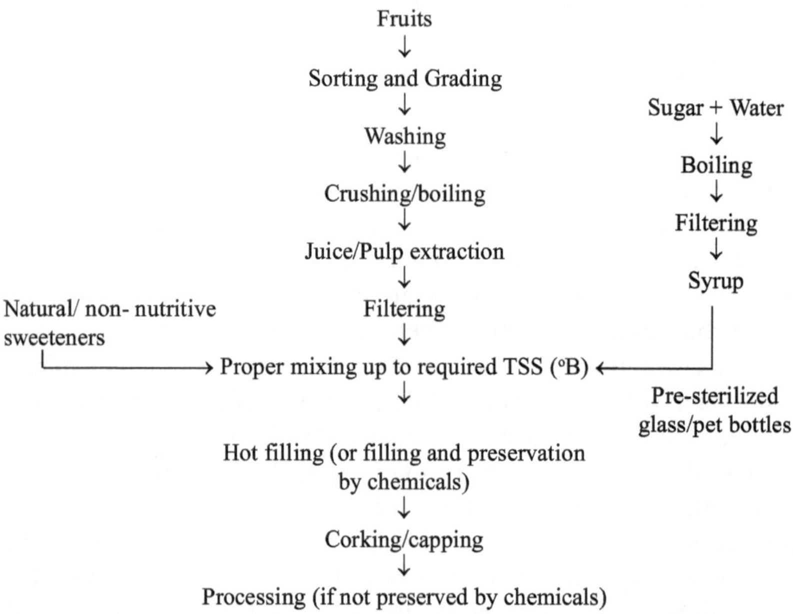

Fig. 2: Unit operations for the preparation of low calorie functional beverages using non-nutritive sweeteners

Table 2: Effect of non-nutritive sweeteners on physico-chemical and overall sensory characteristics of plum spiced squash and bitter gourd RTS beverage

Treatment	% sweetness equivalent used		TSS (°B)	TA (%)	TS (%)	AA (mg/100g)	OAA
	Sucrose	NNS					
Plum spiced squash							
T_1(control)	100	0	40.65	0.98	31.16	0.78	7.30
T_2	75	25 (Saccharin)	30.95	0.93	26.06	0.75	7.22
T_3	75	25 (Cyclamate)	32.10	0.92	27.15	0.78	7.28
T_4	75	25 (Sorbitol)	38.50	0.95	28.78	0.76	7.00
T_5	50	50 (Saccharin)	24.50	0.91	18.24	0.72	6.85
T_6	50	50 (Cyclamate)	25.00	0.94	18.11	0.76	7.15
T_7	50	50 (Sorbitol)	38.95	0.96	19.25	0.75	7.08
T_8	25	75(Saccharin)	17.80	0.91	11.12	0.73	6.62
T_9	25	75(Cyclamate)	18.50	0.93	11.17	0.77	6.63
T_{10}	25	75 (Sorbitol)	39.50	0.94	12.20	0.77	7.15
T_{11}	0	100(Saccharin)	7.25	0.91	3.92	0.72	6.53
T_{12}	0	100(Cyclamate)	8.50	0.94	3.96	0.74	6.56
T_{13}	0	100 (Sorbitol)	42.50	0.95	6.18	0.78	7.03
CD at 5%	0.88	NS	0. 40	NS	0.15		
Bitter gourd RTS beverage							
T_1(control)	100	0	12.50	0.39	8.60	0.68	6.25
T_2	75	25 (Saccharin)	9.20	0.40	7.20	0.68	6.18
T_3	75	25 (Sorbitol)	12.60	0.39	7.18	0.69	6.24
T_4	50	50 (Saccharin)	6.00	0.39	4.60	0.70	6.02
T_5	50	50 (Sorbitol)	15.00	0.40	5.00	0.68	6.23
T_6	25	75(Saccharin)	4.08	0.42	2.45	0.70	5.96
T_7	25	75 (Sorbitol)	17.60	0.40	2.65	0.70	6.25
T_8	0	100(Saccharin)	0.35	0.40	0.30	0.69	5.76
T_9	0	100 (Sorbitol)	15.50	0.39	0.66	0.68	6.25
CD at 5%	0.46	NS	0.20	NS	0.08		

TSS= Total soluble solids; TA= Titratable acidity; TS= Total sugars; AA= Ascorbic acid; OAA = Overall sensory acceptability conducted on 9-point Hedonic scale (9 = Liked extremely; 1 = Dislike extremely)

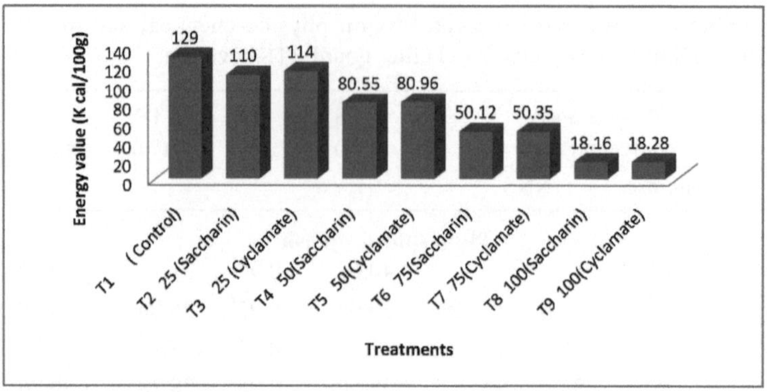

Fig. 3: Effect of different proportions of non-nutritive sweeteners on the energy value of plum spiced squash

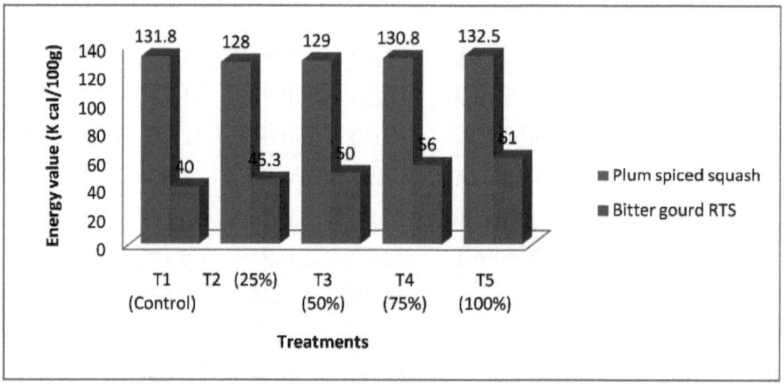

Fig. 4: Effect of different proportions of sorbitol on the energy value of plum spiced squash and bitter gourd RTS drink

Source: Barwal et al. (2002); Sharma et al. (2003); Barwal et al. (2005)

Some research on development of *stevia* sweetened fruit beverages have also been reported by various authors (Chang and Cook 1983; Savita et al., (2004; Kumar et al., 2007; Panpatil and Polsa, 2008; Thakur, 2010; Sharma et al., 2013b). Savita et al., (2004) has recommended up to 75 per cent substitution of sugar with stevioside in different products. Whereas, Tadhani and Subhash (2009) reported that the occurrence of bitter after taste of stevioside which becomes more evident with increasing sweetness proportion from 25 to 100 per cent limited its use up to 75% level in different beverages. They reported 60 per cent sugar replacement in lemon juice, 50 per cent in tea and 75 per cent in coffee with pure steviosides. In another study, Thakur (2010) has replaced sucrose with stevia extract while preparing low calorie aonla based health beverage. The entire process used for preparation of this product is shown in Fig-5.

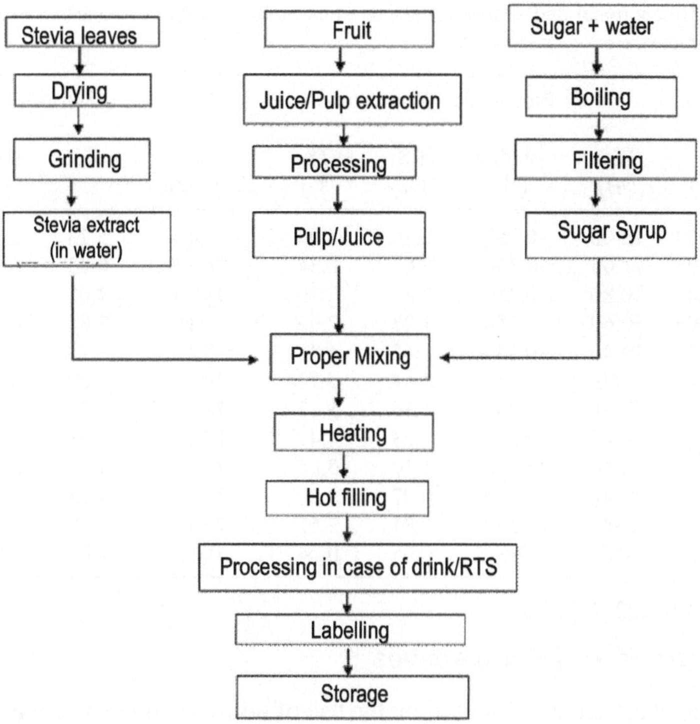

Fig. 5: Unit operations for preparation of stevia sweetened low calorie fruit based functional beverages

It was reported that the beverage sweetened with stevioside having 60 per cent sucrose + 40 per cent stevia extract was adjudged the best on the basis of physico-chemical and sensory quality characteristics (Table-3). Further, Sharma et al., (2013b) have made efforts to prepare low calorie bitter gourd based appetizer using stevioside. On the basis of sensory evaluation, appetizer prepared with 25% sucrose+75% stevioside sweetness was adjudged the best by the panelists. The developed product had 44.5 Kcal/100 g energy value compared to 153.5 Kcal/100g recorded in standard beverage.

Table 3: Biochemical and sensory characteristics of aonla RTS beverages sweetened with stevia extract

Treatments (Sucrose+ Stevia extract %)	Biochemical Parameters					Sensory Parameters		
	TSS (°B)	Acidity (%)	RS (%)	TS (%)	AA (mg/100g)	Colour	Taste	OAA
T_1 (100+0)	12.00	0.350	2.14	8.98	78.15	5.73	6.15	5.60
T_2 (90+10)	11.00	0.350	2.01	8.34	78.15	6.43	6.34	6.15
T_3 (80+20)	10.80	0.350	1.88	7.70	78.14	7.01	6.76	7.55
T_4 (70+30)	10.60	0.350	1.77	7.06	78.15	7.84	7.35	7.50
T_5 (60+40)	10.40	0.350	1.62	6.41	78.14	8.77	7.81	8.50
T_6 (50+50)	9.80	0.350	1.49	5.77	78.15	6.89	6.44	6.53
T_7 (40+60)	9.20	0.350	1.36	5.13	78.16	6.35	6.34	6.03
T_8 (30+70)	6.00	0.350	1.23	4.49	78.16	5.91	6.32	5.30
T_9 (20+80)	5.00	0.350	1.10	3.84	78.12	4.20	6.24	5.04
T_{10} (10+90)	4.00	0.350	0.97	3.20	78.14	4.15	5.87	3.89
T_{11} (0+100)	3.00	0.350	0.84	2.56	78.15	4.08	5.63	3.82
$CD_{0.05}$	0.08	NS	0.05	0.04	0.02	1.02	0.85	0.50

Source: Thakur (2010)

Conclusions and Perspectives

The present article was focused on the use of non-nutritive sweet-eners for the development of low calorie fruit and vegetable beverages. The article provided information on the nutritional as well as functional value of these products. The results showed that calorie value of the product could be reduced up to 50% in case of *saccharin* and *cyclamate*; 70% in case of *sorbitol*. Further, beverages prepared with 75% stevioside sweetness was adjudged the best followed by those sweetened with 50% stevioside sweetness. The calorie value of the stevia sweetened product was reduced up to 60% at 75% sweetness compared to the standard sample. Hence, the manufacture of diabetic-friendly, low calorie fruit and vegetable beverages would provide an alternative assortment to diabetics, obese and health conscious people. Therefore, it should be emphasized that policymakers must know the cost-effectiveness of such interventions, and it would be desirable implementing them for the entire population.

References

1. Barwal, V.S. (1999). Development and significance of low calorie fruit products. The Ind.J. Nutr. Dietet, 36(3):378- 383.
2. Barwal. V.S. and Kalia, M. (1997). Studies on development of reduced sugar apple jelly. Acta Alimentaria, 26(2):107-115.
3. Barwal, V.S. Sharma, R. and Kaushal, B.B.L. (2002). Use of sorbitol for the preparation of low calorie seasoned squash. Journal of Food Science and Technology, 39 (4): 413-417.

4. Barwal, V.S. Sharma, R. and Singh, T.K. (2005). Development and evaluation of dietetic bitter gourd RTS drink. Journal of Food Science and Technology, 42 (2): 202-205.
5. BeMiller, J.N. (1992). Sweeteners, nutritive. In: Encyclopedia of Food Science and Technology, ed. Y.H. Hui. pp. 2487-2490. John Wiley and Sons, Inc., New York.
6. Bright, G (1999). Low calorie sweeteners from molecules to mass markets. In: Cortia A, ed. Low calorie sweeteners: present and future: World Reviev of Nutrition and Dietetics, 85:3-8.
7. Chang, S.S. and Cook, J.M. (1983). Stability studies of stevia and rebaudioside A in carbonated beverages. Journal of Agriculture and Food Chemistry, 31:409-412.
8. Das, K. and Dang, R. (2005). Stevia - a sugar substitute for diabetic patients. Crop Research Hisar, 30(2):305-308.
9. Dias, F.F. (1999). Sorbitol and other sugar alcohols in the food industry. Indian Food Packer, 18(4):229-237.
10. Din, A. Bukhari, S.A.H. Salam, A. and Ishfaq, B. (2011). Development of functional and dietetic beverage from bitter gourd. International Journal of Food Safety, 13:355-360.
11. Grenby, T.H. (1991). Intense sweeteners for the food industry- An overview. Trends in Food Science and Technolology, 2(1): 2-6.
12. Grover, J.K. Yadav, S. and Vats, V. (2002). Medicinal plants of India with anti-diabetic potential. Journal of Ethnopharmacol, 81:81-100.
13. Homler, B.E. (1984). Properties and stability of aspartame. Food Technol. 38(7): 50-55
14. Kumar, S. Jha, Y.K. and Singh, P. (2007). Stevia a natural potential source of sugar replacer. Beverage Food World, 34(7):70-71.
15. Moczara, C. Borgulyab, G. Kovacsc, E. and Rurikc, I. (2012). Could primary care dietary intervention combined with lifestyle changes be effective in the cardiovascular prevention? Acta Alimentaria, 41(2):248–256.
16. Pandey, A. and Nigam, P. (1987). Non-nutritive sweeteners in food systems. Indian Food Industry, 6(4):157-165.
17. Panpatil, V.V. and Polsa, K. (2008). Assessment of stevia (Stevia rebaudiana) natural sweetener: A review. Journal of Food Science and Technology, 45(6):467-473.
18. Polasa, K. (1995). Current status of saccharin-An appraisal. Journal of Food Science and Technology, 32(4):267-277.
19. Savita, S.M. Sheela, K. Sunanda, S. Shankar, A.G. Parama, R. and Srinivas, S. (2004). Health implications of Stevia rebaudiana. J Human Ecology, 15(3):191-194.
20. Sharma, Rakesh. Barwal, V.S. and Kaushal, B.B.L. (2003). Preparation and evaluation of dietetic plum seasoned squash. Beverage Food World, 30(2):38-40.
21. Sharma, R. Barwal, V.S. Joshi, V.K. (2013a). Development of low calorie dietetic beverages from fruits and vegetables using non-nutritive sweeteners. In: Proceedings of National Seminar on Indian Agriculture: Present Situation, Challenges, Remedies and Road Map (August 04-05, 2012), organized by the Youth for Sustainable Development in collaboration with CSK HPKV, Palampur (HP), 154-157.
22. Sharma, R. Joshi, V.K. Tandon, D. and Sharma, K.R. (2013b). Development of antioxidant rich low calorie functional bitter gourd appetizer. In: Souvenir of National Conference on Bioactive Compounds and Functional Foods in Health and Disease Management (15-16 Nov, 2013), organized by the Department of FST, NIFTEM at Kundli (Hry), 176.
23. Shikha, M. and Bajwa, U. (2011). Effect of fat and sugar substitution on the quality characteristics of low calorie milk drinks. Journal of Food Science and Technology, DOI:10.1007/s13197-0100216-9.
24. Singh, S.D. and Rao, G.P. (2005). Stevia: the herbal sugar of 21st century. Sugar Technology, 7(1): 17-24.
25. Tadhani, M.B. Subhash, R. (2009). Preliminary studies on Stevia rebaudiana leaves:

proximal composition, mineral analysis and phytochemical screening. Journal of Medical Science Pakistan, 6 (3): 321-326.

26. Temple, N.J. (2000). Antioxidants and disease: more questions than answers. Nutrition and Research, 20:449-459.
27. Thakur. M. (2010). Use of non-nutritive sweeteners for the preparation of aonla RTS beverage. M Sc Thesis, Dr YSP University of Horticulture and Forestry Solan (HP), 189p.
28. Umbelino, C.D.C. Garcia, D. and Bolini, H.M.A. (2005). Determination of the relative sweetness and acceptability of cyclamate/saccharin blend, aspartame, sucralose and stevia extract as compared to sucrose in mango juice. Acta Alimentaria, 368: 106–110.
29. Verdi, J.R. and Hood, L.L. (1993). Advantages of alternative sweeteners blends. Food Technology, 47(6):1996-1198.
30. Wang. L. and Weller, C.L. (2006). Recent advances in extraction of nutraceuticals from plants. Trends in Food Science and Technollogy, 17: 300-312.
31. WHO (2010). Global strategy on diet, physical activity and health. Available online: http://www.who.int/dietphysicalactivity/strategy/eb11344/strategy_english_web.pdf.

CHAPTER 22

Utilization of Germination Potential in the Development of Nutritional Snacks from Composite of Wheat, Bengal Gram and Green Gram

Sakshi, Jyotika Sharma and C.S. Riar

Abstract

The present study was conducted to develop ready to eat extruded snack food based on germinated flour of wheat, Bengal gram and green gram. A Central Composite Rotatable Design (CCRD) for five variables at five levels was used for experimental combinations and optimization of the process parameters. Experimental design with green gram flour (5 to 20%) and bengal gram flour (15-35%) in wheat flour, moisture content (14-18%), screw speed (200-300 rpm) and barrel temperature (126–134°C) as independent variables having 32 different combinations of snack formulations that were studied using response surface methodology to investigate the effect of these variables on product responses (Lateral expansion, bulk density, Water absorption index, water solubility index and sensory characteristics). Sensory analysis was carried out of all extrudates for texture, color and overall acceptability. Multiple regression equations were obtained to describe the effects of each variable on product responses. The optimum operating conditions for development of germinated wheat, green gram and Bengal gram flour extrudate were 132°C of barrel temperature, 275 rpm of screw speed, 17 % of feed moisture, 6.5% of green gram flour and 23.1% of Bengal gram flour. The responses predicted by the software for these optimum process conditions resulted lateral expansion of 169.59%, bulk density, 0.263g/cm^3; water solubility *index, 19.08; water absoption index, 13.88g/g; and overall acceptability value of 7.38.*

Keywords: Germination, Composite flour, Extrusion, Sensory analysis

Introduction

Germination

Germination is the process in which a plant or fungus emerges from a seed or spore and begins growth. The most common example of germination is the sprouting of a seedling from a seed of an angiosperm or gymnosperm. Germination is induced by rehydration of the seed, which increases both respiration and metabolic activity thus allowing the mobilization of primary and secondary metabolites (Limami et al., 2002). Germination induces the synthesis of hydrolytic enzymes. Significant changes occur in seed during germination in biochemical and physical aspects (Obatolu, 2002) and the total nutrition value is improved (Badau et al., 2005; Jingjun et al., 2010). Flour from germinated seeds has been reported to have better nutritional properties than flours from nongerminated cereals (Lorenz, 1980).

There are many benefits of germination. Vitamin-C is elaborated, phosphorus availability is increased, and lysine and tryptophan are synthesized (Dulby et al.1976). Both starch and protein are partially degraded making for better digestibility (Ram et al. 1979). Some of the flatus factors are also degraded (Rao and Belavady 1978). The elaboration of amylases results in a lower viscosity and thinning down of the starch slurry (Brandtzaeg et al. 1981). Germination of food grains improved the bio-accessibility of iron but not that of zinc (Hemalatha et al., 2007). There is also an overall improvement in the flavour profile. In view of these multiple benefits use of germinated grains, to a larger extent, should be advocated. The germination, however, should not be prolonged as otherwise too much root formation and metabolic weight loss of grain may become a problem.

Germination decreased starch, thereby raising the level of the soluble sugars. Starch digestibility was increased appreciably in green grams. Cooking may gelatinize starch and germination may mobilize starch, resulting in improved digestibility of starch by α-amylase (Grewal and Jood, 2009).

Legumes are high-protein crops that have contributed to the human diet for centuries. They also provide a large amount of available carbohydrates, dietary fiber, water soluble vitamins and minerals. However, pulses contain antinutritional factors such as trypsin inhibitors, α-galactosides and phytic acid, which can negatively affect their nutritional value and that, is the reason why legume seeds have to be processed prior to their consumption (Augustin et al., 1989). Germination has been suggested as an inexpensive and effective technology for improving the quality of legumes by enhancing their nutritional value, and nowadays are becoming very popular in the western countries.

Cereal grains form a major source of dietary nutrients for all people, particularly those in the developing countries. However, the nutritional quality of cereal grains and sensory properties of their products are inferior due to lower protein content, deficiency of certain essential amino acids, lower protein and starch availabilities, presence of certain anti-nutrients, and the coarse nature of the grains. The consumption of sprouted cereals is becoming popular in various parts of the world. Sprouting of grains for a limited period causes increased activities of hydrolytic enzymes, improvement in the contents of certain essential amino acids, total sugars, and B-group vitamins, and a decrease in dry matter, starch, and anti-nutrients. The digestibility of storage proteins and starch are improved due to their partial hydrolysis during sprouting. The magnitude of the nutritional improvement is, however, influenced by the type of cereal, seed quality, sprouting conditions. Sprouting is easy and can be done without sophisticated equipment. Untreated seeds of good quality and high germination percentage are placed in an environment of adequate water, a desirable temperature, and a certain composition of gases in the atmosphere for several days for sprouting. The sprouts can be kept for a few days to over a week under refrigeration. They can be used in many different foods including breakfast items, salads, soups, casseroles, pasta, and baked products. Sprouting of grains causes increased enzyme activity, a loss of total dry matter, an increase in total protein, a change in amino acid composition, a decrease in starch, increases in sugars, a slight increase in crude fat and crude fiber, and slightly higher amounts of certain vitamins and minerals (Lorenz et al., 1997).

According to Shipard (2005) sprouts provide a good supply of Vitamins A, E and C plus B-complex. Like enzymes, vitamins serve as bioactive catalysts to assist in the digestion and metabolism of feeds and the release of energy. They are also essential for the healing and repair of cells. However, vitamins are very perishable, and in general, the fresher the feeds eaten, the higher the vitamin content. The vitamin content of some seeds can increase by up to 20 times their original value within several days of sprouting. Mung bean sprouts have B-vitamin increases, compared to the dry seeds, of - B_1 up 285%, B_2 up 515%, B_3 up 256. Compared with mature plants, sprouts can yield vitamin contents 30 times higher."

Pulses represent one of the most important food categories that have been extensively used as staple foods to cover basic protein and energy needs throughout the history of humanity. In addition to their low lipid and high dietary fiber content, emerging evidence stresses the importance of pulses as carriers of several constituents of potential biological importance, including enzyme inhibitors, lectins, phytates, oxalates, polyphenols, saponins and phytosterols. Investigations in humans suggest that pulses may contribute to human health and wellbeing, mostly through prevention of coronary heart disease and possibly diabetes. The mechanisms responsible for this apparently protective role may include a favorable influence on blood lipids and glucose. The nutritional value of pulses, which are a key component of the traditional Mediterranean diet, is not generally recognized and is frequently under-appreciated. Legumes or pulses play a key role in the acceptability of monotonous diets in many parts of the world; therefore it is important that consumer preferences and acceptability are considered in legume breeding programmes.

Germinated wheat sprouts are an effective and economical tonic to improve general health. The seeds become living food as soon as germination takes place and valuable enzymes are activated. Once eaten, these enzymes act as catalysts that perform important functions in the body, such as: to aid metabolism, neutralize toxins, cleanse the blood, and provide energy for innumerable bodily functions.

The enzyme diastase, activated by germinating the wheat, transforms starch into absorbable nutrients. These enzymes, also aid the digestion of other foods, and can benefit anyone who suffers with digestive and assimilation problems. By eating wheat sprouts, we also get the benefit of the wheat germ, a rich source of vitamin E (usually removed in flour milling). Vitamin E is an essential nutrient for fertility, healthy skin, hair, glands, kidneys, muscles (especially the heart), circulation and the nervous system; also for building red blood cells and improving absorption of fats, protein, vitamins A, C and iron. Wheat germ (the seed embryo), in the early stages of plant growth, is considered one of the best remedies to help overcome a tendency toward miscarriage and birthing prematurely.

The sprouting process starts a beneficial modification of various nutritional elements. Research undertaken at the University of Minnesota, USA, showed that sprouting increases the nutrient density of food. At 3 days of sprouting, much of the original carbohydrate is converted to natural sugars, making it less mucus forming. During the 3 days of sprouting, the vitamin E content can increase 300%, vitamin C increase 600%, and the B vitamins have been found to increase from 20% to 1200%, with B17, the 'anti-cancer vitamin' 100% more than in un-sprouted seed. The protein content increases by 300% in the third day of sprouting compared to the un-sprouted wheat seed. Sprouted wheat, being a raw grain, should be eaten in moderate amounts, as in the early stages of germination (1-4 days) there is insufficient time to convert all raw starch into sugars, for ease of digestion.

Extrusion

Extrusion is defined as "shaping by force through a specially designed opening often after previous heating of the material." Extrusion is the continuous forming of plastic or soft materials through a die. Several types of extruders include ram or piston types and screw or worm types (Harper, 1981).

Extrusion cooking is an important and popular food processing technique classified as a high temperature-short time process to produce fiber rich products (Vasanthan et al., 2002). Harper was the first to study extrusion processing of food material (Mercier et al., 1989). Extrusion combines a number of unit operations i.e. mixing, cooking, shearing, puffing, final shaping and drying in one energy efficient rapid continues process (Harper, 1989). In the extruder, food mix is thermo mechanically cooked to high temperature, pressure and shear stress which is generated in the screw-barrel assembly. The cooked melt is then texturized and shaped in the die (Arhaliass et al., 2003). Extrusion technology is very useful from the standpoint of nutritional value as nutrient losses are low than other thermal processing methods (Moscicki et al., 2003). It offers several advantages over other types of cooking processes, such as faster processing time and significant reduction in energy consumed, which consequently results in lower price for the final product.

The fast changing life style of the consumer demands convenience in terms of saving time and energy. Development of ready to eat products add convenience, saves time and labour, provides hygienic products of standard and uniform quality with enhanced shelf life. Extrusion cooking technology being used increasingly in the food industries for the development of new products, such as cereal-based snacks including dietary fiber, baby foods, and breakfast cereals and modified starch from cereals (Sebio and Chang, 2000). Extrusions cooking of food ingredients bring gelatinization of starch, denaturation of protein, modification of lipid and inactivation of enzymes, microbes and many antinutritional factors (Bhattacharya and Prakash, 1994). It is becoming more and more popular in the production of snack foods due to their technological advantages over the traditional food processing techniques (White, 1994).

Snack foods have become an integral part of the eating habits of the majority of the world's population and are prepared from natural ingredients or components to yield products with specified functional properties (Reid, 1998). In UK the growth of the extruded snack food market has been extensive (Anonymous, 1998), and is now worth between $2.1 to $2.5 billion per annum, in terms of volume, this equates to 486,000 tons of snack products purchased each year (Datamonitor, 2004). India is now world's largest market for snack food with worth of 1530 crore (www.fao.org).

Cereal grains are generally used as major raw material for development of extruded snack foods due to their good expansion characteristics because of their high starch content. Physical characteristics of an extruded snack product such as expansion, hardness and density are important parameters in terms of the consumer acceptability of the final product, as well as its functional properties (Launay and Lisch, 1983). However, extruded food products produced from cereals tend to be low in protein and having a poor biological value due to their limited essential amino acid contents (Iqbal, et al., 2006). Many snack foods have low nutrient densities and are often fortified with proteins. Nutritious snack foods can be produced by incorporation of legumes, vegetables, and fruits into the formulation (Thakur and Saxena 2000).

Green Gram

Green gram also known as mung bean, mung, mongo, moog dhal (in Bengali), mash bean, munggo, is the seed of Vigna radiate which is native to Bangladesh, India, and Pakistan. The split bean is known as moong dal which is green with the husk, and yellow. Green gram (Vigna radiata L.) is one of the most important food legumes grown and consumed in India. It is a good source of carbohydrates, proteins and minerals. Digestibility and availability of nutrients are, however, limited due to presence of toxic compounds including proteolytic enzyme inhibitors. The consumption of mung proteins can fulfil the essential amino acids requirements with the exception of the sulphur-containing amino acids (Khalil, 2006).

Bengal Gram

Bengal gram also called as chick pea, it shares 13.96% of total India's production. It is the major legume crop. Average world production of Bengal gram is at 9.313 million tons (FAO database). India is the largest producer of Bengal gram and constitutes more than 50% of world production. Other major countries are Pakistan, Turkey, Australia, Myanmar, Canada, Ethiopia, Mexico and Iraq. India is also major consuming country and many of the Indian dishes are made with chick pea flour. Bengal gram is third largest pulse crop in India.

Bengal gram is a protein rich pulse. It contains about 23 percent protein, which is almost three times that of cereals. Bengal gram supplies a major share of protein requirement of vegetarian population of the country. Bengal gram is mainly consumed in the form of flour i.e. Besan, which is an essential supplement of cereal based diet. Bengal gram is a valuable source of protein, minerals and vitamins and occupies a very important place in human nutrition. Ganganagar, Hanumangarh, Bikaner, Churu, Jaipur, Bundi, Tonk, Sikar and Alwar are major producing areas in India.

Wheat

India is one of the main wheat producing and consuming countries of the world.Globally, wheat is the leading source of vegetable protein in human food, having higher protein content than either maize (corn) or rice, the other major cereals. In terms of total production tonnages used for food, it is currently second to rice as the main human food crop, and ahead of maize, after allowing for maize's more extensive use in animal feeds.

It contains Manganese, Phosphorus, Magnesium and Selenium in very large quantities. Rich in zinc, copper, iron and potassium, however, calcium is also present in small amounts. It is rich in Vitamin B_6, niacin, thiamin, folate, riboflavin and pantothenic acid. Vitamin-E and Vitamin-K are also present in small but considerable amounts. Wheat has a calorific value of 339.0 per 100 gm. Being a grain, it is very appropriate in calories and hence, filling as a food. Germinated wheat contains 2 or 3 times more vitamin-B than common wheat. This way, the germinated seeds are used for treating gastro-intestinal and cardiovascular maladies, skin diseases or hepatic and breathing affections. It also has the property of balancing the body weight.

Extruded foods are commonly consumed as snacks by every age group and theses are generally produced by using untreated grain flour, but in this study the main objective is to use germinated grain flour, i.e pre-treatment will be given to grains so as to increase their nutritional value. Also, the raw material is choosen as such they are commonly available in the market and are good source of proteins. Although, green gram and Bengal gram are good sources of protein but germination will further helps in increasing digestibility and reduction in anti-nutritional factors. On the basis of above facts the present research was focused on to evaluate the, the effect of germination on the composition of Wheat, Bengal gram and Green gram, development of extruded product from the germinated flour (wheat, Bengal gram and green gram) and optimization of the extrusion process by using Response Surface Methodology and finally the characterization and evaluation of the developed extruded product.

Materials and Methods

Procurement of Raw Material

Good quality raw materials of wheat, green gram and Bengal gram were procured from the local market of Barnala, Punjab. All these grains were freshly harvested and were having good germination power. After procurement, these grains were cleaned in laboratory by using instruments like aspirator, grader etc. and also by manual observation. The grains were packed in air tight containers and stored in the lab for further use

Germination of grains

Wheat

Germination of wheat was carried out as per the method described by (Sadana and Chabra, 2004). Wheat grains were cleaned and soaked separately in water (seed to water ratio of 1:5 w/v) for 12 h at room temperature (4 30°C). The soaked grains were germinated in clean plates lined with wet muslin cloth for 24 h at 30°C with frequent spraying of water, the sprouts were rinsed in water, dried at 55-60°C and pulverized to fine powder (sieved through 60μ sieve). Unprocessed samples of wheat were also pulverized to fine powder and stored in plastic containers for further use.

Bengal Gram

Germination of Bengal gram was carried out as per the method described by Sadana and Chabra (2004). Bengal gram seeds were cleaned and soaked in potable tap water for 12 h at room temperature and grain to water ratio was 1:3 (1 part grain and 3 parts water). The soaked samples were allowed to germinate for 48 h in an incubator at 30°C after placing in muslin cloth with frequent spraying of water. The sprouts were rinsed in distilled water, dried in tray drier at 60°C till constant weight, then pulverized these germinated and un-germinated samples of Bengal gram into fine flour, which was sieved through 60μ sieve and stored in plastic containers for further use.

Green Gram

Germination of Green gram was carried out as per the method described by Akaerue and Onwuka, (2010). Green gram grains were cleaned and soaked in potable tap water for 12 h at room temperature and grain to water ratio was 1:3 (1 part grain and 3 parts water). The soaked samples were placed in between muslin cloth and allowed to sprout for 24 h in an incubator at 25°C. The seeds were kept wet thought germination by spraying them with distilled water frequently. The sprouted green grams were rinsed twice in distilled water and dried in tray drier at 50-60°C till constant weight. Then dried germinated and un-germinated samples of Green gram were subjected to pulverized into fine flour, which was sieved through 60μ sieve and stored in plastic containers for further use.

Chemical analysis of germinated and Un-germinated flours of Wheat, Bengal Gram and Green Gram.

Moisture, ash, total carbohydrate, reducing sugar, starch, Protein, amylose, fat content and crude fiber were estimated as per the AOAC (2000) methods.

Physical characteristics of flours of germinated and Un-germinated grains of Wheat, Bengal Gram and Green Gram.

Water Binding Capacity (WBC)

WBC of the sweet potato starches was determined using the method described by Yamazaki, (1953). A suspension of 5g starch (dry weight basis) in 75ml distilled water was agitated for 1h and centrifuged at 3000rpm for 10min. The free water was removed from the wet starch, which was then drained for 10min. The wet starch was then weighed.

$$\text{Water binding capacity (\%)} = \frac{\text{Weight of residual starch}}{\text{Weight of sample}} \times 100$$

Water Solubility Index (WSI)

The supernatant from the water absorption index determination was decanted into pre-weighed moisture dish and evaporated to constant weight in a precision oven at 103°C. Water solubility index (WSI) was calculated as the percent weight of the original sample weight recovered (Iwe, 1998).

$$\text{WSI (\%)} = \frac{\text{Weight of dry solid in supernatent}}{\text{Dry weight of sampla}} \times 100$$

Oil Binding Capacity (OBC)

Oil binding capacity of the sweet potato starches was determined using the method described by Yamazaki, (1953). A suspension of 5g starch (dry weight basis) in 75ml oil was agitated for 1h and centrifuged 3000 rpm for 10 min. The free oil was removed from the wet starch, which was then drained for 10 min. The residue was then weighed.

$$\text{Oil binding capacity (\%)} = \frac{\text{Weight of residual starch}}{\text{Weight of sampla}} \times 100$$

Experimental design

Response surface methodology (RSM) was adopted in the experimental design as it emphasizes the modeling and analysis of the problem in which response of interest is influenced by several variables and the objective is to optimize this response (Montgomery, 2001). The main advantage of RSM is reduced number of experimental runs needed to provide sufficient information for statistically acceptable results. A five-level, five-factor central composite rotatable design was employed (Myers, 1976).The base flour of the extrudates was of germinated wheat to which the other two legume flours were mixed with different proportions Table-1. The independent variables selected for the experiments were

a) Green gram flour (g) : 0, 5, 10, 15, and 20

b) Bengal gram flour (g) : 15, 20, 25, 30, and 35

c) Moisture content (%) : 14, 15, 16, 17, and 18

d) Barrel temperature (^{0}C) : 126, 128, 130, 132 and 134

e) Screw speed (rpm) : 200, 225, 250, 275 and 300

Experimental designs in coded form for response surface analysis were selected as given in Table-2. For the experiments, response variables were Bulk density, Lateral expansion, Water absorbtivity index, Water solubility index, and Sensory evaluation. The five levels of the process variables were coded as "2, "1, 0, 1 and +2 (Montgomery, 2001) and design in coded (x) form and at the actual levels (X) is given in Table-3.

Sample preparation

Blends were prepared by mixing germinated Wheat flour (10.6%, wb), germinated Bengal gram flour (9.8%, wb), and germinated Green gram flour (9.5%, wb) together in five different feed combination according to design of experiment. Highest and lowest levels of wheat flour were 85% and 55% respectively. All blended samples were prepared in 100gm batch. These blends were chosen according to preliminary tests without jamming of extruder and for acceptable product's physical characteristics. The blended samples were conditioned to 14-18% (wb) moisture with calculated amount of water and mixed thoroughly to adjust feed moisture to desired level in the domestic food processor as mentioned above with mixer attachment for 20 min. These blends were then passed through the 2mm mesh screen for proper mixing or breaking the lumps if any formed due to addition of moisture. Blended mixture were sealed into high density polyethylene bags and placed in container. Samples were allowed to equilibrate at 4°C for 24h and prior to extrusion samples were allowed 3h to reach room temperature (Altan et al., 2008). The moisture content of all samples was estimated using oven method (Ranganna, 2003).

Extrusion experiments

Extrusion trials were performed using a co-rotating twin-screw extruder (Basic Technology Pvt. Ltd. Kolkata, India). The main drive was provided with 7.5 HP motor (400 V, 3 ph, 50 cycles). The output shaft of worm reduction gear was provided with a torque limiter coupling. The screw configuration that was a standard design for processing cereals and flour-based products was used. Barrel length to diameter ratio (L/D) was 8:1.The barrel of the extruder received the feed from a co-rotating variable speed feeder. The barrel was provided with two electric band heaters and two water cooling jackets. A temperature sensor was fitted on the front die plate which was connected to temperature control make placed on the panel board. The die plate of the die fixed by a screw nut tightened by a special wrench provided. The initial experimental temperature was reached within 30 min and samples were then poured in to feed hopper and the feed rate was adjusted to 4 kg/h for easy and non choking operation. The die diameter was selected at 4 mm as recommended by the manufacturer for such product. The barrel zone temperatures were kept constant at 100°C throughout the experiments but die temperature varied according to the experimental design. Samples were then poured in to feed hopper and the feed rate was adjusted to 4 kg/h for easy and non choking operation. Extrudates were cut with a sharp knife and left to cool at room temperature for about 20 min.

Table 1: Values of independent variables at five levels of the CCRD design

Independent variable	Un-coded	Levels in coded form				
		-2	-1	0	+1	+2
Green gram flour(g)	X1	0	5	10	15	20
Bengal gram flour(g)	X2	15	20	25	30	35
Die temperature(°C)	X3	126	128	130	132	134
Screw speed (rpm)	X4	200	225	250	275	300
Feed moisture (%)	X5	14	15	16	17	18

Table 2: Experimental design in coded form for response surface analysis

Coded Variables					Combinations	Replications	No. of Experiments
X1	X2	X3	X4	X5			
±1	±1	±1	±1	±1	16	1	16
±2	0	0	0	0	2	1	2
0	±2	0	0	0	2	1	2
0	0	±2	0	0	2	1	2
0	0	0	±2	0	2	1	2
0	0	0	0	±2	2	1	2
0	0	0	0	0	1	6	6

Code '0' is for centre point of the parameter range investigated,'±1' for factorial points, and '±2' for star points;
X1-Green gram flour(%), X2-Bengal gram flour(%), X3-Die temperature(°C) , X4- Screw speed(rpm) and X5- Feed moisture(%).

Table 3: Experimental combinations in coded and un-coded levels for cereal and pulses based extruded snacks.

Coded	Variables					Un-coded Variables				
S. No	X1	X2	X3	X4	X5	X1: Green Gram (% on flour basis)	X2: Black Gram (% on flour basis)	X3: Temp. (°C)	X4: Screw Speed (rpm)	X5: Moisture (%)
1	-1	-1	-1	-1	1	5	20	128	225	17
2	1	-1	-1	-1	-1	15	20	128	225	15
3	-1	1	-1	-1	-1	5	30	128	225	15
4	1	1	-1	-1	1	15	30	128	225	17
5	-1	-1	1	-1	-1	5	20	132	225	15
6	1	-1	1	-1	1	15	20	132	225	17
7	-1	1	1	-1	1	5	30	132	225	17
8	1	1	1	-1	-1	15	30	132	225	15
9	-1	-1	-1	1	-1	5	20	128	275	15
10	1	-1	-1	1	1	15	20	128	275	17
11	-1	1	-1	1	1	5	30	128	275	17
12	1	1	-1	1	-1	15	30	128	275	15
13	-1	-1	1	1	1	5	20	132	275	17
14	1	-1	1	1	-1	15	20	132	275	15
15	-1	1	1	1	-1	5	30	132	275	15
16	1	1	1	1	1	15	30	132	275	17
17	-2	0	0	0	0	0	25	130	250	16
18	2	0	0	0	0	20	25	130	250	16
19	0	-2	0	0	0	10	15	130	250	16
20	0	2	0	0	0	10	35	130	250	16

(*Contd.*)

Coded	Variables					Un-coded Variables				
S. No	X1	X2	X3	X4	X5	X1: Green Gram (% on flour basis)	X2: Black Gram (% on flour basis)	X3: Temp. (°C)	X4: Screw Speed (rpm)	X5: Moisture (%)
21	0	0	-2	0	0	10	25	126	250	16
22	0	0	2	0	0	10	25	134	250	16
23	0	0	0	-2	0	10	25	130	200	16
24	0	0	0	2	0	10	25	130	300	16
25	0	0	0	0	-2	10	25	130	250	14
26	0	0	0	0	2	10	25	130	250	18
27	0	0	0	0	0	10	25	130	250	16
28	0	0	0	0	0	10	25	130	250	16
29	0	0	0	0	0	10	25	130	250	16
30	0	0	0	0	0	10	25	130	250	16
31	0	0	0	0	0	10	25	130	250	16
32	0	0	0	0	0	10	25	130	250	16

Extrusion of samples

The twin screw extruder was kept running for 30 minutes to stabilize the set temperatures and samples were then poured in to feed hopper and the feed rate was adjusted to 4 kg/h for easy and non choking operation. The die diameter of 4 mm was selected as recommended by the manufacturer for such product. The product was collected at the die end and packed in already numbered zipped lock packs and kept for proper storage

Stabilization of moisture

All the samples were kept in high precision (± 0.1°C) incubator (Macro Scientific works, New Delhi) at 60°C for 12 hour duration for the stabilization of moisture.

Evaluation of extrudates characteristics

Lateral expansion: The ratio of diameter of extrudate and the diameter of die was used to express the expansion of extrudate (Ainsworth, 2006). Six lengths of extrudate (approximately 120 mm) was selected at random during collection of each of the extruded samples, and allowed to cool to room temperature. The diameter of the extrudates was then measured, at 10 different positions along the length of each of the six samples, using a vernier caliper. Lateral expansion (LE, %) was then calculated using the mean of the measured diameters:

$$LE = \frac{(\text{diameter of product-diameter of die hole})}{\text{diameter of die hole}} \times 100$$

Bulk density (BD, g/cm^3) was calculated according to the method of (Stojceska et.al. 2008).

$$BD = \frac{4m}{\pi d^2 L}$$

Whereas: m, is mass (g) of a length L (cm) of extrudate with diameter d (cm)

Water absorption index (WAI) and water solubility index (WSI)

WAI and WSI were determined according to the method developed for cereals (Stojceska et.al. 2008). The ground extrudate was suspended in water at room temperature for 30 min, gently stirred during this period, and then centrifuged at 3000g for 15 min. The supernatant was decanted into an evaporating dish of known weight. The WAI was the weight of gel obtained after removal of the supernatant per unit weight of original dry solids. The WSI was the weight of dry solids in the supernatant expressed as a percentage of the original weight of sample.

$$\text{WAI}(g/g) = \frac{\text{Weight gain by gel}}{\text{Dry weight of extrudate}}$$

$$\text{WSI}(\%) = \frac{\text{Weight of dry solid in supernatent}}{\text{Dry weight of extrudate}} \times 100$$

Sensory characteristics of extrudates

Sensory analysis was conducted on five samples with tomato pomace levels of 5%, 10%, 15%, 20%, and 25%. Thirty panelist were asked to assess the expanded snacks for flavor acceptability, and to mark on a Hedonic Rating Test (1 – Dislike extremely, 5 – Neither like nor dislike and 9 – Like extremely) in accordance with their opinion for taste, texture, color and overall acceptability.

Statistical analysis of responses

The responses (bulk density, lateral expansion, WAI, WSI, and sensory evaluation) for different experimental combinations were related to the coded variables (x_i, i=1, 2, 3 and 4) by a second degree polynomial equation

$$Y = \beta_0 + \beta_1 x_1 + \beta_2 x_2 + \beta_3 x_3 + \beta_4 x_4 + \beta_{11} x_1^2 + \beta_{22} x_2^2 + \beta_{33} x_3^2 + \beta_{44} x_4^2 + \beta_{12} x_1 . x_2$$
$$+ \beta_{13} x_1 . x_3 + \beta_{14} x_1 . x_4 + \beta_{23} x_2 . x_3 + \beta_{24} x_2 . x_4 + \beta_{34} x_3 . x_4 + \varepsilon$$

The coefficients of the polynomial were represented byβ_o(constant), $\beta_{1'}$ $\beta_{2'}$ $\beta_{3'}$ β_4 (linear effects); $\beta_{12'}$ $\beta_{13'}$ $\beta_{14'}$ $\beta_{23'}$ $\beta_{24'}$ β_{34} (interaction effects); $\beta_{11'}$ $\beta_{22'}$ $\beta_{33'}$ $\beta_{44'}$ (quadratic effects); and ε (random error). Data were modeled by multiple regression analysis and the statistical significance of the terms was examined by analysis of variance for each response.

Analysis of data

A complete second order quadratic model employed to fit the data and adequacy of the model was tested considering R^2 (the coefficient of multiple determination, a measure of the amount of variation around the mean explained by the model), Adj R^2 (a measure of the amount of variation around the mean explained by the model, adjusted for the number of terms in the model), predicted R^2 (a measure of how good the model predicts a response value) and Fischer's F-test. Coefficient of determination R^2 is defined as the ratio of the explained variation to the total variation and is measure of the degree of fit (Haber and Runyon, 1977). It is also the proportion of the variability in the response variables, which is accounted for by the regression analysis (Mclaren et al., 1977). When R^2 approaches unity, the better the empirical model fits the actual data. The smaller the value of R^2, the less relevance the dependent variables in the model have in explaining the behavior variation. The models were then used to interpret the effect of various predictors (terms) on the response. The analysis of variance (ANOVA) tables were generated and the effect and regression coefficients of individual linear, quadratic and interaction terms were determined. The significances of all terms in the polynomial were judged statistically by computing the F-value at probability (p) of 0.01 or 0.05. The regression coefficients were then used to make statistical calculation to generate contour maps from the regression models. Optimization of process parameters was done by partially differentiating the model with respect to each parameter, equating to zero and simultaneously solving the resulting functions. Design Expert 6.0 was used for this purpose and contour plots were developed for selected parameters. Design Expert (version 6.0, by STAT-EASE Inc., USA) was used for the analysis of the data.

Results and Discussions

Germinated wheat flour, Green gram flour and Bengal gram flour were prepared and phsico-chemical analysis was carried out. Extruded snack product was developed from Germinated Wheat flour, Germinated Green gram flour and Germinated Bengal gram flour. Germinated green gram flour and germinated Bengal gram flour was added to wheat flour at different levels. Designed experiments were conducted to investigate the effect of feed composition, moisture content, screw speed and barrel temperature on product responses: like, lateral expansion ratio, bulk density, water solubility index, water absorption index, and sensory characteristics. A second order polynomial model for the dependent variables was tested to fit the experimental data and validity of the model was checked using coefficient of determination (R^2), Adj R^2, model F-value, Adequate precision and lack of fit F-value. The results thus obtained are reported and discussed below.

Changes in Physico-Chemical Composition

In this project three different grains were taken viz. wheat, Bengal gram and green gram and comparatively different analysis were carried to study the changes in chemical composition after germination of all these three grains. The mature grain when harvested and stored, in a dormant stage, it is characterized by dehydration and a dramatic decrease of metabolic activity. Germination is induced by rehydration of the seed, which increases both respiration and metabolic activity thus allowing the mobilization of primary and secondary metabolites (Limami et al., 2002). Therefore, the biochemical composition between un-germinated and germinated kernels is different.

Wheat

The moisture content of un-germinated and germinated flour samples of Wheat varied between 8 % to 11 % (wb) as shown in Table-4. Mean moisture content was increased after germination. This finding is similar to the results reported by Khatoon and Prakash, (2006) in germinated legumes (soya bean). As germination proceeds, seeds took up water from the surrounding in order for the metabolic process to commence. Dry seeds absorb water rapidly, influenced by the structure of the seed. The increase in water uptake with time is due to the increasing number of cells within the seed becoming hydrated (Nonogaki et al., 2010).

Ash content of above flour decreased from 1.64 % to 1.39 % (wb) (Table-4). Ash content was significantly decreased in germinated wheat, parallel to observations of Ohtsubo et al., (2005); Khatoon and Prakash, (2006). Wang et al., (1997) reported that the differences in ash content after soaking for a specific time was due to decreased ash content. The decrease in ash content represents loss in minerals due to rootlet and washing of the wheat in water to reduce the sour smell during the period of germination (Tatsadjieu et al., 2004).

Table 4 indicated that the total carbohydrate content of wheat as affected by soaking, germination and heating treatments. The soaking, germination and heating treatments given to wheat grains decreased the total carbohydrate contents 76.23 % to 72.15 %.This was because of active respiration process during soaking and germination. Srivastva et al., (1988) also reported an increase in amylase activity during soaking and subsequent germination of pigeon pea. The results of soaking, germination and heating related with decrease in total carbohydrate and increase in sugars (total, reducing and non-reducing) in wheat are in agreement with those of Dogra et al., (2001). The biological reactions involved in the germination are dependent on the temperature, because the enzymes involved have ranges of action that will retard or accelerate seed germination (Bewley and Black, 1994). The velocity in which the enzymes carry out their functions will interfere in the degrading of the reserved carbohydrates and ATP production, which will be used in forming proteins and other metabolites that originate tissues and cellular compounds that culminate with the protrusion of the rootlets and consequent seedling established (Mayer and Poljakoff-Mayber, 1989).

The protein and fat contents as affected by soaking, germination and heating are given in Table 4. The germination of wheat had an increasing effect on protein content 12.29 % to 14.3 %. Studies done by Khatoon and Prakash (2005), Ghavidel and Prakash (2007) found that total protein increased after germination process. Bau et al. (1997) assumed that the increased was due to synthesis of enzyme proteins or a compositional change following the degradation of other constituents. Fat content was decreased in germinated wheat from 1.09% to 0.95%. Similar results occurred in study by Ghavidel and Prakash (2007) reported a decrease in fat content with increase in the time of germination. This is because fat was used as the major source of carbon for seed growth (Bau et al., 1997). Hahm et al. (2008) also suggested that fatty acids are oxidized to carbon dioxide and water to generate energy for germination.

Crude fiber was increased in germinated wheat from 0.95 % to 1.59 % (Table 1). Martin-Cabrejas et al. (2008) reported that the effect of germination on crude fiber was dependent on type of legumes. In germinated rice, the amount of crude fiber was contributed by the presence of bran layer, an outer layer of rice that contained fiber (Ohtsubo et al., 2005). Study by Azizah and Zainon (1997) demonstrated that crude fiber was decreased in soaked peanut and mung bean, but conversely increased in soaked rice and soybean. This indicates that germination process affect the level of crude fiber during the period of soaking before the actual phase of germination. Therefore, Studies have shown that intake of legumes have many health effects in controlling and preventing metabolic diseases such as diabetes mellitus and coronary heart diseases (Slavin et al., 1997; Liu et al., 1999). Combination of whole grain and legume powder in coronary artery disease patients without diabetes mellitus can reduce fasting levels of glucose and insulin (Jang et al., 2001). Germination meanwhile, altered the biochemical composition of legumes. Decrease of carbohydrate level is beneficial to diabetes mellitus patients. In addition, increase of total dietary fiber and decrease of fat content can give benefit to people with cardiovascular disease and hypercholesterolemia.

Table 4 indicated that the starch content of wheat decreased from 69.42 % to 62.70 % during soaking, germination (96hrs) and heating. The increased α- amylase and β-amylase activities correspond with the decrease in the starch content, increase in amylose content from 12.94% to 15.3% and decrease in amylopectin from 56.40 % to 47.4 % in germinated wheat flour and these results are in agreement with those of Sharma et al., (2007). Palmer, (1989) reported that the germination of cereal grains caused extensive changes in the structure and composition of major macromolecular components of the grains. Since gelatinization rates are related to starch composition and structure, the extent of starch modification during grain germination may influence rates of gelatinization. Therefore, the small increase in gelatinization rates for starches from the out-of-steep through to the four day germinated grains at 50°C assay temperature may suggest small improvements in starch degradability due possibly to differences in the degree of modification of starch structure and composition.

Bengal Gram

Table 5 indicated that the moisture content of un-germinated and germinated flour samples of Bengal gram varied between 8.95 % to 9.49 % (wb) means moisture content was increased significantly after germination same as in case of wheat. Our results corroborates earlier reports of increased moisture content during germination of chick pea and other legumes seeds (Khatoon and Parkash, 2006; Khalil et al., 2007). Germination of Bengal gram led to decrease in ash content of germinated flour (2.79) than un-germinated flour (2.85) as shown in Table 5. These results were in agreement with those reported earlier by several workers (Gopalan et al., 1989). Leaching out of solid matter during pre germination soaking process could be the reason for signiûcant reduction of mineral matter on germination. The further insigniûcant decrease of ash after dehulling could be contributed to removal of hull portion, which may have some amounts of minerals.

Table 4: Chemical composition of un-germinated and germinated flours of wheat

↓ (%) Samples →	Un-germinated Wheat Flour	Germinated Wheat Flour
Moisture	8.24	10.12
Protein	12.29	14.3
Ash	1.64	1.39
Crude fiber	0.95	1.59
Carbohydrate	76.23	72.15
Fat	1.09	0.95
Starch	69.42	62.70
Amylose	12.94	15.3a
Amylopectin	56.40	47.4

Table-5, indicated that the total carbohydrate and reducing sugar contents of Bengal gram flour as affected by soaking, germination and heating treatments. The soaking, germination and heating treatments given to Bengal gram grains decreased the total carbohydrate contents 57.68 % to 54.49 %.This was because of active respiration process during soaking and germination. On the other hand, soaking, germination and heating increased the reducing 2.52% to 2.82%. Bengal gram contains maltose, glucose and fructose but during germination the level of germination get increased because of conversion of maltose and fructose to glucose, due to activities of α-amylase and β-amylase enzymes, which increase with soaking and subsequent germination. Srivastva et al. (1988) also reported an increase in amylase activity during soaking and subsequent germination of pigeon pea. The results of soaking, germination and heating related with decrease in total carbohydrate and increase in sugars (total, reducing and non-reducing) are in agreement with those of germination effect study by Tatsadjieu et al. (2004) and germination effect study on soyabean by Dogra et al. (2001).

The protein and fat contents in Bengal gram as affected by soaking, germination and heating are given in Table-5 The germination had an increasing effect on protein content 18.38 % to 22.36 %. Increase in protein content was also noted by Camacho et al. (1992) during germination of beans, lentils, chick pea and pea's seeds. Ohtsubo et al. (2005) found an increment in crude protein of germinated brown rice. Results, regarding the effect of sprouting on the proximate composition of Bengal gram seeds, agree with Obizoba (1991) who reported increase in % moisture, % crude protein and % ash. According to him, total nitrogen, total non-protein nitrogen; protein nitrogen, true protein nitrogen also increased with sprouting. Khatoon and Prakash (2006), also noted increase in the percent protein in germinated grains. Bau et al. (1997) assumed that the increased was due to synthesis of enzyme proteins (for example, proteases) by germinating seed or a compositional change following the degradation of other constituents. A further explanation was done by Nonogaki et al. (2010) where they noted that protein synthesis occurred during imbibitions and that hormonal changes play an important role in achieving the completion of germination.

The fat content in present study decreased from 3.94 % to 3.82% after germination. These results were in line with Chung et al. (1998) who noted significant losses in lipid content during canola sprouting. The decrease in fat content of seed could be due to total solid loss during soaking prior to germination (Wang et al., 1997).

Crude fiber was increased in germinated Bengal gram from 4.2 % to 5.3 % (Table-5). Muyanja et al., (2001) reported that the crude fiber content increases during germination of millet grains. In germinated rice, the amount of crude fiber was contributed by the presence of bran layer, an outer layer of rice that contained fiber (Ohtsubo et al., 2005). Chung et al. (1998) reported that in barley (but not in canola) sprouting was associated with significant increase in crude fiber from 3.75% in un-sprouted barley to 6% in 5 days sprouts due to synthesis of structural carbohydrates such as cellulose and hemi-cellulose, a major constituent of cell walls. In a similar study conducted on soybean. Therefore, Studies have shown that intake of legumes have many health effects in controlling and preventing metabolic diseases such as diabetes mellitus and coronary heart diseases (Slavin et al., 1997). Combination of whole grain and legume powder in coronary artery disease patients without diabetes mellitus can reduce fasting levels of glucose and insulin (Jang et al., 2001). Germination meanwhile, altered the biochemical composition of legumes. Decrease of carbohydrate level is beneficial to diabetes mellitus patients. In addition, increase of total dietary fiber and decrease of fat content can give benefit to people with cardiovascular disease and hypercholesterolemia.

Table 5 indicated that the starch content of Bengal gram decreased from 44.62 % to 39.64 % during soaking, germination (96 hrs) and heating. The increased á- amylase and β-amylase activities correspond with the decrease in the starch content, increase in amylose content from 7.36% to 9.31% and decrease in amylopectin from 35.3 % to 32.28 % in germinated Bengal gram flour and these results are in agreement with those of Sharma et al., (2007). Palmer, (1989) reported that the germination of cereal grains caused extensive changes in the structure and composition of major macromolecular components of the grains. Since gelatinization rates are related to starch composition and structure, the extent of starch modification during grain germination may influence rates of gelatinization. Therefore, the small increase in gelatinization rates for starches from the out-of-steep through to the four day germinated grains at 50°C assay temperature may suggest small improvements in starch degradability due possibly to differences in the degree of modification of starch structure and composition.

Green gram

The moisture content of un-germinated and germinated flour samples of Green gram varied between 8.06 % to 9.8 % (wb) as shown in Table-6 means moisture content was increased after germination. This finding is similar to the results reported by Khatoon and Prakash, (2006) in germinated legumes. As germination proceeds, legumes took up water from the surrounding in order for the metabolic process to commence. Dry legumes absorb water rapidly, influenced by the structure of the legume. The increase in water uptake with time is due to the increasing number of cells within the seed becoming hydrated (Nonogaki et al., 2010).

Ash content of un-germinated Green gram flour decreased from 4.03 % to 3.88% (wb) during germination as shown in Table-6. Ash content was significantly decreased in germinated green gram. These results were in agreement with those reported earlier by several workers (Savage, 1988; Gopalan et al., 1989). Leaching out of solid matter during pre germination soaking process could be the reason for significant reduction of mineral matter on germination. The further insignificant decrease of ash after dehulling could be contributed to removal of hull portion, which may have some amounts of minerals.

Table 5: Chemical composition of un-germinated and germinated flours of Bengal gram

(↓ (%) Samples →	Un-germinated Bengal gram Flour	Germinated Bengal gram Flour
Moisture	8.95	9.49
Ash	2.85	2.79
Crude fiber	4.2	5.3
Carbohydrate	57.68	54.49
Reducing sugar	2.52	2.82
Starch	44.62	39.64
Amylose	7.36	9.31
Amylopectin	35.3	32.28
Protein	18.38	22.36
Fat	3.94	3.82

Table 6 indicated that the total carbohydrate and reducing sugar contents of Green gram as affected by soaking, germination and heating treatments. The soaking, germination and heating treatments given to Green gram grains decreased the total carbohydrate contents 59.87 % to 55.23 % in germinated flour of Green gram. This was because of active respiration process during soaking and germination. On the other hand, soaking, germination and heating increased the reducing 1.5% to 3.1%, due to activities of α-amylase and β-amylase enzymes, which increase with soaking and subsequent germination. Srivastva et al., (1988) also reported an increase in amylase activity during soaking and subsequent germination of pigeon pea. The results of soaking, germination and heating related with decrease in total carbohydrate and increase in sugars (total, reducing and non-reducing) in soyabean are in agreement with those of Dogra et al., (2001). The biological reactions involved in the germination are dependent on the temperature, because the enzymes involved have ranges of action that will retard or accelerate seed germination (Bewley and Black, 1994).

The protein and fat contents as affected by soaking, germination and heating are given in Table 6. The germination of Green gram had an increasing effect on protein content 21.53 % to 23.93 %. Khader, (1983) also reported increase in protein content after germination. Several other workers also observed increased protein content after germination in legumes (Ghavidel et al., 2007; Osman 2007). The fat content in present study decreased from 1.6 % to 1.4 % after soaking, germination (96hrs) and heating. On germination, there was a signiûcant decrease of fat content, which could be due to total solid loss during soaking prior to germination (Wang, et al,. 1997) or use of fat as an energy source in sprouting process. Similar results of decrease in fat content after soaking, Crude fiber was increased in germinated Green gram from 4.05 % to 5.95 %. In germinated rice, the amount of crude fiber was contributed by the presence of bran layer, an outer layer of rice that contained fiber (Ohtsubo et al., 2005). Therefore, Studies have shown that intake of legumes have many health effects in controlling and preventing metabolic diseases such as diabetes mellitus and coronary heart diseases (Slavin et al., 1997; Liu et al., 1999). Combination of whole grain and legume powder in coronary artery disease patients without diabetes mellitus can reduce fasting levels of glucose and insulin (Jang et al., 2001). Germination meanwhile, altered the biochemical composition of legumes. Decrease of carbohydrate level is beneficial to diabetes mellitus patients. In addition, increase of total dietary fiber and decrease of fat content can give benefit to people with cardiovascular disease and hypercholesterolemia.

Table 6 indicated that the starch content of Green gram decreased from 60.45 % to 52.42 % during soaking, germination (96 hrs) and heating. The increased α- amylase and β-amylase activities correspond with the decrease in the starch content during germination the amylolitic enzymes generated transform starch into fermentable sugar. Palmer, (1989) reported that the germination of cereal grains caused extensive changes in the structure and composition of major macromolecular components of the grains. Since gelatinization rates are related to starch composition and structure, the extent of starch modification during grain germination may influence rates of gelatinization. Therefore, the small increase in gelatinization rates for starches from the out-of-steep through to the four day germinated grains at 50°C assay temperature may suggest small improvements in starch degradability due possibly to differences in the degree of modification of starch structure and composition. Changes such as these have been reported by Glennie et al., (1984) in the structure of sorghum endosperm starch during germination and are also consistent with similar observations on starch from germinating barley grains (MacGregor and Matsuo, 1982).

Table 6: Chemical composition of un-germinated and germinated flours of Green gram

(↓ (%) Samples →	Un-germinated Green gram Flour	Germinated Green gram Flour
Moisture	8.06	9.8
Ash	4.03	3.88
Crude fiber	4.05	5.95
Carbohydrate	59.87	55.23
Reducing sugar	1.5	3.1
Starch	60.45	52.42
Amylose	5.5	5.04
Amylopectin	55.4	46.9
Protein	21.53	23.93
Fat	1.6	1.4

Changes in Physicochemical Properties

WAI of un-germinated grains flour of Wheat, Bengal gram and Green gram increased from 2.7 % to 2.75 %, 2.89 % to 2.94 % and 1.76 % to 2.13 % respectively after soaking (12 hrs), germination (96 hrs) and heating. The WAI of germinated grains got increased due to increased protein content during germination since protein get increased the protein structure then hold maximum water by binding water molecules in its structure so therefore WAI increase after germination, these results are in agreement with the results obtained by Cira-Chavez et al., (2009).

Table 7 indicated that WSI (water solubility index) of un-germinated flour of Wheat, Bengal gram and Green gram increase from 17 % to 27 %, 15 % to 17% and 18% to 25 %, respectively after soaking (12hrs), germination (96 hr) and heating of grains flour, with the fact that during the germination process the carbohydrate content decreased as a result of hydrolysis by the amylase enzymes. The increase in WSI with germination is of significance since it gives an indication that germination can be used to increase the amount of soluble materials, such as starch and amino acids, which can be easily digestible (Pelembe et al., 2000).

OAC (Oil Absorption Capacity) of all three un-germinated grain flours of wheat, Bengal gram and green gram were decreased after 12 hrs soaking, 96 hrs germination and heating, from 1.27% to 1.28%, 1.48% to 1.51% and 1.56% to 1.58 % respectively. In faba-bean flour protein content increases, WAI increases and OAC decreases after germination in the same trend is also exist for above all the grain so; these results are totally in agreement with Rahma, (1988).

Table 7: Physical (flour) properties of un-germinated and germinated flours of wheat, Bengal gram and green gram

Samples	Wheat (u)	Wheat (g)	Bengal gram (u)	Bengal gram (g)	Green gram (u)	Green gram (g)
WAI (%)	2.7	2.75	2.89	2.94	1.76	2.13
WSI (%)	17	27	15	17	18	25
OAC (%)	1.27	1.28	1.48	1.51	1.56	1.58

u= un-germinated, g=germinated

Effect of process variables on product lateral expansion

Expansion is the most important physical property of the snack food. Starch, the main component of cereals plays major role in expansion process Kokini et al. (1992). The measured expansion of germinated wheat flour and Bengal gram flour and green gram flour extrudates varied between 136.66% and 220%.

Tables 8 shows the coefficient of the model and other statistical attributes of various responses used in the formulations. The Model F-value of 1382.60 implies the model is significant. In this case $x_1, x_2, x_3, x_4, x_5, x_1^2, x_2^2, x_3^2, x_4^2, x_5^2$, $x_2 x_3$, and $x_2 x_5$ are significant model terms at ($P<0.05$). Lack of fit is not significant, R^2 and Adjusted R^2 values of the model are 0.9989 and 0.9981 ,respectively.The Adequate precision value of 162.447 indicates that the model can be used to navigate the design space as it is greater than 4.0 (Montgomery, 2001). Considering all the above criteria, the following response model was selected for representing the variation of lateral expansion ratio for further analysis.

$$LER = 169.59+4.19.x_1+3.12.x_2+14.57.x_3+9.12.x_4+6.92.x_5 -1.16x_1^2 -0.87.x_2^2 +4.45x_3^2 -0.71.x_4^2 +2.73x_5^2 + 7.53x_2.x_3 -0.72.x_2.x_5 \quad (1)$$

Where x_1, $x_2 x_3$, x_4 *and* x_5 are the coded values of % Green gram flour,% Bengal gram flour, temperature °C , screw speed rpm and moisture content %, respectively.

Following observations can be made from equation (1), the coefficient of x_1(%GGF), x_2 (%BGF), x_3 (temperature), x_4 (screw speed) and x_5 (moisture content) are positive, therefore increase in feed composition, extrusion temperature and screw speed will increase lateral expansion ratio. Out of these, feed composition is most significant, next to that is temperature. Since coefficient of x_4^2 is negative, a maximum expansion will occur in the range of screw speed selected for the study.

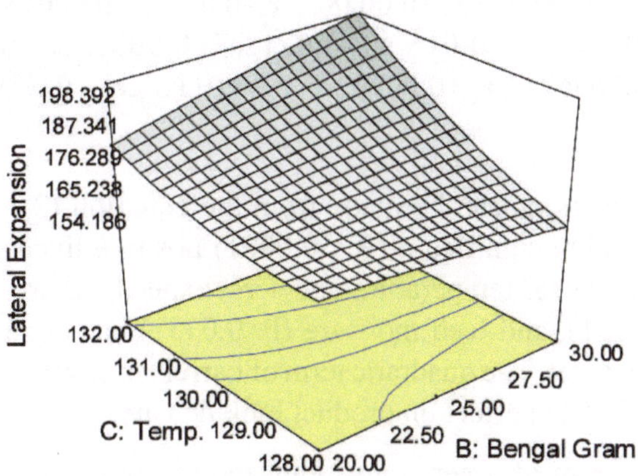

Fig.1: Response surface plot for the variation of lateral expansion ratio of extrudate as a function of barrel temperature and feed composition

F-values for square term of barrel temp and moisture content (x_3^2 *and* x_5^2) were 896.70 and 336.62 and p value was 0.0001and 0.0001 ($P< 0.05$) respectively, validating that these terms are significant. Since coefficients of x_3^2 *and* x_5^2 are positive; they will show the concave shaped variation with the lateral expansion ratio.

F-value for the interaction term $x_2.x_3$ is 1399.43 and P-value is <0.0001. Table 8 shows that this interaction term is significant. From the Fig.-1 it can be observed that lateral expansion ratio decreases with decrease in temperature and increases with increase in % Bengal gram flour.

Effect of process variables on product Bulk Density

Bulk density is a major physical property of the extrudate products. The bulk density, which considers expansion in all direction, ranged from 0.118 to 0.384 g/cm^3. The Model F-value of 248.97 implies the model is significant. In this case x_1, x_2, x_3, x_4, x_5, x_1^2, x_2^2, x_3^2, $x_1.x_2$ are significant model terms at ($P<0.05$). The "Lack of Fit F-value" of 4.17 implies the Lack of Fit is not significant. There was only a 0.01% chance that a "Model F-Value" this large could occur due to noise. The fit of model was also expressed by the coefficient of determination R^2, which was found to be 0.9957. The Adj R^2 was 0.9917 and Adequate Precision was 75.219 showed an adequate signal. A ratio greater than 4 is desirable and hence this model may be used to navigate the design space (Montgomery 2001).

Following regression equation was selected to represent the variation of Bulk density (BD) (g/cm^3) with independent variables.

$$\text{Bulk density} = 0.26 + 0.015.x_1 + 0.002833.x_2 - 0.043.x_3 - 0.006417.x_4 + 0.002417.x_5 + 0.014.x_1^2 + 0.030.x_2^2 - 0.015.x_3^2 - 0.021.x_4^2 + 0.003.x_1.x_2 + 0.00375.x_1.x_4 + 0.00637.x_1.x_5 - 0.00962.x_2.x_3 + 0.00762.x_2.x_5 - 0.011.x_3.x_4 - 0.003.x_3.x_5 \quad (2)$$

Following observations can be made from the equation (2), bulk density of extrudate had highly significant ($P < 0.0001$) positive linear effect of green gram flour (x_1) ,Barrel temperature and screw speed, where as bengal gram flour % ($P<0.0151$), and feed moisture ($P<0.0337$) had significant effect at 95% confidence level. The quadratic term of barrel temperature(x_3^2) had highly significant ($P<0.0001$) effect on product bulk density.

The interaction term of green gram flour and screw speed (x_1x_4) and barrel temperature and feed moisture (x_3x_5) had a significant effect at 95% confidence level with P-value 0.0096 and 0.0317 respectively. In this case x_1, x_2, x_3 x_4, x_5, x_1^2,x_2^2, x_3^2, x_1x_2 and x_2x_3 were the significant model terms. The coefficient of x_1,x_2 and x_5 is positive. Therefore, increase in green gram flour and Bengal gram flour and feed moisture may increase the bulk density of product whereas increase in barrel temperature and screw speed may reduce the bulk density of extrudate (Fig. 3). Since coefficient of x_3^2 is negative, a maximum bulk density will be in the range of temperature considered in the study.

In this study moisture content showed a highly significant effect on the bulk density of extrudates. As the feed moisture content value increased, there was a sharp increase in extrudate bulk density (Fig.2). The dependence of bulk density and expansion on feed moisture would reflect its influence on elasticity characteristics of the starch based material. Increased feed moisture content during extrusion would change the amylopectin molecular structure of the material reducing the melt elasticity thus decreasing the expansion, but increasing the density of extrudate (Ilo et al., 1999;). Barrel temperature was found to have negative significant effect on product bulk density. Increasing temperature decreased bulk density as in agreement with Ilo et.al. (1999), Singh et al., (2007) and Altan et al., (2008). Higher temperature provides higher potentioal energy for flash-off of super –heated water from extrudates as they leave the die. With higher barrel temperatures, the extrudates exiting the die lose more moisture and become lighter in weight (Koksel et. al., 2004).

Quadratic terms of barrel temperature (x_3^2) had negative coefficient showing convex shaped variation in bulk density with these variables. Quadratic terms of feed composition (x_1^2) and (x_2^2) had positive coefficient showing concave shaped effect on bulk density. The sign of all the significant square terms are opposite to the expansion terms as expected. In this study interaction term of feed composition and feed moisture (x_1x_5) was found to be positively significant ($P < 0.05$). Therefore we can assume that BD will show convex shape variation with the change in value of variables. It reveals that higher product bulk density will obtain when increase feed composition and feed moisture. Other significant ($P<0.05$) interaction term of feed moisture and barrel temperature had negative coefficient.

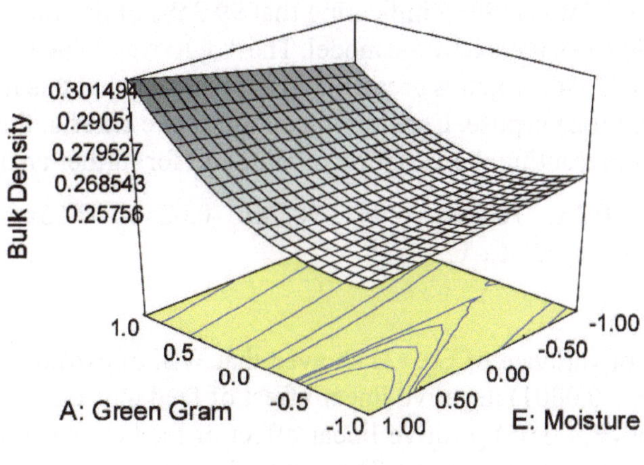

Fig. 2: Response surface plots for the variation of bulk density as function of feed composition (green gram flour) and feed moisture at center level of screw speed and barrel temperature.

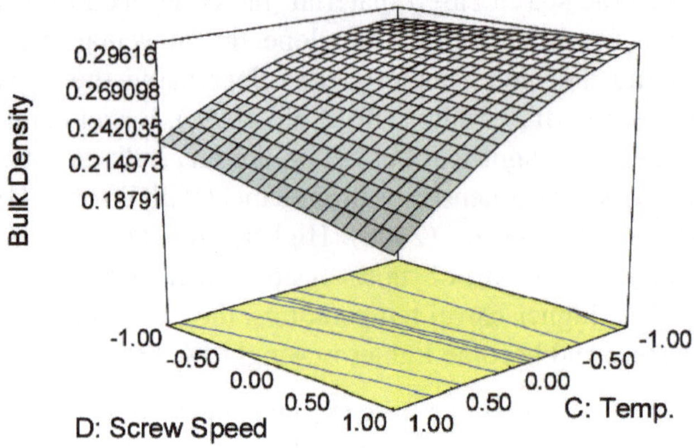

Fig. 3: Response surface plots for the variation of bulk density as function of barrel temperature and screw speed at central level of feed composition and moisture content.

Effect of process variables on product water solubility index (WSI)

WSI used as an indicator of degradation of molecular components. The WSI values ranged from 16.98% to 22.56% for the extrudates. ANOVA and Regression model fitted to experimental results of WSI (Table-8) shows that Model F-value of 318.96 was highly significant ($P < 0.0001$) whereas lack of fit F-value of 0.1384 was not significant ($P > 0.05$). The coefficient of determination, R^2 was 0.9935 indicating that 99.35% of the variability of the response could be explained by the model. The Adj R^2 was 0.9903 and Adequate Precision was 77.042 which is greater than 4, justifying that this model can be used for prediction purpose. Considering all the above criteria, the model was selected for representing the variation of WSI and for further analysis.

$$WSI = 19.09 + 0.36x_1 + 0.28x_2 + 0.94x_3 + 0.74x_4 - 0.025\,x_5 - 0.16x_4^2 + 0.13x_1x_4 + 0.61\,x_2x_4 + 0.16x_2x_5 + 0.36x_3x_4 \qquad (3)$$

The analysis of variance of Eq. (3) showed that WSI of extrudate had highly significant ($P < 0.0001$) negative linear affect of feed moisture (x_5), followed by significant ($P < 0.05$) positive linear effect of feed composition(x_1), (x_2), temperature (x_3) screw speed (x_4). The quadratic term of screw speed (x_4^2) and had significant ($P < 0.05$) negative effect. The interaction term of feed composition (green gram) and feed moisture (x_1x_5) and feed composition (green gram) and barrel temperature had highly significant ($P < 0.0001$) positive effect on product WSI showing concave shape variation with the change in value of variables. Other interaction terms x_2x_5 and x_3x_4 were also significant ($p<0.05$) with negative relation with product WSI.

In this case terms, x_1, x_2, x_3, x_4, x_4^2 x_1x_4, x_2x_4, x_2x_5, x_3x_4 were significant.

Negative coefficient of moisture content of WSI model supports that the value will decrease significantly ($P < 0.05$) with increasing feed moisture content. Similar effects have been reported for rice based extrudates by Ding et al. (2005) and for maize, finger millet extrudates by Onyango et al. (2004). It was suggested that increasing WSI is caused by greater shear degradation of starch during extrusion at low moisture conditions. The WSI increased significantly ($P < 0.05$) with increasing screw speed (Fig-4). The increase in WSI with increasing screw speed was consistent with the results reported for corn meal and corn and wheat extrudates (Mezreb, et al., 2003). Who reported that the increase of screw speed induced a sharp increase of specific mechanical energy, the high mechanical shear degraded macromolecules, and so the molecular weight of starch granules decreased and hence increased WSI.

WSI increased significantly ($P < 0.05$) with increasing green gram and Bengal gram flour proportion in extrudate. Similar findings were reported by Altan et al. (2008) in barley flour and tomato pomace extrudates. WSI is parameter that reflects the degradation suffered by the components of the fiber (Larrea et al., 2005). So, the highest WSI values may be due to disintegration of starch granules and low molecular compounds from extrudate melt during extrusion. This may cause an increase in soluble material. WSI determines the amount of free polysaccharide or polysaccharides released from the granule after addition of excess water (Altan et al., 2008). Increase in feed moisture content significantly ($P<0.050$) decreases WSI. Increase in feed moisture increases the plasticity of feed thus minimizes the chance of formation of small fraction polymer.

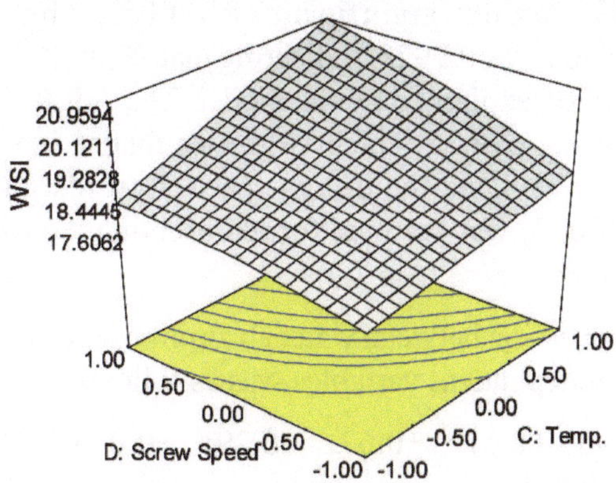

Fig. 4: Response surface plot for water solubility index as a function of screw center value of feed composition and feed mosture

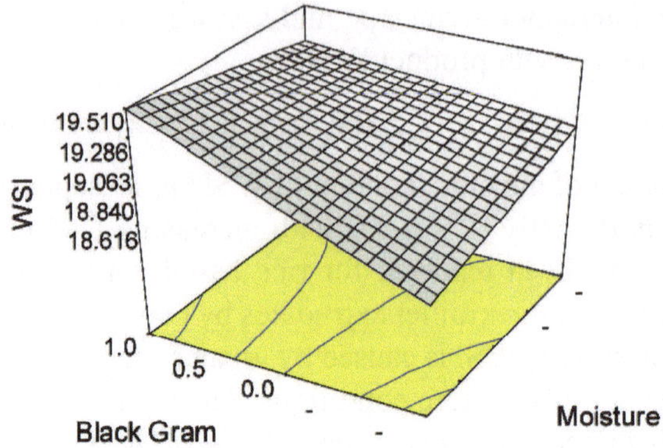

Fig. 5: Response surface plot for water solubility index as a function of feed speed and temperature at compostion and moisture at center value of screw speed and temperature.

In this experiment the WSI increased significantly with increasing barrel temperature. Gutkoski and El-Dash (1999) observed similar behavior in extruded oat products. In this study coefficient of interaction term of x_1x_4, x_2x_4, x_2x_5 and x_3x_4 were found to be positively significant therefore; they will show concave shape variation with the change in value of variables. Response surface plot (Fig.-5) showed that WSI decreased with increase in moisture, whereas increased with an increase in screw speed and temperature

Effect of process variables on product water absorption index (WAI)

WAI measures the amount of water absorbed by starch that can be used as an index of gelatinization. The WAI ranged from 12.13 to 15.19 g/g for extrudate. Regression model fitted to experimental results of WAI (Table 8) shows that Model F-value of 291.03 was highly significant ($P < 0.0001$) whereas lack-of-fit F-value of 3.43 was not significant ($P > 0.05$). The coefficient of determination, R^2 which was 0.9928, indicating that 99.28 % of the variability of the response could be explained by the model. The Adj R^2 was 0.9854 and Adequate Precision was 69.625 which is greater than 4 indicating that the model can be used to navigate the design space. Considering all the above criteria, the model (Eq. 4) was selected for representing the variation of WAI and for further analysis.

Multiple regression equation between water absorption index (g/g) of extrudate and coded levels of independent variables was as follows:

$$WAI = 13.89+0.044x_1 +0.11x_2 +0.67x_3 +0.29x_4 +0.036x_5 -0.21x_1^2 -0.23x_2^2 -0.24x_1x_2-0.051x_2x_4 -0.041x_3x_4 \quad (4)$$

Table 8: Analysis of variance results of equation

Response	Source	Sum of squares	df	Mean squares	F-value	P-value	R^2
LER	Regression	10766.57	12	897.21	1382.60	<0.0001**	0.9989
	Lack of fit	9	14	0.64	0.96	0.5662	NS
	Pure error	3.33	5	0.67			
	Residual	12.33	19	0.65			
	Total	**10791.23**	**50**				
	Regression	0.097	15	0.00647	248.97	< 0.0001*	0.9957
	Lack of fit	0.0003571	11	0.000034	4.17	0.0634	NS
	Pure error	0.00004083	5	0.00002599			
	Residual	0.000415	16	0.00000816			
	Total	**0.09781293**	**47**				
WSI	Regression	48.82	10	4.88	318.96	< 0.0001*	0.9935
	Lack of fit	0.029	16	0.018	2.70	0.1384	NS
	Pure error	0.033	5	0.0066			
	Residual	0.32	21	0.015			
	Total	**49.202**	**52**				
WAI	Regression	16.73	10	1.67	291.03	< 0.0001*	0.9928
	Lack of fit	0.11	16	0.00691	3.43	0.0894	NS
	Pure error	0.010	5	0.00201			
	Residual	0.12	21	0.00574			
	Total	**16.97**	**52**				
Overall acceptability (OA)	Regression	5.14	8	0.64	63.63	< 0.0001*	0.9568
	Lack of fit	0.099	18	0.00548	0.21	0.9945	NS
	Pure error	0.13	5	0.027			
	Residual	0.503735	15	0.033582			
	Total	**5.872735**	**46**				

*significant at $P < 0.05$, df: degrees of freedom

The analyses of variance of Eq. (4) shows that WAI of extrudate had highly significant (P < 0.05) positive linear affect of green gram flour(x_1), Bengal gram flour (x_2), barrel temperature (x_3),screw speed (x_4) and feed moisture(x_5). The quadratic term of feed composition had significant (P<0.05) negative effect on WAI. It shows that variation in WAI will be concave shaped in graph. Other quadratic terms were not significant (P > 0.05).

The interaction term of feed composition (Bengal gram) and screw speed (x_2x_4) had significant (P < 0.001) negative coefficients indicating concave shape variation with the change in value of variables. Other interaction terms found to be not significant (P > 0.05). In this case x_1, x_2, x_3, x_4, x_5, x_1^2, x_2^2, and x_2x_4 were significant model terms. Positive coefficient of feed composition and feed moisture content of WAI model supports that with increasing in feed moisture content, the WAI value will increase. Higher moisture favors gelatinization thus decreasing product WAI. The negative coefficients of temperature indicate that increasing in temperature will decrease the WAI value of the product. A decrease in WAI with increasing temperature was probably due to decomposition or degradation of starch (Pelembe et. al., 2002). Ding et al., (2006) also stated that the WAI decreases with increasing temperature if dextrinization or starch melthing prevails over the gelatinization phenomenon. Positive coefficient of feed composition indicates that increase in green gram flour and Bengal gram flour proportion will increase the WAI of the product. Screw speed had significant positive effect on WAI which shows that increase in screw speed will increase WAI.

The coefficient for interaction term of feed composition (green gram) and screw speed (x_2x_4) is significant and negative thereby it will show concave shape (Fig.-6) variation with the change in value of variables. Figure 7 shows that there will be higher value of WAI at low level of green gram flour irrespective

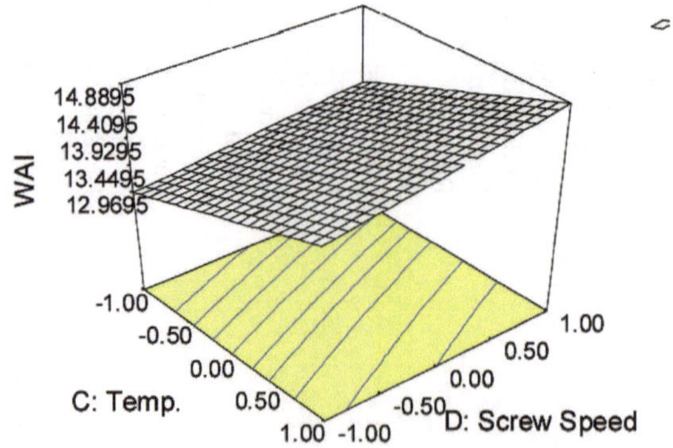

Fig. 6: Response surface plot for water absorption index as a function of feed composition and screw speed at center value of moisture content and temperature.

screw speed level. Increase in feed composition (green gram) is decreasing WAI and increase in feed moisture is increasing WAI. Fig-7 shows that higher value of WAI will obtain at higher level of moisture and lower level of green gram.

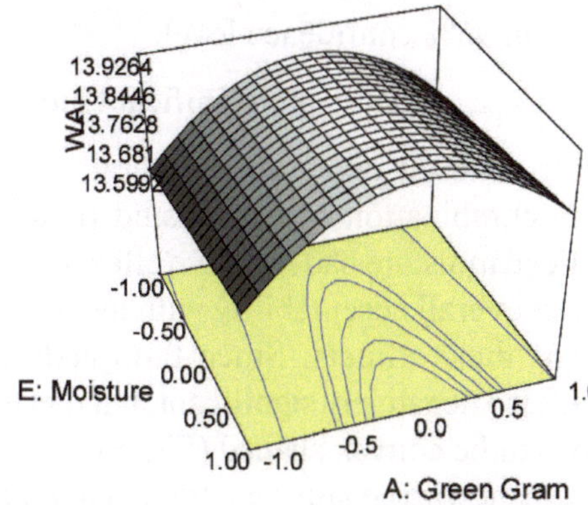

Fig. 7: Response surface plot for water absorption index as a function of feed composition and and feed moisture at center value of temperature and screw speed.

Effect of process variables on product overall acceptability (OA)

Sensory evaluation indicates the acceptability of the product. In present study, overall acceptability was analyzed on the basis of product appearance, taste, texture and color. Hedonic scale was used to find the different aspect of sensory evaluation. The overall acceptability of the product ranges from 6.12 to 7.75 in the extrudates prepared from combination of germinated wheat flour, green gram flour and Bengal gram flour. The coefficient of the model and other statistical attributes of overall acceptability are shown in (Table 8 and 9). The Model F-value of 63.63 implies the model is significant. In this case x_3, x_5, x_1^2, x_2^2, x_1x_5 *and* x_3x_5, are significant model terms at ($P<0.0001$). Analysis of data resulted non–significant "Lack of Fit F-value" 0.21. R^2 value 0.9568, Adj R^2 0.9417 and A deq precision 30.167 shows that the resulting model can be used to navigate the design space. Considering all the above criteria, the following model was selected for further analysis.

$$OA=7.39+0.037.x_1-0.025.x_2+0.100.x_3-0.16.x_5-0.18.x_1^2-0.31.x_2^2+0.11.x_1.x_5+0.16.x_1.x_5 \quad (5)$$

Following observation can be made from equation (5), Coefficients of the linear terms x_2, x_5 are negative which indicates that increase in Bengal gram flour and feed moisture will decrease the overall acceptability. Coefficients of terms

x_1 *and* x_3 are positive indicating that increase in green gram flour and temperature will increase the overall acceptability. Linear term of screw speed was not significant at 95% confidence level. Quadratic terms x_1^2 and x_2^2 have negative coefficient. The interaction term of feed composition and moisture (x_1x_5) had significant positive effect at 95% confidence level.

In this case $x3$, x_5, x_1^2, x_2^2, x_1x_5, and x_3x_5 are significant model terms.

Increase in feed moisture caused decrease in overall acceptability of the extrudate prepared from combination of germinated flours of wheat, green gram and Bengal gram. Feed moisture had negative effect on product expansion and grittiness. Decrease in overall acceptability with increase in feed moisture may be the reflection of these effects. Since the quadratic term of feed composition x_1^2 and x_2^2 is also negatively significant. It indicates that the overall variation in product OA will be convex shaped (Fig. 8). As the quadratic term of x_1^2 and x_2^2 is also significant with negative coefficient it indicates that product OA will be very low with further increase in Bengal gram flour and green gram flour (Fig. 9). Temperature had significant positive linear effect on product OA. It indicates that increase in barrel temperature increases product OA. This may be due to higher expansion of the extrudate. However with further increase in temperature caused slight decrease in product OA. Feed composition had significant effect only at quadratic level showing convex shaped variation on product OA.

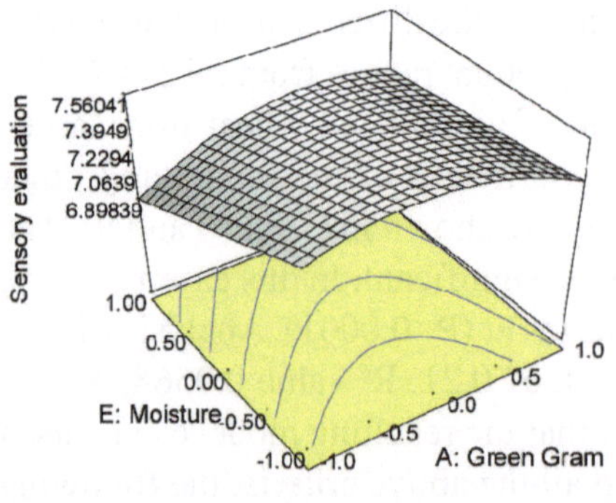

Fig. 8: Response surface plot for overall acceptability as a function of feed composition and and feed moisture at center value of temperature and screw speed.

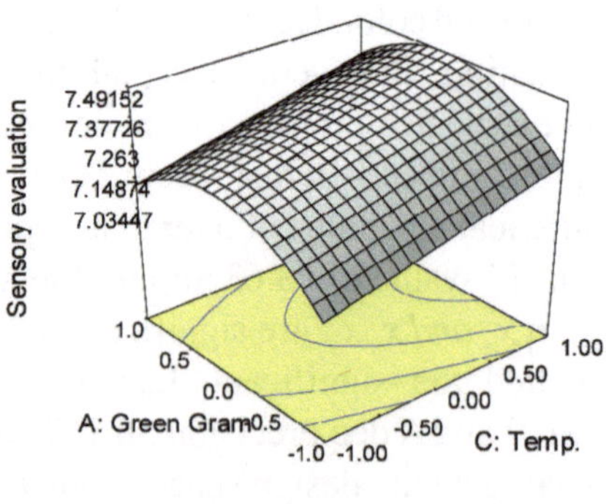

Fig. 9: Response surface plot for overall acceptability as a function of feed composition and and temperature at center value of moisture and screw speed.

Optimization

A numerical multi-response optimization technique was applied to determine the optimum combination of germinated flour of green gram, Bengal gram in feed composition, feed moisture, and screw speed and barrel temperature for the production of extrudate containing germinated flour of wheat, green gram and Bengal gram.

The assumptions were to develop a product which would have maximum score in sensory acceptability so as to get market acceptability, maximum expansion, minimum bulk density, and minimum hardness. Therefore, among responses, these parameters were kept within range. Under these criteria, the un-coded optimum operating conditions for development of germinated wheat, green gram and Bengal gram flour extrudate were 132°C of barrel temperature, 275 rpm of screw speed, 17% of feed moisture, 6.5% of green gram flour and 23.1% of Bengal gram flour. The responses predicted by the software for these optimum process conditions resulted lateral expansion of 169.59%, bulk density, 0.263g/cm^3; water solubility index, 19.08; water absoption index, 13.88g/g; and overall acceptability, 7.38. Table 9 shows the different condition of the constraints for optimization.

Table 9: Multi response optimization constraints of extrudate

Parameters	Goal	Lower limit	Upper limit	Lower Weight	Upper Weight	Importance
Green gram	is in range	-1	1	1	1	3
Bengal gram	is in range	-1	1	1	1	3
Temperature	is in range	-1	1	1	1	3
Screw speed	is in range	-1	1	1	1	3
Moisture	is in range	-1	1	1	1	3
Lateral expansion	Maximize	136.66	220.023	1	1	3
Bulk density	Minimize	0.118	0.384	1	1	3
WSI	Minimize	16.98	22.56	1	1	3
WAI	Maximize	12.13	15.19	1	1	3
Overall Acceptability	Is in range	6.12	7.75	1	1	3

Verification of results

The suitability of the model developed for predicting the optimum response values was tested using the recommended optimum conditions of the variables and was also used to validate experimental and predicted values of the responses. The results of the different physic-chemical properties of extrudate are shown in Table 10.

Table 10: Predicted and actual values of the responses at the optimized condition

Responses	Predicted	Actual
Lateral expansion	169.59	167.92
Bulk density	0.263	0.271
WSI	19.08	20.20
WAI	13.88	14.64
Overall acceptibility	7.38	7.35

References

1. Ainsworth, P. Þenol Ýbanog¡lu, Emir Ay°e Özer, and Andrew, P. (2006), Physical and sensory evaluation of a nutritionally balanced gluten-free extruded snack, Journal of Food Engineering, 75: 469-472.
2. Akaerue. B.I. and Onwuka, G.I. (2010). Evaluation of the yield, protein content and functional properties of Mung bean [Vigna radiata (L.) Wilczek] protein isolates as affected by processing. Pakistan Journal of Nutrition, 9(8):728-735.
3. Altan, A. McCarthy, K.L. Maskan, M. (2008). Twin-screw extrusion of barley-grape pomace plends: Extrudate characteristics anddetermination of optimum processing conditions. Journal of Food Engineering, 89:24-12.
4. Antu, Grewal. Sudesh, Jood. (2009). Chemical composition and digestibility (in vitro) of green gram as affected by processing and cooking methods, Nutrition and Food Science, 39(4):342-349.
5. Arhaliass, A. Bouvier, J.M. and Legrand, J. (2003). Melt growth and shrinkage at the exit of the die in the extrusion cooking process. Journal of Food Engineering, 60:185-192.
6. Augustin, J. Klein, B.P. (1989). Nutrient composition of raw, cooked, canned, and sprouted legumes. In: Matthews RH (ed) Legumes: chemistry, technology, and human nutrition. Dekker, New York, 187-217.
7. Badau, M.H. Nkama, I. and Jideani, I.A. (2005). Phytic acid content and hydrochloric acid extractability of minerals in pearl millet as affected by germination time and cultivar. Food Chem., 92: 425-435.
8. Bau, H.C. Villaume, J.N. Mejean, L. (1997). Effect of germination on chemical composition, biochemical constituents and antinutritional factors of soya bean (Glycine max) seeds. Journal of Science Food and Agriculture, 73(1):1-9.
9. Bewley, J.D. and Black, M. (1994). Seeds: Physiology of Development and Germination. In Seeds: Germination,Structure, and Composition, p. 1-2. Plenum Press, New York
10. Bhattacharya, S. and Choudhury, G.S. (1994) Twin screw extrusion of rice flour: effect of extruder length to diameter ratio and barrel temperature on extrusion parameters and product characteristics. Journal of Food Processing and Preservation, 18:389-406.
11. Brandtzaeg, B. Malleshi, N. G. Svanberg, U. Desikachar, H.S.R. Mellander, O. (1981). Dietary bulk as a limiting factor for nutrient intake in pre-school children. III. Studies of malted flours from ragi, sorghum and green gram. Journal of Tropical Pediatrics, 27: 184-189.
12. Camacho L, Sierra C, Campos R, Guzman E, and Marcus, D. (1992). Nutritional changes caused by the germination of legumes commonly eaten in Chile. Arch Latinoam Nutrition, 42(3):283-90.
13. Chung, T.Y. Nwokolo, E.N. Sim, J.S. (1998). Compositional and digestibility Changes in sprouted barley and canola seeds. Plant Foods Human Nutrition, 39(3):267-78.

14. Ding, Q.B. Paul, A. Plunkett, A. Tucker, G. and Marson, H. (2006). The effect of extrusion conditions on the functional and physical properties of wheat based expanded snacks. Journal of Food Engineering, 73:142-148.
15. Dogra, J.Y.S. Dhaliwal and Kalia, M. (2001). Effect of soaking,germination,heating and roasting on the chemical composition and nutritional quality of soybean and its utilization in Indian leavened products.Journal Food Science and Technology, 38:453-457.
16. Dulby, A. and Tsai, C.Y. (1976). Lysine and Tryptophan Increases during Germination of Cereal Grains. Cereal Chemistry, 53:222-224.
17. Finney, P.L. (1983). Effect of germination on cereal and legume nutrient changes and food and feed value. Recent Advances in Phytochemistry, 17:229-305.
18. Ghavidel, R.A. and Prakash, J. 2007 The impact of germination and dehulling on nutrients, antinutrients, in vitro iron and calcium bioavailability and in vitro starch and protein digestibility of some legume seeds. LWT, Food Science and Technology, 40(7):1292-1299.
19. Glennie, C.W. (1984). Endosperm cell wall modiûcation in sorghum grain during germination. Cereal Chemistry, 61:285–289.
20. Gopalan, C. Sastri, B.V. R. and Balasubramanian, S.C. (1989). Nutritive value of Indian foods. Hyderabad, India: National Institute of Nutrition, Indian Council for Medical Research.
21. Haber, A. and Runyon, R. (1977). General statistics (3rd ed.). Reading, MA: Addison-Wesley.
22. Harper, J.M. (1989). Food extruders and their applications. American Association of Cereal Chemists, 1-15.
23. Hemalatha, S. Platel, K. Srinivasan, K. Influence of germination and fermentation on bioaccessibility of zinc and iron from food grains. Department of Biochemistry and Nutrition, Central Food Technological Research Institute, Mysore, India.
24. Iqbal, A. Khalil, I.A. Ateeq, N. and Khan, M.S. (2006). Nutritional quality of important food legumes. Food Chemistry, 97:331-335.
25. Jang, Y. Lee, J.H. Kim, O.Y. Park, H.Y. and Lee, S.Y. (2001). Consumption of whole grain and legume powder reduces insulin demand, lipid peroxidation, and plasma homocysteine concentrations in patients with coronary artery disease: randomize controlled clinical trial. Arteriosclerosis, Thrombosis and Vascular Biology, 21:2065-2071.
26. Khalil, A.W. Zeb, A. Mahmood, F. Tariq, S. Khattak, A.B. Shah, H. (2007). Impact of germination time on comparative sprout quality characteristics of desi and Kabuli type chickpea cultivars (Cicer arietinum L). LWT-Food Science and Technolology, 103:115-120.
27. Khatoon, N. and Prakash, J. (2006). Nutrient retention in microwave cooked germinated legumes. Food Chemistry, 97(1):115-121.
28. Kokini J. Chang C. and Karwe M. (1992). Food Extrusion Science and Technology. Marcel Dekker Inc. New York, pp. 631–652.
29. Larrea, M.A. Chang, Y.K. and Martinez Bustos, F. (2005). Some functional properties of extruded orange pulp and its effect on the quality of cookies. Lebensm.Wiss.u. Technology, 38: 213-220.
30. Launay, B. and Lisch, L.M. (1983) Twin screw extrusion cooking of starches: flow behaviour of starch pastes, expansion and mechanical properties of extrudates. Journal of Food Engineering, 52:1746–1747.
31. Limami, A.M. Rouillon, C. Glevarec, G. Gallais A. and Hirel, B. (2002). Genetic and physiological analysis of germination efficiency in maize in relation to nitrogen metabolism reveals the importance of cytosolic glutamine synthetase. Plant Physiology, 130:1860-1870.

32. Lorenz, K. (1980). Cereal sprouts: Composition, nutritive value, food application. CRC Crittical Rev. Food Sci.Nutr., 13: 353-385.
33. Lorenz, K., 1980. Cereal sprouts: Composition, nutritive value, food application. CRC Critical Review: Food Science and Nutrition, 13:353-385.
34. MacGregor, A.W. and R.R. Matsuo (1982). Starch degradation in endosperms of barley and wheat kernels during initial stages of germination. Cereal Chem. 59:210-216.
35. Mclaren C.G Bartolome V.I. Carrasco M.C. Quintana L.C. Ferino M.I.B. Mojica J.Z. Olea A.B. Paunlagui L.C. Ramos C.G and Ynalvez M. A. (1977). Experimental design and data analysis for agricultural research, vol. 1. Los Banos. Laguna: International Rice Research Institute.
36. Mercier, C. and Feillet, P. (1975). Modification of carbohydrate component by extrusion cooking of cereal product. Cereal Chemistry, 52:283-297.
37. Mezreb, K. Goullieux, A. Ralainirina, R. and Queneudec, M. (2003). Application of image analysis to measure screw speed influence on physical properties of corn and wheat extrudates. Journal of Food Engineering, 57:145-152.
38. Montgomery, D.C. (2001) Design and Analysis of Experiments. John Wiley and Sons, Inc, New York, pp.427-500.
39. Moscicki, L. (2003). Effect of screw conûguration on quality and SME value of corn extrudate. Teka Commission of Motorization Power Industry in Agriculture, vol. III, pp.182–186.
40. Myers, R. H., Montgomery, D. C. and Anderson-cook, C. M. (1976). Response surface methodology: Process and product optimization using Designed Experiments. WILEY publications.
41. Nonogaki, H. Bassel, G.W. and Bewley, J.W. (2010). Germination-still a mystery. Plant Science doi:10.1016/j.plantsci.2010.02.010.
42. Obatolu, V.A. (2002). Nutrient and sensory qualities of extruded malted or unmalted millet/soybean mixture. Food Chem., 76: 129-133.
43. Obizoba, I.C. (1991). Effect of sprouting on the nitrogenous constituents and mineral composition of pigeon pea (Cajanus cajan) seeds. Plant Foods Human Nutrition, 41(1):21-26.
44. Ohtsubo, K. Suzuki, K. Yasui, Y. Kasumi, T. (2005). Bio-functional components in the processed pregerminated brown rice by a twinscrew extruder. Journal of Food Composition and Analysis, 18(4):303-316.
45. Onyango, C. Noetzold, H. Bley, T. and Henle, T. (2004). Proximate composition and digestibility of fermented and extruded uji from maize finger millet blend. Lebensm.Wiss u.Technology, 37:827-832.
46. Pelembe L. A. M. Erasmus C. and Taylor J. R. N. (2002). Development of a protein rich composite sorghum-cowpea instant porridge by extrusion cooking process. Lebensmitted-Wissenschaft Untersuchung Technology, 35:120-127.
47. Ram, P.C. Lodha, M L. Srivastava, K.N. Tyagi, R.S. Joginder Singh, and Mehta. S.L. (1979). Improving Nutritive Value of Maize (Zea mays L) By Germination." Journal of Food Science and Technology, 16:268-260.
48. Ranganna. S. (2003). Handbook of Analysis and Quality Control of Fruit and Vegetable Products, 2nd Ed., Tata McGraw Hill Publications. Co., New Delhi, India.
49. Rao, P.U. and Belavady, B. (1978). Oligosaccharides in pulses: varietal differences and effects of cooking and germination. Journal of Agricultural and Food Chemistry 26:316-319.
50. Reid, R. (1998). Hot and fruity: the future for snack flavours. Food Reviews, 25:15–21.
51. Sadana, B. and Chhabra,C. (2004). Development and sensory evaluation of low cost weaning food formulation.Journal of Human Ecology, 16:133-136.

52. Savage, G. P. (1988). The composition and nutritive value of lentils (Lens culinaris). Nutrition Abstract Review, 58(5):319–343.
53. Sebio, L., and Chang, Y.K. (2000). Effects of selected process parameters in extrusion of yam flour on physicochemical properties of extrudates. Nahrung, 44:96–101.
54. Singh, B. Sekhon, K. S. and Singh, N. (2007). Effects of moisture, temperature and level of pea grits on extrusion behaviour and product characteristics of rice. Food Chemistry, 100: 198-202.
55. Slavin, J. Jacobs, D. and Marquart L. (1997). Wholegrain consumption and chronic disease: protective mechanisms. Nutritional of Cancer, 27(1):4–21.
56. Stojceska. V, Ainsworth. P. Andrew, P. Esra, I. Senol, I. (2008). Cauliflower by-products as a new source of dietary fibre, antioxidants and proteins in cereal based ready-to-eat expanded snacks. Journal of Food Engineering. 87:554–563.
57. Tatsadjieu, N.L. Etoa F.X. and Mbofung, C.M.F. (2004). Drying Kinetics, Physicochemical and Nutritional Characteristics of "Kindimu", a Fermented Milk-Based-Sorghum-Flour. The Journal of Food Technology in Africa, 9(1):17-22.
58. Thakur S. and D.C. Saxena (2000). Formulation of Extruded Snack Food (Gum Based Cereal-Pulse Blend): Optimization of Ingredients Levels Using Response Surface Methodology. Lebensmittel-Wissenschaft und- Technologie, 33:354-361.
59. Vasanthan, T. Gaosong, J. Yeung, J. and Li, J. (2002). Dietary fiber profile of barley flour as affected by extrusion cooking. Food Chemistry, 77:35-40.
60. Wang, N. Lewis, M.J. Brennan, J.G. and Westby, A. (1997). Effect of processing methods on nutrients and antinutrional factors in cowpea. FoodChemistry, 58:59-68.
61. White, G. (1994). Defining the true meaning of snacks. Food Technology International Europe, 2:115-117.
62. Yamazaki, W. T. (1953). An alkaline water retention capacity test for the evaluation of cookie baking potentialities of soft winter wheat flours. Cereal Chemistry, 30:242-246.

Unit-IV: Cereals and other Food Grains as a Source of Functional Foods and Nutraceuticals

Chapter 23

Pre-Gelatinized Cereal Flours: Processing, Characteristics and Functionality

Gurkirat Kaur, Savita Sharma and Baljit Singh

Abstract

The concern of consumers for convenience and healthy foods involves many parameters like functional and nutraceutical properties, ease to prepare, good quality, economics as well as good nutritional profile of the food. This trend of convenience has been increasing in today's world because of the societal changes like the aging of population, the increasing entry of women into the labor force, the change to a multi-cultured society and the increasing polarization between poor and rich. Value addition to the traditional sources is a medium to bring variety products into the market. Cereal starches possess a wide application in the food industry as an additive but the native starch is not widely used in food industry due to its poor functional properties. Pregelatinized starches and flours have become important in processed foods because the functional properties of the starches and flours are improved over the native starches and cereal flours. Modification is a process which alters the structure of the starch in the wheat endosperm by affecting the hydrogen bonds in a controllable manner. Modification of starch and flours is generally done by several methods such as chemical degradation, physical alteration or enzymatic transformation. The primary objective of modification is to reduce the hot viscosity of the starch paste so that higher concentrations of starch can be dispersed without excessive thickening. The pregelatinized flours offer a tremendous number of functional benefits to variety of foods such as bakeries, snacks, beverages as well as nutritional foods. The various functional benefits of pre-gelatinized starch/flour as an additive are classified as fat replacer, texture improver, stabilizer, emulsifier, thickening agent and for freeze-thaw stability.

Keywords: Pregelatinization, modification, functionality, cereal flour.

Introduction

The current tendency of customers is to use practical foods which are easy to prepare and provides nutritional and health benefits. The trend of convenience has been increasing because of the societal changes like aging of population, increasing entry of women into labour force and the change to a multi-cultured society. The concept of convenience can be achieved by carrying out the modification processes. Cereal starches possess a wide application in the food industry as an additive but the native starch is not widely used in food industry due to its poor functional properties. Pregelatinized starches and flours have become important in processed foods because the functional properties of the starches and flours are improved over the native starches and cereal flours (Rubin, 2010). The pregelatinized flours offer a tremendous number of functional benefits to variety of foods such as bakeries, snacks, beverages as well as nutritional foods. The various functional benefits of pregelatinized starch/flour as an additive are classified as fat replacer, texture improver, stabilizer, emulsifier, thickening agent and for freeze-thaw stability (Alexander, 1995).

Modification is a process which alters the structure of the starch in the wheat endosperm by affecting the hydrogen bonds in a controllable manner. Starch is a biopolymer that occurs naturally as water insoluble and birefringent granules. Each granule consists of concentric growth rings of alternating amorphous and semi-crystalline composition. The semi-crystalline growth rings contain stacks of amorphous and crystalline lamellae. The crystalline lamellae consist of chain sections of amylopectin that form double helices, while other chain sections of amylopectin form connections between the helices (Waigh et al., 2000). The primary objective of modification is to reduce the hot viscosity of the starch paste so that higher concentrations of starch can be dispersed without excessive thickening.

Starch of the wheat flour can be modified/pregelatinized can be done by several methods such as chemical degradation, physical alteration, genetic modification or enzymatic transformations. Extrusion technology is the widely used and new technique for carrying out the modification process. It is a powerful processing operation, which utilizes high temperature, pressure, and shear force to produce highly expanded, low density products with unique texture properties. Extrusion processing of starchy materials is known to cause substantial changes in the physical constitution of the starch granules (Charboniere et al. 1973). Starch modification, which involves the alteration of the physical and chemical characteristics of the native starch to improve its functional characteristics, can be used to tailor starch to specific food applications.

Pre-gelatinization is done to reduce the hot viscosity of the starch paste so that high concentrations of starch can be dispersed without excessive thickening (Thomas and Atwell, 1999). Native starch cannot be gelatinized in cold water. Native starch is not widely used in the food industry due to its poor functional properties; therefore most starches currently incorporated into foods have to be pregelatinized

Methods of Pregelatinization

There are several methods by which modification can be carried out. Chemical modifications such as cross-linking and conversion, etc. (Takahashi et al., 1989), and physical treatments such as high pressure treatment (Douzals et al., 1998), microwave heating (Lewandwicz et al., 2000), heat-shear extrusion processing (Bryant et al., 2001) and warm-water treatment (Yamamoto and Shirakawa, 1999) are widely used to change starch properties.

Chemical Modification

Conversion: Conversion is a process that is used to reduce the viscosity of raw starches. Its main objectives are to allow the use of starches at higher percentages, increase the water solubility, control gel strength, or modify the stability of starch. Typically, native starch cannot be used at much more than 6% solids because it imparts such a high viscosity. However, in the confection industry in a product such as a soft gum candy, a low-viscosity starch at high solids is needed to obtain the desired gel structure and set. Thus, a converted starch is required. Methods of conversion include acid hydrolysis, oxidation, dextrinization, and enzyme conversion

***Crosslinking**:* This is a treatment whereby small amounts of compounds that can react with more than one hydroxyl group are added to the starch polymers. Crosslinking yields starch granules with increased resistance to overcooking and other variations in processing conditions. Its main purposes are to provide a short texture to cooked dispersions of starch and to impart resistance to processing conditions such as temperature, acid and shear.

Stabilization: This is a process whereby blocking groups are reacted with starch polymers to inhibit retrogradation, the alignment of polymers that causes a change in the structure of the food product. Inhibiting retrogradation imparts textural, and freeze-thaw stability, thus prolonging the shelf life of the food product. This modification is most important in frozen foods, since retrogradation of starch polymers is accelerated at cold temperatures, leading to an opaque, gelled, and/or chunky texture with eventual syneresis or "weeping" of liquid from the gel.

Physical Methods

Extrusion processing

Extrusion technology is widely used to make the starch based extruded products with low cost and high efficiency. Degree of starch gelatinization is dependent on the parameter of extrusion. Modified flours must have specific and tailored functional properties to meet the increased changing demands of convenient finished products. To achieve this target, use of highly flexible manufacturing techniques are to be identified. Extrusion cooking as a modern High temperature short time (HTST) method, meets these requirements in an ideal manner. Because of its low cost, versatility and continuous processing capability, it is one of the promising techniques used to modify the functional and digestible characteristics of cereal grains (Camire et al., 1990). Extrusion has made manufacturing more effective by bringing together several processing steps in a single, continuous unit.

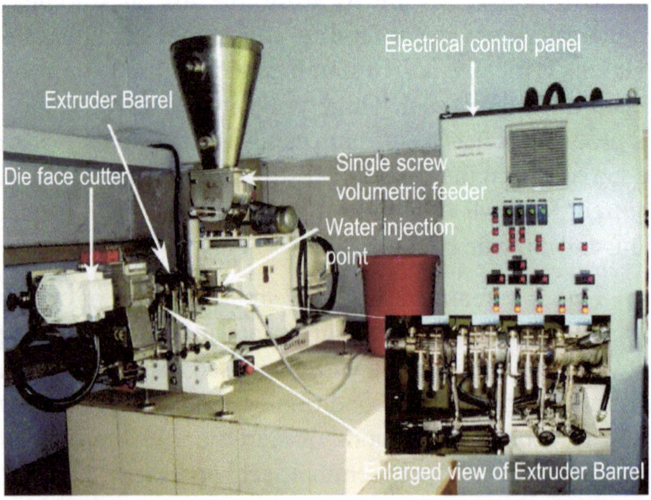

Fig. 1: Co-rotating intermeshing twin screw extruder.

The results of extrusion are gelatinization of starch, denaturation of proteins, inactivation of many native enzymes and antinutritional factors, reduction of microbial counts, and improvement in digestibility and biological value of proteins. Gelatinization of starch is phase transition of starch granules from an ordered to a disordered state during heating with excess water (Babic et al., 2009). The process of gelatinization causes substantial changes in both chemical and physical nature of granular starch due to the rearrangement of intra and intermolecular hydrogen bonding between water and starch molecules resulting in the collapse or disruption of molecular orders within the starch granules. This causes irreversible changes in the starch properties including loss of organized structure of starch, granule swelling, loss of birefringence and crystallinity (Anastasiades et al., 2002).

Characteristics of Pregelatinized Flours

Functional properties

The physico-chemical properties of starches such as swelling, solubility, and light transmittance have been reported to be affected significantly by extent of modification. The change in these properties upon modification depends on the type of modification. Chemical modifications such as acetylation and hydroxypropylation increase, while cross-linking has been observed to decrease (depending on the type of cross-linking agent and degree of cross-linking) the swelling power and solubility of starches from various sources. It has been reported that the waxy starches show an increased swelling power upon acetylation because of the presence of mainly amylopectin with a more open structure than in non-waxy starch; this allows rapid water penetration, and increased swelling power and solubility.

Morphological properties

Starch modification involves physical, chemical and biochemical phenomena on the surface of contacting phases. Microscopy (light and SEM) has played an important role in increasing understanding of granular structure of modified starches. It has been used to detect structural changes caused by chemical modifications and the most substituted regions in starch granules. Many of the less affected modified granules developed a depression that resulted later in slight fragmentation, indentation, and the formation of a deep groove in the central core region along the longitudinal axis in highly affected granules. These granules appeared as folded structures with their outer sides drawn inwards, giving the appearance of a doughnut. Moreover, these altered regions were observed to be apparently larger in large size granules compared with small size granules in all the potato starches examined; this may be attributed to differences in the native granule architecture and fragility.

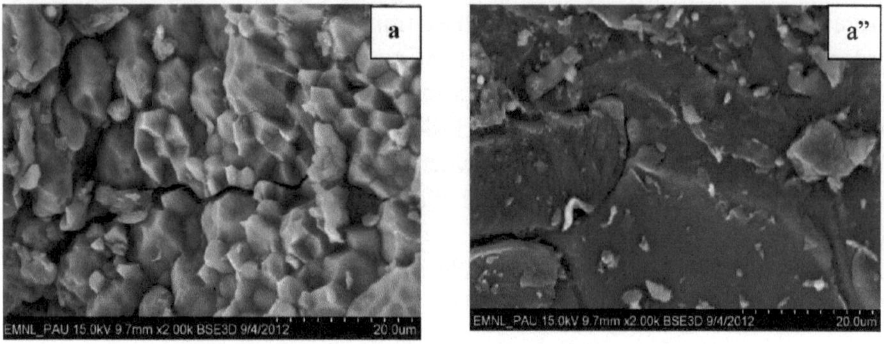

Fig. 2: Effect of pregelatinization on the morphology of cereal flours (a- Native rice flour, a”- Modified rice flour

Thermal properties

Starch gelatinization is the collapse (disruption) of molecular orders within the starch granule manifested in irreversible changes in properties such as granular swelling, native crystalline melting, loss of birefringence, and starch solubilization. It is an important starch functional property that varies with respect to the composition (amylose to amylopectin ratio, phosphorus, lipids, proteins and enzymes, etc.), the molecular structure of amylopectin (unit chain length, extent of branching, molecular weight and granule architecture (crystalline to amorphous ratio), granule morphology and size distribution of starches. DSC is the most common technique used for detecting both first order (melting) and second order (glass) thermal transitions. DSC studies have shown that modification alters thermal transition temperatures (onset (To); peak (Tp); conclusion (Tc) and the overall enthalpy (“Hgel) associated with gelatinization.

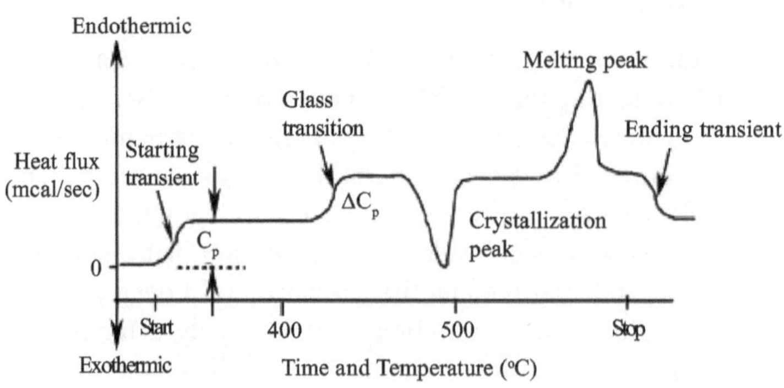

Fig. 3: Typical DSC scan. The heat capacity of the sample is calculated from the shift in the baseline at the starting transient. “H is calculated from the area under the peaks.

Functionality of Modified flours

The Modified Flours offer a tremendous number of functional benefits to variety of foods such as bakeries, snacks, beverages as well as nutritional foods. Modified flours act as a versatile utilitarian ingredient and have a potential to act as a raw material, intermediate product and final product due to its definite characteristics in variety food products. The various functional benefits of modified starch/flour as an additive are classified as fat replacer, texture improver, stabilizer, emulsifier, thickening agent and for freeze-thaw stability (Alexander 1995). Examples of the applications in selected food products will also be included as follows:

***Fat Replacer*:** Starch also can be modified by hydrolysis to form a fat replacer. Some well-known fat replacer from starch base is maltodextrin beside some other starch hydrolysis products. It is widely used for low-fat butter spread/ margarine, low fat mayonnaise, low fat milk type products and low -fat ice cream (Sajilata and Singhal, 2004).

Low-fat yogurt

Some recent studies have been done in replacing fat with modified starch. Castilla et al., (2003) prepared three commercial fat replacers consisting of whey protein concentrate (WPC), microparticulated whey protein (MWP) and modified tapioca starch (MTS), along with the preparation of seven reduced-fat yogurts from reconstituted milk. They found that yogurts with WPC and blends of WPC and MWP possessed textural characteristics that resemble those of full-fat yogurt (FFY), whereas yogurt with MWP showed lower tension and firmness but higher cohesiveness. Yogurt with MTS showed higher firmness than FFY. Blends of carbohydrate–protein-based fat replacer resulted in yogurts that are less dense, firm and adhesive, but more cohesive than FFY.

Low-fat cheese

Functionality of modified tapioca starch and lecithin as fat mimetic in Feta cheese was studied by Sipahioglu et al., (2000). Cheeses were made without any fat mimetic, 1% modified tapioca starch, 0.2% lecithin, and a combination of 0.5% tapioca starch and 0.1% lecithin. Low-fat Feta cheeses (9–16% fat) were prepared from bovine 1.6–2.2% fat milk and were ripened for 45 days before analysis.

Low-fat Mayonnaise

In the production of low-fat mayonnaise, Mun et al., (2009), has used 4-α-glucanotransferase (4αGTase) enzyme to modify the structural properties of rice starch in combination with xanthan gum. They concluded that the usage of 4αGTase-treated rice starch as a viable fat replacer in mayonnaise is feasible.

Texture Improvers

Several studies have been carried out on the potential use of modified starches as texture improver in the food industry. It can provide crispness of crackers and biscuits, viscosity breakdown resistant for can foods, desired chewiness for certain foods and improvement in the quality of extruded products.

Bread making

Starch is a major component of bread making and plays an important role in texture and quality of the dough and bread. Modified starches were developed to decrease the undesirable properties of native starch, which will affect the properties of dough and quality of bread. Bread crumb prepared with phosphorylated cross-linked tapioca starch has a dry feel, whereas bread that was produced from flour substituted with native, hydroxypropylated and acetylated tapioca starches have a tacky texture (Miyazaki et al., 2005, 2006).

Crispy snack

High amylose cornstarch is used in extruded and fried snack products to obtain crisp, evenly browned product. High amylose starches or slightly degraded waxy pregels can be used to give crunchy or crispy mouth feel because they can expand to tender structure when heated or dried. By incorporating the modified starch into the dough composition and by controlling the water absorption index of the starch-based materials, snacks can have a high degree of mouth melt, less waxiness, improved texture and increased crispiness (Sajilata and Singhal, 2004).

Salad mayonnaise

The usage of modified starch as a water-soluble gelling agent to ensure stability of mayonnaise in high-shear emulsion especially for low-fat content of salad dressings and New Zealand-style mayonnaises were reported by Depree and Savage (2001). In these products, much of the rheology of the product depends on the gelling agent rather than the nature of the emulsion.

Conclusion

The developed modified cereal flour has the requisite functional properties over the native cereal flours to be used as a versatile additive in food products. The processed flour can be evaluated for its functionality by adding the flour as convenience ingredient into different food products. Modified cereal flour has potential application in food industry as a raw material, intermediate product and food additive as it has the specific and tailored functional properties.

References

1. Alexander, R. J. (1995). Pregelatinized Starches-What are they all about? Cereal Foods World, 40:769-70.
2. Anastasiades, A. Thanou, S. Loulis, D. and Stapatoris, A. (2002). Rheological and physical characteristics of pre-gelatinized maize starches. Journal of Food Engineering, 52:57-66.
3. Babic, J. Subaric, D. Milicevic, B. Ackar, D. Kopjar, M. Tiban, N. N. (2009). Influence of trehalose, glucose, fructose, and sucrose on gelatinisation and retrogradation of corn and tapioca starches. Czech Journal of Food Science, 27:151-57.
4. Bryant, R. J. Kadan, R. S. Champagne, E. T. Vinyard, B. T. and Boykin, D. (2001). Functional and digestive characteristics of extruded rice flour. Cereal Chemistry, 78:131-37.
5. Camire, M. E. Camire, A. and Krumhar, K. (1990). Chemical and nutritional changes in foods during extrusion.Food Science and Nutrition, 29:35-57.
6. Castilla, O.S. Calleros, C.L. Carter, E.J.V. (2003). Microstructure and texture of yogurt as influenced by fat replacers.Internationa Dairy Journal, 14(2):151-59.
7. Charboniere, R. Duprat, F. and Guibolt, A. (1973). Changes in various starches by extrusion cooking. II. Physical structures of extruded starches. Cereal Science Today, 18: 286.
8. Depree, J.A. and Savage. G.P. (2001). Physical and flavour stability of mayonnaise. Trends in Food Science and Technology, 12:157-63.
9. Douzals, J. P. Perrier, Cornet, J. M. Gervais, P. and Coquille. J. C. (1998). High-pressure gelatinization of wheat starch and properties of pressureinduced gels. Journal of Agricultural Chemistry, 46:4824-29.
10. Lewandowicz, G. Jankowski, T. and Fornal, J. (2000). Effect of microwave radiation on physico-chemical properties and structure of cereal starches.Carbohydrate Polymer, 42:193-99.
11. Miyazaki, M.R. Hung, P.V. Maeda. T. and Morita, N. (2006). Recent advances in application of modified starches for breadmaking. Trends in Food Science and Technology, 117:591-99.
12. Miyazaki, M.R. Maeda, T. and Morita, N. (2005). Gelatinization properties and bread quality of flours substituted with hydroxypropylated, acetylated and phosphorylated cross-linked tapioca starches for wheat flour. Applied Glycoscience, 52:345-50.
13. Mun, S. Kim, Y.L. Kang, C.G. Park, K.H. Shim, J.Y. and Kim, Y.R. (2009). Development of reduced-fat mayonnaise using 4áGTase-modified rice starch and xanthan gum. Biology Macromolecule, 44(5):400-07.
14. Rubin, C. (2010). Extrusion technology for the production of modified flours. Journal of Food Marketing and Technology, 120-22.
15. Sajilata, M.G. and Singhal, R.S. (2004). Speciality starches for snack foods. Carbohydrate Polymer, 59:131-51.
16. Sipahioglu, O. Alvarez, V.B. Solano-Lopez, C. (2000). Structure, physico-chemical and sensory properties of feta cheese made with tapioca starch and lecithin as fat mimetics. International Dairy Journal, 9(11):783-89.
17. Takahashi, S. Maningat, C.C. and Seib, P. A. (1989). Acetylated and hydroxypropylated wheat starch: Paste and gel properties compared with modified maize and tapioca starches. Cereal Chemistry, 66:499-506.

18. Thomas, D.J. and Atwell, W.A. (1999). Gelatinization, pasting and retrogradation. In Thomas D.J. and Atwell A W (Eds.), Starches. St Paul: American Association of Cereal Chemists.
19. Waigh, T.A. Gidley, M.J. Komanshek, B.U. and Donald, A.M. (2000). The phase transformations in starch during gelatinization: a liquid crystalline approach. Carbohydrate Research, 328:165-79.
20. Yamamoto, A. and Shirakawa, K. (1999). Annealing of long-term stored rice grains improves gelatinization properties. Cereal Chemistry, 76:646-49.

CHAPTER 24

Standardisation and Preparation of Accelerated Fermentation of 'Idli' Batter Using Soy Residue Okara

Kishor Pawar, Deepika Kamble, Dnyaneshwar Holkar and Vinod Palve

Abstract

The aim of this work was to reduce the natural fermentation period of 'idli' from the conventional 14 hr to10 hr by adding underutilized okara for the preparation of 'idli'. Black gram was partially substituted with soy residue okara in the ratio of (1:1). After 14 hr of natural fermentation, the pH and total acidity of control 'idli' batter was 4.55 and 0.68% and that of okara fortified 'idli' batter was 4.57 and 0.56%, respectively. The amount of CO_2 released by the control and okara fortified batter was 12.5% and 22.5% respectively. The viable count of yeast and mold, lactics and mesophilic bacteria in control and okara batter increased with time reaching 8.4 and 9.6, 8.2 and 7.5, and 8.6 and 9.2 log10 cfu/g, respectively at the end of 10 hr of natural fermentation. Okara fortified 'idli' was soft and spongy compared to control 'idli'.

Keywords: Batter, natural fermentation time, fortification, 'idli', soy residue okara

Introduction

Idli is a traditional cereal/legume-based naturally fermented steamed product with a soft and spongy texture which is highly popular and widely consumed as a food item in India. Idli makes an important contribution to the diet as source of protein, calories and vitamins, especially B-complex vitamins, compared to the raw unfermented ingredients. Idli is also known as "Rice cake" is a traditional food of India. It is a favourite breakfast food in south India with spongy texture, attractive appearance, appetizing taste and flavour to get with its easy digestibility and good nutritive value contribute to its increasing popularity in all parts of India and also in other countries (Radhakrishnamurthy, 1962; Nazni and Shalini, 2010).

Soybean, (Glycine max) although is superior to the legumes commonly used in India like black gram, red gram and bengal gram in nutritive value, its use is limited due to the presence of antinutritional components like trypsin inhibitors and hemagglutinins (Ramakrishnan, et al., 1996). Fortification of soybean in fermented foods is advantageous as increase in digestibility and nutritive value of fermented foods. Soybean exerts many health beneficial effects, like reducing menopausal symptoms, reduction in bone loss, cancer, antioxidant activity due to presence of its polyphenols mainly isoflavones (Tsangalis et al., 2002). Natural fermentation of soybean brings desirable changes in taste and texture and may result in the breakdown of some of the antinutritional factors.

During soymilk production, about 70% of soybean solids and 80% of soybean proteins are extracted and the residue solids make up okara. From 1 kg of dry soybeans about 1.0 to 1.1 kg of fresh okara is produced. A high protein content of 27–32% (on dry basis) makes okara a potential source of low cost plant origin protein for human nutrition (Chan and Ma, 1999). Areas such as preparation methods, nutritive value and microbiology of idli have been extensively studied and well documented (Balasubramanium et al., 2006). In the present study, the effect of fortification of 'idli' batter with okara on natural fermentation rate, physicochemical parameters and microbial growth in 'idli' batter and, texture and sensory attributes of 'idli' were studied.

Material and Methods

Preparation of okara

Soybean was purchased from a local market. Hundred grams of cleaned soybeans were soaked in 800 ml water for 12–14 h, rinsed 2 or 3 times, dehulled manually and ground with 400 ml of water in an electrically operated blender for about 2–3 min. Additional water was added to make up the volume of Soy slurry to 800 ml (Water to bean ratio was 8:1) (Jama and Varadaraj, 1999). The soy slurry was filtered through a double layered cheese cloth to obtain soymilk and the okara was separated. The wet okara was dried in an oven at 45°C for 10–12 h and stored in airtight containers.

Preparation of control 'idli' batter

Rice (oryza sativa) and black gram dhal (Phaseolus mungo) were purchased from local market for the entire study. Rice soji and black gram dhal was taken in the ratio of 3:1 (Rekha and Gijaylakshmi, 2011). Dehusked split black gram dhal (30 g) was washed twice, soaked in 120 ml water for 4 h at room temperature (28±2°C) and ground separately with required quantity of water into a batter of desirable consisitency, using an electrically operated blender. Rice soji (90 g) was washed with water and mixed with dhal batter. The batter was dispensed into steel container and allowed to ferment for 14h at room temperature. After every 2 h of natural fermentation pH, acidity and the amount of CO_2 released were measured. At the end of natural fermentation, the fermented batter was distributed into hollow depressions in steaming pans and steam cooked for 10 min.

Preparation of okara fortified 'idli' batter

Idli batter was prepared using three different proportions of rice soji, black gram and okara. To standardize ideal ratio, batters were prepared with different combinations of rice soji, black gram dhal and okara (3:0.5:0.5, 3:1:1 and 3:1:1.5). This was allowed to ferment for 10 h at room temperature. Total acidity, pH, acidity and the amount of CO_2 released were measured after every 2 h of natural fermentation time. At the end of natural fermentation, the fermented batter was distributed into hollow depressions (moulds) pans and steam cooked.

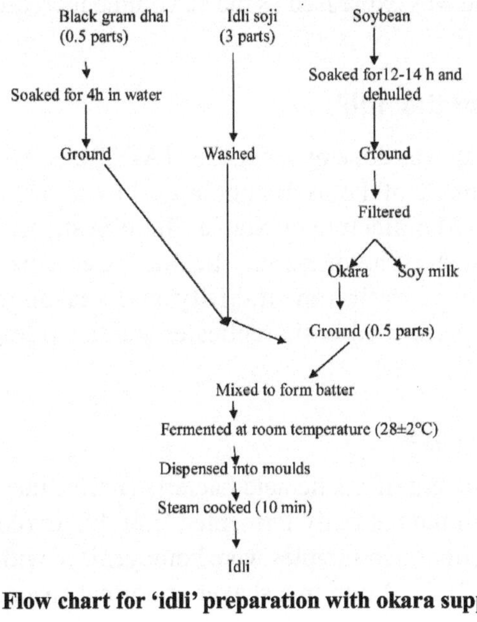

Fig. 1: Flow chart for 'idli' preparation with okara supplementation

Analysis of batter

Titrable acidity and pH

The pH of the samples was measured using a pH meter. To determine titrable acidity, 10 g of fermented batter was taken in a 100 ml conical flask to which 20 ml of distilled water was added. After adding 3–4 drops of phenolphthalein, the contents were mixed well and titrated against 0.1 N NaoH to an end point of pale pink colour and expressed as % lactic acid produced (AOAC, 1999).

Measurement of CO_2

Measurement of CO_2 was done using acid salt solution. Acid salt solution was prepared by dissolving 200gm Sodium Sulphate in one liter Sulphuric Acid (5% v/v). Methyl red indicator was added to this solution to impart a red colour. The set up was made of three bottles. One of which contained batter for fermentation and other two had equal level of acid salt solution. The decrease in acid salt solution level was recorded as percent evolution of $CO_{2.}$

Measurement of increase in batter volume during natural fermentation

A 50 ml of batter was transferred to a sterile measuring cylinder of 250 ml capacity, covered with aluminum foil and incubated at room temperature. The increase in volume at 0 and 10 h of natural fermentation was recorded and the raise in batter volume was expressed as the % volume increase over the initial volume.

Texture analysis of the 'idli'

Texture analysis of idli was done by using the 'TA-XT plus Texture Analyzer'. Installed at Department of Food Technology, Laxminarayan Institute of Technology, Nagpur. Manufacture by Stable Micro Systems, UK. Among the several textural parameters, hardness and stickiness were selected to represent the results because of their repeatability and reasonable variations. Measurements were performed in six replicates and the average was reported in Newtons.

Microbiological analysis

To determine viable count of Lactic acid bacteria (LAB), mesophilic bacteria and yeast and mold of the naturally fermented 'idli' batter (control and okara fortified batter), 10 g of dough samples were homogenized with 90 ml of sterile diluents (0.85% NaCl) for 2 min in a shaker at normal speed. Ten fold serial dilutions were prepared and pour plated on MRS agar for the enumeration of LAB. Spread plate technique was employed to determine the counts of total mesophilic bacteria, and yeast and molds using Nutrient agar (NA) and Potato dextrose agar (PDA).

Sensory evaluation of the 'idli's

Idli is prepared with different batters were subjected to sensory evaluation by semi trained judges using 9-point Hedonic scale on the basis of sensory attributes fluffiness, compactness, sponginess, firmness, sticky, beany, fermented, salty, sour, bitter after taste. The scores for all the attributes were tabulated and the mean values were calculated. These mean scores represented the panel's judgment about the sensory quality of the samples.

Results and Discussion

Effect of okara fortification on pH and Titrable acidity

The changes in pH of okara fortified and unfortified 'idli' batter are represented in Fig-2. The pH of control batter at 0 h was 6.35 which decreased to 4.55 at the end of 14 h of natural fermentation and in okara fortified batter the pH at 0 h was 6.31 and 4.57 at 14 h of natural fermentation.

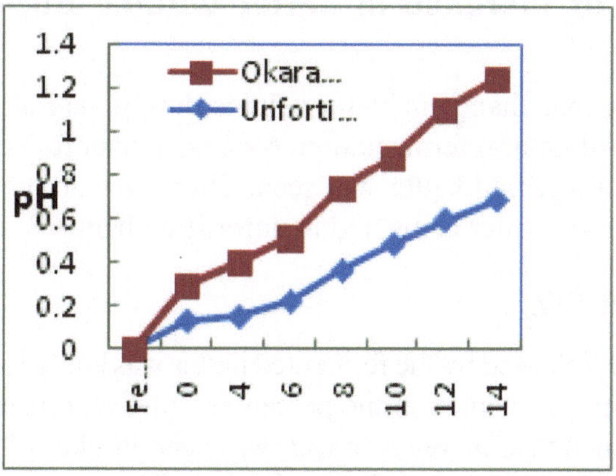

Fig. 2: Effect of natural fermentation period on pH of Control 'idli' batter and okara fortified batter

With the progress of natural fermentation, increase in the acid content of batter during natural fermentation was observed in both control and okara fortified 'idli' batter (Fig-3). The acidity increased from 0.13 to 0.68 in control batter and from 0.16 to 0.56 in okara fortified batter over a period of 14 h of natural fermentation. The change in pH and acid content can be attributed to the microbial growth especially of lactics producing lactic acid, which lowers the pH and leavens the batter. The role of LAB is to reduce the pH of the batter to an optimum level (4.4–4.5) for yeast activity. This accounts for fall in pH and rise in acid content of batter as the natural fermentation progresses.

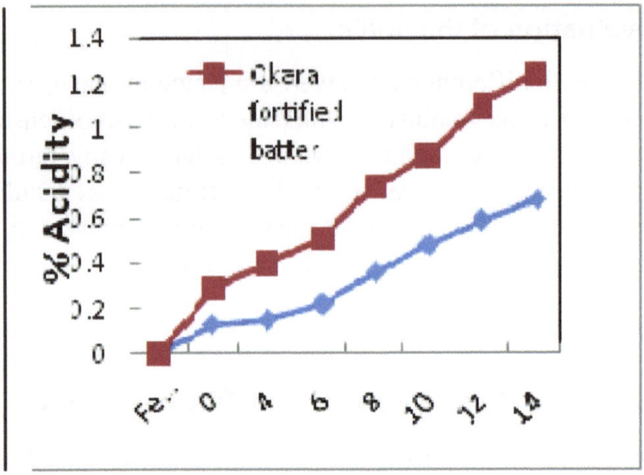

Fig. 3: Effect of natural fermentation period on total acidity of control 'idli' batter and okara fortified batter

Measurement of increase in batter volume during natural fermentation

There was a noticeable change in batter volume during natural fermentation. At the end of 10 h of natural fermentation, 44% raise in unfortified batter and 70% raise in okara fortified batter was seen. There was an increase of 26% raise in okara fortified batter compared to unfortified batter.

Measurement of CO_2

The amount of CO_2 released by the fermented batter was analyzed from 0 to 14 h. At 10 h of natural fermentation the percentage of CO_2 released in control was 4.0 and a tremendous increase in CO_2 was seen in okara fortified batter (20.0). Further at 14 h (at the end of natural fermentation), percentage of CO_2 was 12.5 and 22.5 in control and okara fortified batter, respectively. The increase in CO_2 production was very significant in okara fortified batter within 10 h of natural fermentation. This is due to the presence of more number of yeasts present in okara fortified batter producing CO_2 as a byproduct during its metabolism. It has also been reported earlier that yeasts are responsible for more than 50% of the CO_2 and two fold increase in the volume of batter (Venkatasubbaiah et al., 1984).

Microbiological analysis

The microbial analysis of both the batter samples presented in Table 1 revealed that both bacteria and yeasts play an important role throughout the process of natural fermentation. The counts of mesophilic bacteria and LAB present naturally in batter showed a progressive increase in their numbers with increase in time. There was a significant increase in yeast and mold count in okara fortified batter compared to control batter. It was seen that the bacterial counts were higher in the okara fortified batter than that of control batter. This is due to the growth of LAB in sourdough and has been found to be enhanced by the presence of amino acid (Gobetti et al., 1984, Soni and Arora, 2000).

Table 1: Microbiological profile of control and okara fortified batter

Parameter	Natural Fermentation Period (h)	Type of batter	
		Control	Okara fortified
Lactic count($\log_{10}$ cfu/g)	0	2.4	3.1
	10	8.2	7.5
Yeast and mold count($\log_{10}$ cfu/g)	0	1.8	2.7
	10	8.4	9.6
Mesophilic bacteria count ($\log_{10}$ cfu/g)	0	3.2	3.3
	10	8.6	9.2

*Results are mean of triplicate

Texture analysis of 'idli'

Hardness is measured as the peak force during compression in the first cycle. Hardness of traditional 'idli' was 226.6734 Newton and 'idli' prepared with okara substituted batter fermented for 10 h 'idli' was 225.1884 Newton, respectively. These values indicated that the control 'idli' offered more resistance to compression than that of okara fortified 'idli'. Thus the okara substituted samples were softer and easy to bite compared to control samples (soft texture of 'idli's is a desirable quality). On the other hand the values for stickiness in case of okara 'idli' were relatively low. The stickiness of traditional 'idli' was 0.13 Newton and okara substituted 'idli' was 0.14 Newton respectively.

Sensory evaluation and standardization of 'idli's

The scores of sensory evaluation of control and okara fortified 'idli' batter are presented in Table 2. For sensory attributes of the final product ('idli'), a 14 h fermented batter prepared in traditional method was taken as control, as it was reported optimum time for natural fermentation of 'idli' batter. It has been reported by Soni and Arora (2000) that yeast involved in the fermentation not only contribute towards gas production, which results in good texture but also towards the sensory qualities of the 'idli'.

Table 2: Sensory profile of 'idli'

Parameters	Types of 'idli'			
	S 1	S 2	S 3	S 4
Fluffiness	8.12	5	7.54	8.22
Cmpactness	7.66	6.23	7.12	7.5
Sponginesso	8.28	4.12	7	8.35
Firmness	8.32	5.98	6.75	8.10
Sticky	6.56	6.59	7.52	6.32
Fermented	8.20	7.88	7.96	8.10
Salty	7.58	6	7.18	7
Sour	7.64	6.4	6.99	7.66

*Sample1-3:1(Control), Sample2-3:1:0.5,
Sample 3- 3:1:1, Sample 4- 3:1:1.5

*Rice soji : Black gram : Soy okara

Nutritive value of developed idlis

Okara fortified idli has more nutritional value than control idli. As shown in Table 3. Due to addition of soy okara the protein content of idli wsa (9.16%) whereas control idli had (6.12%) protein. Also the fiber content was more in okara fortified idli i.e. (2.01%) and (0.622%) in control idli respectively. The moisture content was nearly same in both idlis. Also fat content was more in okara fortified idli (2.87%) than control idli (0.54%). Carbohydrates were significantly more in control idli (52.10%), whereas okra fortified idli had (42.29%) carbohydrates.

Table 3: Proximate Composition of Control 'idli' and okara fortified 'idli'

Parameters	Control idli	Okara fortified idli
Moisture (%)	23.51	22.62
Protein (%)	6.12	9.16
Fat (%)	0.54	2.87
Ash (%)	0.23	0.54
Fiber (%)	0.622	2.01
CBH (%)	52.10	42.29

Conclusion

The addition of soy residue okara to the 'idli' batter accelerated the natural fermentation rapidly and shortened the fermentation time. Fortification of okara to 'idli' batter has a beneficial effect in terms of higher amount of gas production and leavening during natural fermentation. Increase in CO_2 production (33.5%) resulted in soft and spongy 'idli' compared to control sample. Thus, the fermentation time was brought down to 10 h from the reported 14 h and with improved quality of the final product. Reduction in the fermentation time of the 'idli' batter is of great commercial significance for large scale 'idli' production (Bharti and Laxmi, 2008) and this can be potentially achieved by addition of soy residue okara. Thus the under-utilized protein rich okara can be used for the preparation of value added fermented products.

References

1. Nazni, P. and Shalini, S. (2010). Standardization and quality evaluation of idli prepared from pearl millet. International Journal of Current Research, 5:084-087.
2. Ramakrishnan, C.V, Parekh, L. J, Akolkar, P.N, and Rao, G. S. (1976). Studies on soy'idli' natural fermentation. Plant Foods for Man, 2:15–33.
3. Tsangalis, D, Ashton, J. E. Mcgill, A.E.J. and Shah, N.P. (2002). Enzymic transformation of isoflavone phytoestrogens in soymilk by β-glucosidase producing Bifidobacteria. Journal of Food Science, 67:3104–13.
4. Chan, W. M. and Ma, C. Y. (1999). Acid modification of proteins from soy milk residue (okara). Food Research International, 32(2): 119-727.
5. Balasubramanium, S. Singh, N. Ilys, S. M. and Wanjari, O.D. (2006). Effect of selected decorticated legumes protein on rheology of maize extrudate pastes. Journal of Food Science and Technology, 43:590–594.
6. Desikachar, H.S.R. Radhakrishnamurthy, R. Rao, G.R. Kadol, S.B. Srinivasan, M. and Subramanyan, V. (1962). Studies on idli fermenteation. Part I. Some accompanying changes in the batter. Journal of Science Industrial Research (India), 19C:168–17.
7. Jama, Y.H. and Varadaraj, M.C. (1999). Antibacterial effect of plantaricin LP84 on foodborne pathogenic bacteria occurring as contaminants during 'idli' batter natural fermentation. World Journal of Microbiol Biotechnology, 15:27–32.
8. Rekha, C.R. and Vijaylakshmi, G. (2011). Accelerated fermentation of 'idli' batter using soy residue okara. Journal of Food Science and Technology, 48(3):329-334.
9. AOAC (1999). Methods of Analysis. Washington DC: Association of Official Analytical Chemists.
10. Venkatasubbaiah, P. Dwarkanath, C.T. and Sreenivasamurthy, V. (1984). Microbiological and physico chemical changes in 'idli' batter during natural fermentation. Journal of Food Science and Technology, 21:59–63.
11. Gobetti, M. Corsetti, A. and Rossi, J. (1994). The sour dough microflora. Interactions between lactic acid bacteria and yeasts: metabolism of amino acids. World Journal of Microbiol Biotechnology, 10:275–279.
12. Bharti, K. and Laxmi, A. (2008). Effect of á"amylase addition on natural fermentation of 'idli'- A popular south Indian cereal-Legume based snack food. Lebensm Wiss Technollogy, 41(6):1053–1059.

13. Soni, S. K. and Arora, J.K. (2000). Indian fermented foods: Biotechnological approaches. Food processing: Biotechnological Application. Asiatech Publishers Pvt Ltd, New Delhi, pp.171.

CHAPTER 25

Bioactive Components and Functional Ingredients in the Traditional Rice Varieties of Temperate Region and their Potential Health Benefits

Farhan Mohiuddin Bhat and C.S. Riar

Abstract

Rice is the most produced and consumed source of stable food in J andK, and is excellent in genetic diversity. Rice production is the principal activity and a major source of income for the State. Traditional rice varieties have an enhancing potential in wide range of nutraceutical and functional foods. Besides possessing medicinal and nutritional properties, the by-products generated from these rice varieties in various post harvest operations are equally valuable. Traditional healers and local farmers have been using these traditional varieties in ayurveda and in curing of various kinds of ailments, such as cooling the body in ayurvedic treatments, improving vocal clarity, curing of boils, swellings and skin blemishes. Extracts of brown rice are also used as energy drink in individuals, patients and in treatment of chronic gastric problems, jaundice, dysenteric complaints and to increase lactation and nutrition to Childs. Sticky glutinous rice has been reported in treatment of stomach upsets, nausea, heart-burn and indigestion. Brown rice extracts have been reported in treatments of breast and stomach cancer and warts.

Introduction

Rice is a major source of staple food of more than two third of world population. About 80% of world rice is produced from the cultivation of Asian rice (Oryza sativa L.) (Tyagi et al., 2004). Agriculture is the mainstay of more than 80% people in Jammu and Kashmir. In the state Jammu and Kashmir, rice represent main source of staple food crop, particularly those of Kashmir's. The acreage under rice cultivation in the valley constitutes about two-third of the total area under the crops in the whole state (JandK). About 75% of states rice is grown in the Kashmir valley. The land under rice cultivation in Jammu region is 159 thousand acres while it is 374 thousand acres under rice cultivation in Kashmir region. The paddy cultivation in the valley is known since ancient times and is grown even at an altitude of 2500 mtrs above sea level. Most of the farmers are earning good amount of their produce in Kashmir valley as well as in the Jammu region. Out of 14 districts of Kashmir region, rice is cultivated in about 09 districts. Out of which 4 districts come under high productivity group (yield more than 2,500 kg/ha) (Gupta et al., 2009).

Rice productivity in the state is high (2.2 Tonns/Ha) compared to the national average productivity of about 1.9 Tons/Ha. Jammu and Kashmir State is rich in rice culture from the ancient times. About 32 varieties of traditional rice are being cultivated in the valley and varied from each other in various attribute such as, variation in dimensions, color, aroma, stickiness and presence of inherited health benefits etc (Rana et al., 2009).

Characteristics of Traditional Rice Varieties

Traditional varieties of paddy and rice possess distinguishable physical, physico-chemical and cooking characteristics, which justified leading to their classification into different categories on the basis of these of parameters. Varieties within a particular group revealed certain desirable traits than others. Variability had been observed in these varieties in husk color, presence of awns, and boldness of seeds among paddy varieties (Figure-1) and in color of rice kernels, size and boldness, elongation ratios, water uptake ratios among rice varieties (Table-1). International Treaty on Plant Genetic Resources for Food and Agriculture have established obligations among the countries 'promote or support as appropriate, farmers and local community's efforts to manage and conserve on-farm their plant genetic resources for food and agriculture'. Some varieties fetch much more revenue in the market than modern varieties due to their unique characteristics like bold grained, sweet nourishing, aromatic, pigmented rice and are still venerated in social, cultural and economic values (Sultan and Subba Rao., 2013). Different groups of traditional rice possess different inherent qualities, and affect the human body differently. Even today, unani and ayurvedic practitioners recommend different rice for treatment of various ailments (Ahuja et al., 2008).

Table 1: Characteristics of traditional rice varieties of Kashmir

Aromatic	Coloured varieties	Bold seeded	Good water uptake ratio	Better Elongation ratio
Mushki Budgi	Zag	Koshkari	Mushki Budgi	Gull Zag
Mushki Kandi	Gull Zag	Kaw quder	Koshkari	Kaw kareed
	Kaw kareed	Tel zag	Mushki Kandi	Mushki Kandi
	Kaw quder	Mushki Budgi	Gull Zag	Tel zag
	Tel zag	Kamad	Kaw kareed	
	Shel kew		Tel zag	
	Gull Zag		Shel kew	

During recent times, both at national level and international levels, the land races are being preserved in the gene banks. The scientists said that they were also checking these land races for some such characteristics which were not available in the existing varieties so as to incorporate the same in the present varieties (Sultan and Subba Rao, 2013). The trend toward hybrid from traditional varieties may be reversing as traditional varieties are experiencing an increasing trend among consumers, due to their incredible health benefits for example the traditional varieties possess low sugar content, making them pleasing choice for consumers who are suffering from diabetics, overweight, or regulating their sugar intake. They have higher amount of glutamiuc acid, fiber and vitamins. Some people also credit traditional varieties with other health benefits, such as giving sensations of cooling in the body, improves vocal clarity, eyesight, fertility and mitigating rashes (Caius, 1986).

Owing to several health-promoting impacts associated with anthocyanins, such as anti-oxidative, anti-inflammatory and anticarcinogenic effects (Shipp and Abdel-Aal, 2010), colored rice is considered as a functional food and food ingredient in many Asian countries (Kim et al., 2008; Sukhonthara et al., 2009). The functional properties of anthocyanins in red and black rice varieties have been verified in numerous nutritional cases (Hou et al., 2010).

Fig. 1: Genetic diversity of traditional rice varieties of high altitude areas of temperate region of India.

Increased concern over the safety of synthetic antioxidants like butylated hydroxyanisole (BHA) and butylated hydroxytoluene (BHT) (Razali et al., 1997; Tsuda et al., 1998) has lead to an increased interest in exploration of effective and economical natural antioxidants. It has been experimented that alteration in ones diet to an increased intake of natural foods having high content of antioxidants, such as tocopherols, polyphenols, flavonoids, vitamin-C, play an important role in reducing the occurrence of chronic diseases. Pigmented and scented rice varieties besides their pleasing taste are also associated with several health benefits. Pigmented rice possess pigments such as Cyanidin-3-O-β-D-glucopyranoside in abundant content (Ryu, Park, and Ho., 1998). which is associated with diverse physiological benefits such as, protection against cytotoxicity (Chen et al., 2005), antineurodegenerative activity (Kim et al.,

2005), inhibition of glycogen phosphorylase (Jakobs et al., 2006), and possessing antioxidant and scavenging activity higher than white rices and hybrid rice varieties (Oki et al., 2005).

Nutritional Facts about Traditional Rice

Traditional varieties of rice possess nutritive value higher than hybrid varieties. Rice represents an important source of carbohydrate for more than two-thirds of the world's population and whole of valley's population, who depend on rice as their staple food. It is a major source of carbohydrate and serves as fuel for the body to carry on vital activities. These traditional rice varieties have lesser content of fat as compared to hybrids and are cholesterol free. Thus represent as an excellent source of food to be included into balanced diet. These are good sources of minerals and vitamins such as niacin, thiamine, iron, riboflavin, vitamin-D, calcium, and possess higher fiber and lesser amount of sugar content, making these the appealing choice for diabetic patients. Rice is free from gluten, thereby making it an essential choice for people having requirements of gluten free diets such as patients suffering from celiac diseases. These varieties are non allergic due to absence of any additives, because these do not require excessive fertilizer and pesticide applications during their developmental and growth stages (Oudhia, 2008).

These rice varieties possess higher amylose content and are rich in resistant starch, which is the starch that cannot be hydrolyzed in the gastrointestinal tract and serves as a substrate for bacterial fermentation. This possesses several health benefits such as reducing the risk of developing type-II diabetes, obesity and cardiovascular diseases by lowering the glycemic and insulin responses. Brown rice contains high amounts of insoluble fiber, which is reported by scientists to protect the body against a variety of cancers. These are also valuable sources of food for those suffering from hypertension due to its low sodium content and are also a fair source of protein containing all eight amino acids (Umadevi et. al., 2012).

Medicinal uses of Traditional Rice

Jammu And Kashmir State has been a rich source of medicinal plants since ancient times, among these traditional rice varieties have been used by local inhabitants in unani, Ayurvedic, system of medicines for generations. In Ayurveda the medicinal values of rice have been described as acrid, tonic, aphrodisiac, oleaginous, diuretic, fattening and useful in biliousness (Caius, 1986). These traditional varieties have been regarded an energetic food for people and were recommended by saints of having the medicinal values to keep juvenile and long life (Table 2). The extracted from the bran of these rice is used for curing neural diseases and also used to cure body pain and eye disorders It is having properties to rectify the basic ills affecting the circulatory respiratory as well as digestive system. Different groups of rice affect the human beings differently, as they possess different inherent qualities to alleviate the three *doshas*, (which according to ayurveda is considered three principles of energy, believed to be circulating in the body and govern physiological balance and activity). Ayurvedic practitioners prescribe different rices for various ailments. The traditional doctors (Vadis) possess profound knowledge of different effects of rice and were particular about their prescription (Watt, 1891).

Table 2: Medicinal uses of traditional rice varieties of temperate regions

S. No	Rice Varieties	Medicinal Use
1.	Kamad	Cooked rice is useful for dysenteric patients
2.	Mushki Budgi	Lowers glycemic index value in patients suffering from diabetes
3.	Koshkari	Cooling effect, health tonic
4.	Zag	Skin infections, increases secretion of milk in women. Red rice yeast is prepared by fermenting yeast Monascus purpurea over red rice. It is said to promote blood circulation, and is used in treating upset stomach, indigestion, bruised muscles, and hangovers
5.	Mushki Kandi	Cooked rice water (starchy water) is given to patients suffering from diabetes
6.	Naaz Bar	Post natal tonic for women
7.	Gull Zag	Used to increase lactation and more nutrition to child
8.	Tcheri Bar	Cooked rice is useful in breathing problems and epilepsy
9.	Kaw kareed	Given to patients suffering from chronic gastric patients, post delivery restoration of size of reproductive organs
10.	Kaw quder	Cooling effect, health tonic, given to weak patients
11.	Tel zag	Cooked rice is given to lactating mother to cure boil caused on scalpel of newly born child
12.	Shel kew	Used to cure Pimples, small boils in Childs, Post-natal tonic for women.

In Ayurveda, the traditional Indian medical system, colored rice has been called *shastika* rice and claims that it can restore imbalances in the human body. Pigmented rice is rich in antioxidants and polyphenols and has two or three times as much zinc and iron as white rice (Ramaiah and Rao., 1953). Which possess the desirable quality to boost, strengthen, regenerate and energize the body (Sensarma, 1989). It is also used as baby food and replaces white rice on special occasion in the state. The red color, varying from light to dark red, is confined to the bran layer. Red Rice (Zag, Tel Zag, Gull Zag, Shel Kew, Kaw Quder) Keeps You Away From Obesity, Diabetes and Cancer. Rice with a red bran layer is called red rice. Susruta (400 BC), Charaka (700 BC), and Vagbhata (700 AD), the well-known vriddha trayi (Trio of Elders) of Ayurveda, considered red rice (Rakta Shali) the best among the other rice varieties, due to desirable property as they had the power to redress the imbalance in the tridosha or humours (the vata, the pitta, and the kapha – are collectively called the tridosha) whose imbalance in the body causes various types of diseases. In recent times, interest in red rices has been revived because of the presence of antioxidants. The antioxidant and scavenging activity of red rice is higher than that of white rice (Shen et al., 1994).

Ancient Ayurvedic treatises laud the red rice as a nutritive food and medicine. They are known to be influential in the treatment of various ailments such as diarrhea, vomiting, fever, hemorrhage, chest pain, wounds, and burns. (Hegde et al., 2013) Coloured rice has been preferred in the past for their special features such as medicinal value and exclusive taste. A large number of these varieties are still grown in various parts of the sate by tribal's and small farmers who are deprived of modern technologies and health care systems, where indigenous rice with their nutritional and medicinal properties are a rich alternative for the same. This indigenous rice variety in state and India seems to contribute tremendously to the health of the women including adolescent girls, lactating mothers and pregnant women (Hegde et al., 2013) Ayurvedic properties of Raktasali (red rice) and their effect on human physiology. Red rice (Raktasali) was the most efficacious in subduing deranged humors, (Kumar, 1988; Krishnamurthy, 1991). It was considered good treatment for fevers and ulcers, Improves eyesight, voice improver, semen enhancer, diuretic, spermatophytic, refrigerant, cosmetic, and tonic and was antitoxic. Pharmological and clinical trials with red rice has shown anti-fungal, anti-bacterial, anti-viral, ant-diarrheal, anti-inflammatory, antioxidant, antitumor, anti-thyroid and anti-hyper cholesterolemic activities. It also stimulates protein secretion besides having radical scavenging effects (Oki et al., 2005).

Aromatic rice provides most benefits of health and nutrition. These are considered healthier as they possess more vitamin and fiber in their outer bran layers. Brown aromatic basmati rice contains 20% more fiber than other brown rice varieties, which prevents the formation of cancerous cells in the body. According to Canadian diabetes association, glycemic index of basmati aromatic rice is lower than other rice varieties, and thus essential for those suffering from diabetes. Ayurveda supports its properties and proved it to be a great healing food. Traditional scented rice varieties have been revealed by scientists to possess higher amount of Fe and Zn and helps in the bioavailability of iron

(Chaudry and Tran, 2001). The need of the day is to commercialize these varieties and promote them through public awareness about their important characteristics.

Present Research

Currently, rice is regarded as a nutraceutical and functional food besides being a staple source of food and primary source of carbohydrate or starch. Its role of having low glycemic index in comparison to other genetically modified rice varieties, which containing complex carbohydrates and regarded as high glycemic index food provides a better novel substitute for diabetic patients. Presence of valuable mineral content, excellent starch characteristics, antioxidant and anti-inflammatory activity represent these unique among cereals sources (wheat, maize etc) starch of this rice is almost completely absorbed by human body (Strocchi and levitt, 1991). Scientifically, it have been proved, that amino acids possessed by these varieties have high biological value, high content of essential fatty acids and selenium, and have anti-hypertension effect. Hence, rice is now regarded as a functional food.

Rice extracts have been validated in inhibiting chloride channels in human body and thus reduce intestinal losses' in acute diarrhea, thereby serving as oral rehydration solutions (ORS) (Goldberg and Saltzman, 1996). These rice based ORS are now preferred than glucose based ORS, as reported by World Health Organization programmes (Gore et.al., 1992). Rice due to its least allergic properties is recommended for patients suffering from irritable bowl syndrome. Colored rice varieties have been reported to possess antioxidant properties (Zhang et.al., 2005) and considered more nutritious, being rich in iron (Fe), Zinc (Zn), and minerals. These rices have been found to reduce atherosclerotic plaque by 50% more than white rice in rabbits (Ling et.al., 2001). Clinical trials conducted in USA have concluded that red rice yeast reduces cholesterol and total triglycerides, providing a novel food based approach to lowering cholesterol (Herber et.al., 1999). Rice thus meet most of the requirements of a good and healthy food and is the only cereal that is eaten as whole grain, which according to Ayurvedic texts is more easily digestible than flour.

Conclusion

The temperate regions in north India have been an abode of traditional rice varieties since prehistoric times and were used recommended for medical uses by traditional healers and local farmers. Theses varieties thus fit into the description of healthy and functional foods. Due to emergence of high yielding varieties these have been replaced at a large pace and are posing the threat of extinction. Thus these varieties need to be conserved and promote them by commercialize and through general public awareness about their medicinal benefits. Considering the presence of flavonoid and anthocyanin content together with highest radical scavenging activity due to the presence of polyphenols in these varieties, these favors the health consciousness in consumers. Clinical validation of these varieties is mandatory, so that these excel in the international market (like the red rice yeast of china having worldwide demand).

References

1. Caius J.F. (1999). The medicinal and poisonous plants of India (Reprint), Pbl. Scientific Publishers, Jodhpur, India.
2. Chaudry R.C and Tran D.V. (2001). Specialty rice's of the world-a prologue. In: specialty rice's of the world: Breeding, Production and Marketing (Chaudry R.C and Tran D.V, eds).FAO,Rome, Italy; and Oxford and IBH Publishing Co.pvt.Ltd, New Delhi, India. pp. 3-14.
3. Chen, P. N. Chu, S. C. Chiou, H. L. Chiang, C. L. Yang, S. F. and Hsieh, Y. S. (2005). Cyanidin 3-glucoside and peonidin 3-glucoside inhibit tumor cell growth and induce apoptosis in vitro and suppress tumor growth in vivo. Nutrition and Cancer, 53:232–243.
4. Goldberg E.D and Saltzman J.R.C. (1996).Rice inhibits intestinal secretion. Nutrition Reviews, 54:36-37.
5. Gore, S.M. Fontaine, O. and Pierce, N.F. (1992). Impact of rice based ORS on stool output and duration of diarrhea: Meta-analysis of 13 clinical trials. British Medical Journal, 304:287-291.
6. Gupta, B. B. Salgotra, R. K. and Bali, A.S. (2009). Status Paper on Rice in Jammu and Kashmir. www.rkmp.co.in.
7. Hedge, S. Yenagi, N. B. and Kasturiba, B.(2013). Indigenous Knowledge of the traditional and qualified ayurveda practitioners on the nutritional significance and use of red rice in medications. Indian Journal of Traditional Knowledge, 12:506-511.
8. Herber,D. Yip, I. Ashley, J.M. Elashoff, D.A. and Go, V.L.W. (1999).Cholesterol lowering effect of a proprietary Chinese red rice yeast-dietry supplement. American Journal of Clinical Nutrition 69:36-37.
9. Hou, Z., Qing, P. and Ren, G. (2010). Effect of anthocyanin-rich extract from black rice (Oryza sativa L. Japonica) on chronically alcohol-induced liver damage in rats, Journal of Agricultural and Food Chemistry, 58:3191-3196.
10. Jakobs, S. Fridrich, D. Hofem, S. Pahlke, G. and Eisenbrand, G. (2006). Natural flavonoids are potent inhibitors of glycogen phosphorylase. Molecular Nutrition and Food Research, 50:52–57.
11. Kim, M.K. Kim, H. Koh, K. Kim, H.S. Lee, Y.S. and Kim, Y.H. (2008). Identification of anthocyanin pigments in colored rice, Nutrition and Research Practice, 2:46-49.
12. Krishnamurthy, K.S. (1991). The Wealth of Susruta. International Institute of Ayurveda, Coimbatore, Tamil Nadu, India. 582.
13. Kumar, T.T. (1988). History of Rice in India. Gian Publishers, Delhi, India.
14. Ling, W.H. Cheng, Q.M. Ma, J. and Wang, T. (2001). Red and black rice decrease atherosclerotic plaque formation and increase antioxidant status in rabbits, Journal of Nutrition,131:1421-1426.

15. Nam, S.H. Choi, S.P. Kang, M.Y. Koh, H.J. Kozukue, N. and Friedman, M. (2006). Antioxidative activities of bran extracts from twenty one pigmented rice cultivars. Food Chemistry, 4:613–620.
16. Oki, T. Masuda, M. Nagai, S. Nishiba, Y. Sugawara, T. Suda, I. and Sato, T. (2005). Radical scavenging activity of black and red rice. Proceedings of world rice research conference, Tokyo and Tsukuba, Japan, 256-259.
17. Ramaiah, K. and Rao, M.V. (1953). Rice Breeding and Genetics. ICAR Science Monograph 19. Indian Council of Agricultural Research, New Delhi, India
18. Rana, J. C. Negi, K. S. Wani, S. A. Saxena, S. Pradheep, K. Kak, A. Pareek, S.K. and Sofi, P. A. (2009).Genetic resources of rice in the Western Himalayan region of India: current status, Genetic Resources and Crop Evolution, 56:963–973.
19. Razali, I. Norhaya, H. and Norasimah, A. S. (1997). Determination of antioxidants in palm oil products by HPLC. Elaeis, 9(1): 25–31.
20. Rossi, A. Serraino, I. Dugo, P. Di Paola, R. Mondello, L. and Genovese, T. (2003). Protective effects of anthocyanins from blackberry in a rat model of acute lung inflammation. Free Radical Research, 37:891–900.
21. Ryu, S. N. Park, S. Z. and Ho, C.T. (1998). High performance liquid chromatographic determination of anthocyanin pigments in some varieties of black rice. Journal of Food and Drug Analysis, 6:729–736.
22. Sensarma, P. (1989). Plants in Indian Puranas. Naya Prokash, Calcutta, West Bengal, India. 193.
23. Shen, M.Q. Zhao. Z.S. Chang, J. Jiao, Kun, H.X, and Wang, Z.H. (1994). Analysis of medicinal components from shangnog black rice, Journal of Shanghai Agricultural Collage, 12:137-139.
24. Shipp, J. and Abdel-Aal, E.S. (2010). Food applications and physiological effects of anthocyanins as functional food ingredients, Open Food Science Journal, 4: 7-22.
25. Strocchi and levitt M.D. (1991). Measurement of starch absorption in humans. Canadian Journal of Physiology and Pharmacology, 69:108-110.
26. Sukhonthara, S. Theerakulkait, C. and Miyazawa, M. (2009). Characterization of volatile aroma compounds from red and black rice bran. Journal of Oleo Science, 58:155-161.
27. Sultan, S. M. and Subba Rao, L.V.(2013). Germplasm collection from last remnants of rice landrace genetic diversity in high altitude areas of Kashmir Himalayas, International Journal of Conservation Science, 4:467-476.
28. Tsuda, T. Watanabe, M. Ohshima, K. Yamamoto, A. Kawakishi, S. and Osawa, T. (1998). Antioxidant components isolated from the seeds of tamarind (*Tamarindus Indica* L.). Journal of Agricultural and Food Chemistry, 42:2671–2674.
29. Tyagi, A.K. Khurana, J.P. and Raghuvanshi, S. (2004). Structural and functional analysis of rice genome, Journal of Genetics, 83:79-99.
30. Umadev, M. Pushpa, R. and Sampathkumar, K.P. (2012). Rice-traditional medicinal plant in India journal of pharmacognosy and phytochemistry, Phyto Journal, 2278- 4136.
31. Watt, G. (1891). A dictionary of the economic products of India. Vol. V. Cosmic Publications, New Delhi, India.
32. Zhang, M.W. Guo, B.J. Chi J.W. Wei, Z.C. Zhang, Y. and Zhang, R.F.(2005).Antioxidant and their correlation with total flavonoid and Anthocyanin contents indifferent black rice varieties, Scientia Agricultural, Sinica, 38:1324-1331.

CHAPTER 26

Amaranth: A Pseudocereal for Extraction of Nutraceutical Components

Narender K. Chandla, Romee Jan, Arti Chauhan and Sukhcharn Singh

Abstract

The amaranth as a pseudocereal has shown its role in the development and diversification of agricultural products and food which is a priority trend for world agriculture. The amaranth has excellent nutrient profile and technological advantage in various food formulations in comparison to other conventional cereals. Recently, current interest in amaranth also resides in the fact that it has a great amount of genetic diversity, phenotypic plasticity and is extremely adaptable to adverse growing conditions, resists heat and drought, has no major disease problem. This crop is among the easiest of plants to grow in agriculturally marginal lands in hilly area. Moreover, these contain significant amount of bioactive components such as phytosterols, squalene and polyphenols along with good amount of micronutrients such as minerals and vitamins. Amaranth consumption has been shown to improve plasma lipid profile in animals. Methanolic extracts of amaranth have demonstrated anti-hyperlipidemic,anti-diarrheic, anti-fungal and anti-malarial properties. Specific polyphenols such as rutin, isoquercetin and nicotiflorin and phenolic acids and amides with antioxidant activity have been found in amaranth seeds. In addition to above, amaranth, seeds are gluten free and thus they are currently emerging as healthy alternatives to gluten containing grains in gluten free diet for celiac people.

Keywords: Pseudocereals, phytosterols, squalene, polyphenols, antioxidant activity, etc.

Introduction

Grain amaranth is a pseudo-cereal with a long cultivation history in Central and South America. America comes into use of this grain at least 6000 years ago. Amaranth (Amaranthus spp.) has been consumed throughout history including by the Inca, Maya and Aztec civilizations, where it was used as a staple food. Recently, an increased interest in amaranth appeared in the 1980s, when the United States National Academy of Sciences performed research on the grain and described its high nutritional value and agronomic potential (Monteros et al., 1998; Ulbricht et al., 2009). This crop is extremely adaptable to adverse growing conditions, resists heat and drought, has no major disease problem, and is among the easiest of plants to grow in agriculturally marginal lands. Whether from the botanical or from the nutrient composition standpoints, Amaranth grain shares characteristics of both a cereal and a leguminous seed. Because its protein content, starch content and its amino acid composition are somewhere in between those of a cereal and a bean, it could be nutritionally considered as a natural mixture of rice and beans (Amaya-Farfan et al., 2005).

Amaranth has been reported protein concentration between ranges of 12.5%, to 17.6. %. It has been found that Amaranth paniculatus and Amaranth cruentus are good sources of flavanoids, especially for rutin, which are mostly produced in the stage of blossoming. The high lysine and methionine concentrations support the high nutritional quality of the grain. Amaranth grain consists of 6 to 9% of oil which is higher than most other cereals. The lipid fraction is unique due to the high squalene content. Amaranth is also a good source of high soluble fiber content 4.2%. Researchers are interested to know the well balanced nutritional profile of pseudocereal. Pseudocereal are important sources of energy due to their starch content and also provide good quality protein, dietary fiber and lipids rich in unsaturated fat (Alvarez-Jubete, Arended and Gallager 2009). Moreover they contain good amount of micronutrients such as minerals, vitamins and significant amount of bioactive components such as phytosterols, squalene and polyphenols. The reason for this interest is their excellent nutrient profile and their technological advantage in various food formulations.

Nutritional Values

The nutritional value of Pseudocereals viz. amaranth has been extensively studied (Martirosyan, 2005). Grain amaranth has higher protein and starch than other cereal grains and has significantly higher lysine content (Lehman, 1996). It has been shown that amaranth leaves are an excellent source of protein, with its maximal accumulation in the blossoming phase (Bressani and Kadoshnnikov, 1989) (17.2–32.6% from dry weight for various samples).

Protein

Protein concentration ranges in between 12.5% to 17.6%, and between 16.09% to 18.19% respectively (Teutonico and Knorr, 1985; Becker et al., 1981). Its methionine contents are around 15.8 mg/g of total protein, whereas its lysine levels are about 55.8 mg/g of total protein. The protein content of the Amaranthus hypochondricus lines ranges between 6.6% – 19.8% (Subhpreet Kaur et al., 2010) which refresh and reconfirm the previous information of protein content in the Amaranth seeds. The high lysine and methionine concentrations support the high nutritional quality of the grain, especially when compared to other grains in which those amino acids are present at limited levels (Teutonico and Knorr, 1985; Bressani, 1989). Although this pseudocereal shares the same classes of storage proteins with other grains, (Kadoshnnikov et al., 2008) have shown that in the ontogenesis of amaranth the proportions of globulins and albumins are reduced, whereas those of glutelins and gliadins are increased. Nevertheless, a low content of storage proteins (42% of the total) has been observed by these authors to occur in amaranth, compared to the high proportions found in wheat and other cereals.

Fats

Amaranth grain consists of 5.4 to 9.77% of oil which is higher than most other cereals (Subhpreet et al., 2010)). Amaranth oil contains approximately 77% unsaturated fatty acids and is high in linoleic acid, which is necessary for human nutrition. The lipid fraction is unique due to the high squalene content. Detailed studies on amaranth grain oil have been researched further in the last 2–3 decades (Lyon, 1987). This is true because amaranth oil contains 77% of polyunsaturated fatty acids, from which 50% comprises the linoleic acid. Furthermore, amaranth oil is characterized by the high content of such biologically active compounds, such as: squalene (up to 8%), tocopherols (to 2%), phospholipids (to 10%), and phytosterols (to 2%). The reported fat content varied from 0.05% to quite high value of 1.8%.

Carbohydrate

Carbohydrate (Starch) constitutes the main component of Amaranth grain and plays an important role in food application viz. thickeners for soups, fat replacer, cookies, snacks, pastas and in health foods due to its high nutritional values (Choi et al., 2004). The average carbohydrate content of the *Amaranthus* spp. is 65%.

Dietary Fiber

Amaranth has high soluble fiber content 4.2% (Early and Early, 1987) and concerning fiber, the amaranth grain is considered a good source of insoluble fiber. Differential milling of amaranth (*Amaranthus* spp.) yields three different granulometric fractions, which are classified according to their composition and properties. The high-fiber fraction can be improved by pneumatic classification until a product with 63.9% insoluble fiber and 6.86 soluble fiber contents is obtained. (Marcýlio et al., 2003) and (Escudero et al., 2004), reported that from the total dietary fiber in *A. cruentus*, 4.2% is soluble fiber.

Nutraceuticals

Nutraceuticals are, as defined by Merriam-Webster Dictionary, foodstuffs which provide health benefits in addition to their basic nutritional value. The term "nutraceutical" is a combination of the words nutrition and pharmaceutical, coined by Stephen DeFelice, a doctor and the founder of the Foundation for Innovation in Medicine (FIM) in 1989. The development of next generation nutraceutical "super foods" or products consist value-addition in the traditional natural diets by the use of non conventional sources viz. pseudocereal.

Functional Values

The Grain amaranth has been suggested as an alternative to marine animals as a natural source of squalene. Although the oil contents of grain amaranth ranges from 5.1% in Amaranthus tricolor to 7.7% in Amaranthus cruentus as compared to other oil-containing grains, high concentrations of squalene were found in total lipids, ranging from 3.6% in Amaranthus hypochondriacus to 6.1% in A. tricolor. The major fatty acids in Amaranthus oil consisted of palmitic acid (19.1-23.4%), oleic acid (18.7-38.9%), and linoleic acid (36.7-55.9%). To create a new food formulation high in protein, fiber, rich in minerals, vitamins with a good amount of nutraceutical leads prove this a super crop among the category of cereals. Amaranth grain is also a competent crop for the isolation of quality oils. The oil content of this crop has squalene which is a one of the functional component by lowering down cholesterol for the preventing CVD.

The nutraceutical usages of amaranth are appraisable, as it provides benefits above and beyond basic carbohydrate, protein and fats. This crop has its edge over the other as a reservoir for the extraction of nutraceutical. Amaranth has also been recognized because of its high vitamin and mineral contents, such as riboflavin, niacin, ascorbic acid, calcium, and magnesium contents. In the last few years, the properties of amaranth in cholesterol reduction, as an anti-oxidant, anticancer, anti-allergic, and antihypertensive agent; and as a food for patients with celiac disease and immune deficiencies, have been assessed in clinical studies. In turn, Lehmann (1996) stated that the decrease in cholesterol would be associated with some soluble component in the lipid fraction of the amaranth, suggesting that squalene and tocotrienols are responsible for the decrease. On the other hand, Grajeta (1997) also observed a significant reduction in blood and liver cholesterol concentrations in rats fed with whole and defatted amaranth flour and attributed the hypo-cholesterolemic effect to the soluble fraction of dietetic fibers present in the flours.

Fig. 1: Amaranth Plant

Fig. 1a: Amaranth Seed

Fig. 1b: Amaranth Cake

Several studies were focused on polyphenols in various Amaranthus species, which resulted in the identification of several phenolic acids, flavonoids, and their glycosides. The seed of grain amaranth is a rich source of iron calcium, magnesium and zinc as well as vitamin riboflavin and ascorbic acid. Therefore the generalized process layout for the extraction of total phenolics and their antioxidant activity has been shown below;

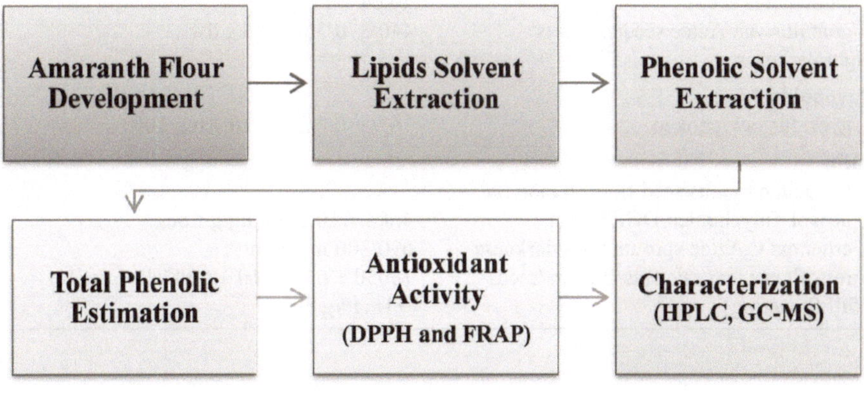

Fig. 2: Process Flow Sheet: for the Extraction of Nutraceutical and Characterization

To create a new food formulation high in protein, fiber, rich in minerals, vitamins with a good amount of nutraceutical leads prove this a super crop among the category of cereals. Amaranth grain is also a competent crop for the isolation of quality oils. The oil content of this crop has squalene which is a one of the functional component by lowering down cholesterol for the preventing heart diseases.

Conclusion

The nutraceutical usages of amaranth are appraisable, as it provides benefits above and beyond basic carbohydrate, protein and fats. This crop has its edge over the other as a reservoir for the extraction of nutraceutical. To create a new food formulation high in protein, fiber, rich in minerals, vitamins with a good amount of nutraceutical leads prove this a super crop among the category of cereals. Amaranth grain is also a competent crop for the isolation of quality oils. The oil content of this crop has squalene which is a one of the functional component by lowering down cholesterol for the preventing CVD.

Table 1: Amaranth Phenolic Components Quantification

Amaranth and Phenolic Constituents	Content
Phenolic Compounds	
A. cruentus seed: raw flours/high-protein flour fraction/cooked/popped/germinated (dried at 30, 60, and 90 æ%C)	5.16/5.89/3.53/4.46/3.04 to 3.68g TAE/kg
A. cruentus: treated vegetables	27.4 to 61.8 GAE/100g fw
A. hypondriacus grain: raw/extruded flour	56.60/69.50 mg GAE/100g dw
Total Phenolic Acids	
A. cruentus var. Aztec seeds/sprouts/light	464/380.7 mg/kg dw
Gallic Acid	
A. paniculatus seeds	40.64 ± 1.1 μg/g
A. cruentus var. Aztec seeds/sprouts light/darkness	440/370/360 mg/kg dw
Flavonoids	
A. hybridus raw/cooked	16.9 ± 0.8/21.1 ± 0.7 mg/100g
Rutin	
A. hypochondriacus seed flour: (Gabriela/ Nutrisol/Tulyehualco/DGETA var.)	4.0/4.7/10.1/5.8 μg/g flour
A. cruentus v. Aztec sprouts light/darkness	690/300 mg/kg dw
A. retroflexus leaves/stems/flowers/seeds (full flowering)	13050 ± 636/3360 ± 99/8725 ± 50/11 ±1mg/kg dw

References

1. Becker. R. Wheeler, E.L. Lorens, K. Stafford, A.Grosjean, O.K. Betschart, A.A. Saunders R.M. (1981). A composition study of amaranth grain. Journal of Food Science, 46:1175–1180.
2. Bertoft, E. (2007). Composition of clusters and their arrangement in potato amylopectin. Carbohydrate Polymer, 68:433-446.
3. Betschart, A. A. Irving, D. W. Shepard, A. D. and Saunders, R. (1981). Amaranth cruentus: Milling characteristics, distributions of nutrients within seed components and effects of temperature on nutritional quality. Journal of Food Science, 46:1181–1184.
4. Breene, W. M. (1991). Food uses of grain amaranth. Cereal Foods World, 36:428-430.
5. Bressani R. (1989). The proteins of grain amaranth. Food Review International, 5:13–38.
6. Bressani, R. Kalinowskw, L. S. Ortiz, M. A. and Elias, L. G. (1987). Nutritional evaluation of roasted, flaked and popped Amaranth candatus. Archivos Latinoamericanos de Nutricio´n, 37:525–531.
7. Correa, P. (1988). A human model of gastric carcinogenesis 1. Cancer Research, 48:3554"3560
8. Escudero, N.L. de Arellano, M.L. Luco, J.M. Gimenez, M.S. Mucciarelli, S.I. (2004). Comparison of the chemical composition and nutritional value of Amaranthus cruentus flour and its protein concentrate. Plant Foods Human Nutrition, 59:15–21.
9. Kadoshnikov, S.I. Kadoshnikova, I.G. Kulikov, Y.A. Martirosyan, D.M. (2008). Researches of fractional composition of protein of amaranth. Current Nutrition Food Science, 4:196–205.
10. Kadoshnikov, Sergey, I. Kadoshnikova, Irina, G. Martirosyan, Danik, M. (2005). Investigation of Fractional Composition of the Protein in Amaranth. In Book "Non-Traditional Natural Resources, Innovation Technologies and Products" Issue 12 Moscow, Russian Academy of Natural Sciences, Moscow; 81-104.
11. Kaul H.E. Aufhammer, W. Laible, B. Nalborczyk, E. Pirog, S. Wasiak, K. (1996). The suitability of amaranth genotypes for grain and fodder use in Central Europe. Die Bodenkultur, 47:173-81.4.
12. Marcýlio, R. Amaya-Farf´an, J. Silva, M.A.A.P. (2005). Avaliac¸˜ao da farinha de amaranto na elaborac¸ao de biscoito tipo "cookie" para cel´ýacos. Brazilian Journal of Food Technology, 8:175–81.
13. Miettinem, T.A. Vanhanen, A. (1994). Serum concentration and metabolism of cholesterol during rapeseed oil and squalene feeding. American Journal of. Clinical Nutrition, 59:356–363.
14. MiroshnichenkoA. V. Gonor, K, V. Kulakovam S.N. Miroshnichenko, L.A. Martirosyan, D.M. (2006). Effect of Amaranth Oil on Lipid Profile of Patients with Cardiovascular Diseases. In Book Functional Foods for Chronic Diseases Edited by: Martirosyan DM. Dallas, USA, pp. 35-45.
15. Pogojeva, A.V. Gonor, K.V. Kulakova, S.N. Miroshnichenko, L.A. Martirosyan, D.M. (2006). Effect of Amaranth Oil on Lipid Profile of Patients with Cardiovascular Diseases. In Book Functional Foods for Chronic Diseases Edited by: Martirosyan DM. Dallas, USA; 2006:35-45.
16. Rodas, B. Bressani, R. (2009). The oil, fatty acid and squalene content of varieties of raw and processed amaranth grain. Arch Latinoam Nutrition, 59:82–7.
17. Sanchez J.M.C. (1990). Amaranth (Amaranthus spp.) as forage. Proc. Fourth Amaranth Conf, Minnesota Ext. Serv, Minnesota Agr, Univ Minnesota, St Paul.
18. Sanders, R. M. and Becker, R. (1984). Amaranthus: a potential food and feed resource. Advances in cereal science and technology, VI. AACC, St. Paul, MN, pp 357-396.

19. Subhpreet, Kaur. Singh, N, Jai Chand Rana (2010). Amaranthus Hypochondricus and Amaranthus Caudatus germplasm: Characteristics of plants grain and flours. Food Chemistry, 123:1227-1234.
20. Teutonico, R. Knorr, D. (1985). Amaranth, composition, properties, and applications of a rediscovered food crop. Food Technology, 39:49–60.
21. Williams, J.T. Brenner, D. Grain amaranth (1995). Amaranthus species In: Williams JT (ed), Underutilized Crops: Cereals and Pseudocereals. Chapman and Hall, London, UK: 129-87. 2. 3.
22. Yorge Soriano Santos (1992). Chemical Characteristics of Amaranth oil. Ciencia, 45, 113-119.
23. Zhao J. and Whistler, R.L. (1994). Spherical aggregates of starch granules as flavor carriers. Food Technol., 48(7), 104-105.

CHAPTER 27

Prebiotics in Cereal Products

Neeraj Gandhi, Suresh Bhise and Amarjeet Kaur

Abstract

Prebiotics are the non-digestible food ingredients that beneficially affect the host by selectively stimulating the growth and/or activity of one or limited number of bacteria in the colon. These are heat resistant components which remain intact during the baking process; this allows prebiotics to be incorporated into every day food choices. Consumption of non-digestible ingredients allows the growth of bio cultures by reaching the intestine unaffected by the digestion process resulting in improvement of digestive health. Prebiotics reach the intestine in an unaltered form and shows the prebiotic effect when there is an increase in the activity of healthy bacteria in the human intestine. The prebiotics stimulate the growth of healthy bacteria such as bifidobacteria and lactobacilli in the gut and increase resistance to invading pathogens. Functional foods increase consumer choice by adding prebiotics to every day food items such as cereals, biscuits, breads, table spreads, drinks, and yoghurts. By continuing to eat and drink common foods, but choosing functional alternatives (such as bread containing prebiotics) dietary requirements can be met, without significant changes to food preferences. Some common example of cereal based products containing prebiotics are oat-based prebiotic drink, Breakstone's and Knudsen live active cottage cheese, prebiotic yogurt- Attune and Muller vitality, multi-fruit drink enrich and flavoured water enrich, prebiotic flavoured water, prebiotic capsules, dried fruit bars, prebiotic powder- Biocare fructo-oligosaccharides, Newtree purplaisir chocolate spread. The claims are constipation relief, suppression of diarrhea and reduction of the risks of osteoporosis, atherosclerotic cardiovascular disease associated with dyslipidemia and insulin resistance, obesity and possibly Type-2 diabetes. At present, claims about reduction of disease risk are only tentative and further research is needed.

Introduction

Currently, there is huge interest in the use of foods which may exert a positive functional effect on our health. Two of these 'Functional Foods' are known as-

- Probiotics
- Prebiotics

Both have a favourable effect on the 'good bacteria' that reside in our digestive system, also known as our 'gut microflora'. These 'good' bacteria live naturally in our intestines and are essential to good health having a number of positive effects, primarily helping our digestive systems work efficiently. We traditionally view bacteria as being 'bad'; however in reality there are relatively only a small number of strains which are pathogenic and most microbes are harmless and contribute to well-being (Gibson 2003).

Probiotics

Probiotics are live strains of 'good' bacteria, which help our digestive system work efficiently, e.g. bifidus and acidopilus. They are found in live yoghurts, powders or specially formulated probiotic drinks which contain one or more of the strains of these 'good' bacteria. With food processing, pollution and antibiotic therapy, numbers of 'good' bacteria occurring naturally in our gut are reduced, so studies have shown, by actively consuming the bacteria, the size of the colonies in the gut can be increased, which improves digestion. Moreover, the numerous studies have also shown that with optimal numbers of 'good' bacteria, the immune system is improved, increasing our ability to fight disease. Probiotics may also have a role in reducing the severity of allergies.

Prebiotics

Prebiotics are certain nutrients and constituents of food which our gut flora feed on, thus increasing their numbers. Prebiotics include fructo-oligosaccharides (FOS) and some other soluble fibres found in pulses, fruit and some cereal products. Thus, prebiotics also help digestion and the immune system. Both probiotics and prebiotics increase the colony size of the gut's natural flora, and more and more people are including them in their diet to promote good health. The bodybuilder can also benefit, as he/she can digest their (large quantity) of food more easily helping to provide a more efficient influx of energy and protein. Pro- and prebiotics also help the immune system keep diseases at bay, so the bodybuilder can train harder and recuperate quicker. Probiotics can be obtained by eating live yoghurts, special powders or probiotic drinks like Yakult, Actimel or LC1. More recently, bodybuilding supplement companies have recognised the benefits and are adding probiotics and FOS to their engineered nutrition meal replacement powders.

Difference

Probiotics are foods which provide the actual 'good bacteria' and Prebiotics are those 'nutrients' which the bacteria feeds on to increase their numbers.The effect of a probiotic may be enhanced by having a prebiotic as a support medium e.g. Milk contains nutrients for lactobacillus and so many probiotic drinks are milk or yogurt-based.

Synbiotics (Eubiotics)

- Synbiotics refers to a product in which a probiotic and prebiotic are combined. Such a combination aids survival of the administered probiotic and facilitates its inoculation into the colon.
- The synbiotic effect may be directed towards two different regions i.e. both large and small intestines.
- Experiments with rats showed that synbiotics protect the organism from carcinogens significantly better than do either prebiotics or probiotics separately (Gallaher and Gill 1999).

Potential Prebiotics

The most powerful and researched prebiotics are FOS (Fructooligosaccharides) and Inulin. Both share the same basic structure i.e. linear chains of fructose units connected by beta (2-1) links.

- **Oligosaccharide-** Found in 36,000 plants and also found in milk. Usually, these are mixtures of saccharides of 3-10 monomers. 16 different types of commercial oligosaccharides are identified. These have been used worldwide. FOS (Frctooligosaccharide) may contain 2-8 units in a chain. These are soluble, sweet and contribute very low calories (2.0kcal/gm). The degree of polymerization varies in the range of 2-29.
- **Inulin-** It is a type of FOS, extracted from chicory root, has a DP upto 60 with the physiological benefits of soluble dietary fibre. It can act as a excellent fat replacer. At high conc., it exhibits gelling properties.

Dutch ingredients firm Sensus has carried out new research to prove the prebiotic effect of native inulin at only 5 grams, allowing food manufacturers to make prebiotic claims on a relatively low dose. Inulin, extracted from the chicory root, is already known to have a prebiotic effect i.e. boosting the growth of beneficial bacteria in the gut but while studies have shown 5g of oligofructose, formed through the partial enzymatic hydrolysis of inulin, to have a prebiotic benefit at 5g, Sensus says it is the first inulin producer to prove the prebiotic effects of native inulin at a dose of 5 grams per day.

- **Resistant Starch:** Functional fibre performing an important role in digestive physiology e.g. high amylose corn starch. It may be divided into 4 groups:
- Seed starch (RS1)
- Granular starch, particularly high amylose and banana starches resistance of which depends, among others, on tendency to swelling, forms the second group (RS2)
- Highly retrograded starch, usually produced either by extrusion cooking or by liberation from cereals by enzymatic digestion, constitute the third group (RS3)
- Chemical modified starch, constitute the fourth, very promising group of resistant starches (RS4)

Others Potential Prebiotics

Pyrodextrine, Galacto-oligosaccharides, Soy-oligosaccharides, Xylo-oligosaccharides, Isomalto-oligosaccharides, Lactulose, Trans-oligo-saccharides, Rafitaline, Frutafit EXL and Fruitofit IQ (Table 1). Also food formulators can use Gums, Arabinogalactane, Lignins to create foods that offer another option for consumers to get more fibre.

Table 1: Production Methods for Prebiotics

Prebiotic	Production methods
Inulin [Fructooligosaccharide (FOS)]	Hot water extraction from chicory root (followed by enzymatic hydrolysis or polymerization of fructose monomers
Galactooligosaccharide (GOS)	Enzymatic lactose transgalactosylation
Xylooligosaccharide (XOS)	Enzymatic hydrolysis of plant xylans liquefied starch
Isomaltooligosacchairde (IMO)	Transglucosylation of liquefied starch
Lactulose	

Cereals as Potential Prebiotics

Traditionally grains, particularly whole grains, are used in nutrition as carriers of macro and microelements, proteins, fibre, and vitamins. In the past 30 years, numerous studies have documented a beneficial role in:

- Controlling blood pressure
- Arterial health
- Insulin level
- Facilitate weight loss, if properly used

The majority of cereal grains appeared to be too readily digested to play an effective role as Prebiotics.

Daily consumption of whole-grain (WG) Wheat exerted a pronounced prebiotic effect on the human gut microbiota composition. As cereals are grown over 73% of the total world harvested area and contribute over 60% of the world food production providing nutrients for human health. The possible applications in functional foods formulations are:

- As fermentable substrate for growth of probiotic microorganisms, especially lactobacilli and bifido- bacteria.
- As dietary fibre promoting several beneficial physiological effects.
- As prebiotics due to their content of specific non-digestible carbohydrates.
- As encapsulation materials for probiotics in order to enhance their stability.

Benefits of Prebiotics

Prebiotic fermentation leads to health benefits, such as:

- Increased fecal biomass.
- Increased stool weight and/or frequency.
- Prebiotics are fermented by bifidobacteria with the production of short chain fatty acids (SCFA), mainly acetate, propionate, butyrate, and hydrogen and carbon dioxide. Production of SCFA leads to lower pH in the colon, which facilitates absorption of Ca, Mg and Zn.
- Lower pH also restricts pathogenic bacteria, thus reducing or eliminating precarcinogenic activity.
- Immune system is improved.

- Prebiotics also have role in reducing the severity of food allergens, food poisoning and reduce effects of food intolerance.
- Also reducing the severity of symptoms in both osteo and rheumatoid arthritis.
- Supplimentation with inulin and FOS reduces the risk of infant diarrhea, constipation, and gut infection.
- As food ingredient, prebiotics reduces fat and carbohydrates in food products and enhancing mouth feel, structure and shelf life.
- Prebiotics not only work symbiotically with probiotics but also could protect probiotics during processing and shelf life through micro-encapsulation (e.g. coating of polyols on bacteria).

Sources of Prebiotics

Prebiotics occur naturally in a variety of fruits and vegetables and grains, especially in:

- Chicory
- Asparagus
- Jerusalem artichokes (Sun root- Hillianthus tuberosus)
- Banana
- Onion
- Garlic
- Tomato
- Barley
- Wheat
- Leeks
- Endive

Also mother's milk contains relatively high levels of prebiotics (3-6 g/l)

Recommended Levels

- With food processing, pollution and antibiotic therapy, the numbers of bacteria living naturally in our gut are reduced.
- Research has shown that there should be consumption of around 5gm of prebiotics daily on the top regular diet for our healthy bifido bacteria to get a boost.

- It may be as high as 15-20 gm/day.
- Allergy, asthma and inflammatory bowel disease could be a consequence of prebiotic deficiency in Western diets.

Role of Prebiotics in Food Industries

- Prebiotics are becoming increasingly popular in food formulations within the United States. They are considered an 'invisible' fibre source allowing food technologists to formulate fibre enriched food without any negative effect on taste or texture.
- Prebiotics can be added to a wider selection of foods than probiotics because they don't have the same storage problems as "live" products.
- As a general rule, only about 15% of all products that contain any prebiotic ingredients make a specific claim statement on the front of the package as to the content or benefits.
- Prebiotics are available in everything from bread and cereal to yougurt, baby milk and flavoured water.

Role of Prebiotics in Cereal Food

- The development of non-dairy prebiotic products is a challenge to the food industry.
- In this respect, prebiotics containing baby foods or confectionary formulations have been developed by a number of leading manufactures.
- Britannia Health Products is introducing a range of breakfast cereals containing Hi-maize prebiotic resistant starch.

Cereal Products Including Prebiotics

There are number of products in the market that promote their prebiotic content and benefits.

- **Prebiotics in Breakfast Products:** A number of leading manufactures within the cereal market worldwide have incorporated prebiotics into products to enhance health claims. These have been successfully formulated into a number of branded cereal products such as:

- **Wholewheat Breakfast Cereal**

Company- Weetabix

Category- Breakfast cereals

Country- France

Product Description- It is said to be rich in fibre and low in fat and the first breakfast cereal with prebiotic fibre. It is said to combine the good taste of whole wheat and the lightness of rice to give a crispy and generous flake. It is available in a 375g pack.

- **Wholegrain Porridge with Prebiotic**

Company- Tesco

Category- Breakfast Cereals

Country- UK

Product Description- Comprising wholegrain oat flakes with prebiotic inulin. The product is free from artificial colours, flavours, sweetners and preservatives and has strict limits on fat, sugar and salt. It can be microwaved in 5 mins., is suitable for vegetarians and is available in 240g pack of twelve 20g sachets.

- **Rice Krispies Multigrain**

Company- Kellogg

Category- Breakfast Cereals

Country- UK

Product Description- A 30g serving provides 2g of inulin. Available in 350g pack.

- **Products with DHA and Prebiotics**

Company- Beech-Nut Nutrition Corp.

Category- Baby foods and dry cereals

Country- N.Y.

Product Description- Contain DHA (An omega-3 fatty acid)+inulin+ brown rice+ fruits and vegetables. The product is packaged in a 7-oz gable-top carton with a twist-off cap and spout.

- **Good Morning and Good Evening Baby Cereal**

Company- Beech-Nut Nutrition Corp.

Category- Baby foods

Country- N.Y.

Product Description- Help babies' digestive health

- **Hipp's Hippness Crisp**

Category- Breakfast Cereal

Country- Germany

- **Prebiotics in Cereal Snack Bars:** Now a days, many people frequently combine snacks as a meal substitute which could result in not getting enough nutrition from their diet. Prebiotics ingredients, such as Beneo, can help deliver a great-tasting bar that consumer won't feel guilty about eating as snack between meals.

- **Cereal Bars**

Company- Schwartauer Werke

Category- Snacks

Country- Germany

Product Description- Corny Activ Dreierlei Nuss is a pack of 6 cereal bars with 3 kinds of nuts. It contains prebiotic dietary fibres and one bar is said to contain 15% of daily recommended intake of fibre. Also available are Yogurt with Soya and Dark Chocolate varieties.

- **Goodness Bar**

Company- Ryvita Company

Category- Snacks

Country- UK

Product Description- Energy bars made from chewy wholegrain rye. The bars are prebiotic and are available in 3 flavours: mixed berry; apple and sultana; and cranberry and raisin. The bars are free from added sugar, added salt, colours, flavours and preservatives. They contain 62 calories, 3% fat and 20% fruit, suitable for vegetarians. This product is available in 138g box of six 23g bars.

- **Nutresan's Active Bran**

Category- Cereal Bar

Country- Argentina

Product Description- Sugar free with added Ca and active fibres which regulate intestinal function and improve natural defenses.

- **Delight Full Snack Bar**
- Help consumers to eat less. The bars contain larch arabinogalactane and other fibres. Each bar offers 2.5g of dietary fibre.

- **Granola Wellness Bar**

Company- Attune Foods

Country- Calif.

Product Description- Each bar contains more than 10 billion probiotics as well as prebiotic inulin.

- **Prebiotics in Bread**
- **Waterburtons Healthy Inside Loaf**

75p, available from all major markets. The first prebiotic loaf of bread. Three slices provide >3rd of the daily prebiotic recommendation.

- **Vitaalbrood Bread**

Sold in Netherlands by Bakkerij Veenhuis. Carries a "healthy colon" claim because of its inulin content.

Other Products

- Oat-based prebiotic drink
- Breakstone's and Knudsen live active cottage cheese
- Prebiotic yogurt- Attune and Muller vitality
- Multi-fruit drink enrich and flavoured water enrich
- Prebiotic flavoured water
- Prebiotic capsules

- Dried fruit bars
- Prebiotic powder- Biocare FOS
- Newtree purplaisir chocolate spread

Current Research

- ORAFTI Active Food Ingredients is also a driving force behind much of the scientific research into the health effects of prebiotic fibres.
- In collaborations with 75 universities and research centers worldwide, the health benefits of prebiotics are documented.
- The application of inulin and oligofructose in functional foods provides key benefits to older people and raises an interesting opportunity for the food industry.

Conclusion

- Consuming prebiotics is an excellent way of promising healthy bacteria that improve digestion and fight potentially harmful bacteria.
- Continuous development of new functional foods in response of science and industry to the increased consumer awareness regarding health and role of foods for improving quality of life.
- Prebiotics fibres are now explored by manufacturers worldwide to respond to an increasing public interest in health and nutrition.
- The U.S. market appears to have awaken to the benefits and functionality of prebiotics.
- Emerging research and supplier innovation continue to enhance the prebiotics available.
- Large distribution and important nutritive value of prebiotics have focused the attention on their use as raw materials for development of food products.
- Foods containing prebiotics and fibre are put into group of products called Nutraceuticals.

References

1. Angelov, A. Gotcheva, V. Kuncheva, R. and Hristozova, T. (2006). Development of a new oat-based probiotic drink. International Journal of Food Microbiology, 112:75-80.
2. Charalampopoulos, D. Wang, R. Pandiella, S. S. and Webb, C. (2002). Application of cereal and cereal components in functional foods: a review. International Journal of Food Microbiology, 79:131-141.

3. Dornblaser, L. (2007). Probiotics and prebiotics: what in the world is going on? Cereal Foods World, 52(1):20-21.
4. Kummel, K. F. and Brokx, S. (2001). Lactitol as a functional prebiotic. Cereal Foods World, 46(9):424-429.
5. Mishra, V. and Prasad, D. N. (2001). Prebiotic foods and their applications in health management. Indian Food Industry, 20(5):59-61.
6. Neill, J. O. (2007). Prebiotics: health and nutrition benefits in functional foods. Cereal Foods World, 52(1):8-11.
7. Ohr, L. M. (2004). Nutraceuticals and functional foods. Food Technology, 58(2):72-75.
8. Ohr, L. M. (2007). Ingredients for good digestion. Food Technology, 83-88.
9. Tamime, A. Y. Saarela, M. Korslund Søndergaard, A. Mistry, V. V. and Shah, N. P. (2005). Production and Maintenance of Viability of Probiotic Micro-Organisms in Dairy Products. In: A Y Tamime (ed.) Probiotic Dairy Products. Blackwell Publishing, Oxford, UK. pp. 44-51.
10. Tomasik, P. J. and Tomasik, P. (2001). Probiotics and prebiotics. Cereal Chemistry, 80(2): 113-117.
11. www.cellscience.com
12. www.tetrapak.com/processing/ads/index.html

CHAPTER 28

Typha angustifolia L.: A Potential Source of Nonconventional Starch

Syed Insha Rafiq, Khalid Muzaffar, Kulsum Jan and D.C. Saxena

Abstract

Demand for starch is increasing day by day and is met by restricted range of crops. In recent years, substantial progress has been made in obtaining starch from non-conventional sources such as sweet potato, buckwheat, sago, water chestnut, which can be used as a substitute for conventional sources. So it is necessary to search for other underutilized and nonconventional sources of starch to meet the increased demand. Typha angustifolia, an aquatic weed containing appreciable amount of starch in its rhizome and roots, can be used for starch isolation to make its use in food and non-food derivatives. In many parts of the world, starch-rich roots of Typha angustifolia are eaten raw or boiled or dried roots are pulverized into a powder, and then, used as a thickener in soups or added to cereal flours to make biscuits and bread.

Keywords: Non-conventional, starch, rhizomes, *Typha angustifolia*, weed.

Introduction

Starch is important food stuff as it accounts for over 30% of the average diet on a dry weight basis and more than 25% on an available energy basis (Galliard, 1987). Among carbohydrate polymers, starch is currently enjoying increased attention owing to its usefulness in different food products. Starch can be used as thickener, an adhesive, binder, encapsulating agent, film former, gelling agent, water binder, texturizer and with numerous other applications both in the food and non-food areas (Mauro, 1996). Non-food applications of starch include in the field of pharmaceuticals, cosmetics, textiles, alcohol-based fuels and adhesives (Othaman et al., 2010).

The world-wide market for industrial starches is expanding and current demand is met by a restricted range of crops, the most important of which include potatoes, maize, wheat and tapioca, rice, finger millet, green gram, chick pea, lotus stem, water chestnut etc. However, these sources act as food material and keeping in mind the food scarcity, we have to look for other non-conventional food sources, which can be used for starch extraction. Among the non-conventional starch sources, weeds are the leading ones. These may include aquatic weeds like Typha species, Acorus calamus, Nympheae species, etc. terrestrial weeds like Amaranth, Chenopodium, Cyperus difformis, Horse chestnut etc. The following is a subjective summary that presents status of knowledge about the non-conventional starch source-Typha species (*Typha angustifolia* L.).

Typha angustifolia L.

Typha angustifolia L. is a perennial rhizomatous aquatic macrophyta from the Typhaceae family. The plant is named as taboa (Brazil), Lesser Bulrush (Britain) and narrow-leaved cattail or cattail (North America and other countries). *T. angustifolia* is characterized by its fast growth and high biomass (Xu et al., 2011).

Distribution and Habitat

Typha angustifolia L. like other Typha species is found throughout the world, from tropical to temperate zones, and from humid to dry climates. It can grow under broad climatic and habitat ranges, which helps it to achieve widespread dominance in a variety of aquatic plant communities (Mitich, 2000; Murkin and Ward, 1980). Typha angustifolia commonly grows in littoral zones of lakes, ponds, and dikes. It is more tolerant to deep water than other Typha species (Grace and Wetzel, 1981; Weisner, 1991).

Plant Description

Typha angustifolia is an erect, perennial rhizomatous aquatic macrophyta, 1-3 m tall with flowers borne on a slender stem. Table-1, shows the scientific classification of Typha angustifolia L. The leaves are long, linear, strongly planoconvex, 3-8 mm wide, occasionally wider, deep green, sheathing at the base. The roots are fibrous, adventitous from rhizomes. The lateral rhizomes are produced at the base of leaves (Grace and Harrison, 1986). Rhizomes besides acting as organs for carbohydrate storage during unfavorable conditions, they also help the plant to survive during winter, play important roles in the propagation and migration of the plant (Cizokova-Koncalova et al., 1992; Asaeda et al., 2005a, b), provide structural support for the roots, provide anchorage during light disturbance (Graneli et al., 1992), and also effect regeneration after heavy disturbance (Archibald, 1995). Typha species show morphological changes with variation of water depths (Grace 1989; Grace and Wetzel, 1981; Waters and Shay, 1992). Typha angustifolia shows changes in rhizome morphology with the changes in water depth to maintain effective aeration (Sharma et al., 2008).

Table 1: Scientific classification of Typha angustifolia

Scientific classification	
Binomial name	*Typha angustifolia* L.
Common Name	Small Reed Mace, Narrowleaf cattail
Kingdom	Plantae
Order	Poales
Family	Typhaceae
Genus	Typha
Species	*T. angustifolia*
Group	Monocot
Duration	Perennial

Life History

Typha angustifolia has a higher allocation to sexual reproduction. Cattails spread both vegetatively and by seed, particularly under drawdown conditions (USDA, NRCS, National Plant Data Center). Typha species generally reproduce by seeds but more extensively by rhizomes as rhizomes grow new shoots quickly by vegetative propagation as compared to seeds by sexual reproduction. Typha species show varied growth behaviour which includes initial leaf production during early spring, followed by flowering stage during mid summer and lastly high clonal growth in the fall. However, the specie is said to have an earlier flowering period with wind pollinated inflorescence. The pollen grain of the specie is said to remain viable for four weeks and upon pollination attach themselves to stigmas of female flowers to form fruit and the fruits with small seeds, mature in august and September (Keddy and Ellis, 1985).

Food Uses

Several parts of the plant are edible, including dormant sprouts on the roots and bases of the leaves, ripe pollen, the stem and the starchy roots (Telias-Dykeman, PA 2009; Schmeda-Hirschmann et al., 1999). The rhizomes of T. angustifolia are characterised by high fibre (17.20 g/100 g of flour) and carbohydrate (67.29 g/ 100 g flour) contents, and are known to be rich in starch granules (Steinbachova-Vojtískova et al., 2006; Asaeda et al., 2008), which can be used by colonic microbiota as substrates

for anaerobic fermentation and the production of short chain fatty acids. The base of the stem where it attaches to the rhizome can be boiled or roasted like potatoes. Cattail pollen is a fine substitute for flours. It is a bright yellow or green color, and gives pancakes, cookies or biscuits a pretty yellow color. The rhizomes and lower stems have a sweet flavor and can be eaten raw, baked, roasted, or broiled. Cattail rhizomes are fairly high in starch content, about 30% to 46% and can be ground into flour (Harrington, 1972). This flour would probably contain about 80 % carbohydrates and around 6% to 8% protein. As cattails occur around the world, it can be a potential source of food for the worlds' population. In many parts of the world, starch-rich roots of *T. angustifolia* (Tanaka, 1976) are eaten raw or boiled or dried roots are pulverized into a powder, and then, used as a thickener in soups or added to cereal flours to make biscuits, bread, cakes etc. Pollens are yet another source of protein used as additive in making bread, porridge etc. The nutrient composition of the rhizome of Typha is presented in Table 2.

Table 2: The nutrient composition of the rhizome of Typha

Composition	Value (%)
Moisture	40
Protein	8
Fat	2
Starch	30
Ash	2
Fibre	18

Medicinal uses

Typha angustifolia and other species of the genus Typha are widely used as medicinal plants. In Brazil, Latin America and North America, the leaves are used as a diuretic, an astringent, a desiccant, a haemostatic agent and a vulnerary. In addition, the rhizomes are used as a diuretic, an astringent and an antimycobacterial. Moreover, the pollen is used in the treatment of scrofula, abscesses and abdominal pain, and the powder of the rhizomes is used to prevent chafing, sores, inflammation, kidney stones and diarrhoea (Duke, 1984).

Conclusion

Rhizomes and shoot of *Typha angustifolia* L. can be exploited as a source of nutritious food products. Besides this, rhizome and roots contain a good content of starch and therefore can be used as potential source of non conventional starch source. Need of the hour is to isolate and characterize *Typha angustifolia* L. starch, to find its use in food and non-food industries.

References

1. Galliard, T. (1987). Starch availability and utilization. In T. Galliard (Ed.), Starch properties and potential (pp. 1–15). Chichester, UK: John Wiley and Sons.
2. Mauro, D.J. (1996). An update on starch. Cereal Food World, 41:776-780.
3. Othaman, Z. Al-Assaf, S. and Hassan, O. (2010). Molecular characterisation of sago starch using gel permeation chromatography multi-angle laser light scattering. Sains Malaysiana. 39:969-973.
4. Xu, J. Li, C. Yang, F. Dong, Z. Zhang, J. Zhao, Y. Qi, P. and Hu, Z. (2011). Typha angustifolia stress tolerance to wastewater with different levels of chemical oxygen demand. Desalination, 280:58–62.
5. Mitch, L. M. (2000). Common cattail, Typha latifolia L. Weed Technology, 14:446-450.
6. Murkin, H. R. and Ward, P. (1980). Early spring cutting to control cattail in a northern marsh. Wildlife Society Bulletin, 8:254-256.
7. Grace. J.B. Wetzel, R.G. (1981). Effect of size and growth rate on vegetative reproduction in Typha. Oecologia, 50:158–161.
8. Weisner S.E.B. (1991). Within lake patterns in depth penetration of emergent vegetation. Freshwater Biology, 26:133–142.
9. Grace, J.B. and Harrison, J.S. (1986). The biology of Canadian weeds, Typha latifolia L., Typha angustifolia L. and Typha xglauca Godr. Canadian Journal of plant Science, 66:361-379.
10. Cizokova-Koncalova, H. Kvet, J. and Thompson, K. (1992). Carbon starvation: a key to reed decline in eutrophic lakes. Aquatic Botany, 43:105–113.
11. Asaeda, T. Fujino, T. and Manatunge, J. (2005a). Morphological adaptations of emergent plants to water flow: a case study with Typha angustifolia, Zizania latifolia and Phragmites australis. Freshwater Biology, 50:1991–2001.
12. Asaeda, T. Hai, D.N. Manatunge. J. Williams, D. and Roberts, J. (2005b). Latitudinal characteristics of below and aboveground biomass of Typha: a modelling approach. Analytical Botany, 96:299–312.
13. Graneli, W. Weisner, S.E.B. and Sytsma. M.D. (1992). Rhizome dynamics and resource storage in Phragmites australis. Wetlands Ecological Management, 1:239–247.
14. Archibald, O.W. (1995). Ecology of world vegetation. Chapman and Hall, London.
15. Grace, J.B. (1989). Effects of water depth on Typha latifolia and Typha domingensis. American Journal of Botany, 76:762–768.
16. Waters, I. and Shay, J.M. (1992). Effect of water depth on population parameter of a Typha glauca stand. Canadian Journal of Botany, 70:349–351.
17. Sharma, P. Asaeda, T. Fujino. T. (2008). Effect of water depth on the rhizome dynamics of Typha Angustifolia. Wetlands Ecological Management, 16:43–49.
18. USDA, NRCS, National plant data center and Idaho plant materials center. Plant Guide. http://plant-materials.nrcs.usda.gov/intranet/pfs.html.
19. Keddy, P.A. and Ellis, T.H. (1985). Seedling recruitment of 11 wetland plant species

along a water level gradient: shared or distinct responses? Canadian Journal of Botany, 63:1876-1879.

20. Telias-Dykeman, P.A. (2009). Edible Wild Plants. Sterling Publishing Co Inc. New York.
21. Schmeda-Hirschmann, G. Razmilic, I. Gutierrez M.I. Loyola, J.I. (1999). Proximate composition and biological activity of food plants gathered by Chilean Amerindians. Economical Botany, 53:177–187
22. Steinbachova-Vojtískova, L. Tylova, E. Soukup, A. Novicka, H. Votrubova, O. Lipavska, H. and Cízkova, H. (2006) Influence of nutrient supply on growth, carbohydrate, and nitrogen metabolic relations in Typha angustifolia. Environmental Expiemntal Botany, 57:246–257.
23. Asaeda, T. Sharma, P. and Rajapakse, L. (2008). Seasonal patterns of carbohydrate translocation and synthesis of structural carbon components in Typha angustifolia. Hydrobiol. 607:87–101.
24. Harrington, H.D. (1972). Western edible wild plants. The University of New Mexico Press. 156 pp.
25. Tanaka, T. (1976). Tanaka's Cyclopaedia of Edible Plants of the World. Keigaku Publishing, Japan.
26. Facciola, S. (1990). Cornucopia - A Source Book of Edible Plants. Kampong Publications
27. Duke, J.A. (1984). The handbook of notheast Indian medicinal plants. Quarterman Pub, Inc, Lincoln, p 212.

CHAPTER 29

Effect of Reducing and Oxidizing Agents on the Rheological, Pasting and Noodle Making Characteristics of Durum and Commercial Wheat Semolina

Ritika Oberoi, Amarjit Kaur and A.K. Bakhshi

Abstract

Noodles are widely consumed throughout the world and their global consumption is second only to bread. In the study, response of reducing agents (cysteine hydrochloride and sodium metabisulphite) and oxidizing agents (potassium bromate and benzoyl peroxide) at 10 and 20 ppm levels on the farinographic and amylographic curve characteristics as well as on the noodle making properties of durum and commercial wheat semolina were investigated according to the AACC (1990) procedure. The farinographic curve characteristics revealed that water absorption was enhanced with increasing levels of reducing agents but decreased with addition of oxidizing agents. However there was no significant change in Dough Development Time in case of durum semolina on addition of reducing agents, but commercial semolina recorded significant changes at 20 ppm level of reducing agent. The best reducing agent for durum and commercial semolina was thus sodium metabisulphite and cysteine hydrochloride respectively. Addition of oxidizing agent potassium bromate at 20 ppm level enhanced stability and was thus chosen as the best oxidizing agent for durum semolina and benzoyl peroxide for commercial semolina. Addition of reducing and oxidizing agents improved yield, appearance, elasticity and extensibility characteristics of instant fresh noodles. The best levels were used in the preparation of noodles on the Italian extruder (NAMAD).

Keywords: Reducing agent, Oxidizing agent, durum, semolina, noodle

Introduction

Noodles are widely consumed throughout the world and their global consumption is second only to bread. Instant noodles are widely consumed throughout the world and it is a fast growing sector of the noodle industry. This is because instant noodles are convenient, easy to cook, low cost and have a relatively long shelf-life. In this study, the response of Reducing (cysteine hydrochloride and sod metabisulphite) and Oxidizing agents (potassium bromated and benzoyl peroxide) at 10 and 20 ppm levels on the farinographic and amylographic curve characteristics as well as on the noodle making properties of durum and commercial wheat semolina were investigated. The best levels of these additives on durum and commercial wheat semolina's were used in preparation of noodles on the Italian extruder (NAMAD). Bakhshi (1998) studied Farinographic curve characteristics of durum semolina. He found a variation of 4.0 – 6.2 minutes in dough development time, 245-340 BU in peak consistency and 10 – 70 BU in mixing tolerance index. Mastromatteo et al., (2013) studied the influence of water content on the rheological, micro structural and sensorial properties of durum wheat bread and found that strong doughs (with low water content), recorded greater resistance to deformation.

Materials and Methods

Response of Reducing agents (cysteine hydrochloride and sod metabisulphite) and Oxidizing agents (potassium bromated and benzoyl peroxide) at 10 and 20 ppm levels on the farinographic and amylographic curve characteristics as well as on the noodle making properties of durum and commercial wheat semolina were investigated. From the farinographic curves, parameters like water absorption, Dough Development Time, Mixing Tolerance Index and Stability were studied. From the amylographic curves, parameters like Gelatinization Time, Gelatinization Temperature, Peak Viscosity, Viscosity at 95°C and Peak Viscosity Temperature of durum and commercial wheat semolina's were studied. The best level of these additives for durum and commercial wheat semolina from farinographic and amylographic results was used in preparation of noodles on the NAMAD extruder.

Results and Discussion

The effect of reducing and oxidizing agents on farinographic and amylographic curve characteristics of durum and commercial wheat semolina is presented in Table-1 and 2 respectively.

Table 1: The effect of reducing agents on farinographic and amylographic curve characteristics of durum and commercial wheat semolina.

Parameter	0 ppm Additives (Control)		10 ppm Cysteine Hydrochloride		20 ppm Cysteine Hydrochloride		10 ppm Sodium Metabisulphite		20 ppm Sodium Metabisulphite	
	Durum Semolina	Commercial Semolina	Durum Semolina	Commercial Semolina	Durum Semolina	Commercial Semolina	Durum Semolina	Commercial Semolina	Durum Semolina	Commercial Semolina
Water Absorption (%)	54.8	49.4	55.0	48.2	54.5	46.0	54.6	48.0	53.8	47.0
DDT (min)	2.0	3.6	2.0	3.5	2.1	3.1	1.9	3.5	1.9	3.9
Stability (min)	15.2	3.8	10.2	3.9	3.3	8.0	10.8	4.1	5.4	3.0
MTI (BU)	12.0	62.0	28.0	45.0	43.0	45.0	45.0	62.0	57.0	42.0
Gel Time (min)	30.0	39.0	31.0	40.0	31.0	40.0	30.0	40.0	36.0	38.5
Gel Temp (°C)	65.0	78.5	66.5	80.0	66.5	80.0	65.0	80.0	74.0	77.8
Peak Viscosity (BU)	520	695	510	710	505	697	555	720	545	730
Peak Viscosity Temp (°C)	89.0	89.75	90.50	91.10	93.50	92.30	89.90	91.10	91.10	90.50
Viscosity at 95°C (BU)	485	680	492	690	502	690	530	710	520	700

Table 2: The effect of oxidizing agents on farinographic and amylographic curve characteristics of durum and commercial wheat semolina.

Parameter	0 ppm Additives (Control)		10 ppm Potassium Bromate		20 ppm Potassium Bromate		10 ppm Benzoyl Peroxide		20 ppm Sodium Benzoyl Peroxide	
	Durum Semolina	Commercial Semolina	Durum Semolina	Commercial Semolina	Durum Semolina	Commercial Semolina	Durum Semolina	Commercial Semolina	Durum Semolina	Commercial Semolina
Water Absorption (%)	54.8	49.4	55.0	48.2	54.5	46.0	54.6	48.0	53.8	47.0
Water Absorption (%)	54.8	49.4	52.2	47.8	51.2	46.4	52.6	48.0	51.6	
DDT (min)	2.0	3.6	1.9	3.5	1.8	3.3	1.9	2.8	1.9	2.9
Stability (min)	15.2	3.8	11.8	04.0	16.6	04.4	13.2	03.4	11.8	11.9
MTI (BU)	12.0	62.0	25.0	62.0	18.0	63.0	30.0	70.0	27.0	70.0
Gel Time (min)	30.0	39.0	32.0	39.5	36.0	39.0	32.0	39.5	32.0	40.0
Gel Temp (°C)	65.0	78.5	68.0	79.3	74.0	78.5	71.0	79.3	68.0	80.0
Peak Viscosity (BU)	520	695	540	710	560	720	550	705	450	710
Peak Viscosity Temp (°C)	89.0	89.75	91.25	92.00	90.50	90.50	89.75	90.95	90.20	91.25
Viscosity at 95°C (BU)	485	680	500	700	520	715	515	695	410	700

Farinographic Water Absorption (FWA)

The durum semolina had 5.4per cent higher farinographic water absorption than the commercial semolina. Bakhshi and Bains (1987) reported that durum in general had higher farinographic water absorption than bread wheat that was attributed to higher damaged starch in durum flour. In control samples durum wheat semolina recorded significantly higher water absorption than commercial semolina. This may be attributed to finer particle size and higher bran content of durum semolina.

The water absorption increased on addition of cysteine hydrochloride at 10 ppm level but decreased slightly on increasing the level to 20 ppm in case of durum semolina. On the other hand, addition of sodium metabisulphite reduced water absorption at both 10 and 20 ppm levels. In case of commercial semolina addition of both the reducing agents significantly reduced water absorption at both 10 and 20 ppm levels.

Addition of oxidizing agents affected water absorption significantly. In durum semolina, addition of both potassium bromate and benzoyl peroxide reduced water absorption such that at 10 ppm level, the decrease was 2.6 per cent while at 20 ppm, the decrease was 3.6 per cent. Similar observations were recorded in case of benzoyl peroxide. In commercial semolina, addition of potassium bromate reduced water absorption by 1.6 per cent and by 3.0 per cent at 20 ppm level. Similarly addition of benzoyl peroxide reduced water absorption by 1.4 per cent and by 3.4 per cent at 20 ppm level. Moore and Herman (1942) also reported decreased water absorption with increasing amount of potassium bromate in the dough.

Dough Development Time (DDT)

For control, durum semolina recorded less dough development time i.e.2.0 min as compared to 3.6 min for commercial semolina. Bakhshi and Bains (1987) studied the rheological properties of durum wheat and reported that dough development time of durum was 0.9-2.1min.

In durum semolina addition of reducing agents recorded non-significant effect on DDT at 10 and 20 ppm levels. In case of commercial semolina however, addition of 20 ppm of cysteine hydrochloride decreased dough development time but 20 ppm of sodium metabisulphite increased dough development time.

In durum semolina addition of 20 ppm potassium bromate significantly reduced dough development time from 2.0 - 1.8 min. However addition of benzoyl peroxide recorded a non-significant change. In commercial semolina, potassium bromate at 20 ppm level and benzoyl peroxide at both the levels reduced dough development time significantly.

Stability

Stability values for control were much higher for durum semolina (15.2 min) as compared to commercial semolina (3.8 min). This could be attributed to the higher uniformity in particle size of durum semolina.

Addition of reducing agents significantly decreased stability in case of durum semolina. However at 10 ppm level of sodium metabisulphite stability was much higher (10.76 min) than that at other levels. In commercial semolina, an entirely different trend was observed. In case of 20 ppm cysteine hydrochloride, there was significant increase in stability while it was reduced to 2.95 min in case of sodium metabisulphite. However at other levels differences in stability were found non-significant. The results indicated that the best reducing agent for durum semolina was sodium metabisuphite, and cysteine hydrochloride for commercial semolina.

Addition of oxidizing agents resulted in significant decrease in stability in comparison to control for durum semolina except at higher levels of potassium bromated where stability was enhanced to 16.6 min. In case of commercial semolina, stability was significantly increased in comparison to control except for 10 ppm benzoyl peroxide where it was reduced to 3.35 min. In case of 20 ppm benzoyl peroxide stability was greatly increased to 11.85 min. The results indicated that in the present situation, potassium bromated was the best oxidizing agent for durum semolina and benzoyl peroxide for commercial semolina. The improving effect of oxidizing agent is closely associated with sulphydryl oxidation in dough (Tsen 1963; Tsen and Hlynka, 1962; Tsen and Hylnka, 1963). During dough mixing, reduced glutathione (GSH) is rapidly oxidized to the disulphide without change in the total glutathione content (Mair and Grosch, 1979). The flour improver potassium bromate enhances the oxidation rate of glutathione. The oxidation of glutathione by the improvers competes with SH/SS interchange reaction of glutathione with the gluten protein. This SH/SS interchange reaction between low molecular –SH- peptide such as reduced glutathione (GSH) and the high molecular protein components of the gluten is one reason responsible for decrease in dough stability (Redman and Ewart, 1967; Kininori and Sullivan, 1968).

Mixing Tolerance Index (MTI)

MTI values for control samples were 12 BU for durum semolina and 62 BU for commercial semolina indicating weakness of commercial semolina dough. This may be attributed to the hardness of durum wheat. Bakhshi (1998) reported a variation of 10-70 BU in mixing tolerance index in case of durum semolina.

Addition of both the reducing agents resulted in significant increase in mixing tolerance index in case of durum semolina. At 20 ppm level of cysteine hydrochloride and sodium metabisulphite mixing tolerance index was significantly higher than control. In case of commercial semolina, mixing tolerance index was significantly reduced in comparison to control on addition of cysteine hydrochloride at both 10 and 20 ppm levels. On the other hand addition of sodium metabisulphite resulted in no change in mixing tolerance index at 10 ppm level while at 20 ppm level there was significant decrease in values of mixing tolerance index. Kaur (1978) reported that there was decrease in mixing tolerance index on addition of cysteine hydrochloride while there was no change o addition of sodium metabisulphite.

Addition of oxidizing agents also resulted in significant increase in mixing tolerance index but the increase was less as compared to reducing agents. Addition of potassium bromate and benzoyl peroxide at 10 ppm resulted in higher values of mixing tolerance index than those at 20 ppm level. In case of durum semolina, the increase was more on addition of benzoyl peroxide than potassium bromate. In commercial semolina, there was non-significant increase in mixing tolerance index on addition of potassium bromated however addition of benzoyl peroxide increased MTI to 70BU at both 10 and 20 ppm levels.

The effect of reducing and oxidizing agents on the farinographic curve characteristics of durum and commercial wheat semolina can be seen in Figure-1 and 2 below.

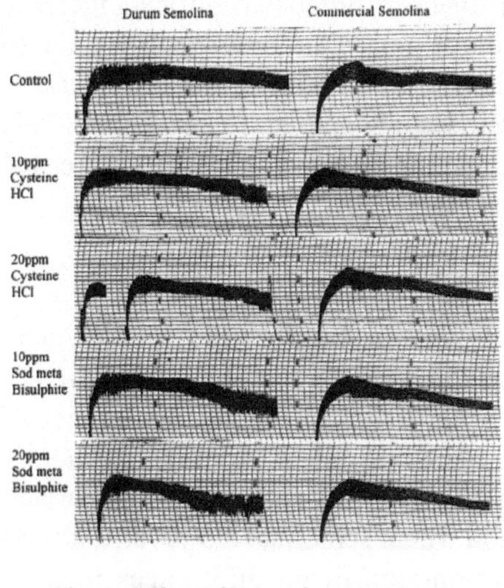

Fig. 1: Effect of Reducing Agents on Farinographic Curve Characteristics

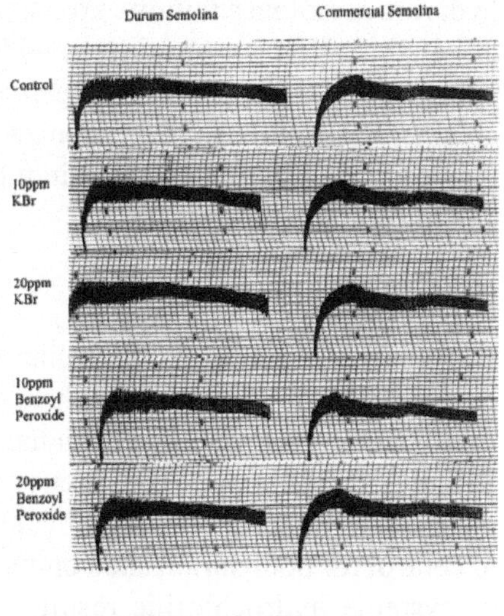

Fig. 2: Effect of Oxidizing Agents on Farinographic Curve Characteristics

Gelatinization time

Gelatinization time was less for control sample of durum semolina (30 min) as compared to commercial semolina (39 min). Addition of reducing agents to durum semolina resulted in significant increase in gelatinization time on addition of 20ppm sodium metabisulphite, however the increase was non-significant at the other levels. In commercial semolina, only addition of 20 ppm sodium metabisulphite resulted in decrease of Gelatinization time, however the decrease was non-significant as compared to control. At other levels, non-significant changes were however observed.

In durum semolina, addition of oxidizing agents resulted in significant increase in gelatinization time on addition of 20 ppm potassium bromate, however at other levels there was non-significant change. In commercial semolina the change in gelatinization time was however found non-significant for both the oxidizing agents at 10 and 20 ppm levels.

Gelatinization temperature (GT)

For control samples of durum and commercial semolina, gelatinization temperatures were 65°C and 78.5°C respectively. Bakhshi and Bains (1987) reported a range of 63.0 – 77.3°C in gelatinization temperatures of Indian durum. In case of durum semolina addition of 20 ppm sodium metabisulphite resulted in significant increase in gelatinization temperature but decrease in case of commercial semolina, however the decrease was non-significant as compared to control. The other levels of reducing agents recorded non-significant changes.

In durum semolina addition of oxidizing agents resulted in significant increase of gelatinization temperature on addition of potassium bromate at 10 and 20 ppm levels and 10 ppm benzoyl peroxide whereas at other levels and in case of commercial semolina the differences were found non-significant. Kaur (1997) reported that addition of potassium bromate and benzoyl peroxide increased gelatinization temperature.

Peak Viscosity

For controls, commercial semolina had more peak viscosity (695 BU) than durum semolina (520 BU). Bakhshi and Bains (1987) reported a range of 425 to 970 BU in peak viscosity of Indian durum. Bakhshi (1998) found a range of 245 to 340 BU in peak consistency of durum semolina. Addition of sodium metabisulphite resulted in significant increase in peak viscosity in comparison to control for both durum and commercial wheat semolina. However, addition of cysteine hydrochloride resulted in non-significant differences in peak viscosity at 10 and 20 ppm levels in case of both durum and commercial semolina in comparison to control.

Addition of oxidizing agents resulted in significant differences in peak viscosity in comparison to control in case of durum semolina. In case of commercial semolina there was significant increase in peak viscosity on addition of 20 ppm potassium bromate while at other levels the differences were non-significant. Patil et al., (1976) reported that improving action of potassium bromate is by oxidation of disulphide group of gluten protein. In addition, bromate also affects the non-protein component such as non-starch carbohydrate.

Peak Viscosity Temperature (PVT)

The temperature at peak viscosity for control was 89°C for durum semolina and 89.8°C for commercial semolina. In durum semolina addition of 20 ppm cysteine hydrochloride resulted in significant increase in peak viscosity temperature at compared to control. However at other levels and other reducing agents non-significant changes were observed for durum as well as commercial semolina.

There were non-significant differences in peak viscosity temperature on addition of oxidizing agents in both durum and commercial wheat semolina.

Viscosity at 95°C- The paste viscosity at 95°C for controls was 485 BU for durum semolina and 680 BU for commercial semolina. In case of durum semolina addition of sodium metabisulphite resulted in significant incrase in the viscosity at 10 and 20 ppm levels, howver the incarese was non-significant in case of cysteine hydrochloride. In case of commercial semolina, there was significant incrase in viscosity at 95°C on addition of 10 ppm sodium metabisulphite, however in case of other levels and for Cysteine hydrochloride, the change was non-significant.

Addition of 20 ppm potassium bromate resulted in significant increase in viscosity in durum and commercial semolina. In case of benzoyl peroxide there was significant increase in viscosity at 95°C at 10 ppm level but at 20 ppm level there was significant decrease. However there were non-significant differences on addition of benzoyl peroxide to commercial semolina. The effect of reducing and oxidizing agents on the amylographic curve characteristics of durum and commercial wheat semolina is shown in Figures 3, 4, 5 and 6 below.

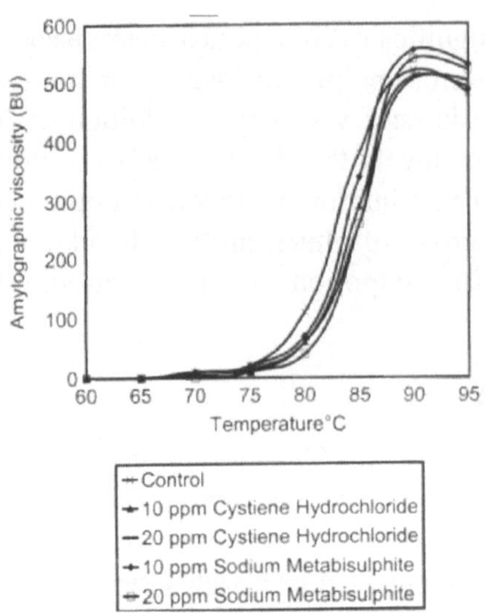

Fig.3: Effect of Reducing Agents on Amylographic Curve Characteristics of Durum Semolina

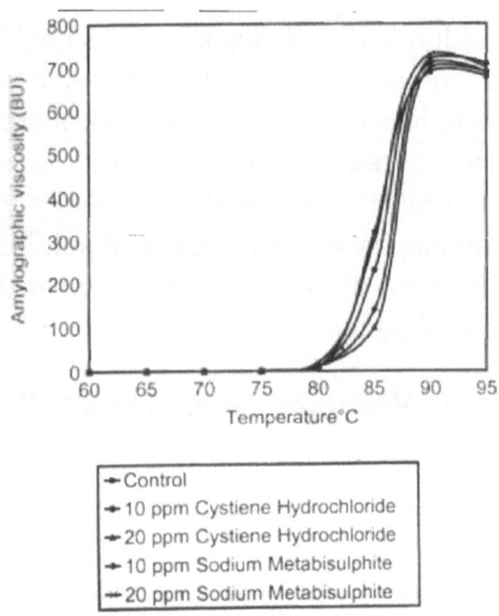

Fig. 4: Effect of Reducing Agents on Amylographic Curve Characteristics of Commercial Semolina

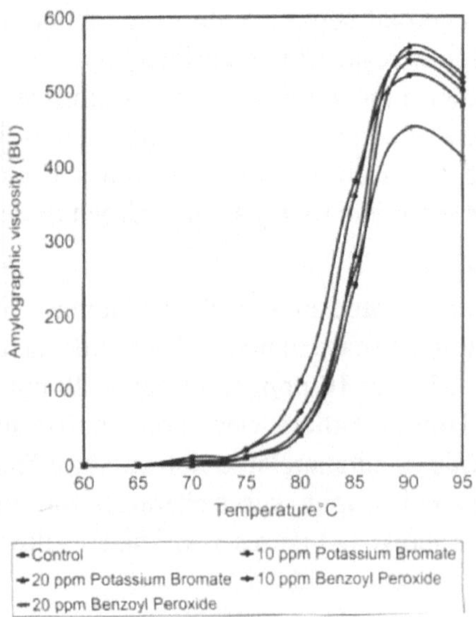

Fig 5: Effect of Oxidizing Agents on Amylographic Curve Characteristics of Durum Semolina

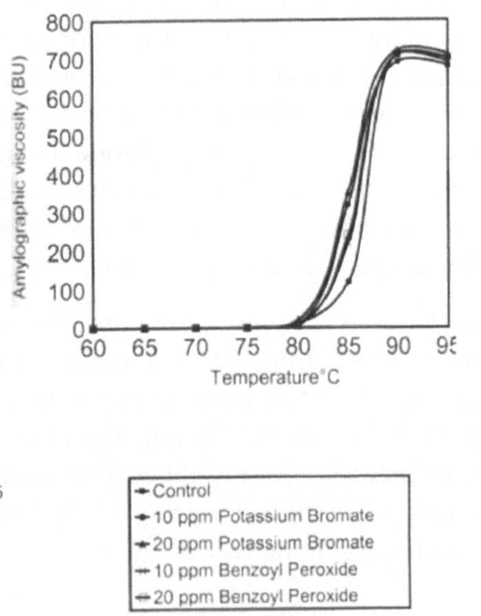

Fig. 6: Effect of Oxidizing Agents on Amylographic Curve Characteristics of Commercial Semolina

Effect of Reducing and Oxidizing Agents on Noodle Quality

Control samples of durum and commercial wheat semolina were made by addition of 2.5 per cent salt and 33 to 35 percent water in case of durum and commercial wheat semolina. Addition of reducing and oxidizing agents improved the elasticity and extensibility characteristics of instant fresh noodles. Moreover the yield was greatly improved. This was done with the idea of standardizing the method of production of instant fresh noodles from durum and commercial wheat semolina, which was the main objective.

Summary

The effect of reducing and oxidizing agents on the rheological, pasting and noodle making characteristics of durum and commercial wheat semolina were studied. The farinographic curve characteristics revealed that water absorption was enhanced with increasing levels of reducing agents but decreased with addition of oxidizing agents. On addition of reducing agents to durum semolina, there was no significant change in DDT but in case of commercial semolina addition of 20 ppm cysteine hydrochloride decreased DDT to 3.1 min and addition of 20 ppm sodium metabisulphite increased DDT to 3.85 min. Addition of oxidizing agents resulted in no significant change in DDT. The present study revealed that best reducing agent for durum and commercial semolina was sodium metabisulphite and cysteine hydrochloride respectively. Addition of oxidizing agents resulted in decrase in stability in comparison to control except in case of 20 ppm potassium bromate where stability was enhanced to 16.6 min. Based on these results, potassium bromate was chosen to be the best oxidizing agent for durum semolina and benzoyl peroxide for commercial semolina. Addition of reducing agents increased MTI in case of durum semolina and decreased in case of commercial semolina. Addition of oxidizing agents also increased MTI. Addition of reducing and oxidizing agents to durum and commercial semolina resulted in non-significant change in Gelatinization time but increased Gelatinization temperature of both durum and commercial semolina. Addition of cysteine hydrochloride resulted in significant change in peak viscosity while addition of sodium metabisulphite resulted in non-significant change in peak viscosity for both the semolina's. Addition of oxidizing agents resulted in increase in peak viscosity in both the semolina's. Addition of reducing and oxidizing agents resulted in non-significant increase in peak viscosity temperature for both the semolina's.

Noodles were prepared from durum and commercial wheat semolina by addition of 2.5 per cent salt and 35 and 33 per cent water respectively. Addition of reducing and oxidizing agents improved the elasticity and extensibility characteristics of instant fresh noodles. Determination of moisture content and water uptake ratio revealed significant differences in control samples of durum and commercial semolina noodles. There were non-significant differences in cooking time in durum and commercial semolina noodles. Durum semolina noodles had significantly higher volume expansion than the commercial semolina noodles.

Conclusions

Based on the results obtained during the investigation it is concluded that the addition of reducing and oxidizing agents improved the yield, appearance, elasticity and extensibility characteristics of instant fresh noodles. The overall acceptability of durum semolina noodles is higher than the commercial semolina noodles.

References

1. Bakhshi, A. K. (1998). Production of export quality wheats. Indian Farming, 48:20-32.
2. Bakhshi, A. K. and Bains, G. S. (1987). Study of the physic-chemical, rheological, baking and noodle quality of improved durum and bread wheats. Journal of Food Science and Technology, 24:217-21.
3. Kaur, K. (1997). To study the effect of improvers on the rheological and bread making properties of indigenous wheats. MSc Thesis, Punjab Agricultural University, Ludhiana.
4. Kaur, M. (1978). Effect of L-cysteine hydrochloride, oxidants, amylases and surfactants on the rheological and bread making properties of Indian wheats. PhD Dissertation, Punjab Agricultural University, Ludhiana, India.
5. Kuninori, T. and Sullivan, B. (1968). Disulphide-sulphydryl interchange studies of wheat flour. II. Reaction of glutathione. Cereal Chemistry, 45:486-95.
6. Mair, G. and Grosch, W. (1979). Changes in glutathione content (reduced and oxidized form) and the effect of ascorbic acid and potassium bromate on glutathione oxidation during dough mixing, Journal of Science Food and Agriculture,30:914-20.
7. Mastromatteo, M. Guida, M. Danza, A. Laverse, J. Frisullo, P. Lampignano, V. and Alessandro, Del Nobile, M. (2013). Rheological, microstructural and sensorial properties of durum wheat bread as affected by dough water content. Food Research Inter, 51(2): 459-466.
8. Moore, C. L. and Herman, R. S. (1942). The effect of certain ingredients and variations in manipulation on the farinographic curve. Cereal Chemistry, 19:568-69.
9. Patil, S. K. Finney, K. F. Shogren, M. D. and Tsen, C. C. (1976). Water soluble pentosans of wheat flour. III. Effect of water soluble pentosans on loaf volume of reconstituted gluten and starch doughs. Cereal Chemistry, 53:347-54.
10. Redman. D. G. and Ewart, J. A. D. (1967). Disulphide interchanges in dough proteins. Journal of Science Food and Agriculture, 18:15-18.
11. Tsen, C. C. (1963).The reaction mechanism of azodicarbonamide in dough. Cereal Chemistry, 40:638-47.
12. Tsen, C. C. and Hlynka. I. (1962). The role of lipids in oxidation of dough. Cereal Chemistry, 39:209-19.
13. Tsen, C. C. and Hlynka, I. (1963). Flour lipids and oxidation of sulphydryl groups in dough. Cereal Chemistry, 40:145-53.

CHAPTER 30

Studies on Isolation of *Bacillus* Spp. and *Thermoactinomycetes* with Cellulase Activity in Relation to Lignocellulosic Post Harvest of Wheat

Vijayta Sharma and Ishpreet Kaur Walia

Abstract

In the present study, a novel work on comparison of physico-chemical characterization and bacterial diversity of wheat straw compost and vermicomposting has been initiated in the tropical region. Wheat straw compost shows lesser pH, EC, TDS, Organic Carbon, Total Nitrogen and Total Phosphorus than vermicompost whereas there was an increase in Total Nitrogen, Total Phosphorus and Electrical Conductivity and a drop in the pH, TDS and Organic Carbon of the vermicompost as the composting progressed from 0, 3, 10, 17, 24 and 31 days. Strains of Bacillus spp. and Thermoactinomycetes were screened from the wheat straw compost, which are ecologically important and have commercial potential for enzyme production. These strains were able to degrade cellulose, hemicellulose and lignin and exhibited cellulase activity. Enterobacter aerogenes, Micrococcus luteus and E.coli were isolated from the gut of Eudrilis eugeniae which help in the transformation of biodegradable waste materials into valuable compost.

Keywords: Composting, vermicomposting, physic-chemical characterization, wheat straw compost, *Eudrilis eugeniae* and waste management.

Introduction

Agricultural waste generated in field and at processing sites is generally discarded without being recycled. If this waste is not properly handled, it may create large environmental hazards. In India, about 320 million tones of agricultural waste are generated annually (Suthar et al., 2005), of which vegetable waste alone is in major proportion. Yang (1997) stated that fruits and vegetables account for almost 15% of total waste produced, comprising 5, 70, 000 metric tons of organic matter with high cellulose content. This waste can be handled in several ways, including direct combustion, landfill, composting and vermin-composting. Among these methods, composting and vermin-composting are considered to be the optimum approach for regulating soil fertility and minimizing the environmental impact.

Composting involves the accelerated degradation of organic matter by microorganisms under controlled conditions, in which the organic material undergoes a characteristic thermophilic stage that allows sanitization of the waste by the elimination of pathogenic microorganisms (Mayende et al., 2006). Vermi-composting involves the bio-oxidation and stabilization of organic material by the joint action of earthworms and microorganisms. Although it is the microorganisms that biochemically degrade the organic matter, earthworms are the crucial agents of the process as they aerate, condition and fragment the substrate, thereby drastically altering the microbial activity. During the process, the nutrients locked up in the organic waste are changed to simple and more readily available and absorbable forms such as nitrate or ammonium nitrogen, exchangeable phosphorus and soluble potassium, calcium, magnesium in worm's gut (Lee, 1985; Atiyeh et al., 2002). Earthworms act as mechanical blenders and by comminuting the organic matter they modify its physical and chemical status by gradually reducing the ratio of C: N and increasing the surface area exposed to microorganisms – thus making it much more favorable for microbial activity and further decomposition (Domínguez et al., 1997). The result of such processes is a biologically stable and humified end product, the compost, which can be used in organic farming (Nakasaki et al, 2005).

The important factors responsible for making good compost and compost stability include Aeration, Temperature, pH, Moisture, C: N Ratio, Particle size, shape of the feedstock (the quality of the microbial "food" and the porosity of the pile. In the agricultural waste composting, lignocellulose accounts for the major part of biomass and consequently, its degradation is essential for the operation of composting (Toumela, 2000; Dixon and Langer, 2006). Lignocellulosic wastes refer to plant biomass wastes that are composed of cellulose, hemicellulose and lignin (Mtui, 2009). In nature, cellulolysis occurs as a result of the combined action of fungi and bacteria with different substrate requirements that shift their biomasss depending on what substrate is being metabolized. Degradation of cellulose is a slow process that is limited by several factors involving cellulases such as concentration, location and mobility of the enzymes. Hemicellulose and lignin content and the degree of crystallinity of cellulose itself also determine the rate at which cellulose is metabolized. Decomposition of lignocellulosic residues is directly mediated by extracellular enzymes (Aira, 2006).

Compost microorganisms that have been identified include *Bacillus* spp., *Actinomycetales, Thermus* spp., *Streptomycetes* spp., *Aspergillus* spp. among others. Interest in these microorganisms has increased due to potential commercial applications which include biodegradation and the production of bioactive compounds such as antibiotics and enzymes. Thermophilic enzymes are important in bioremediation process at higher temperatures in the treatment of wastes have the advantage of increased solubility of the substrates (Mayende, 2006). The sustainance of Agri-biotechnological approach is to transform the biodegradable residues by rhizospheric isolates in association with earthworm population. It is a well documented device to explore the potentials and perspectives of vermin-culture in relation to biodegradable wheat post harvest residues (wheat straw).

Materials and Methods

Collection of samples and Preparation of vermicompost bed

Around 1 kg of wheat compost sample was collected from the Department of Microbiology, Punjab Agricultural University, Ludhiana. Suitable bedding was prepared in the wide plastic tub with approximately 7 kg of fresh cow dung, approximately 2-2.5 kg of the waste vegetable paste (prepared of the chopped waste of spinach, potato, peas, coriander, cabbage, onion, garlic, Trigonella foenum-graecum (methi) and reddish), wheat husk (which was used as bulking agent) and the earthworm spp. *Eudrilus eugeniae* was released into the respective layers. Bed was covered with the gunny bag to prevent the earthworms from ants and to prevent the drying of the vermicompost. Water was sprinkled after every three days to maintain adequate moisture range up to 60-70% at 27°C. Proper turning was done after every week to maintain proper aeration conditions for the earthworms and to get homogenous compost (Nagavallemma et al., 2006). The composting was carried out for 31 days.

Physico-chemical analysis

The physico-chemical characters of wheat compost and that of vermicompost samples were monitored during vermicomposting. Vermicompost samples were collected undestructively at 0, 3, 7, 14, 24 and 31 days and processed for analyses of pH, EC, TDS, Organic Carbon, Total Phosphorus and Total Nitrogen. The pH, EC and TDS of samples were recorded by a digital pH meter and conductivity meter, respectively (Rajor and Anita (2006ab). The Organic Carbon of the samples was measured by Walkley-Black method (Walkley and Black, 1934). Total Nitrogen and Total Phosphorus were measured by the methods given by Murugesan and Rajakumari (2005); Rajor (2006) respectively.

Isolation of *Thermoactinomycetes* and *Bacillus* spp. from wheat compost

Thermoactinomycetes and *Bacillus* spp. were isolated from wheat compost by dilution plating. Tryptic soy agar [Pancreatic digest of Casein, 15 g; Papaic digest of Soyabean meal, 5 g; NaCl, 5 g; Agar, 15 g; distilled water, 1 litre (pH 7.3 ±0.2)], was used for the isolation of both *Thermoactinomycetes* and *Bacillus* spp. 1 gram of compost sample was suspended in 9 ml of sterile distilled water and dilutions were made up to 10^{-8}. Using micropipette, 0.1 ml of the compost suspensions from last 4 dilutions plated onto the media plates and plates were incubated at 50°C for 4-5 days for *Thermoactinomycetes* and 1-2 days for thermophilic *Bacillus* spp. After incubation, thebacterial colonies were identified by cellular morphology and biochemical characteristics.

Assay of Cellulase activity

Colonies appeared on selective agar were further used for cellulase activity testing. Isolates were plated on Mendels- Reese agar medium [$(NH_4)_2SO_4$, 1.40 g; KH_2PO_4, 2 g; $CaCl_2$, 0.30 g; $MgSO_4$, 0.30 g; $FeSO_4$, 0.005 g; $MnSO_4$, 0.0016 g; Proteose peptone, 1.0 g; $ZnCl_2$, 0.0017 g; $CaCl_2$, 0.002 g; Cellulose, 10 g; Urea, 0.3 g; distilled water, 1 litre (pH 7.3 ± 0.2)], and cultivated at 50°C for 4 days. 0.1% of Congo red was then sprayed on the colonies retained on the plates for 15 minutes and the plates were destained with 1M NaCl for 15 minutes. Formation of clear zone was observed (5). Cellulase containing crude enzyme was prepared by centrifugation of the Mandels- Reese broth at 5000 rpm at 4°C for 20 minutes. Supernatant was considered to be the enzyme source and was added to 1 ml of 1% Carboxymethylcellulose (CMC) in 0.05M citrate buffer (pH 4.8). Dinitrosalicylic acid (DNS) method was applied to determine cellulase activity. Carboxymethylcellulose (CMC) was used as the substrate. Glucose standard was used to plot the standard curve. Measurement of glucose released from CMC was expressed as Unit (U), in which, U= mg of glucose equivalents released min^{-1} ml^{-1} crude enzyme (Aneja, 2006).

Assay of Lignocellulose Degradation

Lignocellulose degradation was assayed by the measurement of Neutral detergent fiber (NDF), Acid detergent fiber (ADF) and Acid detergent lignin (ADL) with the procedures outlined by Van Soest et al. (1995). Hemicellulose was estimated as the difference between NDF and ADF. Cellulose was estimated as the difference between ADF and ADL. Lignin was estimated as the difference between ADL and ash content.

Isolation of bacterial species from the gut of *Eudrilus eugeniae*

The gut contents (4-5.5 cm of gut ranging from 18-185 segments in *Eudrilus eugeniae)* were dissected out using sterile blade and the contents were transferred to 1 ml of sterile distilled water into a sterile test tube. It was vortexed and dilution plated from 10^{-3} to 10^{-5} on nutrient agar. The plates were then incubated at 37°C for 24 hours (Parthasarthi et al., 2007). The isolates were characterized by morphological, cultural and biochemical characteristics.

Results and Discussion

Physico-chemical characterization of Wheat compost and Vegetable vermin-compost

Changes in pH

A slightly acidic pH of 6.75 has been observed in the wheat compost whereas there has been a slow decrease in the pH during vegetable vermicomposting. A drop in pH has been observed from the slightly alkaline to slightly acidic phase (Table 1). The initial and final drop in pH at the mesophilic stage may be due to the action of microorganisms on the fraction ofcarbohydrates present in the organic matter leading to the release of organic acids. The pH subsequently increased at the thermophilic stage probably due to the production of ammonia. The observations are in conformity with the results obtained by Jadia and Fulekar (2008).

Changes in Electrical Conductivity

Electrical Conductivity (EC) of wheat compost has been found to be 2.39 dS m^{-1} and a slight increase in electrical conductivity has been observed during the period of vermin-composting at regular time intervals and presented a higher value of salinity than wheat compost (Table-1). The increase in EC during the period of vermin-composting processes is probably due to the degradation of organic matter releasing minerals such as exchangeable Ca, Mg, K, and P in the available forms i.e. in the form of cations in the vermin-compost. Similar results have been reported by Pattnaik and Reddy (2009) but the results are in contrast with results of Nath et al., (2009).

Changes in Total Dissolved Solids

Total dissolved solids content (TDS) in wheat compost has been observed to be 1.93 ppm. TDS content in the vegetable vermin-compost has been found to be decreased with the regular intervals of time (Table 1). The decrease in TDS content may be due to the action of microorganisms on organic matter transforming the waste into compost.

Table 1: Physico-chemical characterization of Wheat compost and Vegetable vermin-compost

Sr. No. Test	Wheat Compost	Vegetable vermin-compost					
		Day 0	Days 3	Days 10	Days 17	Days 24	Days 31
1. pH	6.75	7.13	7.10	7.05	7.29	6.90	6.82
2. Electrical Conductivity(µS)	2.39	2.50	2.52	2.58	2.66	2.74	2.81
3. Total Dissolved Solids (ppm)	1.93	2.89	2.69	2.56	2.49	2.43	2.45
4. Organic Carbon (µgD g)	0.62	1.24	1.18	1.01	0.89	0.76	0.69
5. Phosphorus (µgD ml)	1.18	0.840	0.880	0.96	1.04	1.12	1.24
6. Nitrogen(%)	50.12	36.10	36.13	46.48	51.80	55.16	58.52

Changes in Organic Carbon, Total Phosphorus and Total nitrogen

Organic carbon in wheat compost has been found to be 0.62 µgD g. A subsequent decrease in the organic carbon has been observed during the vermicomposting of the vegetable waste which shows the decomposition of waste by microbial population.The results are in comparison with those of Jadia and Fulekar (2008). Total phosphorus in wheat compost has been found to be 0.321 µgD ml and an increase in the total Phosphorus has been observed at the end of the vermicomposting as compared to the initial substrate. The Phosphorus content may be increased due to the enhancement of phosphatase activity and the physical breakdown of waste that results in greater mineralization. The increased trend of total Phosphorus in the vermicompost is in consistence with the findings of Padmavathiamma et al., (2008) and Pattnaik and Reddy (2009). Total nitrogen in wheat compost has been found to be 50.12% and there has been an increase in the Nitrogen content of the prepared vermicompost analyzed at regular intervals of times. The increase in Nitrogen content may be due to the mineralization of the organic matter containing proteins and conversion of ammonium-nitrogen into nitrates and the release of nitrogenous products of earthworm metabolism such as excretory products, mucus, enzymes. The results are similar in comparison with those of Jadia and Fulekar (2008); Padmavathiamma et al., (2008); Pattnaik and Reddy (2009).

Qualitative determination of *Thermoactinomycetes* and *Bacillus* spp. Isolated from Wheat compost

Morphological studies of the isolates revealed that both of the isolates I and II were Gram positive with the Gram stain and Isolate-II also appeared to be branched and filamentous when stained with lactophenol cotton blue for Hanging drop method. Colonies of Isolate-I were circular, flat, opaque, pale yellow with irregular margins and colonies of Isolate-II were spherical, flat, translucent, yellow with entire margins. Isolate-I was further characterized by the biochemical tests as shown in Table 2. From the above morphological, cultural and biochemical characterization of the bacterial Isolate-I, the strain characters were similar to *Bacillus cereus* and with the above criterion of identification of staining, the bacterial Isolate-II was identified as a *Thermoactinomycetes* spp.

Table 2: Biochemical characterization of Isolate-I

Sr. No.	Test	Result
1.	Indole production test	-
2.	Methyl Red test	-
3.	Citrate Utilization test	-
4.	Starch Hydrolysis	+
5.	Catalase test	+
6.	Cellulase Production Test	+
7.	Gelatin Hydrolysis	+
8.	H_2S Production Test	-
9.	Glucose Fermentation Test	-
10.	Sucrose Fermentation Test	-
11.	Lactose Fermentation Test	-

Measurement of carboxymethyl cellulase activity by reducing sugar determination

Thermophilic Bacillus and Thermoactinomycetes released specific amount of glucose with regard to their cellulase activities at 50°C. Cellulase produced by Thermophilic Bacillus released 0.077 U ml-1 of glucose and Thermoactinomycetes released 0.048 U ml-1 of glucose. The results were in contrast with the results of Mayende et al. (2006) for Thermophilic Bacillus and Aboul-Enein et al., (2010) for Thermoactinomycetes.

Lignocellulose degradation

Cellulose degradation

Cellulose has been found to be 10% degraded in the wheat compost whereas cellulose was partially degraded during the initial stage of composting. Then the degrading ratio was almost unaltered due to high temperature followed by the large decomposition during the temperature falling period (Table 3). The degradation pattern was similar with the degradation studied by Aira et al., (2006) and Yu et al., (2007).

Table 3: Ligno-cellulose degradation

Sr. No.	Degradation (%)	Wheat compost	Vegetable Vermi-compost					
			Day 0	Days 3	Days 10	Days 17	Days 24	Days 31
1.	Cellulose	10.0	4.0	4.0	4.0	8.0	6.0	14.0
2.	Hemicellulose	4.0	20.0	34.0	20.0	20.0	18.0	12.0
3.	Lignin	8.0	3.0	6.0	6.0	12.0	12.0	11.0

Hemicellulose degradation

Hemi-cellulose was found to be degraded up to 4% in the wheat compost. There has been a high degradation in the initial stage of composting. Then the degrading ratio was almost constant during the thermophilic period and then decreased consequently (Table 3). The degradation pattern of the vermicompost is similar with the pattern of Aira et al. (2006) and is in contradiction with those given by Yu et al. (2007). Cellulose and Hemicellulose degradation is the result of the combined action of actinomycets, non filamentous bacteria and fungi over the vermicompost and earthworms enhance the rate of degradation as cellulose is a part of the diet of earthworms.

Lignin degradation

Lignin was degraded up to 8% in wheat compost and the degradation of lignin in vermicompost was quite different from that of cellulose and hemicellulose. Lignin was slightly decomposed during the initial stage of composting. When the composting reached the thermophilic phase, lignin was greatly degraded until the temperature began to fall. The degradation pattern was similar with the degrading pattern as given by Yu et al., (2007).

Isolation of bacterial species from the gut of *Eudrilis eugeniae*

Three bacterial species i.e. Isolates III, IV and V respectively, were isolated from the gut of *Eudrilis eugeniae*. Morphological characterization of the isolates revealed that Isolates-III and V were Gram negative rods whereas Isolate-IV was Gram positive coccus bacteria. Colonies of Isolate-III were small, thick, white and glistening, irregular in shape with entire margins. Colonies of Isolate-IV were smooth, circular, yellow pigmented with entire margins. Colonies of Isolate-V were small and smooth, white, circular having entire margins. The isolates were further characterized by the biochemical tests as shown in Table 4. From the above morphological, cultural and biochemical characterization, Isolate-III had shown similar characters of *Enterobacter aerogenes*, Isolate-IV had shown similar characters of *Micrococcus luteus* and Isolate-V had shown similar characters of *E.coli.* The isolation of these bacterial species was in correspondence with the isolation of the bacterial species by Parthasarthi et al., (2007).

Table 4: Biochemical characterizations of gut bacterial isolates

Sr. No.	Test	Isolate- III	Isolate- IV	Isolate- V
1.	Indole production	-	-	+
2.	Methyl Red	-	-	+
3.	Voges- Proskauer	+	-	-
4.	Citrate Utilization	+	+	-
5.	Starch Hydrolysis	-	-	-
6.	Catalase	+	-	+
7.	H2S Production	-	-	-
8.	Lactose Fermentation	AG	-	AG
9.	Sucrose Fermentation	AG	-	A
10.	Glucose Fermentation	AG	-	AG
11.	Urease	-	+	-

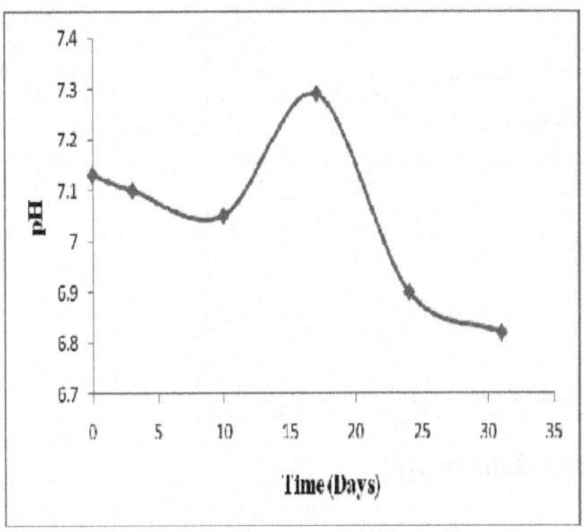

Fig. 1: pH of the vermin-compost samples

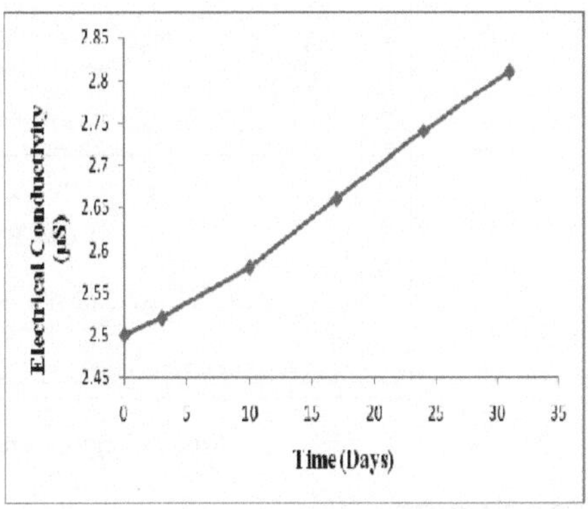

Fig. 2: EC of the vermin-compost samples

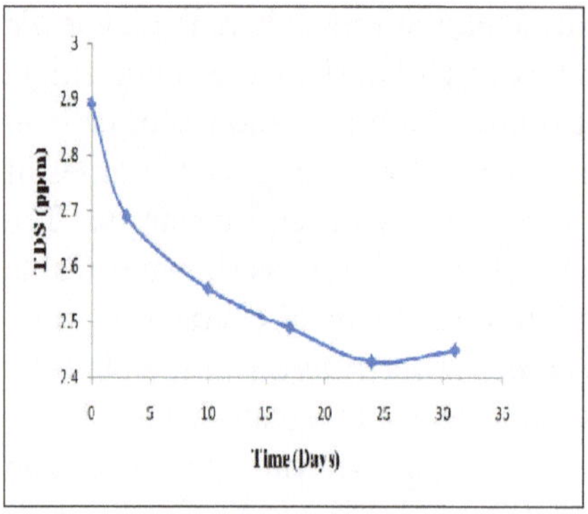

Fig. 3: TDS of the vermin-compost samples

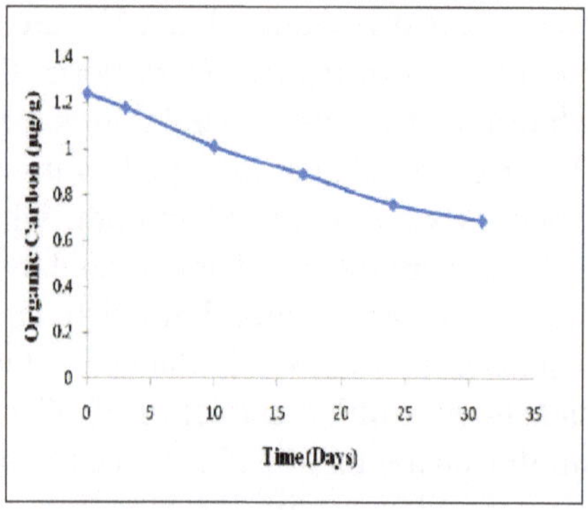

Fig. 4: Organic Carbon of the vermin-compost samples

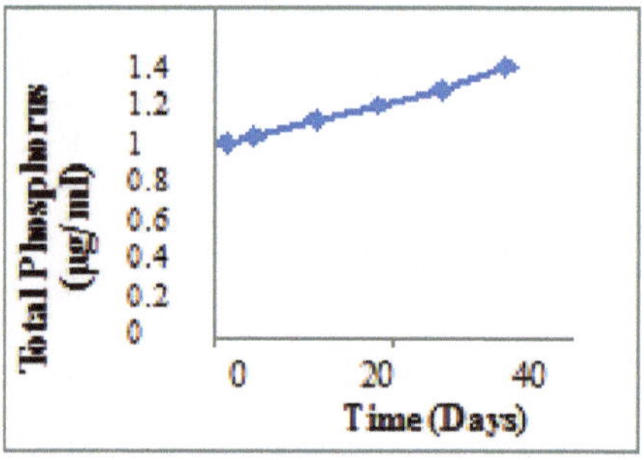

Fig. 5: Total Phosphorus of the vermin-compost samples

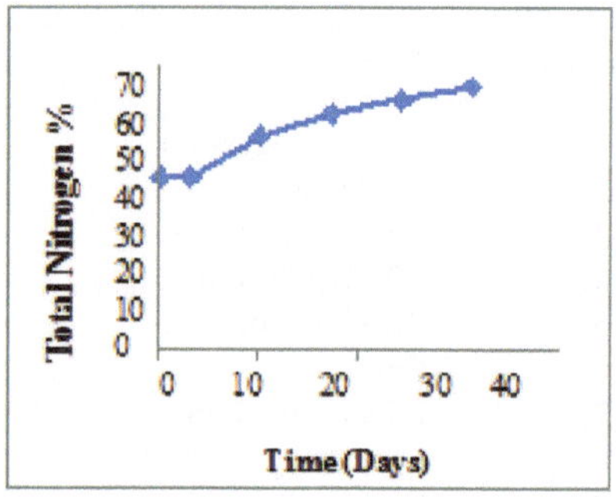

Fig. 6: Total Nitrogen of the vermin-compost

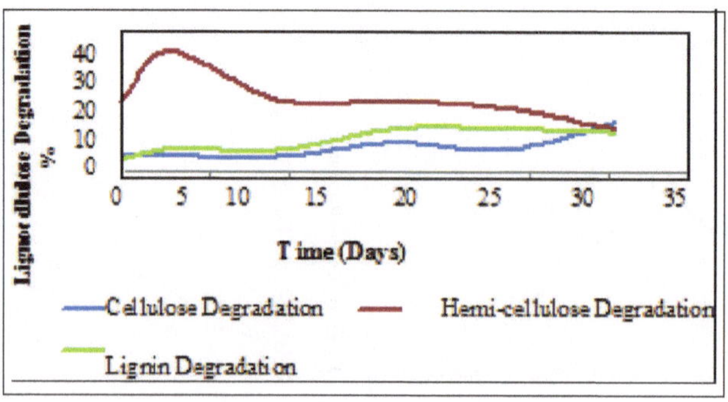

Fig. 7: Ligno-cellulose Degradation

Conclusion

In the present study, it is concluded that among composting and vermin-composting, vermin-composting is much more efficient and eco-friendly technology for the decomposition of organic wastes. It is evident from the results that significant increase and decrease in the physico-chemical characters of the vermicompost by the use of *Eudrilis eugeniae* is due to the vermic activity in laboratory conditions. Vermi-compost produced *Eudrilis eugeniae* showed higher nutrient content as compared to the sole compost. Thus, vermin-composting was proved to be a better technology than sole composting and may be preferred for the management and nutrient recovery from the agricultural waste.

References

1. Aboul-Enein, A. Abou-elalla, F. Serour, E. and Hussein, T. (2010). Purification and characterization of anovel thermoactive cellulase from thermophilic actinomycetes isolated from soil sample of Egypt. Journal of Academic Research, 2(1):81-86.
2. Aira, M. Monroy, F. and Dominguez, J. (2006). Eisenia fetida (Oligochaeta, Lumbricidae) activates fungal growth, triggering cellulose decomposition during vermicomposting. Journal of Microbiplogy and Ecology, 52:738-746.
3. Aneja, K.R. (2006). Experiments in microbiology, plant pathology and biotechnology. In: Measurement of endoglucanase or carboxymethyl cellulase by reducing sugars determination. New age international (P) limited, publishers. pp. 457-458.
4. Atiyeh, R.M. Lee, S., Edwards, C.A. Arancon, N.Q. and Metzger, J.D. (2002). The influence of humic acids derived from earthworm processed organic wastes on plant growth. Journal of Bioresources Technology, 84(1): 7-14.
5. Chang, C.C. Ng, C.C., Wang, C.Y. and Shyu, Y.T. (2009). Activity of cellulase from Thermoactinomycetes and Bacillus spp. isolated from Brassica waste compost. Science and Agriculture (Piracicaba, Braz.). 66(3): 304-308.
6. Dixon, N. and Langer, U. (2006). Development of a MSW classification system for the evaluation of mechanical properties. Journal of Waste Management, 26:220-232.
7. Domínguez, J. Edwards, C. A. and Subler, S. (1997). A comparison of composting and vermin-composting. Biocycling, 4:57–59.
8. Jadia, C.D. and Fulekar M.H. (2008). Vermicomposting of vegetable waste: A biophysicochemical process based on hydro-operating bioreactor. Journal of African Biotechnology, 7(20):3723-3730.
9. Lee, K.E. (1985). Earthworms: Their Ecology and Relationships with Soils and Land Use. Academic Press, London, UK.
10. Lung, A.J. Lin, C.M., Kim, J.M. Marshall, M.R. Nordstedt, R. Thompson, N.P. and Wei, C.I. (2001). Destruction of Escherichia coli O157:H7 and Salmonella enteritidis in cow manure composting. Journal of Food Protection, 64:1309–1314.
11. Mayende, L. Wilhelmi, B.S. and Pletschke, B.I. (2006). Cellulases (CMCases) and polyphenol oxidases from thermophilic Bacillus spp. isolated from compost. Journal of Soil Biology and Biochemistry, 38(9):2963-2966.
12. Mtui, G.Y.S. (2009). Recent advances in pretreatment of lignocellulosic wastes and production of value added products. Journal of African Biotechnology, 8 (8):1398-1415.
13. Murugesan, A.G. and Rajakumari, C. (2005). Soil Analysis. In: Environmental Science and Biotechnology. Theory and Techniques. MJP Publishers. pp. 277-279.
14. Nagavallemma, K.P., Wani, S.P., Stephane, L., Padmaja, V.V., Vineela, C., Babu, R.M. and Sahrawat, K.L. (2006). Vermicomposting: Recycling wastes into valuable organic fertilizer. Global Theme on Agrecosystems Report No.8. An open access journal by ICRISAT. 2(1): pp.1-12.

15. Nakasaki, K., Nag, K., Karita, S. (2005). Microbial succession associated with organic matter decomposition during thermophilic composting of organic waste. Journal of Waste Management and Research, 23:48-56.
16. Nath, G., Singh, K. and Singh, D.K. (2009). Chemical analysis of vermicomposts/ vermiwash of different combinations of animal, agro and kitchen wastes. Australian Journal of Basic Application and Science, 3(4):3672-3676.
17. Padmavathaimma, P.K. Li, L.Y. and Kumari, U.R. (2008). An experimental study of vermi-biowaste composting for agricultural soil improvement. Journal of Bioresources Technology, 99:1672-1681.
18. Pattnaik, S. and Reddy, M. V. (2009). Nutrient status of vermicompost of urban green waste processed by three earthworm species- Eisenia fetida, Eudrilus eugeniae, and Perionyxexcavates. Journal of Applied and Environtal Soil Science, 2010: 1-13.
19. Parthasarthi, K., Ranganathan, L.S., Anandi, V. and Zeyer, J. (2007). Diversity of microflora in the gut and casts of tropical composting earthworms reared on different substrates. Journal of Environmental Biology, 28(1):87-97.
20. Rajor, Anita. (2006). In: Environmental Methods for General and Environmental Chemistry. In: pH. Standards publishers distributors; 1705- B, Nai Sarak, Delhi. pp. 174-178.
21. Rajor, Anita. (2006). In: Environmental Methods for General and Environmental Chemistry. In: Electrical Conductivity. Standards publishers distributors; 1705- B, Nai Sarak, Delhi. pp. 172-173.
22. Rajor, Anita (2006). In: Environmental Methods for General and Environmental hemistry. In: Total Phosphorus. Standards publishers distributors; 1705- B, Nai Sarak, Delhi. pp. 85-87.
23. Suthar, S.S., Watts, J. Sandhu, M. Rana, S. Kanwal, A. Gupta, D. and Meena, M.S. (2005). Vermicomposting of kitchen waste by using Eisenia foetida (SAVIGNY). Asian Journal ofMicrobiology Biotechnology and Environmental Science, 7:541-544.
24. Toumela, M. Vikman, M. Hatakka, A. and Itavaara, M. (2000). Biodegradation of lignin in a compost environment: a review. Journal of Bioresources and Technology, 72:169-183.
25. Van Soest, P. J. Rovertson J. B. and Lewis, B. A. (1991). Methods for dietry fiber, neutral detergent fiber, and nonstarch polysaccharides in relation to animal nutrition. Journal of Dairy Science, 74:3583-3597.
26. Walkley, A., Black, I. A. (1934). An examination of the Degtjareff method for determining soil organic matter, and proposed modification of the chromic acid titration method. Journal of Soil Science, 37:29-38.
27. Yang, S.S. (1997). Preparation and characterization of compost. Journal of the Biomass Energy Society of China, 16:47-62.
28. Yu, H. Guangming, Z. Hongli, H. Xingmei, X. Renyou, W. Danlian, H. Guohe, H. and Jianbing, L. (2007). Microbial community succession and lignocellulose degradation during agricultural waste composting. Biodegradation. 18: 793-802.

CHAPTER 31

Nutraceutical and Functional Potential of Quinoa (*Chenopodium quinoa* Willd)

Khan Nadiya Jan, Ishrat Majid, D.C. Saxena and Sukhcharn Singh

Abstract

Quinoa is a pseudo-cereal which belongs to family Chenopodiaceae. Quinoa can act as highly potential crop for areas where ecological conditions do not support other food crops because of its rusticity, tolerance to frost, salinity, drought and ability to grow in marginal soils. Recently, much attention is focused on quinoa for people with celiac disease and allergic to gluten. Quinoa contains high-quality protein, complete with all essential amino acids and is a good source of carbohydrates, lipids, vitamin C, vitamin E (natural antioxidants) and adequate amount of important micronutrients such as minerals. Carbohydrates of quinoa possess excellent freeze–thaw stability and can be used as thickener in frozen foods. Quinoa also consists of bioactive components such as phytosterols, squalene, fagopyritols and polyphenols which act as powerful anti-oxidants in-vitro and thus could prevent degenerative diseases such as coronary heart disease, atherosclerosis and Alzheimer's disease. Fibre accounts for 6% of the grains total weight, and quinoa intake therefore promotes intestinal transit, regulates cholesterol, stimulates the development of beneficial bacterial flora and helps prevent colon cancer. The quinoa grain can be used to combat liver disease, angina and cystitis. The radical scavenging ability and Angiotensin I converting enzyme (ACE) inhibition activity of quinoa is higher than that of amaranth, rice and wheat. Recent studies have identified significant amounts of ecdysteroids, kaempferol, quercetin, two phytoestrogen - daidzein and cenisteína in quinoa. Ethanolic extract of quinoa was observed to have Antioxidant and antihypertensive activity.

Keywords: Chenopodium quinoa, Celiac, Bioactive Components, Antioxidant, etc.

Introduction

Chenopodium quinoa Willd (quinoa) is an annual dicotyledonous plant that belongs to Andean region. Quinoa seeds are flat; oval-shaped that are usually pale yellow but can range in color from pink to black. Size of quinoa grain is small and if dried can be stored for years (Repo-Carrasco et al., 2001). Quinoa is referred by FAO as a potential crop for future food security because of its wide range of adaptability and diversified tolerance to biotic and abiotic stresses (FAO, 1998; Repo-Carrasco etal., 2003). Its remarkable features include tolerance to low soil nutrition, extreme drought, and high salinity and frost hazards. One cultivar of quinoa was found to grow successfully with peak salinity of 57 mS cm^{-1}(Jacobsen, 2003). Leaves of quinoa also serve as nutritious herb similar to that of spinach. Quinoa can be used to make soup, porridge, muffins, bar and flakes. Quinoa flour can be used in combination with other cereal flours as fortifying agent because of its high nutritional profile. Additional benefits include its low glycemic index and absence of gluten for this purpose it is a favourite diet of people suffering from hypertension and celiac disease. During a search for grain having complete dietary attributes to feed astronauts, quinoa was selected as potential crop for NASA's Controlled Ecological Life Support System (CELSS) in order to abolish CO2 production and to synthesize oxygen and food for the prolonged space mission (Schlick and Bubenheim, 1996). Quinoa seeds can serve as substitute to rice when boiled with almost double quantity of water. (Galwey, 1989; Popenoe et al., 1989). Consumption and demand of quinoa is increasing everywhere. Quinoa consumption in USA, is 1500 tones 90% of which is imported from Latin America (Jacobsen, 2003). Recently quinoa has gained much interest at national and international level because of its nutritional value that it offers to consumers looking for healthy and nutritious food and also for consumers who suffer from muscular or circulatory problems and some kind of allergic reactions. Quinoa has gained importance because of the quality of its proteins with balanced set of amino acids. Grain yields vary from 1 to 3 t ha^{-1}, depending on the variety and the level of cultivation technology (Carmen, 1984).

Plant Description

It is an annual dicotyledonous plant usually standing 0.5-2.0 m high. With large panicles bearing 1.5-2.5 mm long seeds produced at the end of the stem as shown in Fig. 1.

Nutritional Value of Quinoa

Quinoa is commonly referred as 'super grain' because of its well balanced nutritional profile as shown in Table 1. United Nations Food and Agriculture

Fig. 1: Chenopodium Quinoa

Organization has termed the nutritional quality of quinoa similar to that of dried whole milk.

Proteins, Lipids and Vitamins

Quinoa contains 8-22% proteins having a balanced amino acid composition similar to that of milk protein (casein). The principal protein of quinoa, the chenopodina was first isolated and characterised in 1993. This protein was separated into two subgroups, chenopodina A and B by electrophoresis. Molecular weight of these subgroups was found to be 32,000-39,000 and 22,000-23,000, which is higher than that for casein. (Brinegar and Goundan 1993; and Ranhotra et al., 1993). Quinoa seed contains sufficient amounts of phenylalanine (brain stimulant), threonine (used for liver detoxification and facilitating absorption of other nutrients). Crude fat in qunoa is 2% to 10%. Quinoa oil consists of essential fatty acids like linoleic acid 50.24%, oleic acid 26.04% and linolenic acid 4.77%. Quinoa seed contains vitamin A in the range of 0.12 to 0.53 mg/100 g, vitamin E 4.60-5.90 mg/100g, thiamine 0.05-0.60 mg/100g, ascorbic acid 0.00-8.50, Niacin 0.16-1.60 mg/100g (Ruales and Nair 1992).

Carbohydrates, Minerals and Fiber

Carbohydrate content in quinoa varies from 67% to 74%. Most important carbohydrate in quinoa is starch which accounts for 58.1-64.2% and is present in granular form (debruin 1964). Quinoa starch has an excellent freeze thaw stability and resistance to retrogradation, therefore it can be an interesting alternative to replace chemically modified starches. (Ahamed et al. (1998)). Quinoa contains high amount of calcium, copper, iron, magnesium and zinc. Fibre accounts for 6% of the total weight of grains. Dietary fiber of quinoa has various positive health effects like improving digestion and reducing the level of the cholesterol in the blood.

Food Application

Quinoa can be consumed as a whole grain or in a processed form. Quinoa flour can be used for preparation of bread and pastry. Quinoa can be utilized in weaning food mixtures. A study was conducted in 1993 by Repo-Carrasco and Li Hoyos in which two dietary mixtures with high nutritional value were formulated: quinoa-kan˜iwa-beans and quinoa-kiwicha-beans. PER value of the mixtures was close to that of casein: 2.36 and 2.59 (casein 2.5).

Table 1: Nutrient composition in Quinoa as per USDA

Constituents	Unit	Amount (Per100 g)
Water	g	13.28
Energy	kcal	368
Protein	g	14.12
Total lipid (fat)	g	6.07
Carbohydrate, by difference	g	64.16
Fiber, total dietary	g	7.0
Minerals		
Calcium, Ca	mg	47
Iron, Fe	mg	4.57
Magnesium, Mg	mg	197
Phosphorus, P	mg	457
Potassium, K	mg	563
Sodium, Na	mg	5
Zinc, Zn	mg	3.10
Vitamins		
Thiamin	mg	0.360
Riboflavin	mg	0.318
Niacin	mg	1.520
Vitamin B-6	mg	0.487
Folate, DFE	µg	184
Vitamin B-12	µg	0.00
Vitamin A, RAE	µg	1
Vitamin A, IU	IU	14
Vitamin E (alpha-tocopherol)	mg	2.44
Lipids		
Fatty acids, total saturated	g	0.706
Fatty acids, total monounsaturated	g	1.613
Fatty acids, total polyunsaturated	g	3.292
Cholesterol	mg	0

References: USDA National Nutrient Database for Standard Reference Release 27 Full Report (All Nutrients) The National Agricultural Library

Quinoa flour can be extruded and drum-dried resulting in the products with good physical, sensorial, and nutritional qualities. Good-quality tempeh (vegetable meat) can be obtained by solid-state fermentation of quinoa with Rhizopus oligosporus Saito (Valencia- Chamorro 2003). Leaves of quinoa can be used in salads because they are rich in vitamins and minerals, especially phosphorous, calcium and iron. Other products include jellied milk and quinoa ham Nieto Soria (1991). Strarch of quinoa has proved to be a better thickener for fillings than wheat, barley and amaranth starch (Lorenz 1990).

Potential of quinoa to food security

FAO has selected quinoa as the crop destined to offer food security in the 21st century. It is a strategic crop having potential to contribute to food security because of its genetic variability, nutritional quality, adaptability to adverse climate and soil conditions, low production cost and ability to show growth in marginal soils (Jacobsen, 2003). It can be grown in areas were ecological conditions do not support other food crops because of its tolerance to frost, salinity, drought and ability to grow in marginal soils. During grain forming stage some quinoa varieties are tolerant to -5 ° C (Espindola 1986). Cultivation of Quinoa can provide an alternative for countries having limited food production, or where the population has no access to protein sources. Important feature of quinoa is that seed, leaves and inflorescences all are sources of high quality protein.

Functional and Nutraceutical Properties

Quinoa has recently been used as a novel functional food because of all its properties. (Abugoch, 2009). It is capable of reducing bad cholesterol and raising good cholesterol (HDL) in body due to its linolenic and linoleic acid content. High percentage of total dietary fibre (TDF) makes quinoa an ideal food for eliminating toxins and waste products that may harm body there by detoxifying it. The radical scavenging ability, of quinoa and Angiotensin I converting enzyme (ACE) inhibition activity of quinoa is higher than that of amaranth, rice and wheat. Quinoa contains greater phenolic compounds compared to other cereals and are thought to be responsible for their antioxidant activity. Antioxidant activity of quinoa was observed to be higher than that of buckwheat. (Asao et al., 2010). Quinoa has been used in traditional medicine since ancient times its stem and leaves cooked with vinegar, oil and pepper increase the blood. Fresh leaves of quinoa consumed either as a soup or dessert, act as a remedy against scurvy. Quinoa has proven as a remedy against Anthrax, herpes, and other skin conditions.

Many functional properties of quinoa have been reported which include decreased serum and liver cholesterol levels (Takao et al., 2005) suppressive effects on increased blood pressure (Ogawa et al., 2001) and antioxidant activity (Nsimba et al., 2008). Quinoa seeds contain many antioxidant compounds like carotenoids, vitamin C, and flavonoids which act as protective agents against cancer, allergy, and inflammatory diseases. A highly unsaturated triterpene Squalene which is the precursor of steroid family is present in ample amount (58.4±0.69 mg/100g) in quinoa and has the ability to inhibit the cholesterol synthesis in liver. Antioxidants like α- Tocopherol (2.1±0.22 mg/100g), β+γ Tocopherol (3.1±0.07 mg/100g) and phytosterols like β-Sitosterol (63.7±4.0 mg/100g), Campesterol (15.6±8.7 mg/100g), and Stigmasterol (3.2±0.1 mg/ 100g) have been detected from the quinoa seed. These phytosterols have biological effects like anticarcinogenic, anti-oxidative and anti-inflammatory effect. Antioxidative flavonoid glycosides and betalains have been isolated from quinoa. (Zhu et al., 2001) Quinoa offers many health benefits like Gastroprotective activity, Anti-inflammatory, Antioxidant activity, reduced chances of developing type 2 diabetes, helps to avoid Celiac disease due to presence of bioactive components as shown in Table 2.

Table 2: Bioactive components present in Quinoa

Bio-active Components	Content	Author
Arabinan	Not Defined	Cordeiro et al., 2012
Ecdysteroids	405 mg/kg (EIA) (Flour) 846 mg/kg (EIA) (Bran)	Kumpun et al., 2011
Phenolics ,flavonoids Carotenoid,Ascorbic acid	86.4 mg GAE/10g,13.9 mg CE/10g 0.4mg/10g, 1.3 mg/10g	Dini et al., 2010
Flavonoid	83 mg GAE/100 g FW	Hirose et al., 2010
Anthocyanins and Polyphenols and	120.4 (mg CGE 100 g^{-1}), 3.75 (mg GAE g^{-1})	Pas′ko etal., 2009
Total polyphenols	Quinoa (Bolivia) 94.3 mg g^{-1} TAE Quinoa (Japan) 148.0 mg g^{-1} TAE	Nsimba et al., 2007
Squalene	58.4 (mg/100 g	Berganza et al., 2003
β-Sitosterol, Campesterol Stigmasterol	63.7,5.6 and 3.2 (mg/100g)	Weihrauch et al., 1978

Conclusion

Quinoa is widely recognised nowadays as a crop having great potential to eradicate hunger, solve the malnourishment problems and having ability to cope with different climates, geographical conditions, low soil fertility, drought and frost. Consumption of this pseudocereal is a useful suggestion for the consumer health, as it is a rich source of antioxidant compounds. It is a crop having the capability to fight against hunger and food insecurity. Faced with the challenge of increasing the production of quality food to feed the world's population in the context of climate change, quinoa offers an alternative for those countries suffering from food insecurity. Currently, there is a need for obtaining high-quality protein concentrates to solve problems of chronic malnutrition affecting rural and urban populations. Further Low-allergy potential of quinoa and its relatively high digestibility, nutritional value and bioactive components makes it a food of special interest in the diet. Apart from food value saponin (by-product) obtained from quinoa can be used in manufacturing of soaps, foam stabilizer in beverages and as an insecticide.

References

1. Abugoch, J. L. E. (2009). Quinoa (Chenopodium quinoa Willd.): composition, chemistry, nutritional, and functional properties. Advances in Food and Nutrition Research, 58: 1-31.
2. Ahamed, T. Singhal, R., Kulkarni, P. Pal, M. (1998). A lesser-known grain, Chenopodium quinoa: review of the chemical composition of its edible parts. Food Nutr. Bull. 19:1, The United Nations University.
3. Asao, M. and Watanabe, K. (2010). Functional and Bioactive Properties of Quinoa and Amaranth, Food Science and Technology Research, 16 :163 –168,
4. Brinegar, C. Goundan, S. (1993). Isolation and characterization of chenopodin, the 11S seed storage protein of quinoa (Chenopodium quinoa). Journal of Agriculture Food Chemistry, 41(2):182–185.
5. Carmen, M. (1984). Acclimatization of quinoa (Chenpodium quinoa, Willd) and canihua (Chenopodium pallidicaule, Aellen) to Finland. Annales Agriculturae Fenniae, v. 23 : 135-144.
6. de Bruin, A. (1964). Investigation of the food value of quinoa and canihua seed. Journal of Food Science, 29:872–876.
7. Espíndola, G. (1986). Respuestas fisiológicas, morfológicas y agronómicas de la quinua (Chenopodium quinoa Willd.) al déficit hídrico. Tesis M.Sc., Colegio de Postgraduados. Institución de Enseñanza e Investigación en Ciencias Agrícolas. Chapingo México. p101.
8. Galwey, N.W. (1989). Quinoa Biology, 36:267-274.
9. Jacobsen, S.E. (2003). The Worldwide Potential for Quinoa (Chenopodium quinoa Willd.). Food Review International, 19:167-177.
10. Lorenz, K. (1990). Quinoa (Chenopodium quinoa) starch: physico-chemical properties and functional characteristics. Starch, 42:81-86.
11. Nsimba, R.Y. Kikuzaki, H. and Konishi, Y. (2008). Antioxidant activity of various extracts and fractions of Chenopodium quinoa and Amaranthus spp. Seeds. Food Chemistry, 106:760-766.
12. Ogawa, H. Watanabe, K. Mitsunaga, T. and Meguro, T. (2001). Effect of quinoa on blood pressure and lipid metabolism in dietinduced hyperlipidemic spontaneously hypertensive rats (SHR). Journal of Japan Society of Nutrition Food Science, 54:221-227.
13. Popenoe, H. S.R. King, J. Leon, L.S. Kalinowski. (1989). Lost crops of the Incas. In: Vietmeyer, N.D. (Ed.), Little Known Plants of the Andes with Promise for Worldwide Cultivation. National Academy Press, Washington.
14. Ranhotra, G. Gelroth, J. Glaser, B. Lorenz, K. Johnson, D. (1993). Composition and protein nutritional quality of quinoa. Cereal Chemistry, 70(3):303–305.

15. Repo-Carrasco, R., Espinoza, C. and, Jacobsen S.E. (2001). Nutritional Value and Uses of Quinoa (Chenopodium quinoa) and Kañiwa (Chenopodium pallidicaule) Proyecto Quinua CIP-DANIDA.
16. Repo-Carrasco, R. C. Espinoza and S.E. Jacobsen. (2003). Nutritional value and use of the Andean crops quinoa (Chenopodium quinoa) and kañiwa (Chenopodium pallidicaule), Food Review International, 19:179-189.
17. Ruales, J. and B.M. Nair. (1993). Content of fat, vitamins and minerals in quinoa (Chenopodium quinoa, Willd) seeds. Food Chemistry, 48:131-136.
18. Ryan E, Galvin K, O'Connor T.P, Maguire A.R, and O'Brein N.M (2007). Phytosterol, Squalene, Tocopherol Content and Fatty Acid Profile of Selected Seeds, Grains, and Legumes. Plant Foods Human Nutrition, 62:85-91.
19. Schlick, G. and D.L. Bubenheim. (1996). Quinoa: Candidate crop for NASA's Controlled Ecological Life Support Systems. In: Janick, J. (ed.), Progress in new crops. ASHS Press, Arlington, VA. Pp. 632-640.
20. Takao, T. Watanabe, N. Yuhara, K. Itoh, S. Suda, S. Tsuruoka, Y. Nakatsugawa ,K. and Konishi,Y. (2005). Hypocholesterolemic effect of protein isolated from quinoa (*Chenopodium quinoa* Willd.) seeds. Food Science Technology and Research, 11:161-167.
21. Valencia, Chamorro (2003). Quinoa.In: CABALLERO B.: Encyclopedia of food science and Nutrition. Vol. 8. Academic Press, Amsterdam: 4895-4902.
22. Zhu, N. Sheng, S. Edmond, D.L. Karwe, M.V. Rosen, R.T. and Ho, C.T. (2001). Antioxidative flavonoid glycosides from quinoa seeds (*Chenopodium quinoa* Willd). Journal of Food Lipids, 8:37-44.

Unit-V: Fruits and Vegetables as a Source of Functional Foods and Nutraceuticals

CHAPTER 32

Functional Potential and Food Utilization of Seaweeds

Hanuman Bobade, Gurkirat Kaur and Savita Sharma

Abstract

Seaweeds are the large marine algae that grow almost exclusively in the shallow waters at the edge of the oceans. Seaweeds are rich sources of many bioactive compounds and nutraceuticals. These factors are considered under two categories: Seaweed phytochemicals and Seaweed polysaccharides. They are rich in soluble dietary fibers, proteins, minerals, vitamins, antioxidants, phytochemicals, and polyunsaturated fatty acids, with low caloric value. The red, brown and green seaweeds have been shown to have therapeutic properties for health and disease management. Usually they are only used as gelling agent and stabilizers in the food and pharmaceutical industries, but current research has revealed their potential medicinal uses against many diseases like cancer, allergy, diabetes, oxidative stress, inflammation, thrombosis, obesity, hypertensive and other degenerative ailments. In line with other natural foods like fruits, vegetables and grains, there has been marked interest within the scientific community to assess which fractions of seaweeds may be linked to the historically observed health benefits. As consumer health and nutrition become more influential in the food industry, the use of seaweed as an ingredient is on the rise in product development. For several centuries there has been a traditional use of seaweeds as food in China, Japan and the Republic of Korea. This review undertakes successful exploration of potential of seaweed gums as an functional ingredient in a number of food applications. This addition has the potential to enhance the nutritional quality of a product and hence human health.

Introduction

For the last few decades, scientists throughout the world have been searching for suitable and nutritional alternatives to conventional food, partly due to the grave food shortage all over the globe due to the ongoing human population explosion. Agricultural yields are increasingly failing to satisfy the need of our daily diet as population increase is geometric whereas crop production increases more or less arithmetically. To meet this challenge, the use of some alternate food source in our diet becomes indispensable. Alternate food literally means some healthy and readily available supplement excluding the conventional staples of our daily diet (Ghosh, 2004). Edible seaweeds are rich in bioactive antioxidants, soluble dietary fibers, proteins, minerals, vitamins, phytochemicals, and polyunsaturated fatty acids (Mohammed et al., 2012). Although previously the seaweeds were only used as gelling and thickening agents in the food or pharmaceutical industries, recent researches have revealed their potential as complementary medicine. The red, brown and green seaweeds have been shown to have therapeutic properties for health and disease management, such as anticancer, anti-obesity, anti-diabetic, antihypertensive, antioxidant, anticoagulant, anti-inflammatory, thyroid stimulating, neuro-protective, antiviral, anti-fungal and antibacterial.

The term seaweed refers to the large marine algae that grow almost exclusively in the shallow waters at the edge of the world's oceans (Jirge and Chaudhary, 2010). They provide home and food for many different sea animals, lend beauty to the underwater landscape, and are directly valuable to man as a food and industrial raw material. Seaweeds are plants because they use the sun's energy to produce carbohydrates from carbon dioxide and water (photosynthesis). They are simpler than the land plants mainly because they absorb the nutrients that they require from the surrounding water and have no need for roots or complex conducting tissues (bcb.uwc.ac.za).

Table 1: Latin and most common names of some of the seaweeds

Latin name	Most common name
Ascophyllum nodosum	Egg wrack
Laminaria digitata	Kombu/Konbu
Laminaria saccharina	Royal or Sweet Kombu
Himanthalia elongate	Sea spaghetti
Undaria pinnatifida	Wakame
Porphyra umbilicalis	Nori
Palmaria palmata	Dulse or Dillisk
Chondrus crispus	Irish moss or Carrigeen
Ulva lactuca	Sea lettuce
Enteromorpha (Ulva) intenstinalis	Sea grass

Source: Morrissey et al. (2001)

Seaweeds are classified into Rhodophyta (red), Phaeophyta (brown) and Chlorophyta (green) marine macroalgae. They are rich in soluble dietary fibers, proteins, minerals, vitamins, antioxidants, phytochemicals, and polyunsaturated fatty acids, with low caloric value. However, like most flora their nutrient contents are affected by external factors such as the geographic location, environmental, season, sampling conditions (Khotimchenko et al., 2005; Renaud and Luong-Van, 2006). Usually they are only used as gelling agent and stabilizers in the food and pharmaceutical industries, but current research has revealed their potential medicinal uses against cancer, allergy, diabetes, oxidative stress, inflammation, thrombosis, obesity, hypertensive and other degenerative ailments.

For several centuries there has been a traditional use of seaweeds as food in China, Japan and the Republic of Korea. As people from these countries have migrated around the world, this custom has moved with them, so that today there are many more countries where the consumption of seaweed is not unusual. Coastal dwellers in tropical climates such as Indonesia and Malaysia have also eaten fresh seaweeds, especially as salad components.

In recent years there has been a strong movement in France to introduce seaweed into the European cuisine, with some success, although it is still regarded as an exotic component of the menu. It has gained more acceptances in regions like California and Hawaii, where communities of Japanese are larger and the taste for seaweeds spreads out into the surrounding population through finding them on restaurant menus and supermarket shelves.

Functional Properties of Seaweeds and their Health Benefits

Studies on seaweed consumption in humans have centered in the areas of the world where reasonable amounts of seaweeds are habitually consumed (particularly South-East Asia). A variety of potential health benefits are related to their antioxidant properties; potentiality of seaweeds against metabolic syndrome; anticoagulant properties; antitumor, anticancer and anti-proliferative properties; endocrine metabolic properties; etc.

Anti-diabetic

Type-2 or non-insulin-dependent diabetes mellitus is associated with insulin resistance and aging, and is expected to reach 10% of the adult population worldwide. A Korean National Health and Nutrition Survey showed seaweed consumption decreased diabetes risk in men (Lee et al., 2010). The pacific edible brown seaweed *Petaloniabinghamiae* extract (PBE) has anti-diabetic insulin-like and insulin sensitizing action in adipocytes of diabetic mammals.

PBE stimulated the differentiation of 3T3-L1 pre-adipocytes that act on cellular glucose uptake in mature adipocytes and increased transcriptional activities of peroxisome proliferator-activated receptor gamma (PPARg). PPARg improves insulin sensitivity in glucose metabolism and represents a link between lipid and carbohydrate metabolism (Rieusset et al., 2002). *Sea tangle* and *E. stolonifera*also have anti-hyperglycemic and antioxidant effects in diabetic mammals. Most seaweeds contain high levels of soluble dietary fibers (carrageenan, agar, alginates), that passively retard digestion and glucose absorption (Baron, 1998).

Anti-obesity

The seaweeds high soluble dietary fiber contents helps low down digestion and calorie absorption. They help moderate appetite by delaying gastric emptying and absorption in the small intestine (Paxman et al., 2008) although these effects are difficult to prove in humans. Fucoxanthin (the main carotenoid in brown seaweed) showed anti-obesity effects especially in obese patients with non-alcoholic fatty liver disease and elevated indices of chronic inflammation (Abidov et al., 2010). The fucoxanthin help reduce body weight by (i) up regulating mitochondrial uncoupling protein 1 (UCP1) gene expression in white adipose tissue(WAT) (ii) leading to increase resting energy expenditure, through uncoupling step during cellular metabolism(iii) oxidation of fatty acids(iv) reduced blood glucose and plasma insulin and (v) improved insulin.

Antitumor, anticancer and anti-proliferative properties

Regular seaweeds consumptions are associated with lower risks of human breast cancers with suppressive effects on the development of both benign and cancer neoplasis (Eskin et al., 1995). Water-soluble polysaccharides, such as laminarans, sulphated polysaccharides and fucoidans, represent some anticancer compounds from seaweeds (Yamasaki et al., 2009). Iodine may be needed by normal breast tissue, because iodine deficient tissues show changes in, among other things, estrogen receptor proteins. Iodine supplementation makes the breast tissue less susceptible to excess iodine, which is over-produced by the ovaries into the blood, under iodine-deficiency state. Thus, iodine acts as an anti-estrogen. Continuous administration of iodine to mammals results in strong and persistent reduction of breast tumors incidence. Selenium acts synergistically with iodine to affect hormone homeostasis and iodine availability. Arachidonic acid and its metabolites, cyclo-oxygenase (COX) and LOX, stimulate aromatase gene expression in breast fatty tissues, and increases intra-mammary estrogen. When Arachidonic acid is iodinated, the iodo-lactones formed inhibit the EGF (epidermal growth factor) receptor, which consequently inhibited Arachidonic acid metabolism. Seaweeds sulfo-lipids and sulfated polysaccharides possess anti-proliferative properties Fucoidan is naturally effective against bile duct cancer, human breast cancer and reduced chemotherapy toxicities in colorectal cancer patients. Iodine rich mekabu seaweed are more potent than the breast cancer chemo drug, 5-fluorouracil, *in vitro* (Funahashi et al., 2001).

Mechanisms reported for the anti-carcinogenic effects of the various seaweed compounds include:

- Cancer induction inhibition
- Hyaluronidase activity inhibition
- Anti-inflammatory activities
- Anti-proliferative activities
- Regulation of mammary gland integrity
- Anti-oxidative
- Apoptosis induction through the expression of transforming growth factor
- Suppression of tumor initiation
- Anti-angiogenic effects (e.g. by fucoidan), that inhibits VEGF165 (vascular endothelial growth factor165) binding to its cell surface receptors
- Inhibition of tumor cells invasion
- Immune response enhancement
- Estrogen production suppression

Anti-microbial properties

The antimicrobial properties of seaweed extracts have been well documented over the years (Gupta et al., 2011; Cox et al., 2010; Kumaran et al., 2010). Seaweeds act as an antimicrobial agent that kills microorganisms or inhibits their growth. The incorporation of seaweed into foods has also been shown to have a preservative effect, particularly with regards to Gram-negative bacteria, reducing the need to add salt. The antimicrobial properties of seaweed extracts have been well documented over the years. However, there would appear to be a lack of published information regarding the antimicrobial properties and preservative effects when seaweed as a 'whole-food' is incorporated into a food matrix. Previous studies suggest that the overall antimicrobial capacity of seaweeds appears to be linked to their antioxidant content. Seaweed

(*Ascophyllumnodosum*, Sea-greens) as a dried whole-food has been incorporated into various products for example processed meat products and bread products. Previous studies suggest that the overall antimicrobial capacity of seaweeds appears to be linked to their antioxidant content (Devi et al., 2008).

It has been observed that the 3% w/w dried seaweed added to the processed meat product results in an overall 0.3 log10 cfu/g reduction in the total viable count over the two month trial period; however, when looking at specific populations of micro-organisms, this antimicrobial activity is particularly effective against Gram negative micro-organisms such as coliforms (showing a 0.7 log10 reduction cfu per gram of product) which is to be expected since Gram negative micro-organisms have a thinner cell wall than Gram positive micro-organisms making them more susceptible to antimicrobial agents. However, there is a significant reduction in the Gram positive lactic acid bacteria (LAB) population over shelf-life (0.5 log10).

Antioxidant properties

An antioxidant is a molecule that inhibits the oxidation of other molecules. The most abundant terrestrial biophenols (flavonoids) have three interconnected rings. Seaweed phlorotannins have up to eight interconnected rings, making them 10-100 times more powerful and more stable as free radicals cavengers than other polyphenols (green tea catechins, have only four rings). The half-life of phlorotannins which is partially fat soluble, is up to 12 h in the body, compared to 30-180 min for water-soluble, terrestrial polyphenols. The total phenol contents and antioxidant capacity varies with species, but generally the green seaweeds (*Caulerpa spp.*) have higher free-radicals cavenging properties (Chandini et al., 2008; Matanjun et al., 2008).

Lipid oxidation increases with disease and aging, and many types of seaweed can reduce lipid oxidation in vivo, under such conditions. Malondialdehyde (MDA), a marker of endogenous lipid per-oxidation, were significantly reduced in the liver, heart and brain of aging mammal by the porphyrans and sulfated galactans from *Porphyra spp*. Seaweed consumption generally increases the endogenous antioxidant enzymes superoxide dismutase (SOD), glutathione peroxidase (GSH-Px) and sometimes catalase activities *in vivo* (Matanjun *et al* 2010; Zhang et al., 2004).

Anticoagulant properties

An anticoagulant is a substance that prevents coagulation (clotting) of blood. The anticoagulant behavior of seaweeds sulfated polysaccharides depends on having preferably two sulfate groups and a glycosidic linkage on the pyranose ring, to enable them to interact with the basic groups in proteins. However, the anticoagulant activities are mainly by anti-thrombin and/or heparin cofactor mediated thrombin inhibition (Colliec et al., 1991).

Utilization of Seaweeds in Food Products

Seaweeds are used in many maritime countries as a source of food, for industrial applications and as a fertilizer. Nori (Porphyra spp.), a Japanese red seaweed, is very popular in the Japanese diet, has a high protein content (25-35% of dry weight), vitamins (e.g. vitamin C) and mineral salts, especially iodine. Microalgae biomass has a chemical composition which varies depending on the algae used. It can be rich in proteins or rich in lipids or have a balanced composition of lipids, sugars and proteins (Raja et al., 2013).

The uses of seaweeds have gone beyond the culinary and nutritional values as a food for human consumption. In the mid 17th century, the Japanese learned to cook "tengusa" (Gelidium amansi) into a gelatin consiste ncy and cool it to make a jelly type food called "tokoroten", which was later found to be used when freeze dried and powered into an agar called "Seaweed gum", suitable as a base for raising bacteria etc, in medical research and other industrial uses. Chapman (1970) elucidated some uses and products of commercial seaweeds. Dhargalkar and Perreira (2005); Dhargalkar and Verlecar (2009) reported more comprehensive uses of seaweeds and their extracts. However, utilization of these seaweeds upon harvesting would be based on the design of appropriate technology for colloid production and processing. This and further socio-economic analyses would clearly depend on the production ecology of the wild, cultured or transplanted seaweeds.

Furthermore, in recent years it is being proven that a variety of organic substance found in seaweeds have medicinal effects on the human body. The useful qualities of seaweed can be grouped into;

- A food with preservation characteristic
- A health food with medicinal quality
- A seaweed gum, a good source of colloid material for industrial use

Table 2: Nutritional value of some edible seaweed

Seaweed	Carbohydrates (%)	Fiber (%)	Calcium (mg/100g)	Iodine (mg/100g)	Vit B_{12} (µg/100g)
Ascophyllum nodosum	13.1	8.8	575.0	18.2	0.131
Laminaria digitata	9.9	6.2	364.7	70.0	0.459
Himanthalia elongata	15.0	9.8	30.0	10.7	-
Undaria pinnatiûda	4.6	3.4	112.7	3.9	0.345
Porphyra umbilicalis	5.4	3.8	34.2	1.3	0.769
Palmaria palmata	10.6	5.4	148.8	10.2	1.84
Ulva sp.	4.1	3.8	325.0	1.6	6.3
Enteromorpha sp	7.8	4.9	104.0	97.9	-

Source: MacArtain *et al* (2007)

Seaweeds can be eaten in a number of ways. Some are eaten raw as salads, some are flavored before eating and some are made into processed foods. However, the fact that so many species are eaten in so many ways can be seen as the result of an early recognition of the fact that, when processed in the right way, they became long lasting and easily stored preserved foods. Here are some of the observations that surely led to the use of seaweeds as processed foods. When dried seaweed is soaked in water again, it regains its original shape and can be eaten with very little loss in nutritional substance or texture.

The common seaweed dish "tsukudani", which is made by cooking seaweed in soy sauce, can be left out for several days without losing its density or without its liquid component separating out. "Dried laver" is made by chopping up the seaweed and setting it out to dry in thin sheets. In this process, the resinous substance that exudes from the seaweed serves to paste the leaf fragments together and give the sheet a shinny external gloss. These food qualities are all due to the functions of the intercellular polysaccharide. The processing method for seaweeds include, drying (Kelp, undaria, laver) drying with flavouring (laver), salt preserving (Undaria, mozuku)" tuskudam" (Kelp, Undaria, laver and hitoeagusa), and other flavoured processing (for all types of seaweeds).

Seaweeds are consumed by coastal people, particularly in East Asia, e.g., Brunei, Japan, China, Korea, Taiwan, Singapore, Thailand, Cambodia, and Vietnam, but also in South Africa, Indonesia, Malaysia, Belize, Peru, Chile, the Canadian Maritimes, Scandinavia, South West England, Ireland, Wales, California, Philippines, and Scotland. In Asia, Nori (Japan), Zicai (China), and Gim (Korea) are sheets of dried Porphyra used in soups or to wrap sushi. Chondrus crispus (commonly known as Irish Moss or carrageenan moss) is another red alga used in producing various food additives, along with Kappaphycus and various gigartinoid seaweeds. Porphyra is a red alga used in Wales to make laver. Laverbread, made from oats and the laver, is a popular dish there. In northern Belize, edible seaweeds are mixed with milk, nutmeg, cinnamon, and vanilla to make a common beverage affectionately called "Dulce" (or "sweet").

Seaweeds are also harvested or cultivated for the extraction of alginate, agar and carrageenan, gelatinous substances collectively known as hydrocolloids or phycocolloids. Hydrocolloids have attained commercial significance as food additives. The food industry exploits their gelling, water-retention, emulsifying and other physical properties. Agar is used in foods such as confectionery, meat and poultry products, desserts and beverages and moulded foods. Carrageenan is used in salad dressings and sauces, dietetic foods, and as a preservative in meat and fish products, dairy items and baked goods.

Seaweeds as Health Foods

The cell wall material and internally stored substances, which seaweeds produce through photosynthesis, are extremely diverse in terms of their chemical composition. This fact gives seaweeds their exceptional value as health foods. Modern nutritional studies have shown that seaweeds contain a great amount of vitamins and trace minerals essential to the human body. Some unique organic substances found in seaweed help in the prevention of the degenerative diseases. For example, the fucosterol found in kelp and Undaria is believed to reduce blood cholesterol and prevent thrombosis in the blood vessels.

Also, experiments with mice have shown that alginic acids are anti tumor agents. Recent studies have also focused attention on edible seaweeds as a valuable source of dietary fibre. The digestive juice of the stomach cannot dissolve the polysaccharide of seaweeds, but they are broken down by the bacterial action in the large intestine. In short, their difficult- to-digest complex carbohydrates, act to stimulate the digestive function of the intestine, thereby invigorating them (Abowei and Ezekiel, 2013).

Seaweed Polysaccharides

Polysaccharides in the seaweed constitute one of the important components thriving food applications of seaweed in different forms. Chief polysaccharides of the seaweed are agar, alginates and carrangeenan. These polysaccharides are also called as seaweed gums and they find many industrial applications in various fields.

Agar

Agar constitutes group of polysaccharides from red-purple algae of the Rhodophyceae class. The agar-yielding species of *Gracilaria*and*Gelidium* grow in the waters along the coast of Japan, New Zealand, South Africa, Southern California, Mexico, Chile, Morocco, and Portugal.

Agar differs from τ- and κ-carrageenans in that the 3,6-anhydro sugar in agar is the L-enantiomer, whereas in carrageens it is the D-enantiomer. Agar is, therefore, a linear polysaccharide built up of the repeating disaccharide unit of (1→3)-linked β-D-galactose and (1→4)-linked 3, 6-anhydro-α-L-galactose residues. Also, in contrast to carrageenans, agar is only lightly sulphated and may contain methyl groups. Methyl groups, when present, occur at C-6 of the (1→3)-linked β-D-galactose or C-2 of (1→4)-linked 3,6-anhydro-α-L-galactose residues. The agar helix is more compact due to the smaller amount of sulphate groups. Agar is a well known thermo-reversible gelling polysaccharide, which sets at 30 to 40°C. Being less sulphated and carrageenans, agar can form strong gels, which are, subject to pronounced syneresis, attributed to strong aggregation of double helices (not weakened by the sulphate groups).

Among the all gums available, agar is the best gelling agent. Its gelling power arise from strong hydrogen bond of the three equatorial hydrogen atoms on the 3,6-anhydro-L-galactose residues in the agarose. Agar gives cold setting gels at ~38°C. However, the melting temperature is much higher, about 85°C. This gives agar gel a very large gelling/melting hysteresis. Agar gel found applications in the baking industry because of its heat resistant characteristics, which help to prevent chipping, cracking, or sweating of icings, toppings, and glazes in baked goods. Agar can form a stiff gel at low concentrations and it is compatible with most of the other gums. Agar gel has a mouth-feel and other attributes similar to gelatin; it is therefore used in kosher gel desserts, pie ûllings, and jelly candies (Izydocrzyk et al, 2005).

Adding a small portion of locust bean gum to some type of agar (e.g., Gelidium agar in the ratio of 1:9) can enhance gel strength and improve gel texture. This is attributed to a synergistic interaction between locust bean gum and agar. The synergistic interactions found applications in the food industry to produce less brittle gels. In addition, agar is a very important microbiological culture medium due to its excellent gelling characteristics and high resistance to microorganisms.

Agar is a food additive of universal use considered in the US as GRAS (Generally Recognized as Safe) by the FDA (Food and Drug Administration). In Europe it is considered an E406 additive. In the Register Service of the Chemical Abstracts it is registered as 9002-18-0 (Armisen and Galatas, 2000).

Agar is utilized in food as a minor ingredient (0.5% to 1.5%) relative to the total weight of the finished products; its nutritional contribution is low because the human digestive system hardly absorbs it. Agar has been consigned a digestibility below 10% of what is ingested. For this reason agar is used to prepare dietetic formulas and foods for diabetics since it does not depend on sugar to gel as with pectins (Armisen and Galatas, 2000).

Alginates

Alginates constitute the primary structural polysaccharides of brown seaweeds (Phaeophyceae). The major species of seaweeds that produce alginates are *Macrocystispyrifera*, grown primarily along the California coast of the USA, south and north-western coasts of South America, and coasts of Australia and New Zealand. Other good sources of alginates are *Laminariahyperborea, Laminariadigitata,* and*Laminaria japonica*, grown along the north Atlantic coast of the USA, Canada, France, and Norway. Alginates can also be synthesized by bacteria, *Pseudomonas aeruginosa*and *Azobactervinelandii* (Milani and Maleki 2012).

Alginates are unbranched copolymers of (1→4)-linked β-D-mannuronic acid (M) and α-L-guluronic acid (G) residues. If the uronic acid groups are in the acid form (–COOH), the polysaccharide, called alginic acid, is water insoluble. The sodium salts of alginic acid (–COONa), sodium alginates, are water soluble. The sequence of mannuronic and guluronic residues signiûcantly affects the physicochemical properties of alginates (Zhao 2008).

Alginates are used as food additives to improve, modify, and stabilize the texture of foods (Giardi *et al* 2010). This is valid for such properties as viscosity enhancement; gel-forming ability; and stabilization of aqueous mixtures, dispersions, and emulsions. Some of these properties stem from the inherent physical properties of alginates themselves, as outlined above, but they may also result from interactions with other components of food product, e.g., proteins, fat, or fibers. As an example, alginates interact readily with positively charged amino acid residues of denatured proteins, which are utilized in pet foods and reformed meats (Draget et al., 2010).

For applications in jams, jellies, fruit feelings, etc, the synergetic gelling between alginates high in guluronate and highly esterified proteins may be utilized. The alginate/pectin system can give thermo-reversible gels in contrast to the purely ionically cross-linked alginate gels. This gel structure is almost independent of sugar content, in contrast to pectin gels, and therefore may be used in low calorie products (Moe et al., 2010).

The only alginate derivate used in food is propylene glycol alginate (PGA). It gives stable solutions under acidic conditions where the unmodified alginate would precipitate. It is now used to stabilize acid emulsions (such as in French dressings), acid fruit drinks and juices. PGA also is used to stabilize beer foam.

Carrageenans

Carrageenans are structural polysaccharides of marine red algae of the Rhodophyceae class. They are extracted mainly from *Chondruscrispus, Euchemacottoni, Euchemaspinosum, Gigartinaskottsbergi*, and *Iradaeala-minarioides*. These red seaweeds grow mostly along the Atlantic coasts of North America, Europe, and the western Paciûc coasts of Korea and Japan (Milani and Maleki, 2012).

Carrageenans are linear polysaccharides whose backbone structure is based on a repeating disaccharide sequence of sulphate esters of (1→3) linked β-D-galactose and (1→4) linked 3,6-anhydro-α-D-galactose. κ-Carrageenans have one sulphate group per repeating disaccharide unit, positioned at C-4 of the β-D-galactopyranosyl residue, whereas τ-carrageenans have two sulphate groups, positioned at C-4 of the β-D-galactopyranosyl residue and C-2 of the 3,6-anhydro-α-D-galactopyranosyl residue. The functional properties of carrageenan gels, such as rigidity, turbidity, and tendency to syneresis (separation of water from gel upon aging), generally decrease with the increasing degree of sulphation in these polymers.

γ-Carrageenans constitute another group of the red seaweed polysaccharides. The repeating disacharide unit in λ-carrageenans consists of β-D-galactopyranosyl residue sulphated at C-2 (instead of C-4 as in τ- and κ-carrageenans) and 2,6-di-O-sulfato-α-D-galactopyranosyl units (instead of 3,6-anhydro-α- D-galactopyranosyl residue). λ-Carrageenans are the most sulphated polysaccharides, carrying three sulphate groups per repeating disaccharide unit. λ-Carrageenans are nongelling polysaccharides used as cold soluble thickeners in syrups, fruit drinks, pizza sauces, and salad dressings (Izydocrzyk et al., 2005).

Carrageenans are used mainly as gelling, thickening, and suspending agents. κ-Carrageenan gives firm, clear, and brittle gels with poor freeze-thaw stability. These gels may be softened by adding locust bean gum through synergistic interactions. κ-Carrageenan forms soft elastic gels with good freeze-thaw stability. The gelation of τ-carrageenan is ionic strength dependent: higher ionic strength promotes formation of junction zones and favors the formation of gels. λ-Carrageenan is a nongelling polysaccharide. There is a lack of 3,6-anhydro group in the (1→4)-linked β-D-galactopyranosyl residues which is required for the initial formation of double helix. λ-Carrageenan is used as a cryo-protectant. A combination of λ-carrageenan with locust bean gum can improve the freeze-thaw behavior of frozen products.

λ-Carrageenan has been used widely in dairy products to prevent whey separation on account of its interactions with casein micelles. It is also used in ice creams as a second stabilizer to prevent the phase separation caused by the incompatibility of locust bean gum and β-casein. λ-Carrageenan is also used as a water binder in cooked meats and as a thickener in toothpaste and puddings.

Water dessert gels and cake glazes are one of the most traditional uses for carrageenan. These products are based on the firm, brittle gels properties of kappa carrageenan with the texture modified as appropriate for elasticity, cohesiveness and syneresis control using iota carrageenan. Similar gels are also used for aspics and gels in canned meats and pet foods and in cooked sliced meats. In these latter products the carrageenan is incorporated to improve moisture retention, cooking yields, slicing properties and mouthfeel and succulence. Beer and wine fining are applications which rely on the protein reactivity of carrageenan (Imeson 2000).

Milk-based puddings were one of the original uses for carrageenan. Extremely low levels of carrageenan, around 100–200ppm, are used to stabilise and prevent whey separation in a number of dairy products. These include milk shake and ice cream mixes, chocolate milks, and pasteurised and sterilised creams (Imeson, 2000).

Table 3: Typical applications of carrageenan

Food Product	Usage level (%)	Function
Dessert gels	0.5-1.0	Gelation
Low calorie gels	0.5-1.0	Gelation
Non-dairy puddings	0.1-0.3	Emulsion stabilization
Syrups	0.3-0.5	Suspension, bodying
pizza sauces	0.2-0.5	Bodying
Whipped toppings	0.1-0.2	Emulsion stabilization
Imitation coffee creams	0.1-0.2	Emulsion stabilization
Pet foods	0.5-1.0	Thickening, suspending, Gelation, fat stabilisation

Seaweeds: As a Whole Food

As consumer health and nutrition become more influential in the food industry, the use of seaweed as an ingredient is on the rise in product development. The rich mineral and trace element content of seaweed compared to terrestrial plant foods can however, impact negatively on its sensory characteristics (MacArtain et al., 2007). As an ingredient of composite foods, it has been shown to be acceptable to consumers when baked into breads (*Ascophyllumnodosum* up to 5% w/w) and added to pasta (up to 10% w/w) (Prabhasankar et al., 2009). Further from these applications, seaweed has been added to low fat meat products where it contributes to water retention and gel formation. Collectively, these results suggest that seaweed may be successfully included as an ingredient in a number of food applications. As dried seaweed is high in dietary fibre, along with a range of other potentially bioactive components, this addition has the potential to enhance the nutritional quality of a product. Habitual consumption of seaweed may offer a nutritionally rich addition to the diet.

Other Uses of Seaweed Gums

Alginates are used in wound dressings, and production of dental moulds. In microbiology research, agar plant-based gum similar to gelatin and made from seaweed - is extensively used as culture medium. Carrageenans, alginates and agaroses (the latter are prepared from agar by purification), together with other lesser-known macroalgal polysaccharides, also have several important biological activities or applications in biomedicine. Seaweed is a source of iodine, necessary for thyroid function and to prevent goitre. However, an excess of iodine is suspected in the heightened cancer risk in Japanese who consume a lot of the plant, and even bigger risks in post-menopausal women. Seaweeds may have curative properties for tuberculosis, arthritis, colds and influenza, worm infestations and even tumors. In Japan, seaweed eaten as nori is known as a remedy for radiation poisoning. Seaweed extract is used in some diet pills. Other seaweed pills exploit the same effect as gastric banding, expanding in the stomach.

Other seaweeds may be used as fertilizer, compost for landscaping, or a means of combating beach erosion through burial in beach dunes. Seaweed is currently under consideration as a potential source of bioethanol. Seaweed is an ingredient in toothpaste, cosmetics and paints. Alginates enjoy many of the same uses as carrageenan, and are used in industrial products such as paper coatings, adhesives, dyes, gels, explosives and in processes such as paper sizing, textile printing, hydro-mulching and drilling.

Health Risks

Rotting seaweed is a potent source of hydrogen sulfide, a highly toxic gas, and has been implicated in some incidents of apparent hydrogen-sulphide poisoning. It can cause vomiting and diarrhoea.

Conclusion

Constantly increasing population and stagnant production of food urges to search for traditional food alternatives. In this perspective, seaweeds found the potential applications in the development of food products. Seaweeds are a large marine alga which grows in shallow water at the edge of oceans. Alginate, carrageenans and agar constitute the important members of gums. Seaweeds provide suitable alternative due to their unique properties and can be used as functional ingredients in the prevention of many chronic diseases. Seaweed contains many phytochemicals and nutraceuticals such as antioxidants, dietary fibers, vitamins, polyunsaturated fatty acids, phenolic acids, etc. Seaweeds have been shown to have beneficial properties for health and disease management, such as anticancer, antiobesity, antidiabetic, and antihypertensive, antioxidant, anticoagulant, anti-inflammatory, antifungal and antibacterial. These seaweed gums find applications in many food products, particularly, in dairy and bakery products for gelation, emulsion stabilization, thickener, texturizer, etc. The seaweeds gums also have potential to use as whole food.

References

1. Abidov, M. Ramazanov, Z. Seifulla, R. and Grachev, S. (2010). The effects of xanthan in the weight management of obese premenopausal women with non-alcoholic fatty liver disease andnormal liver fat. Diabetes, Obesty and Metabolism, 12:72-81.
2. Abowei, J. F. N. and Ezekiel, E. N. (2013). The potentials and utilization of seaweeds. Science and Agriculture, 4:58-66.
3. Ameson, A. (2000). Carrageenan. In: Phillips G O and William P A (ed.) Handbook of hydrocolloids. Woodhead Publishing Ltd., Cambridge, U.K.
4. Armisen, R. and Galatas, F. (2000). Carrageenan. In: Phillips G O and William P A (ed.) Handbook of hydrocolloids. Woodhead Publishing Ltd., Cambridge, U.K.
5. Baron, A. D. (1998). Postprandial hyperglycemia and á-glucosidase inhibitors. Diabetes Research and Clinical Practice, 40:51-55.
6. bcb.uwc.ac.za. Resources: Ireland, University of Galway. Available from: http://www.bcb.uwc.ac.za/envfacts/seaweeds/ind ex.htm.
7. Chandini, S. K. Ganesan, P. and Bhaskar, N. (2008). In vitro antioxidant activities of three selected brown seaweeds of India. Food Chemistry, 107:707-13.
8. Chapman V J (1970). Seaweeds and their Uses. 2nd Edn, Metheun, London, pp: 304.
9. Colliec, S. Fischer, A. M. Tapon-Bretaudiere, J. Boisson, C. Durand, P. and Jozefonvicz, J. (1991). Anticoagulant properties of a fucoidan fraction. Thromb Research, 64:143-54.
10. Devi, K. P. Suganthy, N. Kesika, P. and Pandian, S. K. (2008). Antioxidant and Antibacterial Properties of Methanolic Extract of Green Seaweed Chaetomorphalinum From Gulf of Mannar: Southeast Coast of India. BMC Complement Alter Medicine, 14:8-11
11. Dhargalkar, V. K. and Pereira, N. (2005). Seaweed: Promising plant of the millennium. Science Culture, 71:60-66.
12. Dhargalkar, V. K. and Verleca,r X. N. (2009). Southern Ocean seaweeds: A resource for exploration in food and drugs. Aquaculture, 287:229-242.

13. Draget, K. I. Smidsred, O. and Skjak-Braek, G. (2010). Alginates from Algae. In Steinbuchel A and Rhee S K (ed.). Polysaccharides and polyamides in the food industry: Properties, Production and Patents. Willey-VCH Verlag GambH and Co. KGA, Weinhein.
14. Eskin, B. A. Grotkowski, C. E. and Connolly, C. P. (1995). Different tissue responses for iodine and iodide in rat thyroid and mammary glands. Biology Trace Element Resesrch, 49: 9-19.
15. Ghosh, D. (2004). Search for Future Viands: Algae and Fungi as Food. Resonance, 6: 33-40.
16. Giardi, M. T. Rea, G. and Berra, B. (2010). Bio farms for nutraceuticals: Functional foods and safety control by biosensors. Landes Biosciences, USA.
17. Imeson, A. P. (2000). Carrageenan. In: Phillips G O and William P A (ed.) Handbook of hydrocolloids. Woodhead Publishing Ltd., Cambridge, U.K.
18. Izydocrzyk, M. Cui, S. W, and Wang, Q. (2005). Polysaccharide Gums: Structure, functional properties and applications. In Raton B (ed.) Food carbohydrates, chemistry, physical properties and applications. Taylor and Francis group, CRC press, USA.
19. Jirge, S. S. and Chaudhary, Y. (2010). Marine: The ultimate source of bioactives and drug metabolite. International Journal of Research in Ayurveda and Pharmacy, 1:55-62.
20. Kumaran, S. Deivasigamani, B. Alagappan, K. Sakthivel, M. and Karthikeyan, R. (2010). Therapeutic importance of sulfated polysaccharides from seaweeds: updating the recent findings. Asian Pacific Journal of Tropical Medicine, 3:977-88.
21. Lee, H. J. Kim, H. C. Vitek, L. and Nam, M. C. (2010). Algae consumption and risk of type 2 diabetes, Korean National Health and Nutrition Examination Survey in 2005. Journal of Nutritional Science andVitaminology, 56:13-18.
22. MacArtain P, Gill C I R, Brooks M, Campbell R and Rowland IR (2007). Nutritional value of edible seaweeds. Nutritional Review, 65:535-43.
23. Matanjun, P. Mohamed, S. Kharidah, M. and Noordin, M. M. (2010). Comparison of cardiovascular protective effects of tropical seaweeds, Eucheumacottonii, Caulerpalentillifera, and Sargassumpolycystum, on high-cholesterol/high-fat diet in rats. Journal of Medical Food, 13:792-800.
24. Milani. J. and Maleki, G. (2012). Hydrocolloids in Food Industry. In Valdez B. (ed.) Food Industrial Processes-Methods and Equipment. Intech, Rijeka, Croatia.
25. Moe, S. T. Draget, K. I. Skjak-Break, G. and Smidsred, O. (2010). In: Stephen A M, Phillips G O and William P A (ed.) Food polysaccharides and their applications. Mercel Dekkar Inc., New York
26. Mohammed, S. Siti, Nanda, H. and Abdul, R.. H. (2012). Seaweeds: A sustainable functional food for complementary and alternative therapy. Trends in Food Science and Technology, 23:83-96.
27. Morrissey, J. Kraan, S. Guiry, M.D. (2001). A Guide to Com-mercially Important Seaweeds on the Irish Coast. Dublin, Bord Iascaigh Mhara/Irish Fisheries Board, pp.5–40.
28. Paxman, J. R. Richardson, J. C. Dettmar, P. W. and Corfe, B.M. (2008). Daily ingestion of alginate reduces energy intake in free living subjects. Appetite, 51:713-19.
29. Prabhasankar, P. Ganesan, P. Bhaskar, N. Hirose, A. Stephen, N. and Gowda, L.R. (2009). Edible Japanese seaweed, wakame (Undariapinnatifida) as an ingredient in pasta: Chemical, functional and structural evaluation. Food Chemistry, 115:501-08
30. Raja, A. Vipin, C. and Aiyappan, A. (2013). Biological importance of marine algae- An Overview. International Journal of Current Microbiology Applied Science, 2:222-227.
31. Rieusset, J. Touri, F. Michalik, L. Escher, P. Desvergne, B. and Niesor, E. (2002). A new selective peroxisome proliferator-activated receptor gamma antagonist with anti-obesity and anti-diabetic activity. Molecular Endocrinology, 16:2628-44.

32. Zhang, Q. Ning, L. Xiguang, L. Zengqin, Z. Zhien, L. and Zuhong Xu. (2004). The structure of a sulfated galactan from Porphyrahaitanensis and its *in vivo* antioxidant activity. Carbohydrate Research, 339:105-11.
33. Zhao, Y. (2008). Polysaccharide gel microspheres for peptide and protein drug encapsulation. Thesis, Purdue University, USA.

CHAPTER 33

Studies on Utilization of Gums for Encapsulation of Various Ingredients Used in Food Industries

Khalid Bashir, Kulsum Jan, Basharat Yousuf and Manjeet Aggarwal

Abstract

Microencapsulation has found great applications in food business. However, the technology remains aloof from being fully exploited. Microcapsules offer food processors a way to shield sensitive food ingredients, mask or preserve flavours and aromas and conversion of volatile food components into easy-handled solid ingredients (like powders, capsules etc). A number of naturally occurring and synthetic polymeric materials have been identified so far that possess the entrapping property and are categorised as GRAS materials. Numerous techniques have been developed for using these entrapping materials to encapsulate the sensitive food ingredients. Controlled release of the food ingredients at proper place and right time could be therefore a key practicality that may be provided by microencapsulation. With strictly fine-tuned and stabilized release characteristic, this method adds worth to the merchandise. The areas of interest in this field include the development of alternative and inexpensive polymers that may be considered natural or mimic the properties of natural polymeric materials, like Gum Arabic, Gellan Gum, Gum Karaya, Gaur Gum and Gum Acacia. Limitations in many of the encapsulation techniques have occurred due to high value of raw materials and production. This review is effectively committed to the, properties of encapsulating agents along with the emphasis on different techniques used for encapsulation in the varied fields like food technology, pharmaceutical, commercial engineering and biotechnology.

Keywords: Gum Arabic, Gellan gum, Gum Karaya, Guar gum, Gum Acacia, Pectin, Starch, Microencapsulation, Dextrin.

Introduction

Microencapsulation can be defined as a technology in which the liquids, volatiles or solid materials (including flavours, enzymes, cells, medicine or other active ingredients) are packed within a continuous polymer film, so that the contents are released at predictable and controlled manner under specific conditions (Desai and Park, 2005). Microencapsulation dates back to early 1930s and this technology has boosted during the past decade due to the advancement in technologies and modifications in the entrapping materials. The entrapping material is called as wall, which may be single or double layered while the ingredients that are entrapped within the wall form the core, which may be composed of just one or several different types of ingredients. The retention of the core is governed by several factors such as its chemical nature including the chemical functionality, relative volatility, polarity, properties of the wall material and the encapsulation technique employed (Savitha et al., 2005). The application of encapsulation technique has increased in the food industry as the encapsulated materials can be protected from moisture, heat, oxygen, free radicals or other extreme conditions, thus enhancing their stability and maintaining viability. Since the matrix structure is modified, the intended site time delivery is achieved by controlling the dosage and the release time of the bioactive contents. In around 1950s the eventual development of ribbons and paper in which dye was entrapped in small gelatine capsules came into existence, the dye was released on impact by a typewriter key or the pressure of a pen or pencil. This was the time that provided a stimulus to food scientists and other concerned bodies to develop techniques for encapsulating enzymes, flavours, drugs etc. (Alagusundaram et al., 2009; Allen et al., 2005).

Encapsulation technology has now served the food industries for over 60 years, the most important being the flavour retention, especially in spices, as most of the volatile components were lost during their grinding (although cryogenic grinding is employed nowadays but encapsulation has its deep roots in processed foods famous because of their flavours). Encapsulation technique (Fig.1) wall material (Fig.2) can be selected depending upon the nature of the food item and its intended use. Among the widely used techniques include extrusion coating, spray drying, spray cooling (spray chilling), liposome entrapment, fluidized bed coating, solvent evaporation, rotational suspension separation,

Microcapsules may range from sub micrometer to several millimetres in size and have a multitude of different shapes that depends on the material and methods used to prepare them. An ample variety of food ingredients that have been successfully encapsulated include: colorants, acids and bases, flavouring agents, preservatives, artificial sweeteners, antioxidants, leavening agents,

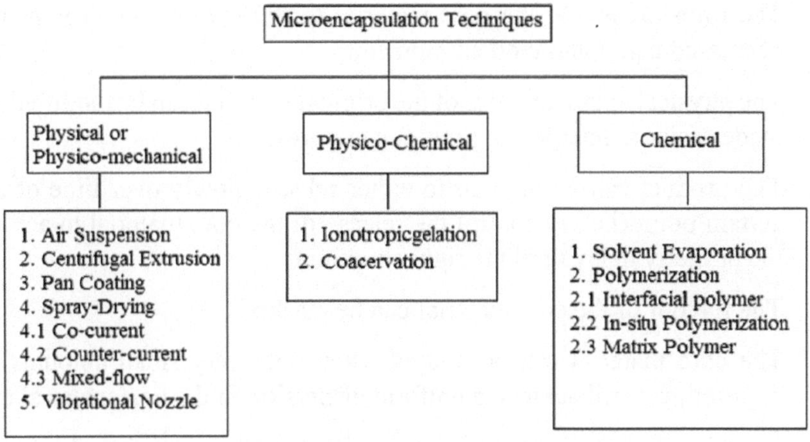

Fig.1: General overview of Encapsulation techniques

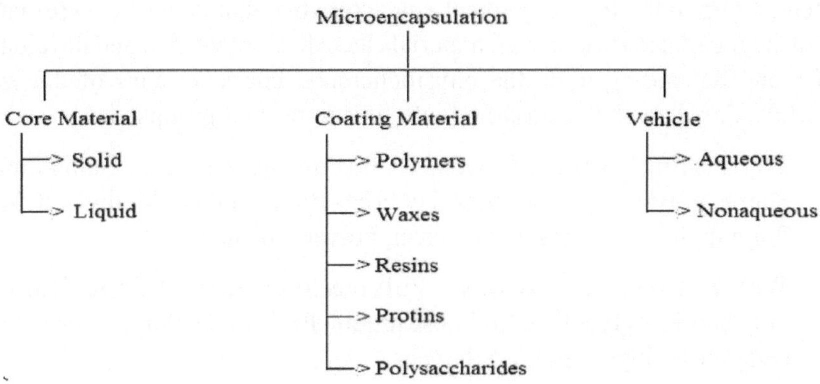

Fig. 2: Type of Core Materials, Coating Materials and Vehicles Used in Microencapsulation

coacervation, inclusion complexation, polymerization, centrifugal extruson. substances with undesirable flavour and odour. The use of encapsulation for aspartame as sweetener and other artificial flavours in chewing gum is well known. Among the wall materials polysaccharides (gums, dextrins, alginates, starch), fats, protein and lipid derivatives have been successfully employed. Different methods exist to release the ingredients from the capsules which can be stage-specific, site-specific, or signalled by changes in temperature, pH, or osmotic shock (Poshadri and Aparna, 2010). The food industries apply the microencapsulation process for a variety of reasons which include:

1. Encapsulation/entrapment protects the core material from degradation by minimizing its reactivity to the outside environment (e.g., heat, moisture, air, and light) like enzyme entrapment.

2. The transfer rate of the core material to the external environment is decreased e.g. volatile oil entrapment.
3. The physical characteristics of the original material can be modified and made easier to handle e.g. probiotic cultures.
4. The product can be tailored to either release slowly over time or at a certain point (i.e., to control the release of the core material to achieve the property delay until the right stimulus).
5. The flavour of the core material can be masked.
6. The core material can be diluted when only very small amounts are required, yet still achieve a uniform dispersion in the host material and
7. It can be employed to separate components within a mixture that would otherwise react with one another.

Depending upon the food ingredient characteristics that is to be encapsulated and its intended use various wall materials have been identified and developed so far and depending upon the physiochemical characteristics of the wall materials, they have been classified into four important groups viz,

1. **Water soluble resins**: Gelatin, Carboxymethylcellulose, Gum Arabic, Starch, Polyvinylpyrro- lidone, Hydroxy-ethylcellulose, Methylcellulose, Polyvinylalcohol, Arabinogalactan, Polyacrylic acid.
2. **Water insoluble resins**: Polymethacrylate, Ethylcellulose, Polylactidecoglycolide, Cellulose nitrate, Polyamide, Polyethylene, Poly (Ethylene Vinyl acetate), Silicones.
3. **Waxes and lipids:** Beeswax, Spermaceti, Paraffin, Stearyl alcohol, Stearic acid, Carnauba, Glyceryl stearates.
4. **Enteric resins:** Zein, Shellac, Cellulose acetate phthalate.

The wall material for encapsulation must possess the properties like: Inert towards active ingredients, non-hygroscopic, low viscosity and economical, soluble in an aqueous media, film-forming, pliant, tasteless and stable. Further it must result in the stabilization of the core material and its controlled release under specific conditions.

The microcapsules formed after the encapsulation process may consist of a single particle or cluster of particles. After isolation from the liquid manufacturing vehicle and drying, the material appears as a free flowing powder. The powder remains suitable for formulation as compressed tablets, hard gelatin capsules, suspensions, and other dosage forms. Not all wall materials meet all the properties needed, so they are often used in combination with other wall materials and modifiers such as oxygen scavengers, antioxidants, chelating agents and surfactants. Generally hydrophilic polymers, hydrophobic polymers (or) a combination of both are used for the microencapsulation process. A number of coating materials have been used successfully (Fabregas and Garcia, 1995).

Wall Materials for Encapsulation

The wall materials employed for the encapsulation process have been classified according to its source and structure (Table 1, Fig. 3). Selecting a suitable coating material for encapsulation decides the physico-chemical properties of the resultant microspheres. The important parameters that are to be taken into consideration while selecting a polymeric wall material include: stabilization, environmental conditions, release characteristics and reduced volatility. The wall material must be able to form a film i.e. cohesive with the core material. It should be chemically compatible, non-reactive with the core material and provide the desired coating properties such as flexibility, strength, optical properties and stability.

Table 1: Classification of natural carbohydrate polymers related to the source

Origin	Carbohydrate polymer	Protein	Lipid
Plant	Polysaccharides (Starch derivatives, Cellulose derivatives), Plant exudates (Gum Arabic, Gum karaya, Mesquite gum, Tragacanth), Plant extracts (Galactomannans, Soluble soybean), Seeds (Locust bean), Roots (Konjac).	Gluten (corn), Isolates (pea, soy).	Fatty acids, Waxes, Glycerides, Phospholipids.
Microbial	Dextran, Chitosan, Gellan	Caseins, Whey proteins, Gelatin	Fatty acids, Waxes, Glycerides, Phospholipids
Marine	Carrageenan (Red seaweed), Alginate (Brown seaweed),		

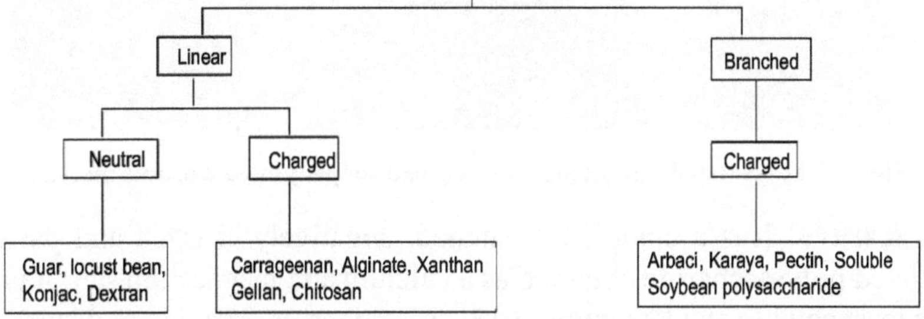

Fig. 3: Classification of carbohydrate polymers related to structural characteristics.

Wall materials are complex polysaccharide macromolecular substances, consisting of oligomers or polymers of diverse chemical structure. Chemical structure of some commonly used wall materials is shown in Fig. 7. Most of the wall materials used for encapsulating the food ingredients are products of gummnosis and are thus commonly referred as plant gums. Gummnosis is the production of gums by certain plant types in response to stress conditions like drought, wound, heat etc, thus acting as a barrier against microorganisms (Nitika et al. 2012). Most of the plant species cultivated for the production of gums generally belong to the family Leguminosae like: Gum arabic (Acacia senegal), Locust bean gum (Ceratoniasiliqua), Guar gum (Cyamopsistetragonolobos), Gum tragacanth (*Astragalus* spp.), Mesquite gum.

Gum Arabic: Gum Arabic also known as Gum Acacia, it is a complex mixture of polysaccharides, oligosaccharides (arabinogalactan) and glycoproteins. It is branched, negatively charged, odorless, tasteless, highly water-soluble and does not alter the odour, colour, and taste of the system to which it is added. β-(1→3)-linked d-galactopyranosyl units form the back of gum acacia and the side chains are composed of two to five β -(1→3)-linked d-galactopyranosyl units, attached to the main chain by 1,6-linkages (Orozco-Villafuerte et al., 2005). Gum Arabic has been successfully used for the encapsulation of various food ingredients like essential oils, food flavours and other essential ingredients (Savitha et al., 2005). Fig. 4 shows encapsulated flavouring compounds in gum Arabic matrix.

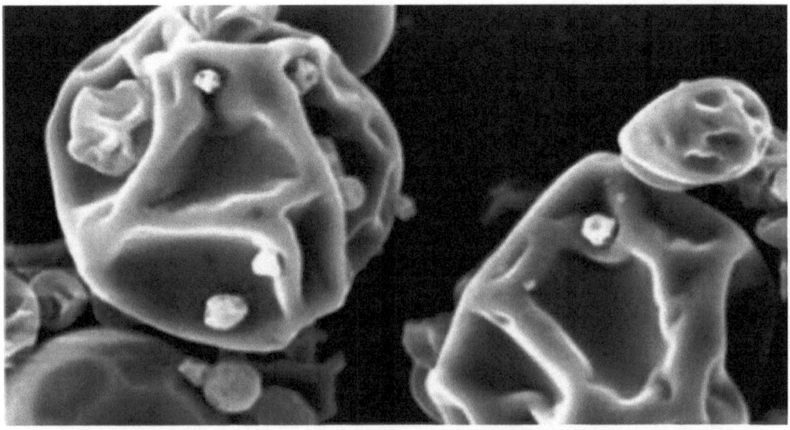

Fig. 4: Encapsulated flavouring agents (bead shaped) in gum arabic matrix.

Gum Karaya: It is a complex, branched, negatively charged and partially acetylated polysaccharide obtained as a calcium and magnesium salt. It is one of the least soluble exudate gums (10 % in cold water, 30% in hot water). The backbone has been identified as α–L- rhamnose and α-D-galacturonic acid residues. The side chains are linked by (1→2)-linkage of β-D-galactoseor by (1→3)-linkage of β-D-guluronic acid to the galacturonic acid of the main chain.

Galactomannans: They are linear, neutral, pseudoplastic solution behaviour gums, also called as Locust bean gum (LBG), tara gum, carob bean gum and guar gum. They consist of linearly (1→4)-linked β-D-mannopyranosyl units with single α-D-galactopyranosyl units connected by (1→6) linkages as side branches (Meer et al., 1980).

Mesquite Gum: They are branched, negatively charged with solubility comparable to GA, but possess steeper increase of the viscosity and has a good film forming capacity. The backbone is formed by (1→3)-linked β-D-galactose residues with (1→6)-linked branches, having L-arabinose (pyranose and furanose rings), L-rhamnose, β-D-glucuronate, and 4-O-methyl-β-D–glucuronateas single sugar or oligosaccharide side chains, with a small amount of proteins (0.7-5.8%) (Orozco-Villafuerte et al., 2005).

Gum Tragacanth: it is a complex mixture of heterogeneous polysaccharide with highly branched, negatively charged and one of the most acid resistant gums, also used as an emulsifier besides a thickening agent even at low concentrations. It has been successfully employed for the formation of films around oil droplets. It consists of two components: water-swellable component (bassorin) and water-soluble component (tragacanthin) (Elias, 1992).

Other Commonly Used Wall Materials

Pectins: These are hetero-polysaccharides with large molecular weight, soluble in water, low viscosity when compared to other gums and possess a pseudoplastic behaviour with complicated gelation behaviour. The basic unit is α-(1→4)-linked D-galacturonic acid. These units may be present as free acid or salt of Na, K, Ca, NH_4. Some of the neutral sugars like D-galactose, L-rhamnose, L-arabinose, D-xylose form a certain part of the polymeric chain (Elias 1992). Pectins display a very complex and non-random structure with linear blocks of homo-poly (galacturonic acid) known as smooth regions and with highly branched blocks of homo-poly (galacturonic acid), called hairy regions (May, 2000).

Soluble Soybean Polysaccharide: It is composed of galacturonic acid, galactose and arabinose, but some of the sugars are also present in the chain at low quantities including glucose, rhamnose andxylose. Galactouronic acid present on the backbone is responsible for the negative charge. They are soluble in both hot and cold water yielding solutions of relatively low viscosity. SSPS have gained attention nowadays because of the following properties: they possess excellent adhesion and film forming properties, no gellation, films are colourless and transparent, prevent oxidation of oils and can be used as emulsifier and stabilizer for emulsions, films formed are highly resistant to tension. SSPS is obtained from the residue left after oil and soy protein extraction from soybean. The powder obtained is then tailored as per the specific use (Maeda, 2000).

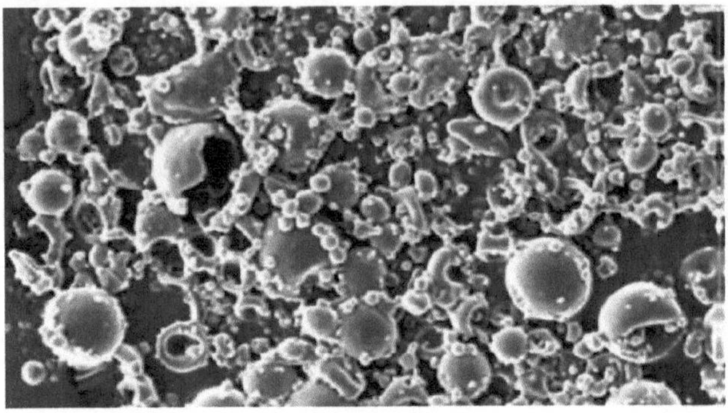

Fig. 5: Encapsulated core material in polysaccharide wall material

Waxes: Among the waxes, Bee wax, Candellia wax and Carnauba wax are popularly used. Waxes that are actually the esters of fatty acids are primarily composed of Triacontyl palmitate (75%), paraffin (15%) and triacontyl cerotinate (10%). Beeswax is obtained from the honeycomb. The melting point of bee wax falls in the range of 62–64°C, with colour varying from nearly white to brownish (Beyer, 1968). Carnauba wax is extracted from palm leaves. Carnauba wax has a melting point of 83°C and is one of the hardest natural waxes. Fig. 6 depicts encapsulated vanillin in carnauba wax.Candellila wax is derived from the leaves of the Candelilla shrub. Candelilla wax is light brown to light yellow and melts in the range of 67–79°C. It is compatible with all waxes (vegetable and animal), glycerides, fatty acids and natural and synthetic resins.

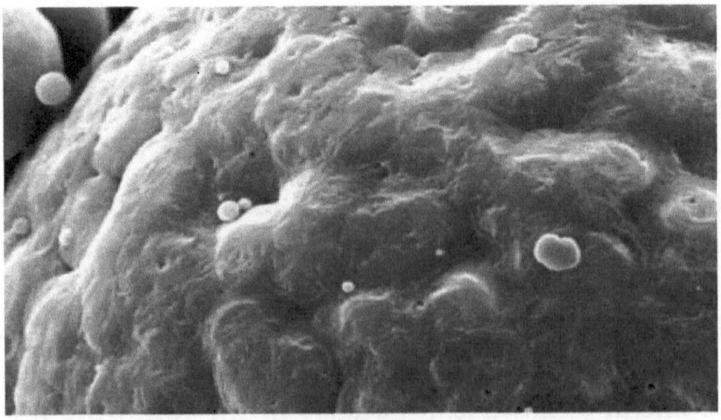

Fig. 6: Encapsulated vanillin in carnauba wax

Glycerides: They have been classified as monoglyceride (monoacylglycerol), diglyceride (diacylglycerol) and Triglyceride (triacylglycerol, triacylglyceride). The fatty acid chains are bounded by ester linkages by covalent bond to aglycerol molecule. In monoacylglycerol, only one fatty acid is attached which can be at the C-1 or C-2. In diacyglycerol, two fatty acids may be attached to any of the three carbon positions and in triacylglycerol all the three fatty acids may be either same or may differ by one or all three. The most common residues are those with 16, 18, and 20 carbons. Mono and diglycerides possess emulsifying properties. The melting point depends mainly on the chemical nature and symmetry of the fatty acid residues. More saturation results in increase in the melting point.

Gluten: It is a complex mixture of gliadins and glutenins and comprise about 80% of the proteins contained in wheat seed. Both components have unusually high levels of amino acids proline and glutamine and are therefore designated as prolamins (Shewry et al., 1986). Gliadins are monomeric gluten proteins with high heterogeneity. It consists of three structurally distinct groups, designated as: α, γ and ω. The α and γ-types are classified as sulphur-rich prolamins whereas the γ-type are called as sulphur-poor prolamins. Glutenin is a heterogeneous mixture of disulfide stabilized polymers of high-molar-mass and low molar mass glutenin subunits. Glutenins have poor water solubility because of high content of non-polar amino acids. Glutenins are rich in glutamine.

Alginate: These are the family of linear anionic polysaccharides, that are the copolymers of (1→4) linked α-l-guluronic acid and β-D-mannuronic acid. They are extracted from brown sea weed and depending upon the extraction process, the proportion and sequential arrangement of the two structural units vary widely.

Carrageenans

They are obtained from red sea weed and include the family of high molar mass sulphated polysaccharides and the individual structure effectively depends upon the source, extraction and purification process employed. The polymer contain alternating (1→3)-linked β-D–galactopyranosyl and (1→4)-linked α-D-galactopyranosyl units. Carrageenans are anionic polyelectrolytes. Carrageenan is about five times more effective in milk than it is in water (Perceval, 1972).

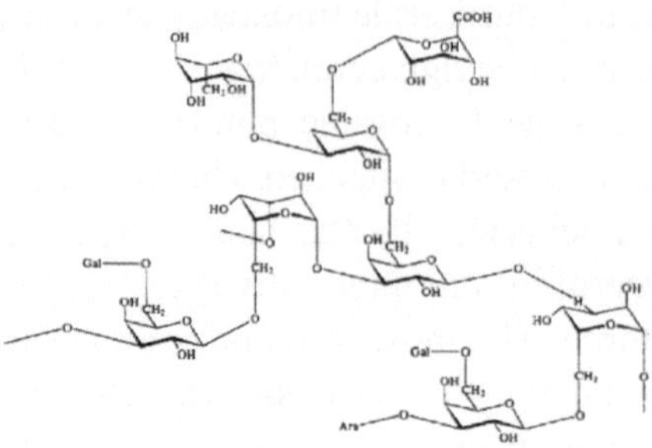

a) Gum Arabic

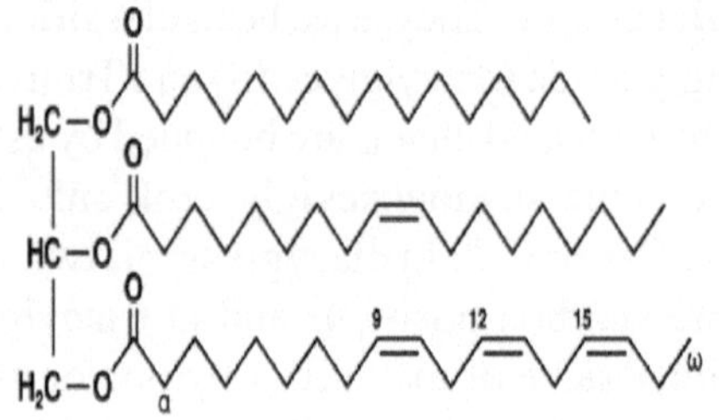

b) Triglyceride

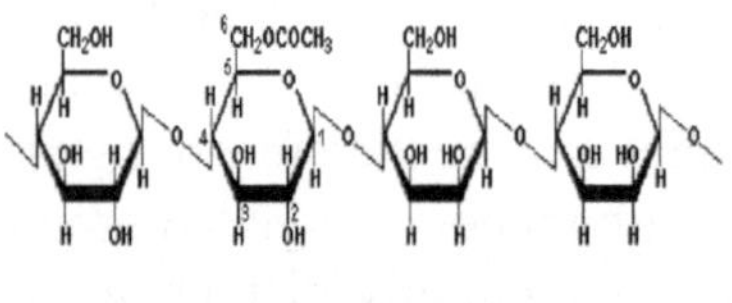

c) Gellan Gum

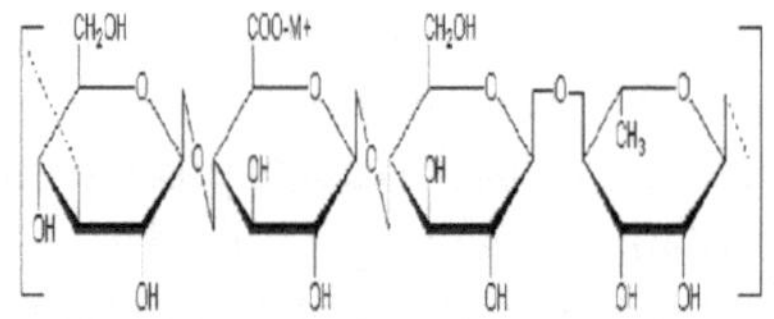

d) Glucomannan

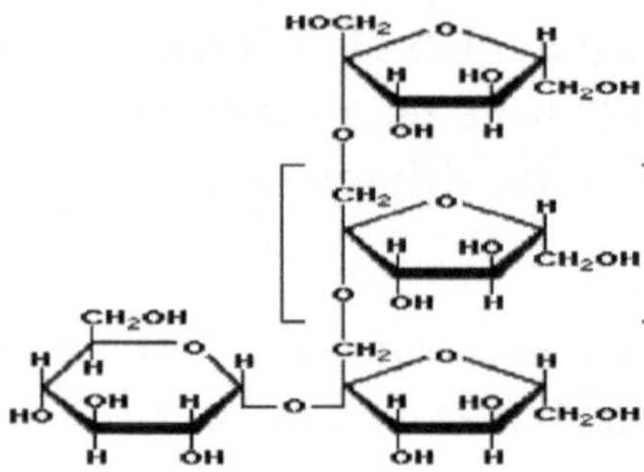

e) Inulin

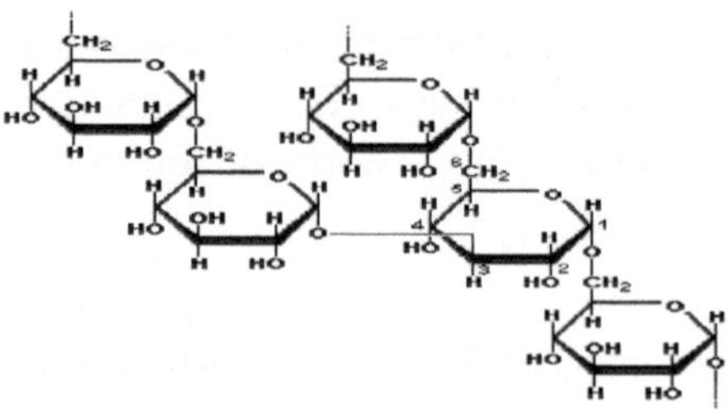

f) Dextran

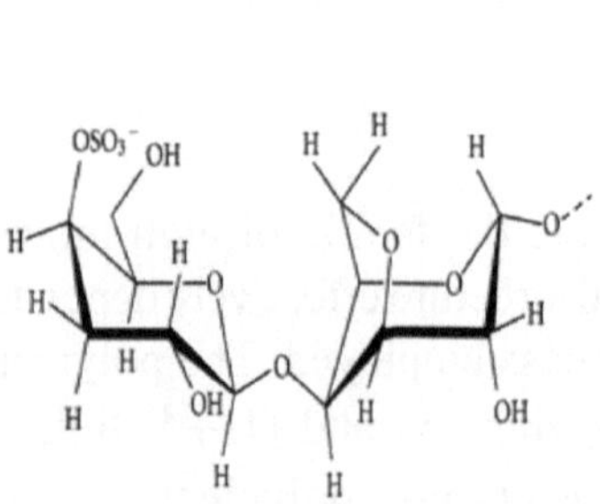

g) Carageenan

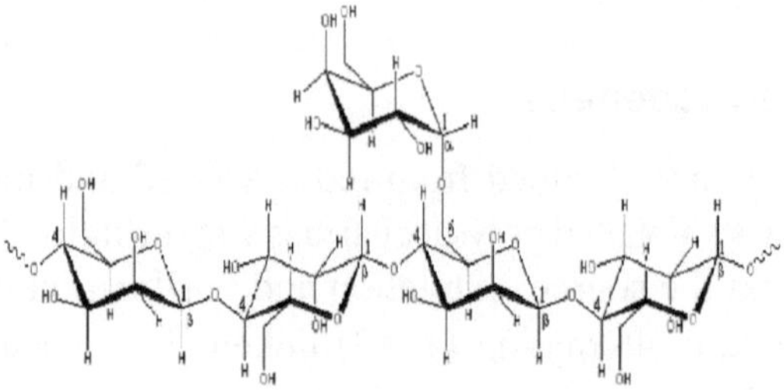

h) Gaur gum

Fig. 7: Chemical structure of various substances commonly used as wall materials

Encapsulation Techniques at a Glance

Spray drying: It is one of the most commonly employed microencapsulation technique in food industries. This technique was developed in 1937 by Boake Roberts. The process was in fact accidently developed when acetone was added by mistake to tomato puree which resulted in colour retention of tomato puree during spray drying. Afterwards, spray drying has emerged as a wonderful technology for the production of encapsulated food flavours (powders), minerals, oleoresins, vitamins, natural and synthetic colorants, fat and oil flavour, aroma compounds and enzymes. Fig. 8. represents encapsulation by spray drying.

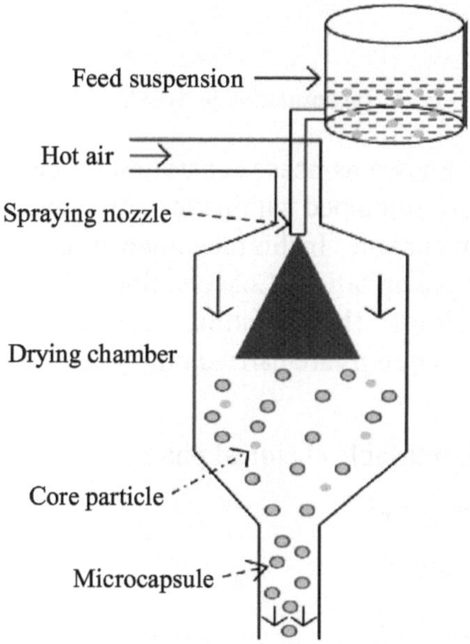

Fig. 8: Encapsulation by Spray drier

The advantages of this technique include being economical and more effective. In spray drying the wall materials like gums, modified starch, malto-dextrin etc are hydrated and then mixed with the core material in the ratio of 4:1. The mixture is then fed into a spray dryer at high pressure and temperature and atomized through a spinning wheel/nozzle. The water is evaporated by high temperature of hot air. The flow of hot air is either parallel to feed, anti-parallel or mixed. The capsules formed are then collected (Gibbs, 1999).

Rapid expansion of supercritical solution (RESS): In RESS process, a supercritical fluid like carbon dioxide along with the core material i.e. active ingredient and the wall material are kept at high pressures and which when released through a small plunger/nozzle to normal atmospheric pressure results in the deposition of wall material (in the form of coating layer) around the active core ingredient due to sudden drop of pressure (Poshadri and Kuna, 2010).

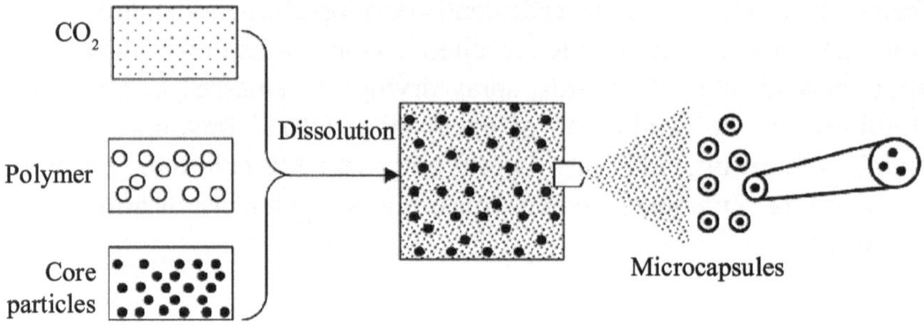

Fig. 9: Encapsulation by RESS

Coacervation: It is also known as phase separation process. In this process the core material is entirely entrapped within the matrix and is thus referred as true microencapsulation technique. In this technique the colloidal and aqueous phases are separated by precipitation. Coacervation can be either Complex (ionic interaction) or simple (hydrophobic interaction). In batch-type coacervation processes three steps are carried out under continuous agitation (Kirby, 1991).

1. Formation of a three-immiscible chemical phase.
2. Deposition of the coating and
3. Solidification of the coating.

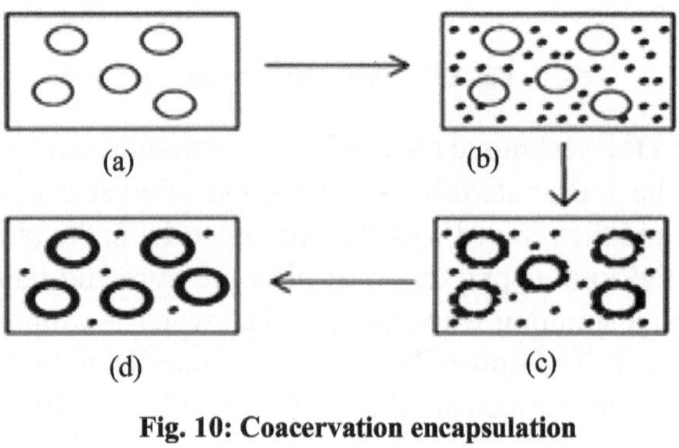

Fig. 10: Coacervation encapsulation

Fluidized-bed: They are used for the encapsulation of solid or porous particles. A liquid coating is sprayed onto the core material particles and the rapid evaporation helps in the formation of an outer layer on the particles. The thickness of the coating can be obtained by regulation of the temperature and flow of the liquid coating material. Depending upon the flow direction of the liquid matrix, it is classified into three types viz, Top spray, Bottom spray and Tangential spray (Korsmeyer et al., 1983).

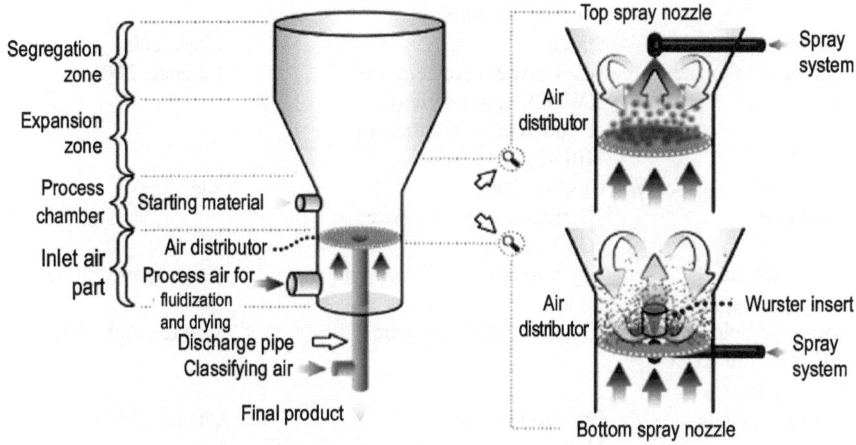

Fig. 11: Fluidized bottom bed spray encapsulation

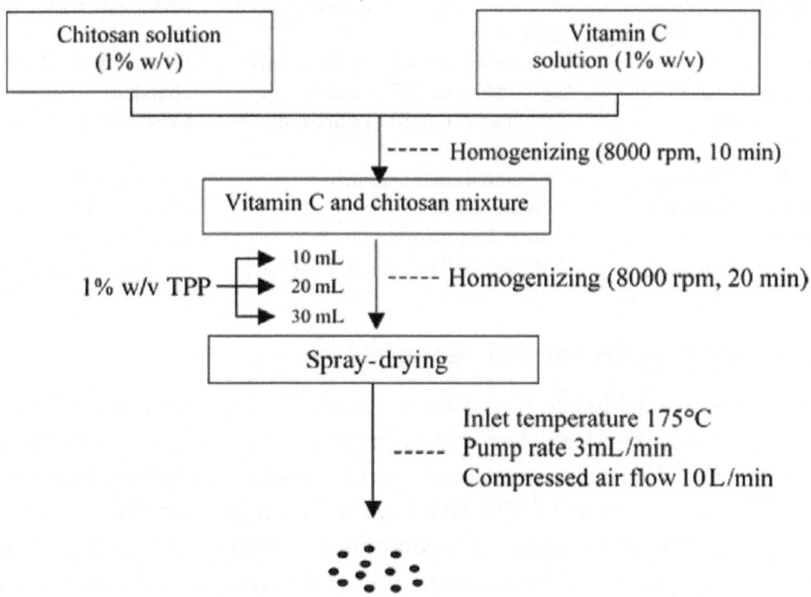

Fig. 12: Vitamin C encapsulation by Chitosan microspheres

Table 2: Important studies on encapsulation of some essential active compounds

Active component	Material for encapsulation	Reference
Vitamin C,Acerola, Synthetic ascorbic acid	a) Maltodextrin of DE20mixture DE20/gum arabic (GA); b) Maltodextrin DE25, GA, or mix of both	a) Righetto, 2006 b) Righetto, 2005
Vitamin E	Starch	Chen, 2004
Conjugated linoleic acid CLA)	Whey protein concentrate (WPC), gum arabic (GA), blend WPC/maltodextrin 10DE (1:1, w/w)	Jimenez, 2006
Vitamin A acetate dissolved in coconut oil	Hi-CAO 100 (starch octenylsuccinate, OSA-starch)	Xie, 2007
Probiotic bacteria (L. acidophilus, B. Lactis)	Alginate/Hi-Maize™ starch	Kailasapathy, 2006
Probiotics (L. acidophilus 547, B. bifidum ATCC 1994, L. casei 01)	Alginate/CaCl2/chitosan	Krasaekoopt , 2006
Allyl isothiocyanate	Gum arabic	Chacon, 2006
Isoflavone, beta-galactosidase	Medium-chain triacylglycerol (MCT), polyglycerol monostearat (PGMS)	Kim, 2006
Lactobacillus sp.	Gum arabic (GA), gellan gum (GG), mesquite gum(MG), and binary mixtures thereof	Fernanandez, 2007
Cardamom oleoresein	Gum arabic, modified starch, malto dextrin	Krishanlal et al., 2005
Bixin	Gum arabic, sucrose, malto dextrin	Barbosa et al., 2005
Black pepper oleoresin	Gum arabic, modified starch	Shaikh et al., 2006
Cumin oleoresin	Gum arabic, modified starch, malto dextrin	Krishanlal et al., 2005
Short chain fatty acids	Gum arabic, malto dextrin	Teixeira et al., 2004
Cardamom essential oil	Mesquite gum	Beristain et al., 2001
Limonene	Gum arabic, gelatin	Kaushik and Roos, 2007
Bifidobacterium lactis	Hydrated Gellan gum, xanthan gum	McMaster et al.,2005

Regulatory Aspects for Wall Materials

All additives are subjected to thorough scientific safety evaluation, prior to use them in foods. Food and Agriculture Organization (FAO) and World Health Organization (WHO) have formed an expert committee at an international level in 1955 known as Joint FAO/WHO Expert Committee on Food Additives (JECFA). JECFA acts as an advisory committee to WHO, CAC, FAO and other government agencies. The fundamental role of the committee is to recommend Acceptable Daily Intake levels and also assess the hazards and risks associated with the consumption of food additives. The ADI approach for additives was initiated by the JECFA in 1961. The procedures follow the collection of data related to the additives and the determination of no-observed-adverse-effect level (NOAEL) using Lethal dose-50 (LD-50) and Lethal concentration (LC-50) concepts and then applying the appropriate safety factors to achieve ADI for humans. The ADI provides a large safety margin and is the amount of a food additive that can be consumed daily over a life time without any adverse effect on health. CAC and JECFA has now came up with a new concept General Standards for Food Additives (GSFA), with the aim of establishing a harmonized, workable, and indisputable international standard for world trade. The wall material found safe is only allowed to be added to the food commodities. Table 2, gives the list of some of the wall materials along with their European numbers that have been allowed by CAC, JECFA and EU.

Table 3: List of some food additives used as encapsulating agents along with their E numbers

Name	E Number	Name	E Number
Alginic acid	400	Sodium alginate	401
Potassium alginate	402	Calcium alginate	403
Ammonium alginate	404	Propylene glycol alginate	405
Agar	406	Carrageenan	407
Locust (carob bean gum)	410	Guar gum	412
Gum tragacanth	413	Gum arabic	414
Xanthan gum	415	Karaya gum	416
Tara gum	417	Gellan gum	418
Curdlan	424	Alpha-cyclodextrin	457
Gamma-cyclodextrin	458	Powdered cellulose	460
Beeswax	901	Candelilla wax	902
Carnauba wax	903	Gum benzoic	906
Acid treated starch	1401	Bleached starch	1403
Enzyme treated starch	1405	Starch acetate	1420

Conclusion

Encapsulation is a useful method to protect sensitive food ingredients until they reach the target organ. Encapsulated bioactive compounds are easier to handle and offer improved stability. This technology can protect sensitive ingredients with an encapsulating agent, improving their stability and increasing their shelf life. Encapsulating agents (wall material) act as a protecting coat against ambient adverse conditions, such as light, humidity and oxygen. A large number of encapsulating agents have been investigated and are already in use for quite a long time. This review is effectively committed to the, properties of encapsulating agents along with the brief emphasis on different techniques used for encapsulation in varied fields like commercial engineering, pharmaceutical, industrial and biotechnology. This review focuses on the far side standard microcapsules to all or any substitute small particulate systems.

References

1. Alagusundaram, M. Chetty, M. S. and Umashankari, C. (2009). Microspheres as a Novel drug delivery system – A review. International Journal of Chemical Technology, 12:526-534.
2. Allen, L. V. Popovich, N. G. and Ansel, H. C. (2005). Pharmaceutical Dosage Forms and Drug Delivery Systems. Delhi, India: BI Pubications, 8, 265.
3. Bernard, F. Gibbs, Selim Kermasha, Inteaz Alli, Catherine N. Mulligan, (1999). Encapsulation in the food industry: a review. International Journal of Food Science and Nutrition, 50(3):213-224.
4. Desai, Kashappa Goud H. and Park, Hyun Jin (2005). 'Recent Developments in Microencapsulation of Food Ingredients, Drying Technology, 23:1361-1394.
5. Elias H-G. (1992). Makromoleküle. Band 2, Technologie. Hüthig and Wepf Verlag, Basel, Heidelberg, New York
6. Fabregas, J. L. and Garcia, N. (1995). Invitro studies on buccoadhesive tablet formulations of hydrocortisone hemisuccinate. Drug Delivery Industrial Pharmacology, 21:1689-1696.
7. Gibbs, B. F. Kermasha, S. Alli, I. and Mulligan, C. N. (1999). Encapsulation in the food industry: a review. International Journal of Food Science and Nutrition, 50:213–224.
8. JEFCA, Chemical and Technical Assessments, available on-line on FAO JECFA website: http://www.fao.org/ag/agn/agns//jecfa_archive_cta_en.asp
9. Kirby, C. J. Whittle, C. J. Rigby, N. Coxon, D. T. and Law, B.A. (1991). Stabilization of ascorbic acid by microencapsulation. International Journal of Food Science and Technolology, 26:437– 449.
10. Kirby, C.J. (1991). Microencapsulation and controlled delivery of food ingredients. Food Science and Technology Today, 5 (2):74–80.
11. Korsmeyer, R. W., Gurny, R. Doelker, E. M. Buri, P. and Peppas, N. A. (1983). Mechanism of solute release from porous hydrophilic polymers. International Journal of Pharmaceuticals, 15:25–30.
12. Maeda, H. (2000). Soluble soybean polysaccharide. In: Phillips, G.O. Williamsm, P.A. (eds) Handbook of hydrocolloids. Woodhead Publishing Limited, Cambridge, England, pp 309–320.
13. May, C. D. (2000). Pectins. In: Phillips GO, Williams PA (eds) Handbook of hydrocolloids. Woodhead Publishing Limited, Cambridge, England, pp. 169–188.
14. Meer, W. (1980). Gum Arabic. In: Davidson RL (ed) Handbook of water-soluble gums and resins. McGraw-Hill, New York, pp. 8-1–8-21.
15. Orozco-Villafuerte, J. Buendía-González, L. Cruz-Sosa. F. and Vernon-Carter, E. J. (2005). Increased mesquite gum formation in nodal explants cultures after treatment with a microbial biomass preparation. Plant Physiol-Biochemistry, 43:802–807.
16. Perceval, E. (1972). Chemistry of Agaroids, Carrageenans and Furcellarans. Journal of the Science of Food and Agricultutre, 23(8):933–940.
17. Poshadri, A. and Aparna, K. (2010). Microencapsulation technology: a review. Journal of Research ANGRAU 38:86-102.
18. Savitha, Krishnan, Amol C. Kshirsagar, Rekha S. and Singhal. (2005). The use of gum arabic and modified starch in the microencapsulation of a food flavoring agent. Carbohydrate Polymers, 62:309–315.

Chapter 34

Studies on Characterization, Processing and Utilization of (*Nymphaea alba* L.) Rhizomes

Kulsum Jan, Shumaila Jan, Syed Insha Rafiq and D.C. Saxena

Abstract

This review presents an idea on the botanical description, medicinal properties and traditional usage of wild, edible and aromatic plant (Nymphaea alba) in the human nutrition. Nymphaea alba Lin, *also known as the White Water lily, White Lotus, or Nenuphar is an aquatic flowering plant of the family Nymphaeaceae. Nymphaea is a weed in India and its production pretty much as good as water chest nut and Nelumbo nucifera. Its rhizome is rich in tannic acid, gallic acid, alkaloids, sterols, flavonoids, glycosides, hydrolysable tannins and high-molecular-weight polyphenolic compounds. It is used as associate aphrodisiac, anodyne, antiscrophulatic, astringent, cardiotonic, demulcent, sedative and anti-inflammatory. Further, it conjointly produces calming and sedative effects upon the nervous system, and is beneficial within the treatment of sleep disorder, anxiety and similar disorders. Moreover, the content of starch in the rhizome is said to be 40%. Extensive work has been carried out on agronomic and phenotypic properties for most plants, but relatively little study has been carried out on the properties which are of direct technical and economic interest to competitive food and non-food industry. Non-conventional food resources and their value addition have attracted attention in the recent years for their potential use as functional ingredients in food formulations, besides food they have also been used for non-food applications. Moreover, Plants provide the predominant ingredients of medicines in most medical traditions. So, there is tremendous potential for the profitable industrial use of these medicinal plants and exploring them for further use.*

Keywords: Non-conventional food resources, Nymphaea alba, Rhizome.

Introduction

More than 35,000 medicinal plant species are being used around the world for medicinal purpose. In developing countries, herbal medicines get the provenance of about 75-80% of the whole population. These plants have reported to have minimal side effects (Desta, 1993; Kapoor, 2005). These plants are used for the treatment of various diseases which dates back to prehistoric period. In a country like India, medicinal properties of some plants like herbs are mentioned in the Rig-Veda which seems to be the earliest records of the use of these plants as medicine (Chitra et al., 2009). The increasing demands of developing countries i.e., balanced nutrition, food security, agricultural development, healthful enhancements and improvement of economy could be met by the exploration of these plant sources. Nymphaea alba (L.) commonly known as White Water Lily is an aquatic herb belonging to the same group of plants i.e., medicinal plants.

This aquatic herb has a perennial rhizomes anchored within the mud. On global level, it is distributed in China, Europe, Southwest Asia, India, North Africa and Russia. In traditional system of medicine, all the parts of the plant have been used for medicinal purposes (Naghma and Sarwat, 2005). Starch, crude fiber, crude proteins and ash are present in rhizome part. An alkaloid nymphaeine is present in all parts of the plant except the seeds, besides tannins and glycosides. Seeds contain starch, fatty oil of seeds contain di, tri and tetraenoic acid (Wehmer, 1949). Water lilies have attracted botanists from at least 200 years. Nymphaeales (Nymphaeaceae) comprise eight genera and approximately 70 species of aquatic plants, with a worldwide distribution in tropical to temperate regions (Valentin et al., 2006).

The genus Nymphaea comprises of approximately 45 species of aquatic herbs with mostly floating leaves, flowers and a perennial rhizomes (Wiersema, 1987). In India, ten species of Nymphaea, both wild (*N. candida, N. alba, N. tetragona, N. nouchali, N. rubra* and *N. pubescens*) and cultivated (*N. Alba, N. caerulea, N. Micrantha, N. marliacea,* and var. *rubra*) have been reported (Kiran et al., 2000). These species are widely distributed from Bengal, Orissa to Kashmir in lakes and ponds (Anindya et al., 2012). Nymphaea alba rhizomes have been used as popular vegetable and exhibit multiple medicinal and nutritional properties. These rhizomes are wealthy supply of proteins and starch (Ying et al., 1986) and may be consumed in roasted form as pickles, dried slices and fried forms. Biochemically, the rhizomes are composed of proteins, fats, carbohydrates and minerals and are sensible supply of energy (Sridhar et al., 2007).

The genus Nymphaea takes its name from the Greek word, numphe which implies virgin or water nymphe and is reputed for the anti-aphrodisiac activity of its members (Dodamani et al., 2012). *Nymphaea alba* L, also known as the European White Water Lily, White Lotus, or Nenuphar, is an aquatic flowering plant of the family Nymphaeaceae. It grows in water from 30-150 cm deep mostly in large ponds and lakes. Most are aquatic herbs, which are widely distributed in India, from Bengal, Orissa to Kashmir in lakes and ponds. The roots have also been used extensively in a Variety of folk medicine. *Nymphaea alba* is rich in phyto-constituents that posses pharmacological actions. But any scientific literature about its antioxidant and analgesic actions using its rhizome part is not documented.

It is widely used in indigenous systems of medicine in India. Its main use is an antiseptic and astringent. It is useful in burning and insomnia. The pulp of the rhizome when applied externally acts as a rubefacient. It is also administered in dysentery, diarrhoea, gonorrhoea and leucorrhoea. The leaves and roots have been used in form of poultice for boils, scrofulous ulcers and inflamed skin; the infusion is used as a gargle for ulcers in the mouth and throat (Madhusudhan et al., 2011).

Plant Description

Height

As it is an aquatic plant, its height varies as the depth of the water varies. These plants can live in water as shallow as 6 inches if the water level is stable or as deep as 15 feet if the water is clear.

Stem

Their stems are cylindrical, long, soft, spongy, flexible and mucous-coated. They anchor the leaves and flowers to the roots. Both the leaves and the flowers have separate stems. Depending upon the subspecies, these stems may or may not have 4-5 purple streaks.

These stems have rounded cross-sections and have 4 hollow air-filled channels or tubes (lacunae) for buoyancy and for the movement of oxygen, carbon dioxide, and methane gases between the leaves and the roots. These gases enter through the younger leaves and exit through the older leaves as shown in Fig. 1.

Fig.1: Stem

Leaves

The White Water Lily only has basal leaves as shown in Fig.2. The leaves may be up to thirty cm in diameter and they take up a spread of 150 cm per plant. Their 1st year leaves are a cluster of pointed leaves upon short stalks and are submerged below the water. Their 2nd and subsequent year leaves are flat, floating, and orbicular. Their oval rounded shape may have evolved to protect it from both the wind and the waves. They have a V-shaped notch or cleft at their base and have entire margins. They are palmately veined and are about 3-12 inches in diameter. The leaves of this plant contain a flavones glycoside, myricitin. A glycoside nymphalin with digitalis like action has been also identified in flowers. Various parts of the plants contain ascorbic acid (Hoppe, 1943). Mature leaves of Nymphaea alba contain high concentrations of phenolics, secondary substances known for their fungistatic properties (Vergeera et al., 1997).

Fig. 2: Leaves

Rootstock

The rhizomes are densely covered with short black hairs. Their rootstocks consist of roots and large, thick, horizontal, ovate, spongy, scaly, and starchy rhizomes. These rhizomes may be 3-4 inches long. They are brown to black and may turn purple when exposed to the air (Fig.3). These rootstalks branch freely and can form new plants. They may break apart to float away and to form new plants elsewhere. During the winter, most of the plant dies back to

Fig. 3: Rootstock

the rootstocks. The roots have also been used externally, as infusions to treat lesions and inflammations associated with mucous membranes and in treatment of a variety of dermatological conditions (Robin, 2001). The dried root and rhizome of the White Water Lily have been used orally to treat gastrointestinal, genital, and bronchial conditions (Naghma et al., 2005).

Flowers

The white Water Lily usually flowers during their 3rd year. These flowers are white or pink, radially symmetrical, about 1½-9 inches wide, and are solitary (Fig.4) and have many small stamens inside. Each flower has numerous upright spreading or overlapping petals arranged in numerous rows with those within the inner rows being narrower and with all petals having their tips being either broadly rounded or tapered. Each flower also has 4 rounded, green or colored sepals that may be red on the outside and white on the inside. The bractless flowers have mostly tetramerous calyces, whereas the other flower organs occur in whorls or rings of many members. The petals are white, blue, red or yellow in various shades.

Fig. 4: Flower

The Fruit

The fruit pod is composed of globular or ovate, leathery, spongy, fleshy, and leathery berries. The fruit is submerged underwater for about 3-4 weeks while the seeds mature (Fig.5).

Seeds

The seeds are about 1/12-1/6 inches long, oblong, starchy, and green. After the seeds are released from the decomposed aril in the fall, they sink to the bottom. These seeds will germinate in the following spring.

Fig.5: Fruit

Edibility

Parts of the White Water Lily are edible to humans. The young, unfurling leaves may be eaten raw or cooked. The raw leaves should be washed, chopped, and added to soups or stews. The rhizomes' tubers may be boiled, roasted, or prepared like potatoes. The unopened flower buds may be cooked or pickled. The ripened seeds which have protein and oil may be boiled, fried, dried and ground into flour, or creamed like corn. All boiling should be done for about 5-10 minutes and within several changes of water. Amount of mineral content is in significant level in Nymphaea alba as described in the (Table-1) given below:

Table 1: The mineral composition of Nymphaea alba

Mineral	K	Fe	Ba	Co	Cu	Mn
Percent	0.024	0.79	0.034	0.034	0.004	0.019

Medicinal importance

All the parts of the plant have medicinal uses in traditional system of medicines. It contains the active alkaloids nupharine and nymphaeine, and is a sedative and an aphrodisiac/anaphrodisiac depending on sources. The root of the plant was used by monks and nuns for hundreds of years as an anaphrodisiac, being crushed and mixed with wine. It is rich in tannic acid, gallic acid, alkaloids, sterols, flavonoids, glycosides, hydrolyzable tannins and high-molecular-weight polyphenolic compounds (Eliana et al., 2008). It is used as an anodyne, antiscrophulatic, astringent, cardiotonic, demulcent, sedative and anti inflammatory. Further, it also produces calming and sedative effects upon the nervous system, and is useful in the treatment of insomnia, anxiety and similar disorders (Adnaik et al., 2009; Robin, 2001). It also produces calming and sedative effects upon the nervous system (Robin et al., 2001; Vergeera et al., 1997; James, 2008). It's anti carcinogenic action and inhibition of renal oxidative stress and hyper-proliferative response were reported by Duke (2008) and Khan et al., (2005).

Alcoholic extract of rhizome containing alkaloids has a mild sedative and spasmolytic action and do not significantly depress the heart in large doses but are reported to have a paralyzing action on medulla (Irvine et al., 1945). It is also said to be useful in improving the voice.

As bulking agents

Bulking agents such as starch are additives that increase the bulk of a food without affecting its nutritional value. Starch or amylum is a carbohydrate consisting of a large number of glucose units joined by glycosidic bonds. This polysaccharide is produced by all green plants as an energy store. It is the most common carbohydrate in the human diet and is contained in large amounts in such staple foods as potatoes (24%), wheat (75%), maize (corn) (72%), rice (86%), and the root vegetables (Eliasson et al., 2004). However, in Nymphaea alba rhizome, it contains 40% starch and 20% sugars (Chittendon et al., 1956).

Conclusion

Plants with good medicinal and aromatic properties are valued for their biological activities which can be justified by the fact that about 80% of the local population still depend on these plants for primary health care. Non-conventional food resources and their value addition have attracted attention in the recent years for their potential use as functional ingredients in food formulations, besides food they have also been used for non-food applications. The present study suggests the uses of Nymphea alba rhizome in different ways like extraction and characterization of starch, its modification and development of starch based products. Moreover, its active ingredients (phenols, antioxidants) suggest a better option to serve it as a important raw material for the development of functional foods.

References

1. Adnaik, R.S. Pai, P.T. Sapakal, V.D. Naikwade, N.S. Magdum, C.S. (2009). Anxiolytic activity of Vitexnegundo Linn in experimental models of anxiety in mice. International Journal of Green Pharm, 3:243-47.
2. Agnieszka, K. (2004). Content of selected chemicals in two protected macrophytes: nymphaea alba and nupharlutea. Sibith. and sm. In relation to site chemistry. Polish Journal of Ecology 52(2):229–232.
3. Anindya, B. Moumita, S. Sarbani, D.R. (2012). *In vivo* evaluation of anti-diarrheal activity of the rhizome of Nymphaea alba (Nymphaeaceae). Orient Pharm Experemnt in Medicine, 12:129–134.
4. Chitra, S. Patil, M.B. Ravikumar. (2009). Preliminary phytochemical investigation and wound healing activity of Allium Cepa Lin (Liliaceae). International journal of pharmacy and Pharmaceutical Sciences, 2(2):167-175.
5. Chittendon, F. (1956), RHS Dictionary of Plants plus Supplement. Oxford University.
6. Desta, B. (1993), Journal of of Ethanopharmacol, 39:129-139.
7. Dodamani, S.S. Sanakal, R.D. Kaliwal, B.B. (2012). Effect of ethanolic leaf extract of Nymphaea odorata on biochemical and oxidative stress parameters of liver and pancreas in alloxan induced diabeticmice, Research Opinions in Animal and Veterinary Sciences, 2:3 151-157.
8. Duke, J.A. (2008). Handbook of Medicinal Plants of the Bible. CRC Press, Boca Raton, FL, USA, pp 302-305.
9. Eliana, R. Ricardo, T. Jose, C. Galduroz, F. Giuseppina, N. (2008). Plants with possible anxiolytic and/or hypnotic effects indicated by three Brazilian cultures - Indians, afrobrazilians, and river-dwellers. Studies in Natural Products Chemistry, 35:549-95.
10. Eliasson, (2004). Starch in Food:Structure, Function and Applications. Wood head publishing. ISBN 978-0-8493-2555-7.
11. James, A.D. (2008). Duke's hand book of medicinal plants of the bible. USA: Taylor and Francis group. pp. 302–5.
12. Kapoor, L.D. (2005). Replica press private pvt,Ltd. India, 86.
13. Khan, N. Sultana, S. (2005). Anticarcinogenic effect of Nymphaea alba against oxidative damage and hyperproliferative response and renal carcinogenis in Wistar rats. Molecular Cell Biochemistry,; 271:1-11.

14. Kiran, A.K. Sai, B. Ammani, K. (2000). Antimicrobial and phytochemical analysis of nymphaea Nauchali leaf extracts. IJRRPAS, 2:2 142-151.
15. Madhusudhan, N. Lakshmi, T. Gowtham, K.S. Ramakrishanan, K. V.G.R. Roy, A. Geetha, R.V. (2011). Invitro antioxidant and free radical scavenging activity of aqueous and ethanolic flower extract of Nymphaea alba. International Journal of Drug Delivery and Research, 3:23-6.
16. Naghma, K. Sarwat, S. (2005). Inhibition of potassium bromated induced renal oxidative stress and hyperproliferative response by Nymphaea alba in Wistar rats. Journal of Enzym Inhibition and Medical Chemistry, 20:275–283.
17. Robin, D. (2001). Nymphaea odorata: White pond lily. Medical Herbalism; 11:6-7.
18. Robin, D. (2001). Nymphaea odorata: White pond lily. Medical Herbalism. Material Medica Pharm, 11:231-233.
19. Sridhar, K.R. Bhat, R. (2007). Lotus–A potential nutraceutical source. Journal of Agricultural Technology, 3(1):143-155.
20. Valentin, G. Philip, M. Evelin, P. Alejandro, R. N. Rolf, R. (2006). Developmental morphology of branching flowers in Nymphaea prolifera. Journal of Plant Research, 119:561–570.
21. Vergeera, L.H. Vander, V.G. (1997). Phenolic content of daylight-exposed and shaded floating leaves of water lilies (Nymphaeaceae) in relation to infection by fungi. Oecologia; 112:481-4.
22. Vergeera, L.H. Vander, V.G. (1997). Phenolic content of daylight-exposed and shaded floating leaves of water lilies (Nymphaeaceae) in relation to infection by fungi. Oecologia. 112:481–4.
23. Wehmer, I. (1949). Chem. Abstract 1933, 27: 5782, 43-135.
24. Wiersema, J.H. (1987). A monograph of Nymphaea subgenus Hydrocallis (Nymphaeaceae). Syst Botany Monogram, 16:1–112.
25. Ying, Y.S. Charls, F. (1986). Gelation properties of Chinese water chestnut and lotus root starch. Journal of Food Science, 51(2): 445-9.

CHAPTER 35

Value Added Green Mango-Mint-Tulsi-Squash Modified with Honey

P.V. Bade, K.R. Kale, A.S. Binnar, V.S. Sadewale

Abstract

Mango pulp is high in prebiotic dietary fibers vitamin c, polyphenols, carotenoids, vitamins and dietary minerals. Mint leaves are known for its medicinal aromatic and flavorsome qualities. Mint act as powerful antioxidant, protecting the body against the formation of cancerous cells and is rich source of polyphenolic compounds. Tulsi also have antioxidant, adaptogenic, anti-inflammatory, antibacterial and immune-enhancing properties. Honey is natural sweetener and has antiseptic and antibacterial properties. Medicinally, honey is non irritant promote rapid growth of healthy tissue and is useful in purities value, bed spores, skin and intestinal disorder, etc. Keeping in mind the beneficial effects of these raw materials, attempt was made to prepare value added drink from it, which may have medicinal value and good sensory acceptability with objective to prepare and evaluate the recipe for value added green mango-mint-tulsi beverage and to study the shelf life of the developed value added beverage. The formulation prepared with partial substitution of honey was more acceptable. Throughout the storage period, the squash prepared with honey scored more in flavor, taste, mouth-feel and overall acceptability than without honey. The cost of production without including labour, bottling and other overhead charges came out to be Rs. 34.5/L for squash without honey and Rs. 59.0/L for squash without honey.

Introduction

There are traditional drinks like aam panna, jaljeera, tulsi tea etc., made from green mango, mint and tulsi, respectively. Blending of two or more fruit juices and their beverages with the addition of spices is convenient alternative for its utilization in order to have some value added fruit drinks, having high quality in respect of both sensory and nutritional aspects. The present research was carried out to prepare value added green mango-mint-tulsi beverage and evaluate its chemical composition changes during storage and sensory perceptions. To study the shelf life of the developed value added beverage.

Materials and Methods

Preparation of raw materials

The raw material procured from the local market includes Green mango pulp (Var. Ramlela), Tulsi leaves paste (Var. Rama) and Mint paste. The material was of good quality and kept under refrigeration temperature until used.

Preparation of squash

Best combination on the basis of organoleptic evaluation was selected for partial substitution of sugar by honey as per flow sheet given in Figure-1

Preparation of sugar syrup (sugar+citric acid+ water) | Green mango-mint-tulsi proportion (as per recipe)

↓

Straining of syrup

↓

Cooling

↓

Mixing (25% pulp, 50% TSS, 1% acidity, 2% black salt)

↓

Homogenization (In colloidal mill)

↓

Addition of sodium benzoate (1g/l squash)

↓

Bottling

↓

Capping

↓

Pasteurizing and cooling

↓

Storage

Chemical analysis

Chemical Constituents analysis of processed products were carried out for total soluble solids using (Abbe's or hand refractometer), Total and reducing sugars as per the method of Hulme and Narain, (1931), Acidity, pH and ascorbic acid were determined according to A.O.A.C. (2005) methods. Total Carotenoids were determined as per the method of Rodriguez-Amaya, (1999). Total chlorophyll content was determined as per the method of Arnon, (1949) whereas total phenols were determined as per the meth of of Amorium et al., (1997).

Storage

The products were stored in glass bottles at room temperature are analyzed for chemical constituents and were scored for sensory attributes after diluting and squash 3-times with water at monthly intervals for three months.

Organoleptic evaluation (9-point hedonic scale)

Thirty panelist were asked to assess the of green mango-mint-tulsi squash for flavor, taste, mouth-feel and overall acceptability, and to mark on a Hedonic Rating Test (1 – Dislike extremely, 5 – Neither like nor dislike and 9 – Like extremely) in accordance with their opinion for taste, texture, color and overall acceptability.

Statistical analysis

Experimental data were analyzed by analysis of variance (ANOVA), and means were compared using Duncan's multiple range tests with significance defined at P0.05. All statistical analysis was performed using commercial statistical package SPSS (16.0, Chicago, IL, USA).

Results and Discussions

A recipe for squash prep ration has been developed using green mango, mint and tulsi as raw materials. Their different ratios were tried and further the best formulation showing high overall acceptability in terms of sensory parameters was selected. The sugar in the best formulation was partially substituted by honey. The squash thus prepared was stored in glass bottles for three months and chemically analyzed for different components when prepared fresh and later on regular interval of 30 days.

Changes in chemical constituents of green mango-mint-tulsi squash during storage at room temperature

During storage, the sugar content was observed to increase slightly; acidity and TSS remained same; whereas chlorophyll, carotenoids, total phenols and

Table 1: Changes in chemical constituents of green mango-mint-tulsi squash with or without honey during storage at room temperature

Treatment	Storage Period Month	TSS (%)	Total Sugar mg/100ml	Reducing Sugar mg/100ml	Acidity (%)	pH	Ascorbic acid mg/100ml	Total Carotenoids mg/100ml	Total Chlorophyll mg/100ml	Total Phenols mg/100ml
Withouthoney	0	50.3	50.2	26.7	1.1	4.4	11.5	2.23	4.06	18.2
	1	52.0	54.6	27.9	1.1	4.5	8.3	2.08	2.47	16.8
	2	51.7	55.7	28.6	1.1	4.5	5.6	1.44	2.34	14.0
	3	52.0	56.5	29.5	1.1	4.5	3.4	1.0	1.90	12.3
With honey	0	50.3	49.9	30.5	1.1	4.4	11.3	2.63	5.11	16.1
	1	51.3	54.8	31.5	1.1	4.4	8.3	2.44	3.33	14.2
	2	51.0	55.1	32.1	1.1	4.5	5.3	1.53	2.48	12.6
	3	51.3	55.4	32.8	1.1	4.5	3.4	1.15	2.01	10.4
C.D. at 5%	Treatment	N.S.	0.28	0.50	N.S.	N.S.	N.S.	0.10	0.30	0.49
	Storage	0.7	0.34	0.70	N.S.	N.S.	0.33	0.14	0.43	0.59
	Treatment + storage	0.9	0.55	0.99	N.S	N.S.	0.47	0.16	0.49	0.66

ascorbic acid contents decreased. The value added squash was diluted there times and the drink was found to have good sensory attributes and nutritive value. During storage it was observed that sugar content increases slightly; acidity and TSS remained significantly unaffected; whereas chlorophyll, total carotenoids, total phenols and ascorbic acid contents decreased significantly. A significant decrease in the ascorbic acid content during storage indicate high oxidation which has to be controlled by efficient storage and packaging methods.

Changes in organoleptic evaluation of green mango-mint-tulsi squash during storage at room temperature

The formulation prepared with partial substitution of honey was more acceptable. Throughout the storage period, the squash prepared with honey scored more in flavor, taste, mouth-feel and overall acceptability than without honey. The overall acceptability did not decrease significantly during the storage period. There was some however change in the colour perceptions as indicated by lower appearance score. The cost of production without including labour, bottling and other overhead charges came out to be Rs. 34.5/L for squash without honey and Rs. 59.0/L for squash without honey.

Table 2: Changes in organoleptic evaluation of green mango-mint-tulsi squash with or without honey during storage at room temperature (35 ± 5°C)

Treatment	Storage Period Months	Color and Appearance	Flavor	Taste	Mouth-feel	Overall acceptability
Without honey	0	7.83	7.33	7.33	7.33	7.47
	1	8.0	8.0	7.67	7.67	7.83
	2	7.67	7.67	7.67	7.67	7.67
	3	7.83	7.83	7.83	7.83	7.83
With honey	0	8.00	8.33	8.33	7.83	8.13
	1	7.83	8.67	8.67	8.33	8.37
	2	7.33	8.33	8.33	8.33	8.07
	3	7.67	8.33	8.33	8.33	8.17
C.D.at5%	Treatments	N.S.	0.54	0.45	0.40	0.24
	Storage	N.S.	N.S.	N.S.	N.S.	N.S.
	Treatment+ Storage	N.S.	N.S.	0.50	0.55	N.S.

Conclusion

During storage it was observed that sugar content increases slightly; acidity and TSS remained significantly unaffected; whereas chlorophyll, total carotenoids, total phenols and ascorbic acid contents decreased. The formulation prepared with partial substitution of honey was more acceptable. Throughout the storage period, the squash prepared with honey scored more in flavor, taste, mouth-feel and overall acceptability than without honey. The cost of production without including labour, bottling and other overhead charges came out to be Rs. 34.5/L for squash without honey and Rs. 59.0/L for squash without honey.

References

1. AOAC (2005). Official Methods of Analysis.Association of Official Analytical Chemists. Washington, D.C. 18th Edition.
2. D'Amato, D. Sinigaglia, M. and Corbo, M.R. (2010). Use of chitosan honey and pineapple juice as filling liquids for increasing the microbiological shelf life of a fruit-based salad. International journal of Food Science and Technology, 45(5):1033-1041.
3. Ranganna. S. (2003). Handbook of Analysis and Quality Control of Fruit and Vegetable Products, 2nd Ed., Tata McGraw Hill Publications. Co., New Delhi, India.

CHAPTER 36

Potential Application of Starch Extracted from Sweet Flag (*Acorus calamus*) in India

Shumaila Jan, Kulsum Jan and D.C. Saxena

Abstract

Starch is a naturally occurring biodegradable polymer and readily available from agriculture. It is widely used in food and non-food applications and isolated from crops such as maize, potatoes, tapioca, rice and wheat. However, in recent years substantial progress has been made on the molecular structure of cereal and tuber starches and their impact on functionality, the following is a subjective summary that presents status of knowledge on isolation of starch from non-conventional source like Acorus calamus (sweet flag). The focus is on potential application of the source as food and non-food derivatives. Extensive work has been carried out on agronomic and phenotypic properties for most tropical crops, but relatively little study has been carried out on the properties which are of direct technical and economic interest to competitive food and non-food industry. Sweet flag (Acorus calamus L.) is an emergent macrophyte, originating in Asia, commonly found in European wetlands. Rhizome is the most important storage organ in Acorus. It contains considerable amounts of carbohydrates and nitrogen-rich compound. This species is growing to 6 feet, with long lanceolate leaves ¾-inch wide with pronounced midrib, aromatic, with stout rhizomes. The major proportion of total non-structural carbohydrates in rhizomes is starch (91-93% and 51-64% in rhizomes and roots, respectively). However, amount of starch varies among the organs as well. Rhizomes contain the highest amount of starch followed by branched roots and finally by unbranched roots. Rhizomes comprises single to compound grains, with clusters present in high frequencies. Non-conventional food resources and their value addition have attracted attention in the recent years for their potential use as functional ingredients in food formulations, besides food they have also been used for non-food applications. So there is tremendous potential for the profitable commercial use of these tropical starches and exploring them for further use.

Keywords: Acorus calamus, Non-conventional food sources, Rhizome, Sweet flag.

Introduction

Acorus calamus Linn., commonly known as Sweet flag belongs to a family Araceae. It is a well-known medicinal plant for centuries, is semi-aquatic herb with creeping rhizomes and sword shaped long leaves, found mostly in marshy places, river banks and lake (Rastogi et al., 1980). India, Pakistan and Ceylon, Acorus calamus is called bach, ghorabach, and safed-bach in Hindi, themepru in Assameese, bach in Bengali, ganghilovaj in Gujarati, bajegida in Kannada, vashampe in Malayalam, vek-hand in Marathi, bhuta-nashini in Sanskrit, vasambu in Tamil, and vasa in Telugu (Jain,1968). It is an alien plant species which forms a shallow and compact network of rhizomes thus prevents the existence of many more other species. Many thick adventitious roots grow out of the ventral side of the rhizome and anchoring the plant deeply in the substrate. Because of its biomass, the rhizome is the most important storage organ in Acorus. It contains considerable amounts of carbohydrates mainly starch and nitrogen rich compounds like Arginine and Asparagine (Vojtiskova et al., 2004). It's also known as Vacha and is one of the oldest herbs in Indian traditional system and has been used as in number of applications like as laxative, diuretic, antioxidant, antimicrobial. It is up to 6 feet tall, aromatic, sword shaped leaves bearing small yellow/green flowers and branched rhizome. It is widely distributed throughout India and Ceylon, in marshes, wild or cultivated, ascending the Himalayas up to 6000 feet in Sikkim, marshy tracts of Kashmir and Sirmoor in Manipur and Naga Hills (Meena et al., 2010 and Bisht, et al., 2010).

Plant Description

This species grows upto 6 feet, with long lanceolate leaves ¾-inch wide with pronounced midrib, aromatic, with stout rhizomes. Spadix is stout about 4 inches long. It is up to 2.5 cm thick as shown in Fig-1, internally the rhizome is whitish pink in color, purplish-brown to light brown externally and pleasantly aromatic, smelling of citrus, although it has a bitter taste.

The inflorescence consists of a leaf-like spathe and a spike-like spadix, produced from the middle of the spathe, which are densely covered with yellow and green flowers. Its leaf scars are brown white and spongy and it possess slight slender roots.

Leaves are thick, erect and are very similar in appearance to the iris but edges are crimped (Fig. 1). These leaves are few in number and size ranges between 0.7 and 1.7 cm wide, with average of 1 cm and distichously alternate (Raja et al., 2009) The sympodial leaf of A. calamus is somewhat shorter than the vegetative leaves. The leaves are free, alternate, green and wavy as shown in Fig 2 These are cylindrical in shape and green in colour and contains multitude of rounded spikes covering it. The fruits are found to be small and berry like with few seeds (Balakumbahan et al., 2010).

Root consists of long creeping roots which spread out just below the surface of the soil (Fig 1,3).Very rarely, flower is grown in this plant, if grown, then it is 3- 8cm long, cylindrical in shape, greenish brown in colour and covered with the multitude of rounded spikes. The flowers are small, sessile and densely packed and 5-10 cm of spadix. The spadix, during expansion, can reach to a length ranging between 4.9 and 8.9 cm. Flowers grows wild in marshy places up to 2000 m altitude in the Himalayas, Manipur, Naga Hills (Fig-4) and in some parts of South India from early to late summer, depending on the latitude (Paithankar et al., 2011).

Fig. 1:

Fig. 2:

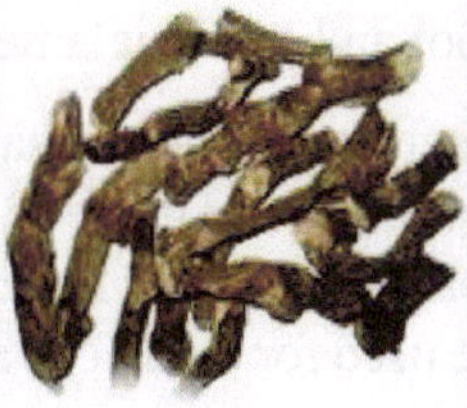

Fig. 3:

Fig. 4:

Distribution

Acorus calamus is a native plant of eastern countries and also it is indigenous to the marshes of the mountains of India. It is cultivated throughout India, ascending to an altitude of about 2200 metres. It is also found in marshy tracts of Kashmir, Manipur, in Naga Hills and in Shirmaur (Himachal Pradesh). It is regularly cultivated in the Koratagere taluka of Karnataka state in peninsular India (Raja et al., 2009).

Medicinal Importance

The roots and rhizomes are used medicinally since ancient times. Acorus calamus is a plant which is rich in alkaloids, phenolics and flavonoids and acts as a potential source of compounds comprising beneficial biological activities (Neha et al., 2012). Most of the present- day western medicines are based on the traditional medicinal plants of European, Mediterranean, and Arabic origin. For drug development, the variety of plants in use around the world may very well represent an enormous treasure (Verpoorte et al., 2000). Acorus calamus possesses an antispasmodic, carminative and anthelmintic property and is also used for the treatment of epilepsy, bronchial catarrh, fever and glandular and abdominal tumours, mental ailments, chronic diarrhoea and dysentery, (Chopra et al., 2002). They are also employed for kidney and liver troubles, rheumatism, sinusitis, eczema and anti-cellular activities (Nadkarni et al., 1982). The main active ingredient of sweet flag rhizomes is held to be their essential oil, whose composition includes various terpenes, sesquiterpenes, and their oxygen-containing derivatives. The essential oil of sweet flag had positive results and acting against microorganisms thus are anti-microbial agents.

The plant has been extensively investigated and a number of chemical constituents from the rhizomes, leave and roots of the plant have previously reported which includes β- Asarone, α- Asarone, elemicine, cis-isoelemicine, cis and trans isoeugenol and their methyl ethers, camphene, P-cymene, bgurjunene, a-selinene, b-cadinene, camphor,terpinen-4-ol, aterpineol and a-calacorene, acorone, acrenone, acoragermacrone, 2-deca –4,7 dienol, shyobunones, linalool and preisocalamendiol are also present.

It is cultivated in many countries to satisfy the demand for its essential oil which is used for flavouring and in the perfumery and pharmaceutical industries (Nadkarni et al., 1982). It is used for fevers, stomach cramps and cholic diseases. Their rhizomes were used for toothache and powdered rhizome for congestion. The rhizome part is also used to treat several diseases like asthma and bronchitis and as sedative. Native tribes treated cough by making a decoction of the plant as a carminative and also for cholic diseases (Modhumita et al., 2006).

Starch

Assemblage (rhizome) is primarily comprised of single to compound grains in which clusters are also present in high frequencies. Grains tend to be quite small, 6 × 4.5 μm, and are highly variable in morphology that is ranging from irregular sub-rounded to sub-angular domes, ovals, kidneys, and elongates. Small clusters having two granules can be recognized when viewed under cross-polarized light as shown in Figs.5 and 6. Forty percent of the rhizome exhibited extinction cross distortion either distally or medially. These characteristics conceal portions of the grain, resulting in weak birefringence (Reichert et al., 1913).

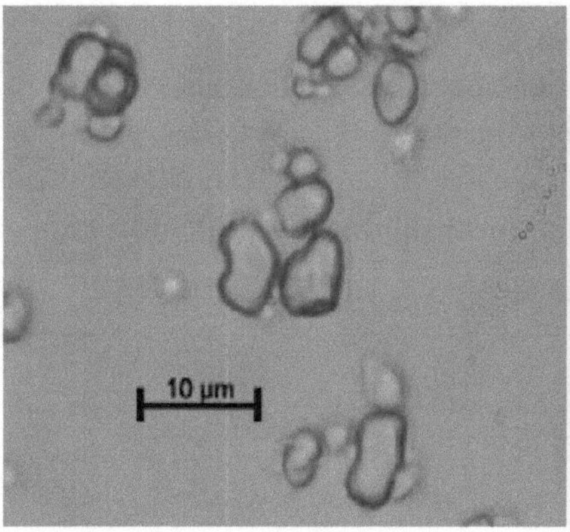

Fig. 5

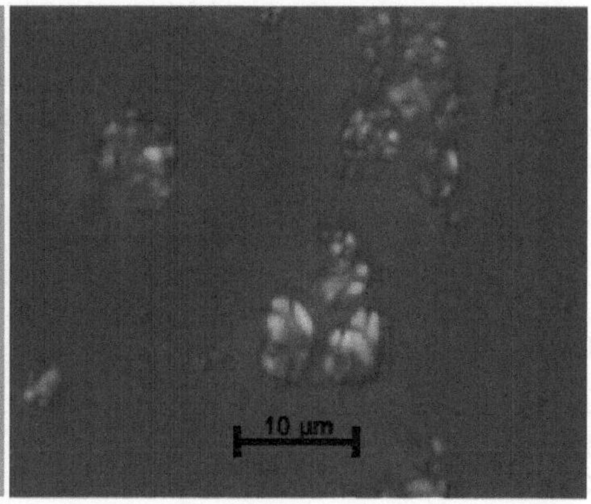

Fig. 6

Nutritional Composition

A high amount of storage compounds is typical for over-wintering organs of perennial plants. Wetland plants make great demands on the amounts of storage compounds in their over-wintering organs due to the prolonged period of oxygen deficiency common in the flooded habitats (Brandle, 1991). The rhizomes and/ or roots of wetland plants mainly store different types of non-structural carbohydrates and lower amounts of proteins and amino acids; lipids are rarely stored (Haldemann and Brandle, 1986; Brandle, 1991).

Sweet flag (*Acorus calamus* L.) is an emergent macrophyte, originating in Asia (Weber and Brandle, 1996), commonly colonising European wetlands (Hejny and Husak, 1978). Rhizome is the most important storage organ in Acorus. It contains considerable amounts of carbohydrates and nitrogen-rich compounds (Weber and Brandle, 1994, 1996). The abundance of storage carbohydrates together with other metabolic features is connected with tolerance of wetland plant species to anoxia (Barclay and Crawford, 1982, 1983).

Sweet flag is high in potassium, which has suggested its use in the relief of illnesses which have been attributed to potassium deficiency, such as asthma, hay fever, hiccups, and muscular dystrophy. The phosphorus and potassium in the plant also aid in pain relief (Bethel, 1972). Harris (1972) remarks that the rhizome stores unused carotene, which is converted to vitamin A and can be used to treat eye problems. The sulfur content in the rhizome has been suggested as favourable for bone growth, blood clotting, muscle metabolism, and to counteract toxic substances in the body (Lust 1974) (Table 1). Describes the biochemical composition of acorus calamus leaves and rhizome tissues

Active Constituents

Phytochemically it has been reported that there are presence of glycosides, flavonoids, saponins, tannins, polyphenolic compounds, mucilage, volatile oil and bitter principle in *A. calamus* (Muthuraman et al., 2011). The plant has been also reported for the presence of glucoside, alka- loid and essential oil containing calamen, clamenol, calameon, asarone and sesquiterpenes.

Table 1: Biochemical composition of acorus calamus leaves and rhizome tissues

Biochemical Parameters	Leaves		Rhizomes	
	mg/g extract	mg% fresh wt	mg/g extract	mg% fresh wt
Lipid				
Glyco-lipids	46.2	28.8	99.7	62.3
Sterols	72.7	47.9	25.0	15.6
Free fatty acids	5.9	3.7	3.3	2.1
Carbohydrates				
Total sugar	29.2	18.2	22.5	14.1
Total reducing sugars	9.8	6.1	10.3	6.5
Sucrose	10.5	6.6	8.5	5.3
Fructose	7.0	4.4	10.5	6.6

(*Sources*: Ganjewala et al., 2011)

It also contains a better glycoside that is a corine along with eugenol, pinene and camphene (Jabbar et al., 2010). The plant has been extensively investigated and a number of chemical constituents from the rhizomes, leave and roots of the plant have previously reported which includes β- Asarone, α- Asarone, elemicine, cisisoelemicine, cis and trans isoeugenol and their methyl ethers, camphene, P-cymene, bgurjunene, α-selinene, β-cadinene, camphor,terpinen-4-ol, aterpineol and a calacorene, acorone, acrenone, acoragermacrone, 2-deca –4,7 dienol, shyobunones, linalool and preisocalamendiol are also present. Acoradin, galangin, 2,4,5-trimethoxy benzaldehyde, 2,5 dimethoxy benzoquinone, calamendiol, spathulenol and sitosterol have been isolated from Acorus calamus (Jayaraman et al., 2010).

Alcoholic extracts of the triploid and diploid of *A. calamus* were characterized by a higher percentage of β-asarone which was the main compound, followed by higher percentages of camphene in case of triploid and many other constituents are also present in both triploid and diploid which are shown in Table 2 and 3 (Cinzia et al., 2005).

Table 2: Active constituents in alcoholic extracts of triploid A. calamus

Active constituents	Percentage (%)
b-asarone	11
Camphene	2.27
e-b-ocimene	3.28
Camphor	1.54
Calarene	1.42
a-selinene	5.02
s-cadinol	2.00

(*Sources*: Cinzia et al., 2005).

Table 3: Active constituents in alcoholic extracts of diploid A. calamus

Active constituents	Percentage (%)
Isoshyobunone	8.62
Bsesquiphellandrene	3.28
Preiso calamendiol	22.81
Acorone	26.33

(*Sources*: Cinzia et al., 2005).

Utilization

Though there is a limited literature available on the application and utilization of starch extracted from the *A. calamus* but there are other utilizations of the whole plant parts which include uses of leaves, root (rhizome) and stem. In Asia, Sweet flag has been used for at least the last 2000 years. The ancient peoples of China used it to lessen swelling and for constipation. In Ayurvedic medicinal practice India, the rhizomes have been used to cure several diseases like fever, asthma and bronchitis, and as a sedative. Native tribes used it to treat a cough, made a decoction as a carminative and as an infusion for cholic diseases.

In Western herbal medicine the herb is chiefly employed for digestive problems such as gas, bloating, colic disorder, and poor digestive function. Calamus helps distended and uncomfortable stomachs and headaches associated with weak digestion. Small amounts are thought to reduce stomach acidity, while larger doses increase deficient acid production, it is a good sedative so that the extract is used for epilepsy, insanity and as a tranquillizer along with valeriana jatamansi and nardostacys grandiflora.

It is an ingredient of any Ayurvedic preparation "Brahmi Bati" (Budhivardhar) which is indicated in epilepsy, coma, and hysteria and in cases of mental retardation; the same uses are prescribed for an Acorus containing Unani drug Ma'jun Baladur" (Balakumbahan et al., 2010). It is also used in the Philippines for rheumatism and memory problems. In Korea, it is an ingredient in a type of moonshine called Immortals' Booze.

Research in China has shown the essential oil in this rhizome to be sedative and neuroprotectant Other virtues of this plant include its mature leaves, which act as an insect repellent when cut up and stored in dry foods. It shows strong effect against filamentous fungi like Trichophytum rubrum and Microsporum gypseum, moderate inhibitory effect against yeast and low against bacteria2ethanolic extract traditionally used for antidiabetis, immuno suppressive, antiproliferative and hypolipidimic (Meena et al., 2010). Presently, the fragrant oils obtained from the rhizome are used medicinally, as flavoring in al-coholic beverages, as fragrant essences in per-fumes and sacred oils, and for insecticidal prop-erties (Motley, 1994)

Conclusion

A. calamus, the versatile plant is the unique source of various types of compounds having wide biological activities. It is concluded here that the rhizome part of the plant can be utilized for the extraction of the starch and is considered a non-conventional source for the extraction of starch. Besides this it has also phytochemical constituents such as α-, β- and γ- asarone, sequesterpenes and acorenone has many biological activities. These compounds are active in antimicrobial, anti-inflammatory, antioxidant, antidiarrheal, antiulcer, antispasmodic, immuno-suppressant and nitrogen inhibitor activity. They are also used in the treatment of simple diseases such as stomach cramps, toothache, colic, fever, throat irritation, and cough and also in the treatment of the severe diseases like nephropathy, chronic diarrhoea, tumors and epilepsy. Thus, the plant plays a significant role in the prevention and treatment of various diseases.

References

1. Balakumbahan, R. Rajamani, K. and Kumanan, K. (2010). Acorus calamus: An overview. Journal of Medicinal Plants Research, 4(25):2740-2745.
2. Barclay, A. M. and Crawford, R. M. M. (1982). Plant-growth and survival under strict anaerobiosis. Journal of Experimental Botany, 33:541–549.
3. Barclay, A. M. and Crawford, R. M. M. (1983). The effect of anaerobiosis on carbohydrate-levels in storage tissues of wetland plants. Annals of Botany, 51:255–259.
4. Bethel, M. (1972). The healing power of herbs. Wil-shire Book Co., Hollywood, CA.

5. Bisht, A. S. Chauhan, M. Dimri, A. Joshi, A. and Ali, M. (2010). Pharmacognostical Evaluation of Leaves of *Acorus calamus* Linn. International Journal of Pharmacy and Pharmaceutical Sciences, 2(2):130-31.
6. Brandle, R. (1991). Flooding resistance of rhizomatous amphibious plants. In Jackson, M.B., D.D. Davies and H. Lambers (eds), Plant Life under Oxygen Deprivation. SPB Academic Publishing, The Hague, Netherlands, 35–46.
7. Chopra, R.N. Nayar, S.L. and Chopra, I.C. (2002). Glossary of Indian medicinal plants. NISCIR, CSIR, Delhi
8. Cinzia, M. Chiara, M. Simone, B. Giovanni, D. and Massimo E. (2005). Maffei. Identification of an EcoRI restriction site for a rapid and precise determination of b-asarone-free *Acorus calamus* cytotypes. Phytochemistry, 66:507–514.
9. Ganjewala, D. Asha D.S. and Ashwani, K.S. (2011). Tissue specific variation in biochemical compositions of *Acorus calamus* (L.) leaves and rhizomes. International Journal of Plant Biology. Volume, 2:4.
10. Haldemann, C. and Brandle, R. (1986). Seasonal-variations of reserves and of fermentation processes in wetland plant rhizomes at the natural site. Flora, 178:307–313.
11. Harris, B.C. (1972). The complete herbal. Barre Pub-lishers, Barre, MA.
12. Hejny, S. and Husak, S. (1978). Higher plant communities. In Dykyjova, D. and J. Kvet (eds), Pond Littoral Ecosystems-Structure and Functioning. Springer-Verlag, Berlin Heidelberg New York, 23–64.
13. Jabbar, A. and Gilania. A.H. (2010). Bronchodilatory effect of Acorus calamus (Linn.) is mediated through multiple pathways,c, Journal of Ethnopharmacology, 131,471–477,
14. Jain, S. K. (1968). Medicinal plants. National Book Trust, New Delhi, India.
15. Jayaraman, R. Anitha, T. and Joshi, V.D. (2010). Analgesic and Anticonvulsant Effects of *Acorus calamus* Roots In Mice International Journal of PharmTechnology Research, 2(1):552-555.
16. Lust, J.B. (1974). The herb book. Bantam Books, Toronto.
17. Meena, A.K. Rao, M.M. Singh, A. and Kumari, S. (2010). Physicochemical and Preliminary Phytochemical Studies on the Rhizome of *Acorus Calamus* Linn. International Journal of Pharmacy and Pharmaceutical Sciences, 2(2).
18. Modhumita, G. (2006). Antifungal Properties of Haem Peroxidase from *Acorus calamus*. Annals of Botany, 98:1145–1153.
19. Motley, T.J. (1994). The Ethnobotany of Sweet Flag, *Acorus calamus* (Araceae). Economic Botany, Vol. 48, No. 4, pp. 397-412.
20. Muthuraman, A. and Singh, N. (2011). Attenuating effect of *Acorus calamus* extract in chronic constriction injury induced neuropathic pain in rats: an evidence of anti-oxidative, anti-inflammatory, neuroprotective and calcium inhibitory effects, BMC Complementary and Alternative Medicine 1-14.
21. Nadkarni, K.M. (1982). Indian Materia Medica, 3rd Edn., Vol. I, Popular Prakashan Pvt. Ltd. Bombay, 12:11-12.
22. Neha, B., Honey, J. Ranjan, B. and Mukesh, B. (2012). Pharmacognostical and preliminary phytochemical investigation of *Acorus calamus* Linn. Journal of Pharmacy, 2(1):39-42.
23. Paithankar, V.V. Belsare, S. L. Charde, R.M. and Vyas, J.V. (2011). *Acorus calamus*: an overview. IJBR 2:10, 518 529.
24. Raja, A. E. Vijayalakshmi, M. and Devalarao, G. (2009). *Acorus calamus* Linn.: chemistry and biology. Research Journal of Pharmacy and Technology, 2(2):256-261.
25. Rastogi, R. P. Mehrotra, B. N. Sinha, S. and Seth, R. (2001). Compendium of Indian medicinal plants (Vol. 3). Central Drug Research Institute and Publications and Information Directorate, New Delhi.

26. Reichert, E.T. (1913). The Differentiation and Specificity of Starches in Relation to Genera, Species, etc. Carnegie Institute, Washington, D.C.
27. Verpoorte, R. (2000). Pharmacognosy in the new millennium: lead finding and biotechnology. Journal of pharmacy and pharmacology, 52(3):253-262.
28. Vojtiskova, L. Munzarova, E. Votrubova, O. Rihova, A. and Juricova B. (2004). Growth and biomass allocation of sweet flag (*Acorus calamus* L.) under different nutrient conditions. Hydrobiologia, 518:9-22.
29. Weber, M. and Brandle, R. (1996). Some aspects of the extreme anoxia tolerance of the sweet flag Acorus calamus L. Folia Geobotanica and Phytotaxonomica, 31:37–46.
30. Weber, M. and Brandle, R. (1994). Dynamics of nitrogen-rich compounds in roots, rhizomes, and leaves of the sweet flag (*Acorus calamus* L) at its natural site. Flora, 189: 63-68.

CHAPTER 37

Preparation of Mouth Freshener (Anardana Goli) From Pomegranate (*Punica granatum* L)

Walunj Amol, Patil Priyanka, Deshmukh Chaitanya and Chothe Deepak

Abstract

Pomegranates are usually consumed as fresh fruit, with only a limited quantity being processed into products like juice, syrup, beverages, wine, confectionary and anardana. The special structure of fruit is one of main factors which limit its industrial processing. Knowing the importance of health benefits of pomegranate and all related aspects, the present study was carried out in order to utilize pomegranate fruit for preparation of pomegranate mouth freshener. Arils of two cultivars like Ganesh and Araktaweredried by two drying methods viz., sun drying (SD) and cabinet drying (CD).Nutritive and palatable mouth freshener was prepared from anardana of two different cultivars as Ganesh (G) and Arakta (A). Production cost of 100g pomegranate mouth freshener is about Rs.55/. It was noticed that moisture content, acidity and ascorbic acid content decreased at the end of storage.

Introduction

Pomegranate (*Punica granatum* L.) is commonly known as 'Anar', belongs to plant family Punicaceae, which includes one genus Punica and two species namely,*Punica protopunica* and *Punica granatum* which is a fruit of tropical and subtropical regions. The most important growing regions are Egypt, China,

Afghanistan, Pakistan, Bangladesh, Iran, Iraq, India, Burma, Saudi Arabia, Malaysia, China, Japan and Russia. Pomegranate is a rich source of antioxidant polyphenols, tannins and anthocyanins.

To utilize low quality pomegranate fruit it needs to be preserved and processed into different products like juice, squash, syrup, jelly, wine, anardana and canned beverages. The pomegranate varieties such as Arakta being a large sized have rind deep red, arils dark blood red, hard seeds, slightly bitter juice. Other variety such as *Ganesh* is a medium sized have soft seeds; pinkish flesh and juice with agreeable taste.

Any mouth freshener is a mixture of nuts, seeds, herbs, and spices which are served after meals in India. Spices are valuable source of a number of natural phytochemicals, which are essential for good health. On the basis of the fruit characteristics the present work was carried out to prepare and assess nutritional quality of anardana, to prepare and assess nutritional quality of mouth freshener and finally to study the effect of storage on quality of mouth freshener.

Materials and Methods

Selection of Raw Material

Two varieties of locally available pomegranate fruits Cv. Ganesh and Arakta were selected for the study. Other ingredients include spices like cardamom, clove and fennel. The reagents used in the testing were of AR grade. All the tests wre carried out in triplicate and the mean of the three values were recorded for statistical analysis.

Preparation Method for Mouth Freshener

The mouth freshener was prepared as per standardized recipe by mixing cardamom, clove, fennel and black salt at an acceptable level as shown in (Figure 1). Pomegranate mouth freshener was prepared after pressing specific quantity i.e. 0.5g of powder, between thumb and finger followed by rolling over palm or by manually operated tablet machine. The prepared mouth freshener looks like flavored balls/tablets. The sample of *Ganesh* variety prepared by cabinet drying scored higher for overall acceptability among other samples may be due to colour of anardana of Ganesh cultivar compared to Arakta cultivar. The pomegranate mouth freshener prepared from Ganesh and Arakta cultivar by using sun drying and cabinet drying contain moisture 14.20 percent, 10.20 percent, 15.00 percent and 10.50 percent respectively.

Aril Powder

↓

Mixing of spices as per standardized recipe

↓

Pressing and rolling with palm/ by hand operated machine

↓

Pomegranate mouth freshener/ tablets

↓

Packing in plastic bottle

Fig. 1: Preparation of Pomegranate Mouth Freshener

Fig. 2: Pomegranate Cv. Ganesh Pomegranate Cv. Arakta

Chemical Analysis of The Product

Anardana and Mouth Freshener were analyzed for, moisture, Acidity, Ascorbic acid (mg/100g), and Total and reducing sugars as per the standard method of AOAC, (2005).

Drying of Arils

Two methods one natural (Sun drying for 3 days) and other mechanical drying using cabinet drying at a temperature of 60⁰C for 8hr were used for dying of arils. In the sun drying a uniform layes was spread under the sun and the arils were turned intermittently to carry out the uniform drying. In the cabinet drier as well as in the sun drying when the product was found dried as required drying was stopped and the products were checked for any discoloration, sorted out and packed in air tight container for further study.

Storage

The products were stored in plastic bottles at ambient temperature. Analysis for chemical constituents was carried out and sensory attributes were checked using the semi-trained panel of the department after familiarizing them to the sensory and mouth-feel attributes of flavour, colour, kinesthetic. The products were analyzed after 15days intervals for two months.

Statistical analysis

Experimental data were analyzed by analysis of variance (ANOVA), and means were compared using Duncan's multiple range tests with significance defined at P0.05. All statistical analysis was performed using commercial statistical package SPSS (16.0, Chicago, IL, USA).

Results and Discussions

The chemical composition of anardana obtained form two varieties of Ganesh and Arakta are shown in Table -1. There was no significant difference between two varieties for the chemical constituents analyzed except for Non- reducing sugars which were higher in Arakta.

Table 1: Chemical composition of Anardana[1]

Sr.No	Parameter	Ganesh	Arakta
1.	Moisture (%)	10.20	10.50
2.	Acidity (%)	1.3	1.5
3.	Ascorbic acid (mg/100g)	6.70	7.00
4.	Total sugars (%)	30.20	30.50
5.	Reducing sugars (%)	29.00	29.52
6.	Non-reducing sugars (%)	1.20^{b}	0.98^{a}
7.	Protein (%)	1.3	1.4
8.	Fat (%)	0.91	0.92

[1]Mean of three were taken.

Mean with different small letters differ significantly (pd"0.05) in the row.

Table 2: Chemical composition of Mouth Freshener[1]

Sr. No.	Parameter	Ganesh		Arakta	
		Sun Drying	Cabinet Drying	Sun Drying	Cabinet Drying
1.	Moisture (%)	14.20[a]	10.20[b]	15.00[a]	10.50[b]
2.	Acidity (%)	1.40	1.30	1.60	1.50
3.	Ascorbic acid (mg/100g)	6.0[b]	6.7[a]	6.5[b]	7.0[a]
4.	Total sugars (%)	30.00	30.20	30.25	30.50
5.	Reducing Sugars (%)	28.50[b]	29.00[a]	29.00[b]	29.52[a]
6	Non-reducing sugar (%)	1.50[a]	1.20[b]	1.20[a]	0.98[b]

[1]Mean of three were taken.
Mean with different small letters differ significantly (pd"0.05) in the row.

Effect of drying on the chemical composition of mouth freshener

Effect of drying methods on the chemical composition of mouth freshener has been presented in Table 2. It is observed that the moisture content of mouth fresher prepared from both the varieties was significantly reduced by mechanical drying as compared to sun drying. The ascorbic acid content was also higher along with reducing sugar contents in the cabinet dried mouth freshener as compared to the sun dried product; however the non-reducing contents were higher significantly in sun dried mouth freshener. The variation in the constituent is due to long drying cycle in case of sun drying which cause oxidization of ascorbic acid content and low drying rate in case of sun drying

Effect of storage at ambient temperature on quality of mouth freshener

Effect of storage at ambient temperature on quality of mouth freshener is shown in Table 3. There was found no significant difference in the chemical constituent of store mouth freshener at ambient storage temperature for 15 to 60 days. It is evident that the ambient temperature is suitable for mouth freshener stored if prepared under hygienic condition and stored in air tight containers.

Table 3: Effect of storage (ambient) temperature on quality of mouth freshener

Sr. No.	Parameter	Fresh	Storage Period in Days			
			15th	30th	45th	60th
1.	Moisture (%)	10.20	10.15	10.10	10.00	9.80
2.	Acidity (%)	1.3	1.28	1.23	1.20	1.18
3.	Ascorbic acid (mg/100g)	6.70	6.50	6.20	6.00	5.80
4.	Total Sugars (%)	30.20	30.22	30.25	30.28	30.30
5.	Reducing Sugars (%)	29.00	29.10	29.25	29.32	29.40
6.	Non-reducing Sugars (%)	1.20	1.12	1.00	0.96	0.90

[1]Mean of three were taken. Mean with different small letters differ significantly (pd"0.05) in the row.

Fig. 3: Dried Arils of Pomegranate cv. Ganesh and Dried Arils of Pomegranate cv. Arakta

Fig. 4: Pomegranate Mouth Freshner cv. Ganesh and Pomegranate Mouth Freshner cv. Arakta

Conclusion

Pomegranate cultivar Ganesh was suitable for preparation of highly nutritive and palatable mouth freshener. During the present research, attempts have been made to utilize anardana prepared by cabinet drying as well as sun drying as a mouth freshener. Production cost of 100g pomegranate mouth freshener is about Rs.55/-. It was noticed that moisture content, acidity and ascorbic acid content decreased at the end of storage. The cheap mouth freshener containing natural ingredients and containing functional ingredients can be prepared used safely as a mouth freshener.

References

1. Berry, S. K. (2005). Pomegranate: A Miracle Fruit. Review article. Indian Food Packer. Nov.-Dec.189-194.
2. Singh, D. Chaudhary, M. Meena, M. L. Wangchu, L. and Dayal, H. (2009). Drying of pomegranate seeds (anardana) under different conditions. ISHS Acta horticulturae 890:II International Symposium on Pomegranate and Minor - including Mediterranean – fruits.

3. Garande, V.K. Masalkar, S.D. Gaikwad, R.S. and Patil, R.S. (2004). Studies on preparation and storage of anardana. Agriculture Science Digest, 24(4):283-285.
4. Kapil, V. (1991) Manufacture of lactose: Effect of processing parameter on yield and purity. Journal of Food Science, 23(3):167.
5. Ranganna. S. (2003) Handbook of Analysis and Quality Control of Fruit and Vegetable Products, 2nd Ed., Tata McGraw Hill Publications. Co., New Delhi, India.

CHAPTER 38

Nutritional and Nutraceutical Potential of Licorice (*Glycyrrhiza glabra*)

Ishrat Majid, Khan Nadiya Jan, Gulzar Ahmad Nayik and Vikas Nanda

Abstract

Due to rapidly increasing knowledge on nutrition and medicine, concepts about food and health has dramatically changed and resulted in consumption of nutraceuticals and natural plant foods. Use of nutritional therapy and phytotherapy have become progressively popular to improve health, and to prevent and treat diseases. Plants have long history of their uses as ingredients of human nutrition and folk medicine, because of availability of huge array of high-value components and multiple-activity bio-actives. "Licorice" means "sweet root" in Greek is a plant of ancient origin, indigenous to Southern Europe and Asia Minor. Most of the supply comes from former USSR, Spain, Turkey, Syria, Iraq and Afghanistan, with total production between 40,000 and 50,000 tonnes per annum. The chemical constituents of the roots include several bioactive compounds, such as glycyrrhizin (~16%), different sugars (up to 18%) ,flavonoids, saponoids, sterols, starches, amino acids, gums and essential oils. A large number of clinical and experimental studies reported its useful biological properties such as antioxidant, immunomodulatory, cardioprotective, anti-inflammatory, anti-viral and anticancer effects. It is useful in anaemia, gout, asthama, ulcers, sore throat, tonsillitis, diabetes, tuberculosis, flatulence, diarrhea, sexual debility, epilepsy, hyperdypsia, fever, coughs, skin diseases, eye diseases, swellings, constipation, rheumatism, acidity, leucorrhoea, bleeding, jaundice, hiccough, hoarseness, bronchitis. Recently, it is reported to have neurological properties such as antidepressant, anxiolytic, and anticonvulsant effects. The most abundant minerals present are calcium, sodium, phosphorus, potassium and iron. Licorice extracts and its principle component, glycyrrhizin, have extensive use in foods,as a natural sweetener and flavoring agent in beverage, confectionery, tobacco and pharmaceutical industry. It is in great demand globally as medicinal and nutritional supplement due to it's wide range of properties, which also indicates that in future the plant could serve as a "lead" for the development of novel agents having good efficacy in various disorders.

Keywords: Nutritional therapy, Phytotherapy, Nutraceutical

Introduction

Plants are the most important sources of human nutrition and medicine. Moreover, concept about food and health has changed a lot due to rapidly increasing knowledge on nutrition and medicine and resulted in a revolution in them. Recently, new concept of nutritional therapy and phytotherapy has emerged that has quickly and widely spread. In order to improve health, and to prevent and treat diseases, consumption of nutraceuticals and natural plant foods has been strongly recommended and thus use of nutritional therapy and phytotherapy have become progressively popular (Patil, 2011). Glycyrrhiza glabra L., a member of the Fabaceae family, is the source of the novel food ingredient is and has a long history of human consumption (EFSA Panel on Dietetic Products, Nutrition and Allergies, 2011).

It has been described as 'the grandfather of herbs' and is one of the most commonly used natural medicines in the world. In fact, the name "licorice" means "sweet root" in Greek (glukus "sweet" and rhiza "root"),(Asl and Hosseinzadeh, 2008) and is used as natural sweetener. (Wang et al., 2000).It is used both as flavouring herb and also as a medicine. It's commercial source is the sweet root of various species of *Glycyrrhiza*, a genus which contains about fourteen species (Jatav et al., 2011), and is a herbaceous perennial native to the Mediterranean region, the Near East, Central Asia, as well as Western Siberia. The genus *Glycyrrhiza* is widely distributed in Eurasia, extending to Australia, North America, and the temperate regions of South America. In Eurasia, around 20 species occur, among them five are confined to Europe. Athough often only of local importance, the roots of many species are sources of liquorice.

In trade, geographical origin is used to distinguish between various types of liquorice: Spanish, Russian, Persian, and Mongolian liquorice, but not to plant species from which the roots are derived. Other highly exploited species are: G. uralensis, that extends from Western to Eastern Siberia, as well as across Central Asia to Mongolia; G. echinata with an area extending from the Balkans across Asia Minor to South-Eastern Russia and Western Siberia; and

G. pallidiflora native to the Far East and China.The plant is cultivated in many instances, e.g. in Italy, Southern France and Spain,and also in Central Asia, Australia, and Brazil (Mansfeld, 1986). It is available in India, Spain, Italy, Turkey, Israel, Syria, Iran, China and Russia where it is commonly called as licorice (Karami et al., 2013). Spain, Turkey, Syria, Iraq Afghanistan and former USSR are the major suppliers. There is only limited area under cultivation and hence a major portion of the commercial supply comes from wild sources. Total production is between 40,000 and 50,000 tonnes per annum (Husain, 1991). It is commonly known as Yashti-madhuh. Madhuka (Sanskrit) Jashtimadhu, Jaishbomodhu (Bengali) Jethimadhu (Gujarat) Jothi-madh, Mulhatti (Hindi) Yastimadhuka, atimaddhura (Kannada) Iratimadhuram (Malayalam) Jeshtamadha (Marathi) Jatimadhu (Oriya) Atimaduram (Tamil) Atimadhuranu, Yashtimadhukam (Telugu) Licorice, Liquorice, Sweet wood (English) (Jatav et al., 2011). This perennial woody herb produce spikes of small pea-like flowers in summer, has dark green leaflets and grows to a height of 1.5 m. It's frost tender foliage dies down in winter before emerging the following spring (Karami et al., 2013). Either peeled or unpeeled stolons (Jatav et al. 2011), and straight pieces of wrinkled fibrous wood (dried roots) of the plant are used, because they are valuable sources of useful chemicals, such as secondary metabolites and enzymes, used by the food and pharmaceutical industries (Karami et al 2013). The extract finds ample use in foods, tobacco and in both traditional and herbal medicine. (Isbrucker et al., 2006).In Western and Eastern countries, it has been used medicinally for more than 4000 years (Saleem et al., 2011).In US there is a high level of use of licorice and glycyrrhizin with an estimated consumption of 0.027–3.6 mg glycyrrhizin/kg/day. Both of these products have been approved by most national and supranational regulatory agencies for use in foods (Isbrucker et al., 2006).

(Roots)

(Plant)

Phytochemistry

Licorice contains sucrose (2.4 to 6.5 per cent), glucose (upto 3.8 per cent), bitter principles, resins, mannite, asparagines (2 to 4 per cent) and fat (0.8 per cent). Glycyrrhizin is the chief constituent of liquorice. (Vispute et al 2011). Glycyrrhizin is beneficial to body and is 50 times sweeter than sugar (Afreen et al., 2005).Root extract showed the existence of triterpenes (glycyrrhizin, glycyrrhetinic acid and liquirtic acid), flavonoids (liquirtin and formononetin) and various other substances upon chemical analysis (Farag et al., 2012). Flavonoids isolated from Glycyrrhizin species are more than 300 in number and belong to various types, including flavanones or flavanols, chalcones, isoflavans, isoflavenes, flavones or flavonols, isoflavones and isoflavanones (Qingying Zhang and Min Ya, 2009). Moreover flavonoid content of the plant is responsible for the yellow color of licorice, which includes liquiritin, isoliquiritin (a chalcone) and other compounds. Significant antioxidant activity is imparted by the isoflavones glabridin and hispaglabridins A and B (Karami et al., 2013). In both tea and infusion forms licorice was found to contain the most abundant minerals like; calcium, sodium, phosphorus, potassium and iron. However, Zinc and copper were detected as trace elements (Table-1). All licorice forms showed presence of essential and non essential amino acids and HPLC analysis of the organic acids also showed the presence of several aliphatic acids such as butyric acids, tartaric acid and acetic acids (Badr et al., 2013).

Table 1: Chemical composition of licorice

Components (%)	Raw herb
Protein	9.15
Fat	0.53
Moisture	6.80
Ash	7.70
Fiber	24.48
Carbohydrate	47.11
Silica	3.56
Elements	mg/100g
Phosphorus (P)	1720
Calcium (Ca)	78
Elements (ppm)	
Sodium (Na)	18580
Potassium (k)	7276
Iron (Fe)	1224
Zinc (Zn)	17.08
Copper (Cu)	11.01

(Contd.)

Components (%)	Raw herb
Amino acids (%)	
Aspartic	0.88
Glutamic	0.50
Threonine	0.32
Serine	0.41
Proline	1.02
Glycine	0.25
Alanine	0.51
Valine	0.44
Isoleucine	0.21
Leucine	0.38
Tyrosine	0.31
Phenylalanine	0.33
Histidine	0.22
Lysine	0.36
Arginine	0.30

Uses

Food additives

In several cases licorice is used as a flavoring additive as well as natural sweetener (Kitagawa, 2002). In addition to imparting sweet taste to foods, Glycyrrhizin has salt-softening and flavor-enhancing properties and is also heat stable.In salty foods such as soy sauce, other sauces, savory snacks, Kamaboko (boiled fish paste), Tsukudani (fish boiled in soy sauce), tsukemono (Japanese pickles), and sausages, it is also used to reduce saltiness (Liu et al., 2000).

Confectionery

Root is used to obtain licorice with water and then vacuum concentrated. All licorice candies are made from this 'block juice'.Licorice juice when combined with sugar, corn syrup, and flour forms a paste that can be moulded into any desirable shape (Davis and Morris, 1991). During18th century licorice extracts for the first time were used for flavoring confectionery products in England and was blended with sugar, flour, and other ingredients to make Pontefract cakes (Neiman, 1959). In most of the licorice candies, the taste is reinforced by aniseed oil in which actual content of licorice varies. However, the sweetness and licorice flavor are well known to consumers of licorice candies. (Asl and Hosseinzadeh, 2008). It is also used as a bakery beverage, dairy, candy and meat flavor etc (Nutr Cancer, 2001).

Liquorice products are also employed in soft drinks, herbal teas, chewing gum and its root chewed as a mouth freshener.In USA and Japan Glavonoid is on the market, as per the information available. Moreover, Glavonoid is intended by the applicant to be marketed as a food supplement and as an ingredient for fruit juices, yoghurts and yoghurt drinks up to a dose of 300 mg per day. The target population is the general adult population. (EFSA Panel on Dietetic Products, Nutrition and Allergies, 2011).

Health Benefits

Licorice has been largely studied for its biological and health benefits, but its long history and biologically active constituents are still interesting to many research groups. It has been traditionally used for respiratory, gastrointestinal, cardiovascular, and skin disorders (Hatano et al., 2000). Various clinical and experimental studies reported its useful biological properties such as antioxidant, immunomodulatory, anti-viral cardioprotective, anti-inflammatory, and anticancer effects (Hardy et al., 2012). Recently, it is reported to have neurological properties such as antidepressant, anxiolytic, and anticonvulsant effects (Cho et al., 2012). Licorice root lessens thirst in dropsies, lowers heat of the blood, helps defluctions of the breast, softens acrimonious humours, temperates salt, promotes urine, and thickens the sanguinary fluid, when too thin'. Root is also helpful for pleurisy, gravel, dysury and intense pain' (Davis and Morris, 1991).

Licorice roots are useful in anaemia, gout, asthama, sore throat, tonsillitis, flatulence, sexual debility, epilepsy, hyperdypsia, fever, coughs, skin diseases, swellings, acidity, leucorrhoea, bleeding, jaundice, hiccough, hoarseness, bronchitis, gastralgia, cephalalgia, ophthalmopathy, pharyngodnia and possess demulcent and expectorant property. It is also used in the treatment of diarrhoea, fevers, fever with delirium and anuria. Glycyrrhetinic acid in licorice imparts minerocorticoid activity and in place of corticosteroids, it is used for the treatment of rheumatoid arthritis, inflammations and Addison's disease. Glycyrrhizin on being broken down in the gut, exerts an inflammatory action similar to hydrocortisone and other corticosteroid hormones (Vispute et al., 2011). Glycyrrhizin is also proposed to make adrenal hormones such as cortisol (the body's primary stress-fighting adrenal hormone) available to the body, by preventing their breakdown (Afreen et al., 2005). Glycyrrhizin also proved useful in the treatment of chronic hepatitis and liver cirrhosis by reducing breakdown of steroids by the liver and kidneys. For relieving pain, discomfort and other symptoms caused by acrid matter in the stomach, licorice is considered as one of the best remedies.

Licorice is known to be best remedy in case of pain, discomfort and other symptoms caused by acrid matter in the stomach. It is also used to nullify the irritating effects of acids in a better way than alkalies. (Vispute et al., 2011). Licorice is reported to have ulcer healing property due to additive effect of increased concentration of prostaglandins in the digestive system that promotes mucus secretion from the stomach, prolongs the life span of surface cells in the stomach and has an antipepsin effect (Aly et al., 2005). Components of licorice root are used for treating hormone-related female problems because they posses estrogenic and anti-estrogenic activity (Jatav et al., 2011). People suffering from obesity can use licorice tea to reduce their body weight as the methanolic extract of licorice showed cytotoxic activities against intestinal carcinoma cell line (Caco-2) and prostate carcinoma cell line (PC-3) with IC50 values of 40 and 40.6µg/ml, respectively. (Badr et al., 2013). Under some circumstances, genotoxic studies have revealed glycyrrhizin to be neither teratogenic nor mutagenic, but may possess anti-genotoxic properties. Many *in vivo* and clinical studies indicated beneficial effects of both licorice and glycyrrhizin consumption including anti-ulcer, anti-viral, and hepatoprotective responses (Isbrucker et al., 2006). Research suggests that licorice may also be beneficial in the treatment of AIDS (Hatano et al., 1988).

Other Uses

Glycyrrhizin as well as powdered Glycyrrhiza roots, Glycyrrhiza extracts, glycyrrhetinic acid, stearyl glycyrrhetinate, pyridoxine glycyrrhetinate, glycyrrhetinic acid 3-b -O-hemisuccinate (carbenoxolone) glabridin-containing glycyrrhiza flavonoids isolated from G. glabra are used in cosmetic preparations owing to their skin-whitening, anti-sensitizing, antiinflammatory (Yokota et al., 1998) skin depigmenting, skin lightening, antiaging, emollient, anti-acne and photoprotection properties (Brooke., 2004). After extracting licorice from roots left over residues are used as fertilizer for mushrooms and as a foam stabilizer in foam type of fire extinguisher (Vispute et al., 2011). In fire extinguishers it is used as a wetting and spreading agent and as an adhesive agent in insecticides, and textile dyes (Lioydia, 1978).Licorice flavonoid oil can be used in domestic animals as a feed additive to decrease abdominal fat accumulation (Badr et al., 2013). Licorice root in the form of ammonium glycyrrhizate also finds use in toothpastes, mouth rinses and other products for the control of periodontal disease (Goldie, 2005).

Side Effects and Toxicity

Licorice is rated "Generally recognized as safe" by the U.S. Food and Drug administration (Moharana, 2008). Licorice supplementation is commonly associated with side effects of hypertension which is probably due to the effect of licorice on the renin-angiotensin-aldosterone system. Licorice saponins bind to mineralocorticoid receptors in the kidneys and potentiate aldosterone action which is known as "pseudoaldosteronism."Symptoms of hypokalemia (potassium loss) and sodium retention, resulting in edema may also be observed in patients, which usually disappear with discontinuation of therapy. However, during the course of treatment no side effects were reported by many of the studies (Zadeh and Kor, 2013). Glycyrrhizinates inhibit the enzyme 11-hydroxysteroid dehydrogenase, responsible for inactivating cortisol as revealed by biochemical studies (Isbrucker et al., 2006).

Glycyrrhetinic acid (GA) that is produced by the metabolism of glycyrrhizin (GL) by gut microorganisms is responsible for inhibition of 11â-hydroxysteroid dehydrogenase type 2 (Kim et al., 2012). Concentrated form of licorice flavonoid oil containing 3% glabridin, regarded as representative for Glavonoid, in a subchronic (90-day) oral toxicity study produced an effect on blood coagulation parameters, evidenced in a prolongation of prothrombin time (PT) and activated partial thromboplastin time (APTT), that resulted in haemorrhage in several organs and tissues and the death of several animals in the highest dose group (1600 mg/kg bw). Inhibition of the synthesis of vitamin K-dependent coagulation factors is responsible for the anticoagulant effect as proposed by additional studies, even though the identity of the responsible substance(s) remains unknown (EFSA Panel on Dietetic Products, Nutrition and Allergies, 2011).

Conclusion

The Licorice extracts and its principle component, glycyrrhizin, have extensive use in foods,as a natural sweetener and flavoring agent in beverage, confectionery, tobacco and pharmaceutical industry. Extract of root can be found in various food as well as herbal preparations that are presently available in market. Presence of a wide range of bio active compounds mentioned in this review support it's great global demand as nutritional and medicinal supplement.Various food uses and health benefits reported in the present review also confirm that in future the plant extracts could serve as a "lead" for the development of nutraceuticals having good efficacy in various disorders.

References

1. Aly, A. M. Al-Alousi Laith. and Salem, H.A. (2005). Licorice: A Possible Anti-inflammatory and Anti-ulcer Drug. AAPS Pharma Science and Technology, 6 (1): 74-82.
2. Afreen, F. Zobayed, S.M. and Kozai, T. (2005). Spectral quality and UV-B stress stimulate glycyrrhizin concentration of Glycyrrhiza uralensis in hydroponic and pot system. Plant Physiology Biochemistry, 43:1074-1081.
3. Asl, M.N. and Hosseinzadeh, H. (2008). Review of pharmacological effects of Glycyrrhiza sp. and its bioactive compounds. Phytotherapy Research, 22:709-724.
4. Brooke, K.D. (2004). Natural Products Insider inside cosmeceuticals nutrilearn.com. Accessed on. Chinese medicine. Pure Applied Chemistry, 74(7):1189–1198.
5. Cho, S. Park, J.H. Pae, A, N. Han, D. Kim, D. Cho N,C. No, K.T. Yang, H. Yoon, M. Lee, C. Shimizu, M. and Baek, N.I. (2012). Hypnotic effects and GABAergic mechanism of licorice (*Glycyrrhiza glabra*) ethanol extract and its major flavonoid constituent glabrol. Bioorganic Medicinal Chemistry, 20:3493–3501.
6. Moharana, D. (2008). Shatavari, Jastimadhu and Aswagandha the Ayurvedic Therapy.
7. EFSA Panel on Dietetic Products, Nutrition and Allergies (2011). Scientific Opinion on the safety of 'Glavonoid', an extract derived from the roots or rootstock of *Glycyrrhiza glabra* L., as a Novel Food ingredient, EFSA Journal, 9(7):2287.
8. Davis, E.A. and Morris, D.J. (1991). Medicinal uses of licorice through the millennia: the good and plenty of it. Molecular and Cellular Endocrinology, 78:1-6
9. Farag, M.A. Porzel, A. and Wessjohann, L.A. (2012). Comparative metabolite profiling and fingerprinting of medicinal licorice roots using a multiplex approach of GC-MS, LC-MS and 1D NMR techniques. Phytochemistry, 76:60–72.
10. Goldie, M.P. (2005). Antioxidants in oral health care: making the connection. International Journal of Dental Hygiene, 3:93–95.
11. Hardy, M.E. Hendricks, J.M. Paulson, J.M. Faunce, N.R. (2012). 18b-glycyrrhetinic acid reduces rotavirus replication in culture. Journal of Virology, 22: 96.
12. Hatano, T. Aga, Y. Shintani, Y. Ito, H. Okuda, T. Yoshida, T. (2000). Minor flavonoids from licorice. Phytochemistry, 55:959–963.
13. Hatano, T. (1988). Anti-human immunodeficiency virus phenolics from licorice. Chemical Pharmaceutical Bulletin, 36:2286-2288.
14. Husain, A. (1991). Economic aspects of exploitations of medicinal plants. In: Akerele O., Heywood, V. and Synge H. (Eds) Conservation of medicinal plants. Cambridge University Press, Cambridge.
15. Hyo, J.K. Ji-Yeon, S. Hwa-Jin. S. Soon, S.L. and Jong, S.K. (2012). Antioxidant activities of licorice-derived prenylflavonoids, Nutrition Research and Practice, 6(6):491-498
16. Jalal, B.Z., and Zahra, M.K. (2013). Licorice (Glycyrrhiza glabra Linn) As a Valuable Medicinal Plant, International journal of Advanced Biological and Biomedical Research, 10:1281-1288.
17. Karami, Z. Mirzaei, H. Emam-Djomeh, Z. Sadeghi, M.A.R. and Khomeiri, M. (2013). Effect of harvest time on antioxidant activity of *Glycyrrhiza glabra* root extract and evaluation of its antibacterial activity, International Food Research Journal 20(5): 2951-2957.
18. Kitagawa, I. (2002). Licorice root. A natural sweetener and an important ingredient in Chinese medicine. Pure Applied Chemistry, 74(7):1189–1198.
19. Gibson, M.R. (1978). Glycyrrhiza in old and new perspectives. Lioydia, 41:348-354.
20. Liu, H.M. Sugimoto, N. Akiyama, T. Maitani, T. (2000). Constituents and their sweetness of food additive enzymatically modified licorice extract. Journal Agricultural Food Chemistry, 48:6044–6047.

21. Mansfeld, R. (1986). Verzeihnis landwirschaftlicher und gartnerischer kulturpflanzen (Ohne Zierpflanzen), vol. 1-4, Springer, Berlin, Germany.
22. Nieman, C. (1959). Licorice. Advances in Food Research, 7:339–381.
23. Wang, Z.Y. Nixon, D.W. (2001). Licorice and cancer. Nutrition Cancer, 39 (1): 1-11.
24. Patil C.S. (2011). Current trends and future prospective of nutraceuticals in health promotion, Bioinfo Pharmaceutical Biotechnology,1:1-7.
25. Chemical analysis of the Chinese herbal medicine Gan-Cao (Licorice). Journal of Chromatography A, 1216:1954-1969.
26. Isbrucker, R.A. and Burdock, G.A. (2006). Risk and safety assessment on the consumption of Licorice root (*Glycyrrhiza* sp.), its extract and powder as a food ingredient, with emphasis on the pharmacology and toxicology of glycyrrhizin. Regulatory Toxicology and Pharmacology, 46:167–192.
27. Saleem, M, M. Mohammad, A. A. Al-Tameemi, J.A. Sulaiman, G.M. (2011). Biological study of the effect of licorice roots extract on serum lipid profile, liver enzymes and kidney function tests in albino mice. African Journal of Biotechnology, 10:12702–12706.
28. Sheetal Vispute and Ashlesha Khopade (2011). GLYCYRRHIZA GLABRA Linn. - “KLITAKA”: A Review, International Journal of Pharma and Bio Sciences, 2(3):42-51.
29. Badr, S.E.A. Sakr, D.M. Mahfouz, S.A. and Abdelfattah, M.S. (2013). Licorice (*Glycyrrhiza glabra* L.): Chemical Composition and Biological Impacts. Research Journal of Pharmaceutical, Biological and Chemical Sciences, 4(3):606-621.
30. Vijay, S.J. Santosh, K.S. Pankaj, K. and Ashish, K.S. (2011). Recent Pharmacological Trends of *Glycyrrhiza glabra* Linn. International Journal of Pharmaceutical Frontier Research, 1(1):170-185.
31. Wang, Z.Y. Athar, M. and Bickers D.R. (2000). Licorice in Foods and Herbal Drugs: Chemistry, Pharmacology, Toxicology and Uses. In Herbs, Botanicals and Teas; Mazza, G.,Oomah, B.D., Eds.; Technomic Publishing Co. Inc.: Lancaster, PA, USA, pp: 321-353.
34. Yokota, T. Nishio, H. Kubota, Y. Mizoguchi, M. (1998). The inhibitory effect of glabridin from licorice extracts on melanogenesis and inflammation. Pigment Cell & Melanoma Research, 11:355–361.

Unit-VI: Utilization of Dairy Food in the Development of Functional Foods and Nutraceuticals

CHAPTER 39

Process Standardization for Low Fat and Low Calorie Kulfi

Jyoti Jha and R. S. Dabur

Abstract

Kulfi is a mixture of fat; sugar and milk solid not fat together with added flavour. So diabetics people on restricted diets cannot consume kulfi having normal ingredients. Now-a-days, consumers demand for low fat/calorie products has significantly raised in an attempt to limit health problems, to lose or stabilize their weight and to work within the frame of a healthier diet. The role of fat replacers and sugar substitutes in the successful manufacture of these products is crucial. Development of sugar free kulfi would certainly ensure diabetic/obese people to enjoy ice cream/ kulfi. The current study is to standardize the kulfi recipe with fat and calorie replacer. Basic formula mix consist of Whole milk 60 percent, SMP 12 percent, Sugar 13 percent, Cream 15 percent and Stabilizer 0.5 percent where the whole milk was replaced by skim milk, cream and sugar was replaced by malto-dextrin as a bulking agent and aspartame respectively alongwith sodium alginate as a stabilizer. The replacers were incorporated at four different levels of 0, 25, 50, 75 and 100 percent to the basic formula mix. Freezing of above kulfi mix was done by traditional method and softy freezer method. Sensory evaluation was carried out by using 9-point hedonic scale through semi-trained judges and results were statistically analyzed. On the basis of sensory scores it was revealed that sugar could be replaced 100% by artificial sweetener, bulking agent (Maltodextrin). It was found from the sensory scores that the milk fat in kulfi can be replaced upto 75 percent by sodium caseinate thereafter increase in caseinate gives prominent flavour of casein and kulfi body becomes thick and hard. It was concluded from the results that 100 percent sugar replacement was found feasible in preparation of low fat and low calorie kulfi. Results of the low fat and low calorie kulfi mix indicated that there was 32.29 percent reduction in the calorific value of experimental kulfi as compared to the control sample (traditional). It was concluded from the results that the kulfi with 75 percent less fat and 100 percent sugar free can be prepared with low cost in comparison with control.

Keywords: Kulfi, frozen milk product, whole milk, skim milk, replacer, fat, calorie, sugar, cream, sodium casienate, malto-dextrin, skim milk powder.

Introduction

Kulfi is a heat desiccated frozen milk product (indigenous ice cream). While the milk is boiling, sugar is added, the product is concentrated at a ratio of approximately 2:1. Malai, crushed nuts and flavouring (commonly rose or vanilla) is added to the cooled concentrate, the mix is placed in moulds (triangular, conical or cylindrical) of various capacities made of galvanised iron sheets or of the earth. The moulds are closed on top by placing a small disc over them, the edges made air-tight with wheat dough (modern moulds are made of plastic, generally conical in shape with screw-cap plastic tops) and the moulds are frozen in a large earthen vessel containing a mixture of ice and salt (1:1). A similar product is malai-ka-baraf, prepared by freezing a mixture of malai, sugar, nuts and essence. The product is plastic in nature, with a crunchy texture and delicate nutty, caramel flavour (Ganguly, 1999). In the west, ice cream and frozen desserts have been tried using aspartame as a sweetener with a variety of bulking agents based on polydextrose, microcrystalline cellulose, etc. (Wolkstein, 1986). The products have lower energy content than standard products, particularly if the sweetener and bulking agent replace some of the milk solids as well as carbohydrates. There is need to standardizing the method of manufacture of these products on a commercial scale.

In India, the growth of ice cream is traced back to a restaurant at Calcutta called "Mangolia" run by an Italian caterer who was the first one to prepare and serve ice cream on western bases. Later, in the year 1956, ice cream machinery was imported by Kwality Ice cream Company, which was the beginning of commercial production and marketing of the product (Lamba, 1998). Ice cream represents one of the most dynamic sector of the dairy industry and for two decades has been the sector where the most value addition has taken place. Ice cream and frozen desserts are delicate, delicious and nutritious food, liked by all the age groups and popular throughout the world. It is the source of important nutrients like fat, protein, carbohydrate, minerals and vitamins. It is stored under refrigerated conditions and consumed in frozen state, the distinction which no other foods can claim. Ice cream has occupied a unique place in the diet of western people and is gaining popularity all over the world.

Now days, there is a trend for health foods. According to recent survey conducted in India, on routine market research, the consumers have to cut back on ice cream consumption because of fear of adverse impact on health (Ramchandran, 2000). A market survey in the US also confirmed a strong interest in low calorie frozen dessert and ice cream. Hence, healthier ice cream for a specific group of people will be the focus point of prominence for the Indian Ice Cream Manufacturers in the coming decade. To satisfy the consumer demands for new, nutritionally acceptable products, many types of new frozen desserts began appearing in the market. Current trend towards lower fat and lower cholesterol alternative will continue. Product development activity will be spurred as new ingredients become available and are approved by statutory body for use. Aspartame already has an impact in the dairy industry. Development of fat substitutes is being eagerly awaited by a group of consumers. The challenges for food technologists will be to develop products meeting these consumers needs with little, if any, compromise to the taste of the product. The range of available products is expected to widen as marketers continue to target specific consumer groups.

The present study is on the replacement of fat and calories in the kulfi, the indigenous frozen milk products, with the skim milk to reduce a percent of fat, aspartame to replace sugar, malto-dextrin to provide bulkiness to the kulfi texture and sodium casienate to stabilize the colloidal mass of fat and bulking agent. All the four ingredients were incorporated to the standard kulfi recipe at four levels i.e 25, 50, 75 and 100 percent with were compared to the control sample. The main focus of this study is to standardize the process for preparing a stabilized product which should be delicious besides lacking fat and sugar calories which is desired by most diabetic patients.

Materials and Methods

The material used and the methods performed to complete this study are:

Procurement of Materials

a) **Skim Milk /whole milk/cream:** Skim milk, whole milk was procured from Dairy Plant of the Department of Animal Products Technology, CCS Haryana Agricultural University, Hisar.

b) **Skim Milk Powder (SMP):** Skim milk powder of NOVA Brand was procured from the local market.

c) **Sugar:** Cane sugar was procured form the local market.

d) **Vanilla Essence:** Vanilla essence of Rose brand was procured from the local market.

e) **Sugar Replacer:** Aspartame Cadilla Ltd. Company was used as sugar replacer and it was procured from the local market.

f) **Fat Replacer:** Sodium caseinate was procured from Spectrochem Pvt. India Ltd., Bombay, India

g) **Bulking agent:** Maltodextrin was used as bulking agent and it was procured from S.A. Pharmachem Pvt. Ltd., Mumbai.

Kulfi Freezing Methods

i) Freezing was done by ice and salt mixture mixed in ratio 1:2 in an earthen pot.

ii) Softy freezer method as followed by Yerriswamy et al. (1983).

All of the samples were analyzed for physico-chemical (Meltdown rates, hardness, pH), chemical (Acidity, Total Solids, Milk solids, Milk Fat, Protein Lactose, Ash content) and sensory characteristics. To minimize the differences, the panelists were acquainted with the quality attributes and the defect generally associated with kulfi. Samples were evaluated for flavor (10), sweetness (10), body and texture (10), colour and appearance (10) and taste (10) using 10 points score card on average basis. The judges were also asked to rate the samples for overall acceptability with maximum score of 10. The experimental samples alongwith controls were stored for 60 days and the sensory evaluation was done at 10 days intervals. The results obtained were analyzed statistically.

Standardization of Process for Low Fat and Low Calorie Kulfi

Control sample

Two control samples were prepared:

C_1 first was prepared by traditional method as followed by Salooja et al. (1982) with minor modification. Freezing was done by ice and salt mixture mixed in ratio 1:2 in an earthen pot. The milk was concentrated to its half of volume and then 130 g of sugar was added and further concentrated to it reached 1000 g.

C_2 Second was prepared by the softy freezer method as followed by Yerriswamy et al., (1983) with minor modifications with following ingredients:

Whole milk (600 ml and cream (150ml) was heated to temperature of 60 °C. Skim milk powder (120g) and cane sugar (130g) was added slowly with continues stirring and temperature was raised to 20°C into hot milk and then cooled down to 40°C before freezing. This kulfi mix was frozen by using softy freezer method.

Experimental samples

Low fat kulfi

To prepare low fat kulfi, fat sources were replaced by non-fat milk sources such as whole milk was replaced by skim milk and cream was replaced by sodium caseinate powder. Both fat replacers were incorporated at the rate of 25, 50, 75 and 100 percent simultaneously while other ingredients amount was kept constant and total volume of kulfi mix was also kept constant in all the formulations as given below in (Table 1).

Table 1: Formulations tried for preparation of low fat kulfi mix

Ingredients (g)	Control		Experiment			
	C_1	C_2	K_1	K_2	K_3	K_4
Whole milk	2000*	600	450	300	150	0
Skimmed milk	0.0	0.0	180	360	540	720
Cream	0.0	150	112.5	75.0	37.5	0.0
Caseinate	0.0	0.0	7.5	15.0	22.5	30.0
Skimmed Milk Powder	0.0	120	120	120	120	120
Sugar	130	130	130	130	130	130
Total	1000	1000	1000	1000	1000	1000

*Whole milk was concentrated till the weight reaches 1000 g
C_1-0 percent replacement (Traditional); C_2-0 percent replacement
K_1-25 percent replacement; K_2-50 percent replacement
K_3-75 percent replacement; K_4-100 percent replacement

Low fat and low calorie kulfi

In this kulfi cane sugar was replaced by aspartame and malto-dextrin and bulking agent at the levels 25, 50, 75 and 100 percent (Table-2).

Table 2: Formulations tried for preparation of low calories kulfi mix*

Ingredients (g)	Control		Experiment			
	C_1	C_2	K_1	K_2	K_3	K_4
Whole milk	2000*	600	300	300	300	300
Skimmed milk	0.0	0.0	480	480	480	480
Cream	0.0	150	75	75	75	75
Caseinate	0.0	0.0	15	15	15	15
Skimmed milk powder	0.0	120	120	120	120	120
Sugar	130	130	98	65	33	-
Sweetener	0.0	0.0	0.117	0.234	0.351	0.468
Malto dextrin	0.0	0.0	32.00	65.00	97.00	129.00
Total	1000	1000	1000	1000	1000	1000

*Whole milk was concentrated till the weight reaches 1000g

Results and Discussions

Effect of fat replacement on kulfi sensory attributes

Kulfi samples prepared by using different formulations were offered for judging to semi trained judges and results are presented in Table 3. Fat provides a creamy texture and a flowing mouth feel to foods. It contributes to the food various organoleptic and rheological characteristics. The attributes contribute to the richness of the food product, resulting in better market value and consumer acceptability.

Table 3: Effect of fat replacement* on the sensory attributes scores of Kulfi

Sensory attributes	Control		Experiment			
	C_1	C_2	K_1	K_2	K_3	K_4
Color and appearance	7.83	7.58	7.16	7.00	7.00	6.66
Body and texture	7.50	7.08	6.91	6.66	6.17	5.42
Flavor	7.25	7.16	7.16	6.92	6.66	4.42
Overall	7.52	7.25	6.94	6.86	6.50	5.17

*Milk fat source i.e. cream was replaced by sodium caseinate and whole milk was replaced by skimmed milk

C_{1-}0 percent replacement (Traditional); C_{2-}0 percent replacement

K_{1-}25 percent replacement; K_{2-}50 percent replacement

K_{3-}75 percent replacement; K_{4-}100 percent replacement

It is evident from the Table values that color and appearance in experimental samples of kulfi was affected by fat replacement with fat replacer Sodium casienate as control samples sample were liked moderately (7.00 – 8.00) while as the fat replacement percentage increases in the experimental samples the judges scores were also reduced to liked slightly (6.00-7.00). There was variation in the body and texture of experimental samples to the control samples as fat being a main component in body and texture in kulfi, it was came down to liked slightly (6.00 – 7.00) by the judges. Control sample C_1 was accepted higher in flavor to experimental samples K_2, K_3 and K_4 except in the case of C_2 and K_1, which were scored at par. It was revealed from the overall data of sensory score that as the level of fat decreases the scores of the sensory attributes were also shown decreasing trend for all the attributes. It was concluded from the results of the sensory score that the milk fat in kulfi can be replaced upto 75 percent by sodium caseinate thereafter increase in caseinate gives prominent flavour of casein and kulfi body becomes thick and hard. Rough texture was also observed at 100 percent replacement, so sensory score decreased below acceptable levels.

Effect of sugar replacement on sensory scores of kulfi

Results are presented in Table-4. The sensory evaluation revealed that colour and appearance was not affected by sugar replacement with sugar free (aspartame) and bulking agent (maltodextrin), even at 100 percent replacement sample was liked moderately (7.00 – 8.00) by the judges for their colour and appearance. Body and texture of kulfi sample was influenced by sugar replacement slightly as judges scored 100 percent replacement sample in the range of liked slightly (6.00 – 7.00). The flavour of experimental samples also influenced slightly by sugar replacement but score was decreased less in comparison to body and texture scores. Finally the judges found acceptable kulfi with 100 percent replacement and overall sensory score was found in the range of liked moderately (7.00 – 8.00). It was concluded from the results that 100 percent sugar replacement was found feasible in preparation of low fat and low calorie kulfi.

Table 4: Effect of sugar replacement on the sensory attributes scores of Kulfi

Sensory attributes	Control		Experiment			
	C_1	C_2	K_1	K_2	K_3	K_4
Color and appearance	8.83	8.58	8.16	8.00	8.00	7.66
Body and texture	8.50	8.08	7.91	7.66	7.17	6.42
Flavor	8.25	8.16	8.16	7.92	7.66	7.42
Overall	8.52	8.25	7.94	7.56	7.50	7.17

C_{1-}Zero percent replacement (Traditional); C_{2-}Zero percent replacement
K_{1-}25 percent replacement; K_{2-}50 percent replacement
K_{3-}75 percent replacement; K_4-100 percent replacement

Effect of storage on the kulfi sensory attributes

To study the shelf life of control and selected experimental formulation the samples were stored at –15°C and offered for the sensory evaluation at 10 days interval upto 60 days. The results pertaining to sensory attributes are presented in Table-5. Sensory score for all the sensory attributes of control as well as experimental sample of kulfi were decreased slightly at each subsequent analysis interval. The color of the control samples showed a variation from liked extremely to liked very much (8.66 -7.68) in comparison to the experimental samples which were liked moderately (7.80-7.50) by the judges during the storage period. Body and flavor of the kulfi had shown the same behavior as for the color attribute. At the end of storage the product was found acceptable on the basis of overall acceptability score in both the cases i.e. control as well as experimental. Judges scores indicated that controls as well as experimental samples were liked moderately (7.00 – 8.00) for all the sensory attributes even at the 60th day of storage.

It was concluded from the results that low fat and low calorie kulfi can be stored upto 60 days without affecting its sensory attributes.

Table 5: Effect of storage (at –15° C) of Kulfi on sensory attributes

Days	Control				Experimental			
	Color	Body	Flavor	Overall	Color	Body	Flavor	Overall
0	8.66	8.48	8.12	8.50	7.80	7.53	7.56	7.66
10	8.62	8.50	8.10	8.60	7.83	7.52	7.41	7.60
20	8.61	8.46	8.10	8.50	7.81	7.51	7.40	7.50
30	8.60	8.46	8.08	8.45	7.79	7.50	7.40	7.50
40	8.60	8.41	8.05	8.45	7.77	7.46	7.41	7.45
50	8.18	8.02	8.01	8.00	7.77	7.43	7.33	7.36
60	7.68	7.88	7.70	7.60	7.50	7.40	7.21	7.20

Conclusion

In the present study of low fat and low calorie kulfi fat of whole milk was replaced by skim milk while the fat of cream was replaced by sodium caseinate. Sugar replacer used was aspartame to reduce the table sugar with bulking agent maltodextrin. Both fat replacer and sugar replacer were incorporated at 25 percent, 50 percent, 75 percent and 100 percent and compared with the control samples.

It was observed from the ingredients list that there were two main sources of fat are whole milk and cream, which has to be replaced to reduce the fat level in the low fat kulfi formulations. For sugar replacement aspartame was used as sweetener and maltodextrin as bulking agent in the place of sugar content. Sensory attributes results indicated that upto 75 percent of replacement was feasible to get acceptable product.

In case of low fat kulfi, sugar replacement did not affect the sensory attributes much as results indicated that even 100 percent replacement was feasible without affecting its overall acceptability. Formulation of low fat and low calorie kulfi were tried by taking the whole kulfi mix 1000g i.e. kept constant and further incorporation of sodium caseinate, aspartame and maltodextrin were calculated according to that at various levels.

Results of the low fat and low calorie kulfi mix indicated that there was 32.29 percent reduction in the calorific value of experimental kulfi as compared to the control samples. Judges stored product indicated that control as well as experimental samples was liked moderately till the end of storage. During storage of low fat low calorie kulfi and control samples had not shown any change in the physico-chemical parameters till the end of storage. It was concluded from the results that the kulfi with 75 percent less fat and 100 percent sugar free can be prepared with low cost in comparison with control. Control as well as experimental samples was found acceptable upto 60 days at –15°C storage.

References

1. Ganguly, P. Bandopadhyay and S. Kumar (1999). Smallholder dairying in the tropics: Processed milk products by: International Livestock Research Institute P O Box 30709, Nairobi, Kenya.
2. Lamba, P.L. (1998). Some facts about ice-cream. Processed Food Industry, 1(6): 11-13.
3. Ramachandran, A. (2000). Hope for diabetic patients. In Proceedings of the scientific session. APICON, Medicine Update, 10(1): 491-495.
4. Salooja, M.K. and Balachandran, R. (1982). Studies on the production of kulfi. Journal of Food Science and Technology, 19(3):116-118.
5. Wolkstein, M. (1986). Dietetic frozen dessert containing aspartame. US Patent, US 4 62 411.
6. Yerriswamy, K.; Atmaram, K.; Natrajan, A.M. and Anantakrishan, C.P. (1983). Quality of kulfi manufactured by different method. Cheiron, 12(3):130-135.

CHAPTER 40

Improving Milk and Milk Products Functionality Through Controlled Release of Herbal Nutraceuticals

Rahul Debbarma, Aditya Madan, Ranjit Singh, Sandeep Singh Rana and Neha Balan

Abstract

India has got a pool of medicinal herbs which are proved to be beneficial in curing various diseases. Their regular consumption not only will improve the health status but also reduce risk of lifestyle diseases. Though simple solutions, associated colloids, suspensions and emulsions are some of the traditional techniques for encapsulating functional agents, micro-encapsulation techniques is the most successfully employed technique for the controlled release of sensitive bioactive compounds. Owing to the targeted delivery of functional components directly to the human gut system, emulsion based delivery system are getting more attention now-a-days. With advancement in science, more complex and novel structures can be created which can successfully deliver these nutraceutical and bioactive ingredients into existing food systems. The technological interventions in emulsion science have opened great opportunities to improve the delivery of bioactive herbal components into milk and milk products. Various milk and milk products functionality can be improved in this way which will improve the gut- microflora within short period of time after consumption.

Keywords: Medicinal herbs, Controlled release, Emulsion, Micro-encapsulation, Nutraceutical

Introduction

About 2000 years ago, Hippocrates correctly emphasized "Let food be your medicine and medicine be your food". Currently there is an increased global interest due to the recognition that "Nutraceuticals" plays a major role in health enhancement. The term "Nutraceutical" was coined by combining the terms"Nutrition" and "Pharmaceutical" in 1989 by Dr Stephen DeFelice, Chairman of the Foundation for Innovation in Medicine (Brower, 1998). "Nutraceutical" is a marketing term developed for nutritional supplement that is sold with the intent to treat or prevent disease and thus has no regulatory definition (Ziesel, 1999). Hence a "nutraceutical" is any substance that may be considered a food or part of a food and provides medical or health benefits, encompassing, prevention and treatment of diseases. Such products may range from isolated nutrients, dietary supplements and diets to genetically engineered "designer" foods, herbal products and processed foods such as cereals, soups and beverages. Presently over 470 nutraceutical and functional food products are available with documented health benefits (Eskin and Tamir, 2005). "Nutraceuticals and functional foods have received considerable interest because of their presumed safety and potential nutritional and therapeutic effects".

Encapsulation may be defined as a process to entrap one substance (active agent) within another substance (wall material). The encapsulated substance, except active agent, can be called the core, fill, active, internal or payload phase. The encapsulating substance is called the coating, membrane, shell, capsule, carrier material, external phase, or matrix (Wandrey et al., 2009; Fang and Bhandari, 2010). In the food industry, encapsulation process can be applied for a variety of reasons. Encapsulation is a useful tool to improve delivery of bioactive molecules (e.g. antioxidants, minerals, vitamins, phytosterols, lutein, fatty acids, lycopene) and living cells (e.g. probiotics) into foods (Wandrey et al., 2009; Vos et al., 2010). In most cases, encapsulation refers to a technology in which the bioactive components are completely enveloped, covered and protected by a physical barrier, without any protrusion of the bioactive components (Vos et al., 2010). Also, encapsulation has been defined as a technology of packaging solids, liquids, or gaseous materials in small capsules that release their contents at controlled rates over prolonged periods and under specific conditions (Desai and Park, 2005). The particles produced usually have diameters of a few nm to a few mm (Wandrey et al., 2009).

One of the most important reasons for encapsulation of active ingredients is to provide improved stability in final products and during processing. For example, probiotics are highly sensitive to variation of pH, mechanical stress, transport conditions, and digestive enzymes in the stomach. Probiotic bacteriaare defined as live microorganisms and bioactive food components with serious health benefits in thehost, if they are present in adequate amounts (Vos et al., 2010). At present, probiotics are the driving force in the design of functional foods, especially in dairy products, maintaining their functional effects for supporting human health. These living cells need to survive the food process, storage, and food intake before they can be useful. Encapsulation increases not only their bioavailability, but more importantly functionality. A choice of an encapsulation system is always crucial, since it has to be efficient and easily incorporated into the food without interfering with the texture and taste of the food. Also, some microbes such as Lactobacilli and Bifidobacteria seem to benefit from the encapsulation matrix during dehydration and lyophilisation (Kim et al., 1996).

India is rich source of medicinal herbs which are proved to be beneficial in curing various diseases. Use of herbs in food systems is recognized as safe by health organizations all around the world. Their regular consumption not only will improve the health status but also reduce risk of lifestyle diseases. Though simple solutions, associated colloids, suspensions and emulsions are some of the traditional techniques for encapsulating functional agents, micro-encapsulation techniques is the most successfully employed technique for the controlled release of sensitive bioactive compounds. Owing to the targeted delivery of functional components directly to the human gut system, emulsion based delivery system are getting more attention now-a-days.

Generally, the dairy products are prepared using w/o (water-in-oil) or o/w (oil in water) emulsion methods. With advancement of modern science and technologies, more and more complex and novel structures can be created which can successfully deliver these nutraceutical ingredients into the existing food components. Novel structures include liposomes, vesicles, nano-layers, multiple emulsions, nano-emulsions, solid-liquid particle etc. Thus, there is a huge opportunity for food and dairy industries to improve upon the delivery of active components into foods, particularly of minerals, vitamins, phytosterols, lutein, fatty acids, lycopene and anti-oxidants. Table 1 depicts several such micro-encapsulation technologies which have been developed for use in the industries which demonstrates potentials for the production of functional foods.

Emulsion Based Delivery System

Plain emulsions can be used as a delivery vehicle for either water soluble and/ orlipophilic active agent in food products (Appelqvist et al., 2007). There are two considerations that must be taken into account when formulating an emulsion for controlled delivery. First, the emulsion system must be (storage) stable right up to the point of application. Second, on application the emulsion should behave in a manner consistent for achieving the desired delivery. In many (but by no means all) cases this equates to the "making and breaking" of emulsions for stability and subsequent delivery. Water soluble food active agent might be encapsulated in water-in-oil (w/o) emulsions or double emulsions of the type w/o/w (Appelqvist et al., 2007). Furthermore, oil-in-water (o/w) emulsions may affect taste (e.g., salt) by changing the aqueous phase volume and thus the concentration of taste molecules in water, and by suppressing contacts of salt with taste receptors. Lipophilic active agent (e.g., aroma, cartonenoids such as lycopene and beta-carotene, plant sterols, vitaminE, dietary fats) might be protected and delivered to consumers via o/w emulsions (Appelqvist et al., 2007).

The use of nano sized vehicles for the protection and controlled release of nutrients and bioactive food ingredients is a growing area of interest to the food science and technology community. The reasons for this are that nano particles might be incorporated into food products easily without sedimentation, without beingnoticed by the consumer, and/or with an enhanced bioavailability (Acosta, 2009). Lipid-based nano encapsulation systems are among the most rapidly developing field of nanotechnology application in food systems. Lipid-based nano encapsulation systems have several advantages, including the ability to entrap material with different solubilities and the use of natural ingredients on an industrial scale (Bummer 2004; Mozafari et al., 2006; Taylor et al., 2005). Lipid-based nano carriers can also be used for targeted delivery of their contents to specific areas within the gastro-intestinal tract or food matrix.

Table 1: Bioactive ingredient that needs to be delivered in foods and potential health benefits from them

Component	Types	Potential health benefits
Fatty acids	Omega-3 fatty acids, conjugated linoleic acid, butyric acid	Coronary heart disease, bone health, immune response disorders, cancer, stroke prevention
Antioxidants	Tocopherols, flavonoids, polyphenols	Coronary heart disease, cancer, and urinary tract disease
Carotenoids	Beta-carotene, lycopene, lutein, and zeaxanthin	Coronary heart disease, cancer, maculardegeneration, and cataracts
Proteins, Peptides, Amino acids	Caseins, lactoferrin, lactoperoxidase, immunoglobulin	Brain functioning, minerals bio-availability, cancer, hypertension
Phytosterols	Stigmasterol, campesterol	Coronary heart disease
Bioactive carbohydrates	Dietary fibers, prebiotics	Coronary heart disease, cancer, diabetes, constipation

(*Contd.*)

Component	Types	Potential health benefits
Bacteria	Probiotics	Intestinal heath, cancer, immunity
Essential minerals	Iron, zinc, calcium, selenium, chromium	Anaemia, growth and development, bone health, diabetes, cancer
Herbs	Puereiatuberosa, Bacopamonniera, Withaniasomnifera, Asparagus racemosus	Medicinal properties

When referring to nanoscale lipid vesicles, the term nanoliposome has recently been introduced (Mozafari et al., 2006) to describe lipid vesicles whose diameter ranges tens of nanometres. These, so called, nanoliposomes have similar structural, physical, and thermodynamic properties as liposomes. The manufacture of nanoliposomes (as of liposomes) requires high energy for the dispersion of lipid/ phospholipid molecules in the aqueous medium. One of the most promising lipid-based nano delivery systems for food applications is the development of nano-sized self-assembled liquids (NSSL) (Garti et al., 2005).

Multiple emulsions are structurally very complex emulsion where, water droplets are dispersed in larger oil droplet which is further dispersed in the continuous water phase (Garti 1997; McClements, 2010). Multiple emulsions are required for the delivery of sensitive ingredients where direct contact with food matrix adversely affects the performance of bioactive components. On the other hand, some compounds like iron, copper etc are very reactive in free state and brings about undesirable flavours to the finished products. Double emulsions (w/o/w) are most suitable for encapsulation of hydrophilic bioactive components; however they may also be used to deliver both lipophilic and hydrophilic bioactive components in the same system (Cournarie et al., 2004). Their double compartment structure can be used to encapsulate substances within the internal water droplets or to reduce the fat content of the emulsion based products (McClements et al., 1999). Hence, multiple emulsions are future of emulsified products in global business.

Structural Components in Emulsion Based Delivery Systems

Depending upon the target applications, final continuous phase and charactristics of bioactive ingradient, the structural components in emulsion based delivery system is selected. The three buiding blocks for such a delivery system: lipid phase, surfactant and biopolymer. A wide range of materials both for core and wall and also various technologies, for controlling the interactions of ingredients in a given food system and, for the manufacturing of microcapsules and microparticles of different size, shape, and morphology are being evaluated.

Double emulsion system

Double emulsion system consists of either water in oil in water (w/o/w) or oil in water in oil (o/w/o) system. In case of w/o/w system, an inner aqueous phase, which is encapsulated in an oil phase, is enclosed within a second aqueous phase (Fig 1). The other one, o/w/o system consists of an inner oil phase, which is encapsulated in an aqueous phase, which is enclosed within second oil phase. Principal applications of such type of emulsion system can be found in encapsulation of lipophilic bioactive components like beta-carotene and omega-3 fatty acids; encapsulation of hydrophilic bioactive components such as vitamin B, immunoglobulins, proteins and amino acids; preparation of a delivery system that contains both lipophilic and hydrophilic bioactive components in the same system.

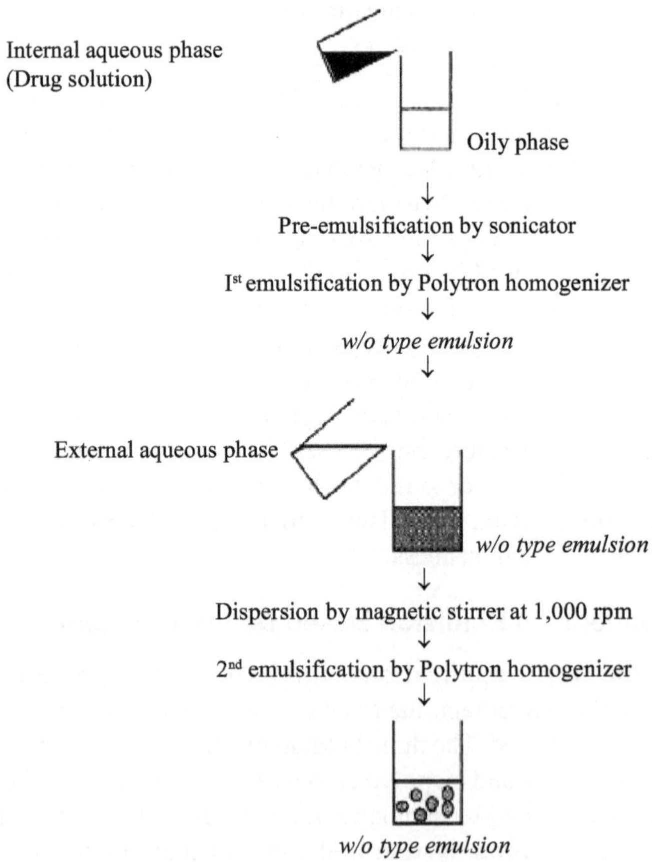

Fig. 1: Schematic diagram of double emulsion w/o/w preparation (Okochi and Nakano, 2000)

Potentially, such kind of w/o/w emulsions have certain advantages over conventional oil-in-water emulsion system for delivery of active components. Functional components could be trapped inside the inner water droplets and released at a controlled rate or in response to specific environmental stimuli, for example, in the mouth, stomach, or small intestine. Active ingredients could be protected from chemical degradation by isolating them from other water-soluble ingredients that they might normally react with. Also, this technique could be employed for producing reduced fat products with no change in sensory and physicochemical properties.

Conclusion

Micro-encapsulation techniques have been successfully employed in the past for the delivery of sensitive bioactive components. Among these, emulsion based delivery systems particularly multiple emulsions (w/o/w) are getting more and more attention now-a-days due to their targeted delivery. They offer significant benefits over the solids matrix materials as they can be used in liquid food systems too. Nevertheless, stability of emulsions in the complex food environment has been a challenging task. But many researchers have been successful to improve stability of these emulsions. Further research in this area is required for controlled and targeted delivery systems, especially in milk and milk based beverages.

References

1. Acosta, E. (2009). Bioavailability of nanoparticles in nutrient and nutraceutical delivery. Current opinion in colloid and interface science, 14(1):3-15.
2. Appelqvist, I. A. Golding, M. Vreeker, R. and Zuidam, N. J. (2007). Emulsions as delivery systems in foods. Encapsulation and Controlled Release Technologies in Food Systems, 41-81.
3. Brower, V. (1998). Nutraceuticals: Poised for a healthy slice of the healthcare market? Nature biotechnology, 16:728-732.
4. Bummer, P.M. (2004). Physical chemical considerations of lipid-based oral drug delivery—solid lipid nanoparticles. Critical Reviews™ in Therapeutic Drug Carrier Systems, 21(1).
5. Cournarie, F. Savelli, M.P. Rosilio, V. Bretez, F. Vauthier, C. Grossiord, J.L. and Seiller, M. (2004). Insulin-loaded W/O/W multiple emulsions: comparison of the performances of systems prepared with medium-chain-triglycerides and fish oil. European journal of pharmaceutics and biopharmaceutics, 58(3):477-482.
6. Desai, K.G. H. and Jin Park, H. (2005). Recent developments in microencapsulation of food ingredients. Drying Technology, 23(7):1361-1394.
7. Eskin, M. and Tamir, S. (2005). Dictionary of nutraceuticals and functional foods. CRC Press.
8. Fang, Z. and Bhandari, B. (2010). Encapsulation of polyphenols– a review. Trends in Food Science and Technology, 21(10):510-523.
9. Garti, N. (1997). Double emulsions—scope, limitations and new achievements. Colloids and Surfaces A: Physicochemical and Engineering Aspects, 123:233-246.

10. Garti, N. Spernath, A. Aserin, A. and Lutz, R. (2005). Nano-sized self-assemblies of nonionic surfactants as solubilization reservoirs and microreactors for food systems. Soft Matter, 1(3):206-218.
11. Kim, K.I. Yoon, Y.H. and Baek, Y.J. (1996). Effects of rehydration media and immobilization in Ca-alginate on the survival of Lactobacillus casei and Bifidobacteriumbifidum.Korean Journal of Dairy Science, 18.
12. McClements, D.J. (1999). Food emulsions: Principles, practices and techniques (Vol. 5). CRC Press.
13. McClements, D. J. (2010). Emulsion design to improve the delivery of functional lipophilic components. Annual review of food science and technology, 1:241-269.
14. Mozafari, M. R. Flanagan, J. Matia Merino, L. Awati, A. Omri, A. Suntres, Z. E. and Singh, H. (2006). Recent trends in the lipid based nanoencapsulation of antioxidants and their role in foods. Journal of the Science of Food and Agriculture, 86(13):2038-2045.
15. Okochi, H. and Nakano, M. (2000). Preparation and evaluation of w/o/w type emulsions containing vancomycin. Advanced drug delivery reviews, 45(1):5-26.
16. Taylor, T. M. Weiss, J. Davidson, P. M. and Bruce, B. D. (2005). Liposomal nanocapsules in food science and agriculture.Critical Reviews in Food Science and Nutrition, 45 (7-8):587-605.
17. Vos, P. Faas, M. M. Spasojevic, M. and Sikkema, J. (2010). Encapsulation for preservation of functionality and targeted delivery of bioactive food components. International dairy journal, 20(4):292-302.
18. Wandrey, C. Bartkowiak, A. and Harding, S.E. (2010). Materials for encapsulation. In Encapsulation technologies for active food ingredients and food processing (pp. 31-100). Springer New York.
19. Zeisel, S.H. (1999). Regulation of" nutraceuticals". Science, 285(5435):1853-1855.

Chapter 41

Functional Dairy Foods: Beyond the Basic Nutrition

Davinder Kaur, Usha Bajwa and Taranpreet Singh

Abstract

A functional food is defined as food that provides health benefit beyond the basic nutrition. Such foods reduce risk of high blood pressure, coronary heart disease, osteoporosis, cancer and dental caries. Consumers' increasing interest for maintaining or improving their health by eating specific food items has led to the development of many new functional foods including dairy products. These functional dairy foods contain some functional ingredients that decrease the absorption of cholesterol, reduce blood pressure and play role in regulation of satiety and have antimicrobial effects. Milk is a good source of fat, protein and minerals. It also contains several intrinsic functional components like melatonin, conjugated linoleic acid etc. in amounts lower than the body requirements. Fermented dairy products contain probiotics with special health benefits. Functional dairy foods could be produced with the addition and/or generation of functional components. Some examples include milk, yogurts and cheese enriched with conjugated linoleic acid or plant stanols/sterols to help reduce the risk of heart disease. Fermented dairy foods possess various nutritional and therapeutic properties. Lactic acid bacteria (LAB) play a major role in determining the positive health effects of fermented milks and related products. Dairy products with probiotics and prebiotics enhance digestive health, prevent colon cancer, help in lowering blood pressure and improve immune functions. Melatonin is important in curing of sleep disturbance and enhancement of alertness. Milk derived peptides have immunological, antimicrobial, ACE-inhibitory activity/antihypertensive, anti-carcinogenic and antioxidant activity. The aspects of functional dairy foods will be presented in detail.

Introduction

Consumers' increasing interest for maintaining or improving their health by eating specific food items has led to the development of many functional foods. A food marketed as "functional" contains added, technologically developed ingredients with specific health benefits. Such foods contain significant levels of biologically active components which have special role to play in health and disease. Functional foods help in reduction of risk of high blood pressure, coronary heart disease, osteoporosis, cancer and dental caries. The functional food comprise conventional foods having naturally occurring bioactive substance (e.g. dietary fiber, phytochemicals etc.) or food enriched with such substances (probiotics, antioxidants) or synthesized food ingredient introduced to traditional foods (e.g. prebiotics).

Various components incorporated to make functional dairy foods are conjugated linoleic acid, phytosterol, probiotic, prebiotic, synbiotic, melatonin and bioactive peptides. These functional dairy foods contain some dynamic ingredients that play a pivotal role in improving digestion, immune system, preventing infections and occurrence of diseases, establishing beneficial micro flora, curing sleep disturbances, increase absorption of minerals and overall provide many health benefits. Foods are not intended to only satisfy hunger and to provide necessary nutrients for human now a day but also to prevent nutrition related diseases and to improve physical and mental well-being. There is an increasing demand of functional foods world over that can be explained by factors like increasing cost of healthcare, the steady increase in life expectancy and the desire of older people for improved quality of their later years (Bhat and Bhat, 2011).

Functional Dairy Foods

Various approaches used to produce Functional Dairy foods are natural synthesis or enrichment and are discussed below.

Enrichment with Conjugated Linoleic Acid

Conjugated Linoleic Acid (CLA) is a complex mixture of different positional isomers of linoleic acid (cis-9, cis-12C18:2) (Aydin, 2005). It prevents cancers, improves immune system, cures diabetics and prevents heart diseases. It is synthesized in the body through the microbial isomerization of dietary linoleic acid in the digestive tract of ruminant animals. The amount of CLA can be increased by grazing grass because grass contain good content of linoleic acid and stearic acid, by giving feed which contain soybean oil and also grains. Synthesis of CLA is more in the presence of light at 530 nm. CLA enriched dairy products are produced by supplementing the basal diet of milch animals with sunflower oil and soybean oil. High CLA yoghurt is produced by using milk derived from animals fed a supplemented diet (Saarela, 2007).

Enrichment with Phytosterols

Phytosterols, which encompass plant sterols and stanols, are steroid compounds similar to cholesterol which occur in plants and vary only in carbon side chains and/or presence or absence of a double bond. Phytosterol are cholesterol lowering and it also help in lowering the occurrence of cancer. Naturally vegetable oil is rich source of phytosterols. Grains, nuts and sesame oil are also good source of phytosterols. In plants, sterols occur in free form and these are insoluble in water and have low solubility in lipid. Earlier phytosterols were dried and added in crystalline form. Now a day these are esterified with fatty acids, which in turns promote higher lipid solubility and thus are easily incorporated into a variety of food commodities (Tasan et al., 2006). Canola oil based margarine, enriched by sitosterol ester is an example of phytosterol enriched functional food. Phytosterols are lipid soluble so margarine is ideal vehicle for these. Yoghurt and ice cream is also a good source of fat but low fat yoghurt enriched with phytosterols can be produced by adding phytosterol esters.

Probiotics

Probiotics are live micro-organisms which when administered in adequate amounts; confer a health benefit on the host. Live probiotic cultures are available in fermented dairy products and probiotic fortified foods. However, tablets, capsules, powders and sachets containing the bacteria in freeze dried form are also available. Health benefits of probiotic products are lactose intolerance, prevention of colon cancer, lowering blood pressure, improving immune function and preventing Infections and diarrhea (Anon, 2013). Bacteria mainly used are strains of the genera Lactobacillus and Bifidobacterium. Probiotics used should be able to survive the passage through the digestive system, to attach to the intestinal epithelia and colonize, to maintain good viability, be non pathogenic and non toxic, capable of exerting a beneficial effect on the host and have a stability of desired characteristics during processing, storage and transportation. Fermented dairy products are produced by using various strains according to products. Probiotic products should contain at least 10 million CFU in the final product. Dairy products (Table 1) which contain probiotics are yogurt having Lactobacillus bulgaricus and Streptococcus thermophilus; cheese containing Lactococci, Lactobacilli or Streptococci ; acidophilus milk contain Lactobacillus acidophilus; kefir having Streptococcus lactis, Betabacterium caucasium; bulgarian buttermilk contain Lactobacillus bulgaricus and kumiss contain Lactobacillus acidophilus or bulgaricus (Saarela, 2007).

Prebiotics

Prebiotics are non-digestible food ingredients that beneficially affect the host by selectively stimulating the growth and/or activity of one or a limited number of bacteria in the colon, and thus improve host health. Health benefits are improvement in bowel functions, increased calcium absorption and prevention of cancer. Some of the oligosaccharides that naturally occur in breast milk are believed to play an important role in the development of a healthy immune system in infants, but these are not considered prebiotics, as they do not act through the intestinal micro flora. Naturally, dairy products contain oligosaccharides. In many dairy products, prebiotics mainly inulin is used. In ice cream inulin is supplemented with sugar as sugar replacer. Addition of inulin, oats and berries in yoghurt, oligosaccharides in cheese and frutooligosaccharide in milk powder could be done to impart prebiotic effect. Inulin is not absorbed at any point after consumption, when ingested orally it has no effect on blood glucose levels, therefore its consumption is good for diabetic patients.

Synbiotics

Synbiotics are a mixture of probiotics and prebiotics that beneficially affects the host by improving the survival and implantation of live microbial dietary supplements in the gastrointestinal tract, by selectively stimulating the growth and/or activating the metabolism of one or a limited number of health promoting bacteria, and thus improving host welfare. The development of new synbiotics typically requires a long-term screening process. Functional benefits of synbiotics are increase in number of ingested bacteria reaching the colon in a viable form, stimulation in the colon growth and implantation of both exogenous and endogenous bacteria, activation of metabolism of beneficial bacteria and production of antimicrobial substances (bacteriocins, hydrogen peroxide, organic acids etc.).

Enhancement of Melatonin

Melatonin is one of important hormone primarily secreted by pineal gland (located in the brain). It is also known as hormone of darkness because production of melatonin by pineal gland is prevented by light and permitted by darkness. The precursor for melatonin formation is tryptophan. Warm milk and banana are good sources of melatonin these contain good amount of

L tryptophan. Melatonin is sensitive to light and its concentration in milk varies with intensity of light (Singh et al., 2011). The use of milk enriched with melatonin or milk products made from it contain sufficient amount of melatonin. The milk obtained during night phase can be processed into milk products, having high melatonin content (Gnann, 2007). Melatonin is important for induction of sleep in healthy people and helps in curing of sleep disturbance, enhancement of alertness and reduces daytime fatigue. It acts as antioxidant or free radical scavenger in oxidative damage conditions and helps in protection of skin from UV light due to its antioxidant properties. It helps in prevention of headache associated with migraine, tension and hypertension.

Table 1: Some examples of dairy products containing both probiotics and prebiotics

Synbiotic Dairy Products	Probiotic	Prebiotic
Ice cream	Lactobacillus acidophilus and Saccharomyces boulardii	Fructo-oligosaccharides
Labneh cheese	Lactobacillus reuteri, Lactobacillus johnsonii and Lactobacillus salivarius	Inulin
Yoghurt	Bifidobacterium lactis, Lactobacillus acidophilus	Inulin
Fermented Milk	L. acidophilus, L. Casei, B. bifidum and B. longum	Peptides, Fructo-oligosaccharides and isomalto-oligosaccharides
Skim Milk Powder	Lactobacillus casei	Maize starch and inulin

Formation of Bioactive Peptides

Milk-derived bioactive peptides (Table 2) are milk components able to influence some physiological functions and finally acting on body health condition. The peptides are released by enzymes in the form of mature bioactive components or the precursors thereof. They are 3 to 20-residue long peptides released during in vivo gastrointestinal digestion, through the action of digestive enzymes or of the microbial enzymes of the intestinal flora, during milk processing (e.g. milk fermentation, cheese production) through the action of microbial enzymes expressed by the microorganisms used as starter and through the action of a single purified enzyme or a combination of selected enzymes. Peptides are formed by hydrolysis of protein molecules. In milk main protein is casein and with coagulation of casein bioactive peptides are produced.

Table 2: Examples of dairy products containing bioactive peptides

Dairy products	Bioactive peptides
Cheddar cheese	αs1 -CN fragments, β-CN fragments
Mozzarella cheese	αs1 -CN, β-CN
Sour milk	β-CN, β-CN , κ-CN
Flavoured milk drink	αs1 -CN
Hydrolysed whey protein isolate	β-LG fragments

Nutraceutical products comprising short bioactive peptides showing in vitro or in vivo antimicrobial, ACE-inhibitory activity and/or antihypertensive and/or antioxidant activity are being considered for possible use by the pharmaceutical industry. The casein hydrolyzates reinforce the body's natural defences or serve as pharmaceutical products for facilitating the control of blood and/or bacterial infections. Recent developments include increasing mineral transport by phosphorylated groups of peptides and increase absorption of chelated calcium, iron, copper, zinc and manganese in the intestine.

References

1. Anonymous Prebiotic fibre http// www.prebiotic.ca.
2. Anonymous (2013). Health benefits of probiotics http// www. Medicine net on health.com.
3. Aydin R. (2005). Conjugated Linoleic Acid: Chemical Structure, Sources and Biological Properties. Turkey J Veterinary and Animal Science, 29:189-195.
4. Bhat Z. F. and Bhat H. (2011). Milk and dairy products as functional foods- A Review. International Journal of Dairy Science, 6(1):1-12.
5. Fitz Gerald R. J. (1998). Potential Uses of Caseinophosphoeptides. International Dairy Journal, 8(5-6):451-457.
6. Gnann T. (2007). Method for the production of milk or milk products with a high proportion of melatonin. Patent CA 2611120 A1.
7. Negro A. (2010). Bioactive peptides from milk proteins: Focusing on peptides displaying Immunomodulatory activity. Università degli studi padova. http://paduaresearch.cab.unipd-.it/2590/1/TESIregazzo.pdf.
8. Saarela M. (2007). Functional dairy products. Wood Head Publishing Limited, Cambridge, England.
9. Singh V. P. Sachan N. and Verma A. K. (2011). Melatonin milk; A Milk of intrinsic health benefits: A Review: International Journal of Dairy Science, 6(4):246-252.
10. Tasan M., Bilgin B, Geçgel U and Demirci A S (2006). Phytosterols as Functional Food Ingredients. Journal of Tekirdag Agriculture Faculty, 3(2):153-159.

Unit-VII: Functional Foods and Nutraceuticals Regulations and Market Prospective

CHAPTER 42

Functional Foods and Nutraceuticals: Regulatory Framework for Protecting Consumers, Promote Fair Trade and Encourage Product Innovation in the Food Industry

Ashok Kumar and Preetinder Kaur

Abstract

Functional foods and nutraceuticals are one of the fastest-growing segments of the food industry. The demand for these foods is driven by health conscious consumer and the growing awareness of the close connection between diet and health, particularly in relation to chronic diseases. Over the past five years (2008-2012), the Indian Nutraceuticals Market has been growing at a rapid pace and is expected to rise to US $ 5 billion by 2015. The rapid growth and development of this new industry combined with the need to protect consumer health and interests, has led to the evolution of a myriad of different regulatory frameworks and policies across multiple markets. Japan was one of the first countries to recognize a different category of healthier food products. The Japanese government formally accepted this category in the 1980s. Other developed countries viz. The EU, The US, Canada etc also regulated this sector during 1990s. In India, the Food Safety and Standards Act 2006 (FSSA) was enacted in 2006 in response to pressure from stakeholders and the food industry, transforming the regulation of food in India. Punjab state, with abundance of food grains and increasing trends in fruits and vegetables, provides options for both the producer and growth promotion in the nutraceuticals *sector through partenship between research centers, private entrepreneurs and indigenous communities. Regulatory frameworks should focus on the protection of consumers, promotion of fair trade and innovation in the sector. This paper discusses functional foods and nutraceuticals as per food regulations prevalent in various countries and highlights how these regulatory frameworks play a major role in transforming the food sector.*

Introduction

The consumers' health consciousness is an important factor driving the food market in the present era. The food industry has reacted to this trend by developing a growing variety of new products with health-related claims, including organic and functional foods that are selected by consumers for their health-promoting properties. The concept of food has undergone a radical transformation during last few years. Food, in addition to its nutritional and sensory properties, also plays an important role in the maintenance of health, on psycho-physical well-being and prevention of certain diseases. Today foods are intended not only to satisfy hunger and to provide necessary nutrients for humans but also to prevent nutrition-related diseases and improve physical and mental well-being of the consumers. Scientific studies over the past decade have shown close connection between diet and health, particularly in relation to chronic diseases, and have encouraged the development of a growing spectrum of products such as nutraceuticals, medifoods, vitafoods. Increased consumer interest in healthier food products is driven by a variety of factors including growing awareness of the link between diet and health, the desire to age 'gracefully' by maintaining good health, greater convenience in meeting nutritional needs and preventing chronic diseases such as diabetes, cancer, cardiovascular diseases and respiratory diseases. Consumers are more conscious of the maintenance of good health through diet, thereby gravitating towards foods that offer additional benefits beyond that provided by conventional food and are demonstrating a willingness to pay a premium for these products. The global market for functional foods (Table-1) and Natural Health Products (NHPs) is estimated to represent approximately 3% of the total food market (Kotilainen et al 2006) nearly 61 billion US$ and the three dominant markets, United States followed by Europe and Japan, contribute over 90% of the total sales. Major functional foods firms in USA and Europe are shown in Table-2 (a) and (b).The functional food and natural health products market has been dominated by developed countries; however, emerging markets such as Brazil, China, and India have come to represent important potential future markets (Nutrition Business Journal, 2007).

Table 1: Estimated Global Market Size for Functional Foods (US$ Billions)

	1997		2000		2005	
	US$ Billions	%	US$ Billions	%	US$ Billions	%
Canada	0.4	1	0.6	1	1.0	1.1
United States	13.6	35	17.4	34	25.7	28.7
Japan	9.6	24.7	12.8	25	22.6	25.2
European Union	12.3	31.6	16.4	32	34.4	38.4
Rest of the World	3.0	7.7	4.1	8	5.9	6.6
Total	38.9	100	51.3	100	89.6	100

Source: Nutritional Business Journal (NBJ 2007a)

Table 2 (a): Major Functional Food Firms in USA

Manufacturer	Sales (Million US$)	Share (%)
PepsiCo	2,397	16.3
Kellogg Co.	2,100	14.3
Coca-Cola Co.	1.591	10.8
General Mills	1,558	10.6
Kraft Foods, Inc.	847	5.8
Group Danone	643	4.4
Nestle USA, Inc.	545	3.7
Ocean Spray	427	2.9
Welch Foods. Inc	330	2.2
Dean Foods	292	2.0
Ferolito Vultaggio & Songs	241	1.6
Red Bull North America Inc	226	1.5

Source: Mintel, 2006

Table 2(b): Major Functional Food Firms in the Europe

Company	Product(s)
Pepsico	Quaker Oats, Gatorade, Tropicana
GSK	Lucozade, Ribena
Danone	Actimel, Activa, Essensis, Senja
Nestle	Calcium Milk, Beta glucans
Unilever	Slim Fast
Muller	Vitality prebiotics and probiotics yoghurt
Yakult	Lactobacillus based drinking yoghurt
Innocent	Fruit Smoothie
Ebro Puleva/Puleva Biotech	Omega-3 enriched milk, Euploy-3

Source: New Zealand Trade & Enterprise, 2007

Table 3(a): Some Examples of Functional Foods in Canada

Functional Components	Source/Food	Potential IIealth Benefits
Carotenoids		
Alpha-carotene/Beta-carotene	Carrots, Fruits, Vegetables	Nutralize free radicals, which may cause damage to cells
Lutein	Green vegetables	Reduce the risk of macular degeneration
Lycopene	Tomato products	Reduce the risk of prostate cancer
Dietary Fibre		
Insolubel Fibre	Wheat bran	Reduce risk of breast or colon cancer
Beta-Glucan	Oats, barley	Reduce risk of cardiovascular disease. Protect against heart disease and some cancers; lower LDL and total cholesterol
Solube Fibre	Psyllium	Reduce risk of cardiovascular disease. Protect against heart disease and some cancers; lower LDL and total cholesterol
Fatty Acids		
Long chain omega-3 Fatty Acids-DHA/EPA	Salmon and other fish oils	Reduce risk of cardiovascular disease. Improve mental, visual functions
Conjugated Linoleic Acid (CLA)	Cheese, meat products	Improve body composition. Decrease risk of certain cancers
High Oleic Canola Oil	Nexera	Reduces the risk of coronary heart disease (CHD)
Fructo-oligosaccharides (FOS)	Jerusalem artichokes, shallots, onion powder	Improve quality of intestinal microflora; gastrointestinal health
Lactobacillus	Yogurt, other dairy	Improve quality of intestinal microflora; gastrointestinal health
Soy Phytoestrogens		
Isoflavones; Diazein Genistein	Soybeans and Soy-based foods	Menopausal symptoms, such as hot flashes; protection against heart disease and some cancers; lower LDL and total cholesterol
Phenolics		
Ligans	Flax, rye, vegetables	Neutrilize free radicals; reduce risk of cancer

Source: Health Canada 2010

Functional foods are one of the fastest-growing segments of the food industry. Functional food is given an additional function by adding new ingredients or more of existing ingredients. The American Dietetic Association (ADA) defines functional foods as foods "that include whole foods and fortified, enriched or enhanced foods have a potentially beneficial effect on health when consumed as part of a varied diet on a regular basis, at effective levels." Functional foods include processed foods fortified with health promoting additives like "Vitamin-enriched" products. Products considered functional generally do not include products where fortification has been done to meet government regulations and the change is not recorded on the label as a significant addition. For example, for fortification of iodine to table salt, vitamin-D to milk, done to resolve public health problems such as goiter and rickets, respectively. Fermented foods with live cultures are considered functional foods with probiotic benefits.

The term nutraceutical was coined from nutrition and pharmaceutical in 1989 by Stephen De Felice, founder and chairman of foundation for innovation in medicine, an American organization which encourages medical health. According to him "a nutraceutical is any substance that is a food or apart of food and provides medical or health benefits, including the prevention and treatment of disease". Such products may range from isolated nutrients, dietary supplements and specific diets to genetically engineered designer foods and herbal products. The concept of nutraceutical was started from the survey in U.K., Germany and France and it concluded that diet is rated more highly by consumer then exercise or hereditary factors to achieving a good health. Table 4 shows some examples of NHPs in Canada.

Table 3 (b): Some Examples of Functional Foods in USA

Functional Food Area	Functional Food Product	Company	Brands
Dairy Products	Probiotic youghurt/kefir	Dannon (Stonyfield Farms	Dan Active (marketed as Actimel outside the US); Activa; YoBaby: All Natural Fruit Blends; YoSelf
		Cool Brands Internationals	Breyers Light! Probiotics Plus Yoghurt
		Lifeway Foods	Lifeway Organic ProBugs Milk Drinks
	Cholesterol-lowering youghurt	General Mills Dannon	Yoplait Healthy Heart Light'n Fit with Fiber
	Active health drinks	Dannon	Actimel
	Fortified milk	Deans; Borden Dairy Farmers; Mayfield Farms; Kemps: Suiza	None are major, market shares distributed among these
Bakery Products	Wholegrain (heart Healthy) and fortified breads	Sare Lee	Heart healthy Plus: Soft & Smooth
	Breads with omega 3 fatty acids	Wegmans Food Markets: The Baker	None are major market shares distributed among these
		Arnold Foods	Arnold Smart & Healthy
	High Fibre Cookies	RD Foods	Right Direction Cookies
		Quaker	Quaker Breakfast Cookies
		Nabisco	Wholegram Chips Ahoy! Wheat Thins and Fig Newtons
Cereal Products	Heart health cereals	Quaker	Take Heart Instant Oatmeal
		General Mills	Cheerios; All Big G Brands
		Kraft's Post	Grape Nuts; Raisin Bran; Shareded Wheat; Toasties; Bran Flakes
		Kellogg	Smart Start brands; Heart to Heart (under Kashi Kellogg subsidiary)
	Calcium fortified cereal bars	Quaker	Chewy

(*Contd.*)

	High fibre cereal bars	Kellogg	Cereal and Milk Bars
	Heart Healthy cereal bars	Nature Valley	Health Heart
	Wholegrain pasta	Barilla	Barilla Plus
		Kraft	Supermac & Chees Pasta and Sauce
Soya Products	Soya Milk	White Wave	Silk
		Odwalla	Odwalla soymilk
Beverages	Fruit juice and juice blends	Tropicana (PepsiCo)	Pure Premium Orange Juice with Calcium and Extra Fiber; Pure premium Essentials (Immunity Defence, Healthy Heart, Healthy Kids, Light & Healthy
		Minute Maid (Coca Cola)	Heartwise; Minute Maid Extra
		Ocean Spray	Cranberry Juice Cocktails
		Ocean Spray	Cranberry Juice Cocktaila
		POM Wonderful	POM Wonderful
	Enhanced waters	Gatorade (PepsiCo)	Propel Fitness Water
		PepsiCo	Aquafina flavoured water; SoBe Life Water
		Coca-Cola	Dasani
Ment, Fish and Eggs	Canned fish with omega-3	Star Kist Chicken of the Sea	None are major, market shares distributed among these.
	Fish oil supplements	Bumble Bee and Leiner Health	
	Omega-3 DHA enriched eggs	Gold Circle Farms	
		Eaggland's Best	

Source: Leatherhead Food International (2006 p 135-153) Klimas et al. (2008)

Table 4: Some Examples of Natural Health Products in Canada

Product/Ingredient Name	Source Material	Uses	Selected Companies/Organization
Conjugated linoleic acids (CLA)	In trancylglycerol form and free fatty acid form, CLA is derived from processed safflower or sunflower oil	May help to support a modest improvement to body composition and a modest reduction of fat mass when used with a program of reduced intake of dietary calories and increased physical activity	Herbal Magic Inc. Jamieson Lab Ltd
Folate	Folic acid/folacin/folate	General: A factor in the maintenance of good health Specific: Helps the body to metabolize proteins Helps to form red blood cells	Nature's Sunshine Products of Canada Ltd.
Selenium	Monophydrate selenium dioxide. Selenium hydrolyzed animal protein chelate. Selenium citrate. Selenium yeast. Sodium selenate	General: A factor in the maintenance of good health. Specific: An antioxidant for the maintenance of good health	1. Flora distributing and Manufacturing Ltd. 2. Pegasus Pharmaceuticals Group Inc.
Pantothenic acid	Calcium DL-panthothenate. D-Panthenol/Dexpanthenol. Solaray ® Pantothenic Acid	General: Maintenance of good health. Specific: Helps the body to metabolize carbohydrates, fats and proteins. Helps in tissue formation. Prevents pantothenic acid deficiency	Au Naturel Inc

India is the most potential market for Nutraceuticals and Dietary Supplement products. In 2010, the Indian nutra industry was estimated at US $ 2 Billion, roughly 1.5 percent of the global nutraceutical industry. Broad segments of Indian nutraceutical industry include Dietary supplement (40%) and Functional food and beverage market (60%). Indian nutraceutical ingredient market is around US $250 million comprising of vitamins and minerals (36%), probiotics (9%), omega 3 fatty acids (5%) and others (50%). Over the past five years (2008-2012), the Indian Nutraceuticals Market has been growing at a rapid pace. Recently, the Organisation of Pharmaceutical Producers of India (OPPI) is rooting for lower customs and central excise duty on nutraceuticals and health supplements in Budget 2013. According to OPPI, basic customs duty (BCD) for health supplements should be slashed to 10% and additional customs duty (ACD)/central excise duty to 6% to make it at par with drugs. This would help in making the products affordable for the common man and help boost the healthcare agenda of the government. In India, functional foods and beverages are expected to see increased consumption over the next five years resulting in this segment garnering greater product share (67%) in the market as opposed to dietary supplements (33%). The total Indian nutraceuticals market in 2015 is expected to be approximately US $ 5 billion.

Shift Towards Functional Foods and Nutraceuticals: The Reasons

The World Health Organization (WHO) identifies nutrition as a significant and manageable determinant of chronic disease, stressing the need for a shift in nutrient intake towards 'healthier' foods (WHO, 2002). Rising health care costs have also spurred research into healthier foods and food ingredients. Significant scientific evidence now demonstrates that some foods and food ingredients provide certain physiological benefits and/or reduction of risk in chronic disease benefits beyond basic nutritional functions. This combination of growing scientific evidence linking the role of diet in maintaining good health and preventing disease, together with increased consumer and industry interest, have raised overall awareness in and the profile of the health food sector.

Functional food refers to food that is intended to be consumed as part of a normal diet and contains ingredients that have the potential to enhance human health or reduce the risk of disease beyond basic nutritional functions (Health Canada, 2009; ADA, 2004; IFIC, 2009; Stein and Rodriguez- Cerezo, 2008). Natural health products (NHPs), also known as nutraceuticals or food supplements or nutritional supplements, are products that have been isolated or purified from food and may include ingredients such as amino acids or vitamins; they can be marketed in the form of pills, powders, capsules or tablets.

NHPs are intended to have a physiological benefit or to provide protection against chronic disease (Health Canada, 1998; EFSA, 2010). Consumption of products from either category provides additional health benefits beyond the supply of basic nutrients.

There has been an explosion of consumer interest in the health enhancing role of physiologically-active specific nutraceuticals. These products include food supplements, dietary supplements, value-added processed foods as well as non-food supplements such as tablets, soft gels, capsules etc. These materials are rich source of bioactive flavonoids, carotenoids, anthocyanins, vitamins, and other polyphenolics. They help in prevention of various diseases through antioxidant activity and reduce the disease risks. Excessive quantities of nutraceutical-rich foods must therefore, be processed and converted into value added nutraceuticals.

The interest in nutraceuticals continues to grow, powered by progressive research efforts to identify properties and potential applications of nutraceutical substances, and coupled with public interest and consumer demand. Nutraceutical is a combination of two words: Nutrition and Pharmaceutical. Nutraceuticals are food products that provide health as well as medical benefits; including the prevention and treatment of disease. Phytochemicals and antioxidants are two specific types of nutraceuticals. Research has proved that foods with phytochemicals may help to provide protection from diseases such as cancer, diabetes, heart disease, and hypertension, e.g. carotenoids found in carrots. Antioxidants may be helpful in avoiding chronic diseases, by preventing oxidative damage in our body.

Over the last 20 years, a number of Nutraceuticals have become available for self medication or for sale. There has been a boom in sale of Nutraceuticals because of:

- Adverse effects of pharmaceuticals
- Increased tendency of patients for self-medication
- Aging population e.g. – arthritis.

Across the globe, populations are aging. Life expectancy continues to rise, as does the contribution made by older individuals to the total population. Also, obesity is now recognized as a global issue as its incidence continues to climb in countries throughout the world. In India nearly 400 million people belong to the middle class and have the disposable income which made them capable to buy nutraceuticals and dietary supplements. It is an inevitable fact that affluence is one of the causes of lifestyle diseases, which nutraceuticals and dietary supplements often address. People can optimize the health-promoting capabilities of their diet by way of supplementation and by consuming foods that have been formulated or fortified to include health-promoting factors.

Classification of Fuctional Foods

The ADA breaks down functional foods into four categories:

1. **Conventional Foods:** Conventional foods are foods which have not been modified by enrichment or fortification; they are still in their natural state. Most whole fruits and vegetables fall into this category because they're rich in phytochemicals such as lycopene and lutein, as well as other beneficial compounds.
2. **Modified Foods:** Modified foods have been enriched, fortified or enhanced with nutrients or other beneficial ingredients. Calcium-fortified orange juice, folic acid enriched breads and margarine enhanced with plant sterols are functional foods that have been modified. Energy drinks that have been enhanced with herbs such as ginseng, as well as other potentially controversial foods, also fall into this category.
3. **Medical Foods:** The Food and Drug Administration (FDA) defines medical food as "food which is formulated to be consumed or administered entirely under the supervision of a physician and which is intended for the specific dietary management of a disease or condition for which distinctive nutritional requirements, based on recognized scientific principles, are established by medical evaluation."
4. **Foods for Special Dietary Use:** Foods for special dietary use are similar to medical foods, but they're available commercially and don't require the supervision of a health care provider.

Conventional functional food sources are

Cereals: Cereals such as wheat and rice are major constituents of the diet of the world's population. Dietary fibre, particularly from cereals, is the major preventive factor against a number of cancers. Increased dietary fibre intake, especially soluble fiber, has also been linked to reduce risk of coronary heart disease. Oat is one of the few foods that have been tested as a dietary source of the cholesterol-lowering soluble fiber β- glucan. Consumption of diets high in soluble fibre β- glucan has been associated with reduction in total and low-density lipoprotein (LDL) cholesterol, thereby reducing the risk of CHD. Insoluble fibre from wheat bran can help reduce the risk of colon cancer.

Legumes: Soy is of the utmost economic importance, is most abundant in phytoestrogens. Soybeans have several anticarcinogens including protease inhibitors, phytosterols, saponins, phenolic acids and phytic acid. The combination of both soy protein and isoflavones are necessary to produce the greatest cholersterol-lowering effect. Soybean protein plays a preventive and therapeutic role in cardiovascular disease.

Blueberries: are often referred to as a super-food. These are flowering plants of the genus Vaccinium with dark-blue berries. Especially in wild species, blueberries contain anthocyanins (flavonoids), the antioxidant pigments and various phytochemicals possibly having a role in reducing risks of some diseases, including inflammation and certain cancers.

Oilseeds: Flaxseed is the most prominent oilseed, leading source of omega-3 fatty acid, α-linolenic acid (ALA). PUFA is important for visual and brain development. Flaxseed has a protective effect against both mammary and colon cancer. Flaxseed is also a good source of dietary fibre known to improve glucose metabolism and decrease risk of colorectal cancer. It also reduces total and LDL cholesterol and platelet aggregation.

Fruits: Lycopene, found in tomatoes has been implicated in risk reduction of prostate cancer as well as other cancers including breast, digestive tract, cervix, bladder, skin and lung. The limonoids present in citrus fruits are the preventive agent against a variety of human cancers. Cranberry juice is helpful in treatment of urinary tract infections. Flavonoids and other phenolics have been shown to possess anti-carcinogenic, anti-inflammatory, anti bacterial, antiviral, anti-allergic and anti-oxidative effects.

Ginger: The active ingredient in ginger is gingerol, a compound which relaxes blood vessels, stimulates blood flow and relieves pain. It's commonly used as a digestive aid and eases motion sickness, nausea and inhibits vomiting. It is helpful for morning sickness and for people who suffering from the side effects of chemotherapy.

Garlic: The medical properties of garlic such as anti bacterial activity, antifungal activity, inhibition of platelet aggregation and reduction of serum cholesterol levels are due to biologically active sulphur compounds. Garlic is helpful in reducing human cancer risk, cancer of stomach, colon and gastrointestinal tract and also cardiovascular diseases.

Cinnamon: is an anti-microbial food that can stop the growth of bacteria, fungi and yeast. It also has anti-clotting and anti-inflammatory properties, which helps to prevent unwanted clumping of blood platelets. Cinnamaldehyde (also called cinnamic aldehyde) has been well-researched for its effects on blood platelets.

Dairy Products: Dairy products have been associated with health benefits for many years containing bioactive peptides, probiotic bacteria, antioxidants, vitamins, specific proteins, oligosaccharides, organic acids, highly absorbable calcium, conjugated linoleic acid and other biologically active components.

Cheese: Cheese is manufactured by various methods of coagulating milk and thus removing liquid called whey, rennet is normally added to this "curd" to make it stabilize. It is a very good source of protein, calcium, phosphorus, iodine and selenium. Calcium helps to protect colon cells from cancer-causing chemicals and also helps to prevent the bone loss.

Various types of Functional Foods

- Functional dairy products include probiotics and cholesterol-lowering yoghurts.
- Bakery products consist primarily of bread and cookies fortified with fibre and omega-3 fatty acids for heart health.
- Cereal products encapsulate calcium fortification, fibre fortification and wholegrain cereals.
- Soy products are dominated by soya milk for reducing the risk of coronary heart disease. Beverage products feature calcium fortification, extra fibre properties and enhanced waters.
- Meat, fish and eggs are mainly fortified with omega 3 and fish oil supplements

Different classes of Nutraceuticals that help in improving the human health

1. Vitamin C, vitamin E, β-Carotene, lycopene (carotenoids), lipoic Acid, glutathione (thiols) play role in cardiovascular disease prevention and inhibition of necrosis; Co-Enzyme Q-10, super oxide dismastase (enzyme), selenium, copper, manganese, zinc (minerals) act as anti-cardiovascular disease in cardiac cells.
2. Polyunsaturated fatty acids (PUFA) such as safflower oil, corn oil, soybean oil, mustard oil, evening primrose oil, flax oil showed protective effects in heart disease and stroke, inflammatory arthritis, inflammatory bowel disease, asthma, cardiovascular disease.
3. Dietary fibers such as oats, dried beans, legumes, chicory as water soluble

fibers, apple, orange, apricot, plum, pine apple contain 18-30% fiber contents. The vegetable sources such as cabbage, carrot, lettuce, onion, tomato containing 9 to 12 % fiber contents showed antioxidant and myocardial cell proliferation inhibitory properties.

Regulatory Frameworks and Policies

The functional food and NHP industry suffers from widespread inconsistencies in terminology and definitions throughout global markets, inconsistencies that are also prevalent in regulatory and policy environments. An industry experiencing rapid growth and development, combined with the need to protect consumer health and interests, has led to the evolution of a myriad of different regulatory frameworks and policies across multiple markets. Different countries have varied regulations and policies regarding both nutraceuticals and functional foods as discussed below:

Japan: Japan was one of the first countries to recognize a different category of healthier food products. In the 1980s the Japanese government formally accepted that food products could positively affect the physiological system of the human body and accepted approved health claims for food (Ministry of Health, Labour and Welfare Website, 2010). These foods were classified as "Food for Specified Health Uses" (FOSHU) (Fig.1). The Japanese government defines FOSHU as food that "is intended to be consumed for the maintenance/ promotion of health or for special health uses by people who wish to control health conditions, including blood pressure or blood cholesterol" (Ministry of Health, Labour and Welfare Website, 2010). This definition applies to both functional foods and natural health products. The FOSHU claim must be approved by the Ministry of Health, Labour and Welfare (MHLW). Specifically, to qualify as FOSHU, a food has to fulfill the following requirements (Ministry of Health, Labour and Welfare Website, 2010):

- Effectiveness on the human body is clearly proven
- Absence of any safety issues (animal toxicity tests, confirmation of effects in the cases of excess intake, etc.)
- Use of nutritionally appropriate ingredients (e.g. No excessive use of salt, etc.)
- Guarantee of compatibility with product specifications by the time of consumption
- Established quality control methods, such as specifications of products and ingredients, processes and methods of analysis"

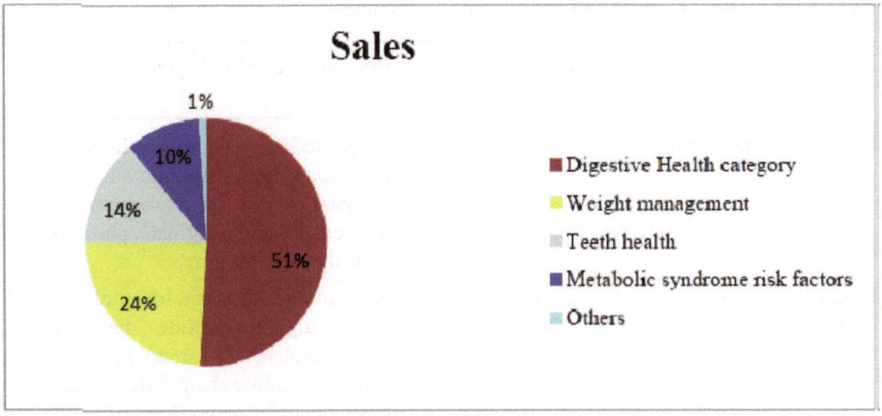

Fig.1: Health Categories of FOSHU (Foods for special health use) in Japan (Baileyl 2008)

Canada: Categories of healthier foods and food derivatives have been evolving over time, with references commonly made to functional food, natural health products, nutraceuticals, novel food, etc

A Functional Food is similar in appearance to, or may be, a conventional food, is consumed as part of a usual diet, and is demonstrated to have physiological benefits and/or reduce the risk of chronic disease beyond basic nutritional functions (Health Canada, 1998).

A nutraceutical is a product isolated or purified from foods that is generally sold in medicinal forms not usually associated with food. A nutraceutical is demonstrated to have a physiological benefit or provide protection against chronic disease (Health Canada, 1998).

A natural health product is a substance or a combination of substances in which all the medicinal ingredients are made up of natural substances. This includes homeopathic medicine or a traditional medicine that is manufactured, sold and represented for use in: the diagnosis, treatment, mitigation or prevention of a disease, disorder or abnormal physical state or its symptoms in human; restoring or correcting organic functions in humans and; modifying organic functions in humans such as modifying those functions in a manner that maintains or promotes health" (Canada Gazette, 2003).

Table 5: FOSHU Functions and Functional Components

FOSHU Function	Number of Products Approved	Principal Function Ingredients
Gastrointestinal health	254	Oilgosaccharides, Lactobacillus, Bifidobacterium, Psyllium husk, indigestible dextrin, wheat bran, low molecular sodium alginate, partially hydrolyzed guar gum
High Cholesterol/ triglyceride level and body fat	117	Soy protein, Chitosan, low molecular sodium alginate, peptides, diacylclycerol, plant sterol/stanol (esters), green tea catechin, middle chain fatty, Psyllium husk, degradation products of globin protein
High blood glucose	71	Indigestible dextrin, L-arabinose, h-wheat albumin
High blood pressure	64	Gamma amino butyric acid (GABA), Peptides
Dental Health	34	Xylitol, polyols, tea Polyphenols, Casein Phosphopeptite and Amporphous Calcium Phosphate (CPP-ACP)
Bone health	26	Soy isoplavone
Anemia	3	Heme iron

Source: Ohama et al., 2006

Australia and New Zealand: There is no universally accepted definition for functional food, although the joint food regulation agency in these two countries has a working definition.

A functional food is defined as being "similar in appearance to conventional food and intended to be consumed as part of a normal diet, but modified to serve physiological roles beyond the provision of simple nutrient requirements" (FSANZ, 2006).

A nutraceutical is defined as a food designed to provide health benefits that may contain ingredients such as specific amino acids or vitamins. Nutraceuticals can be marketed in the form of pills, capsules or tablets (FSANZ Website, 2001).

Russia: Food products with health benefits (functional food, and NHPs) are known as Biologically Active Food Supplements (BAFS). The Russian Federation defines BAFS as nutritive substances and minor food components used to ameliorate deficiencies, decrease risk of debilitating diseases, and improve the quality of life (Zawistowski, 2008). The food products and its components should be natural or identical to natural. They should be consumed along with food and incorporated into food. BAFS are not regarded as medicines and are not meant to treat or diagnose diseases. Biologically Active Food Supplements in Russia are equivalent to nutraceuticals in Canada, however, in Russia the definition used for NHPs also applies to functional foods and novel foods.

Korea: the Health Functional Food Act (HFFA, 2004) defines food products with health benefits (functional foods and NHPs) as healthier food products.

Healthier food products are defined as food supplements containing nutrients or other substances (in a concentrated form) that have a nutritional or physiological effect whose purpose is to supplement the normal diet. A healthier food product in Korea is similar to a NHP in Canada. The Korean Health/ Functional Food Act that came into effect in 2004 requires these products to be marketed in measured doses, such as in pills, tablets, capsules, and liquids (Ministry of Health and Welfare, Korea, 2004). The definition used for NHPs and functional foods also applies to novel foods.

The United States (US): In United States (US), the Food and Drug Administration (FDA) does not provide a legal definition for functional food. However, a number of working definitions have been developed by different organizations including the American Dietetic Association (ADA), the International Food Information Council (IFIC) and the Institute of Food Technologies (IFT).

- A **functional food** is any food that moves beyond necessity (basic nutrition) to provide additional health benefits that may reduce disease risk and/or promote optimal health (ADA, 2004; IFIC, 2009).
- **Functional foods** are, "foods and food components that provide a health benefit beyond basic nutrition (for the intended population). These substances provide essential nutrients often beyond quantities necessary for normal maintenance, growth, and development, and/or other biologically active components that impart health benefits or desirable physiological effects"(Institute of Food Technologies)
- **Dietary supplement** is a product that is intended to supplement the diet and contains any of the following dietary ingredients: vitamin, mineral, amino acid, etc (US Dietary Supplement Health and Education Act: DSHEA, 1994). Dietary supplements would be classified as nutraceuticals or NHPs in Canada. Dietary Supplements must also conform to physical criteria, being 'intended for ingestion in pill, capsule, tablet, powder or liquid form' and should not be represented for use as a conventional food or as the sole item of a meal or diet (DSHEA, 1994).

- **Medical foods** are defined as "a food which is manufactured to be consumed or administered entirely under the supervision of a physician and which is intended for the specific dietary management of a disease or condition for which distinctive nutritional requirements, based on recognized scientific principles, are established by medical evaluation"(Orphan Drug Act,1988). To be considered a medical food, a product must, at a minimum, meet very specific criteria: the product must be a food for oral or tube feeding; the product must be labelled for the dietary management of a specific medical disorder, disease, or condition for which there are distinctive nutritional requirements; and the product must be intended to be used under medical supervision (FDA, 2010).
- **Foods for Special Dietary Use** are defined as "a particular use for which a food purports or is represented to be used, including but not limited to the following: 1. Supplying a special dietary need that exists by reason of a physical, physiological, pathological, or other condition...; 2. Supplying a vitamin, mineral, or other ingredient for use by humans to supplement the diet by increasing the total dietary intake. 3. Supplying a special dietary need by reason of being a food for use as the sole item of the diet...." (FDA, 2009). Examples include infant foods, hypoallergenic foods such as gluten-free foods and lactose-free foods, and foods offered for weight reduction.
- **Novel food**: The FDA does not have a formal definition for novel food. These foods are often referred to as new and emerging food/biotechnology products, and a premarket notification with the FDA is usually recommended FDA (FDA, 2005).

The European Union (EU): In the late nineties, research into functional ingredients was showing promising prospects for the use of such ingredients in foodstuffs. Mainly due to safety concerns, these new scientific developments were accompanied by an urgent call for legislation. The 1997 Green Paper on Food Law gave a new impetus to the foundation of European Food Law. The European Commission 2000 White Paper on Food Safety announced some 80 proposals for new and improved legislation in this field. A General Food Law Regulation was established, laying down the principles of food law and the creation of an independent Food Authority endowed with the task of giving scientific advice on issues based upon scientific risk assessment with clearly separated responsibilities for risk assessment, risk management and risk communication. Its aim was to "provide the basis for the assurance of a high level of protection of human health and consumer interest in relation to food, taking into account in particular the diversity in the supply of food including traditional products, while ensuring the effective functioning of the internal market. It establishes common principles and responsibilities, the means to provide a strong science base, efficient organisational arrangements and procedures to underpin decision-making in matters of food and feed safety". As the Regulation applies to all foodstuffs, its general principles also cover foods with added functional properties (e.g. 'functional foods', 'nutraceuticals', dietetic foods and food supplements). However, there is no regulatory framework for 'functional foods' or 'nutraceuticals' in EU Food Law. The rules of the general food law Regulation are applicable to all foods. In addition, legislation on dietetic foods, on food supplements or on novel foods may also be applicable to functional foods depending on the nature of the product and on their use. Finally, the two proposals on nutrition and health claims and on the addition of vitamins and minerals and other substances to foods, which are currently in the legislative process, will also be an important factor in the future marketing of 'nutraceuticals' in Europe. The cornerstone of EU legislation on food products, including functional foods and nutraceuticals is 'safety'.

- **Food** (or 'foodstuff') is defined as "any substance or product, whether processed, partially processed or unprocessed, intended to be, or reasonably expected to be ingested by humans. 'Food' includes drink, chewing gum and any substance, including water, intentionally incorporated into the food during its manufacture, preparation or treatment. 'Food' shall not include: feed; live animals unless they are prepared for placing on the market for human consumption; plants prior to harvesting; medicinal products; cosmetics; tobacco and tobacco products; narcotic or psychotropic substances within the meaning of the United Nations Single Convention on Narcotic Drugs, 1961, and the United Nations Convention on Psychotropic Substances, 1971; residues and contaminants".
- **Medicinal products:** are defined as: "Any substance or combination of substances presented as having properties for treating or preventing disease in human beings; or any substance or combination of substance which may be used in or administered to human beings either with a view to restoring, correcting or modifying physiological functions by exerting a pharmacological, immunological or metabolic action, or to making a medical diagnosis".
- **Food supplements** are defined as "foodstuffs the purpose of which is to supplement the normal diet and which are concentrated sources of nutrients or other substances with a nutritional or physiological effect, alone or in combination, marketed in dose form, namely forms such as capsules, pastilles, tablets, pills and other similar forms, sachets of powder, ampoules of liquids, drop dispensing bottles, and other similar forms of

liquids and powders designed to be taken in measured small unit quantities".

Risk analysis: This regulation firmly establishes scientific risk analysis at the basis of the decision-making process in matters of food law. It defines this process through a number of definitions and lays down the responsibilities for each of the respective stages of the risk analysis process.

- **Hazard:** means a biological, chemical or physical agent in, or condition of, food or feed with the potential to cause an adverse health effect.
- **Risk:** means a function of the probability of an adverse health effect and the severity of that effect, consequential to a hazard.
- **Risk analysis:** means a process consisting of three interconnected components—risk assessment, risk management and risk communication.
- **Risk assessment:** means a scientifically based process consisting of four steps—hazard identification, hazard characterisation, exposure assessment and risk characterisation; risk assessment is to be based on the available scientific evidence and undertaken in an independent, objective and transparent manner.
- **Risk management:** means the process of weighing policy alternatives in consultation with interested parties, considering risk assessment and other legitimate factors, and, if need be, selecting appropriate prevention and control options; The responsibility of risk management lies with the European Commission (EC), the Council and the European Parliament (EP).
- **Risk communication:** means the interactive exchange of information and opinions throughout the risk analysis process as regards hazards and risks, risk-related factors and risk perceptions, among risk assessors, risk managers, consumers, feed and food businesses, the academic community and other interested parties, including the explanation of risk assessment findings and the basis of risk management decisions.

The Food Supplements Directive focuses only on vitamins and minerals. The Directive forces that maximum amounts of vitamins and minerals present in food supplements shall be set, taking into account the upper safe levels of vitamins and minerals established by scientific risk assessment based on generally accepted scientific data, taking into account, as appropriate, the varying degrees of sensitivity of different consumer groups; and intake of vitamins and minerals from other dietary sources. It also specifies that when the maximum levels will be set, due account should also be taken of reference intakes of vitamins and minerals for the population.

One good example of the application of the risk analysis process: Firstly, the Upper Tolerable Safe Intake Levels (UL) of the vitamins and minerals are set by the Scientific Committee on Food (SCF). The UL is the intake level that can be consumed daily over a lifetime without being likely to pose any risk to health according to available evidence. Secondly, the maximum levels are established for the use in food supplements based on the established UL and the intake of the vitamins and minerals from other food sources including water.

Another example in the Food Supplements Directive for managing risk through scientific risk assessment is the establishment of a positive list of nutritional substances that are allowed in food supplements. The list includes substances that may be added for specific nutritional purposes in foods for particular nutritional uses. European legislation does not consider 'functional foods' or 'nutraceuticals' as specific food categories.

Foods for Particular Nutritional Uses: One category of products that has firmly established legal frameworks are "Foods for Particular Nutritional Uses" (also called PARNUTS or dietetic foods). It offers a good example of a legislation that balances scientific risk assessment, consumer protection, manufacturer responsibility and market innovation in a rather proportionate way. Originally the European Legislator foresaw nine categories of PARNUTS foods that would become subject of a specific Directive. Today, five of them have been established:

- Infant Formulae and Follow-on Formulae (Directive 91/321/EEC) (originally two different Directives were foreseen).
- Processed Cereal-based Foods and Baby Foods for Infants and Young Children (Directive 96/5/EC).
- Foods Intended for Use in Energy-restricted Diets for Weight Reduction (Directive 96/8/EC).
- Foods for Special Medical Purposes (Directive 1999/21/EC).

This PARNUTS legislation is applicable to some 'functional foods' or 'nutraceuticals' if such foods are intended for particular nutritional uses. This basically means that such foods must have a special composition or manufacturing process, must be clearly distinguishable from foodstuffs for normal consumption, must be suitable for their claimed nutritional purposes and marketed in such a way as to indicate such suitability. With "particular nutritional use" is understood that the product must fulfill the particular nutritional requirements of certain categories of persons whose digestive processes or metabolism are disturbed; or of certain categories of persons who are in a special physiological condition and who are therefore able to obtain special benefit from controlled consumption of certain substances in foodstuffs.

India: Palthur et al., (2009) examined the regulatory framework concerning functional food and nutraceuticals (FFN) in India. The study examines the development of a new FFN regulatory regime, reviews speculation on its implementation and makes recommendations on approaches to regulating FFN in India. Prior to2006, no single regulatory body and or body of food law regulated the food sector in India. In 2006 the government of India responded to pressure from stakeholders and the industry by enacting the Food Safety and Standards Act 2006 (FSSA), transforming the regulation of food in India.

The effective implementation of the FSSA faces several challenges, including the inability of small and medium scale firms to identify and comply with regulations due to lack of capacity and access to information; the regulatory vacuum that will be created once the old regime is repealed and before the new FSSA takes effect; and education and competency requirements for firms and regulators (Palthur et al., 2009).

Various definitions proposed for different types of foods are:

- **Foods for special dietary uses/functional foods/ nutraceuticals/health supplements** as: "foods which are specially processed or formulated to satisfy particular dietary requirements which exist because of a particular physical or physiological condition or specific diseases and disorders......wherein the composition must differ significantly from the composition of ordinary foods of comparable nature....... does not claim to cure or mitigate any specific disease, disorder or condition (except for certain health benefit or such promotion claims) as may be permitted by the regulations made under FSSA" (Palthur et al 2009; FSSA).
- **Functional foods** are defined as foods that encompass potentially healthful products, including any modified food or food ingredient that may provide a health benefit beyond that of the traditional nutrients it contains (Ministry of Women and Child Development, Government of India, 2010).
- **Novel foods**: These refer to an article of food for which safety standards have not been specified but are not unsafe. These products should not contain any food or ingredient that is prohibited under the Food Safety Standards Act 2006 (Food and Safety Standards Act 2006 Chapter-IV, article 22). India has no specified definition for natural health products.

Factors that may affect the implementation of the new regulatory framework in India (Palthur et al., 2009)

- Lack of qualified technical personnel;
- Inadequacy of skills of existing personnel;
- Lack of well-equipped laboratories for analysis.

Recommendations to ensure successful implementation of the new regulatory regime (Palthur et al., 2009)

- Development of overarching objectives for regulation;
- Greater collaboration between departments and agencies;
- Seeking industry input on choice and design of regulatory instruments;
- Greater consideration of non-prescriptive types of regulations;
- Sufficient resources for an effective infrastructure to support the regulatory process;
- Harmonization with major trading partners;
- Design of legislations that provides for more regulatory flexibility;
- A comprehensive examination that evaluates using agreed upon principles;
- Ex-post evaluation of regulation; and commitment to implement required changes".

The study concludes that the functional food and nutraceuticals sector has a very important role in the future health of the population in India through the provision of less expensive, regulated, innovative products and technologies which may lead to a reduction in public health care costs.

The Food and Agriculture Organization (FAO) of the United Nations: in 2007 provided a working definition for functional food: "foods which are intended to be consumed as part of the normal diet and that, contain biologically active components which offer the potential of enhanced health or reduced risk of disease" (FAO, 2007). The FAO does not have formal definitions for natural health products or novel foods.

Individual countries have developed their own regulatory frameworks and policy agendas concerning labelling and approval of functional food and NHPs, novel food application procedures, nutritional contents, health claims and the enforcement of these regulations. The regulatory framework in the United States, though comparatively restrictive, is more favourable/supportive to the development and marketing of functional foods compared to that of Canada. Legislation pertaining to the approval, labelling and advertising of dietary supplements, which may be applicable to certain functional foods, was well-defined. In order to patent a product with the United States Patent and Trademark Office, an inventor has to only show that the invention is "useful" and "new" and not necessarily that it works (operative). In contrast, in Canada, an inventor must provide information on the "novelty" (be the first in the world), "utility" (functional and operative) and show inventive "ingenuity" on the product before it can be patented with the Canadian Intellectual Property Office (CIPO, 2010). The regulatory situation in the EU was found to be limited and less restrictive than that of Canada in terms of the commercialization and marketing of functional foods and nutraceuticals due to the fact that adoption and enforcement of functional food legislation varies among the member states.

Current areas of research related to functional foods and nutraceuticals can be:

- Technologies that lead to the development of more stable functional food formulations. Technologies that improve the taste and smell of the products.
- Technologies that remove grainy textures of fibres added to beverages.
- Inclusion of probiotics in beverages and baked goods.
- Inclusion of healthy oils in dairy products.

Conclusion

Functional foods and nutraceuticals offer great potential to improve health and/or help prevent certain diseases when taken as part of a balanced diet and healthy lifestyle. The remarkable growth of the functional foods and natural health products industry can be attributed to consumers' ever increasing interest in eating healthier diets. The timeliness of this interest, combined with a number of factors, has spurred the industry's development and expansion. Scientific evidence in support of some foods and food ingredients providing certain physiological benefits over and above that of conventional foods, and growing public awareness of the link between diet and health, have opened up a new area of research and development in the food sector thus bringing fresh challenges and opportunities for the industry, consumers and policy makers.

This subject of health claims is becoming increasingly important and there is broad consensus that there needs to be a regulatory framework that will protect consumers, promote fair trade and encourage product innovation in the food industry. Regulatory frameworks should focus on the protection of consumers, promotion of fair trade and innovation in the sector. There is a need for greater harmonization and uniformity in nutrition descriptions, disease-risk reduction health claims, structure/function claims; and terminologies internationally. Thus there should be a clear distinction between functional and health claims (disease risk reduction) as well as the international harmonization of claims. The communication of health benefits to consumers is also of critical importance so that they have the knowledge to make informed choices about the foods they eat and enjoy. No matter how impressive-looking a functional food's name or package claim, the Nutrition Facts label must be checked before making decisions regarding consumption. Fortified junk foods are still junk.

Punjab state, with abundance of food grains and increasing trends in fruits and vegetables, provides options for producer and to promote growth in the functional foods and nutraceuticals sector through partnership between research centers, private entrepreneurs and indigenous communities. However, the success requires sufficient proof to establish the health claim and capacity to accurately market functional food to consumers in high end product. These products can find demand in the domestic market and for export. Further studies are required to establish the most critical bottlenecks in production system and identify opportunities with the potential rural employment and competitive advantage for small scale farmers as producers of functional foods and nutraceuticals.

References

1. Azzurra Annunziata and Paola Pascale (2009). Consumers' behaviours and attitudes toward healthy food products: The case of Organic and Functional foods. Paper presented at the 113th EAAE Seminar "A resilient European food industry and food chain in a challenging world", Chania, Crete, Greece, on September, pp. 3-6.
2. Canada Intellectual Property Office (CIPO) (2010). Canadian Patent Database, Available online at :http://brevets-patents.ic.gc.ca/opic-cipo/cpd/eng/introduction.html.
3. Coppens Patrick, da Silva Miguel Fernandes and Pettman Simon (2006). European regulations on nutraceuticals, dietary supplements and functional foods: A framework based on safety. Toxicology 221: 59–74.
4. Food and Safety Act – India (2006) General Provisions as to Articles of Food, http://www.drugscontrol.org/food%20safety%20and%20standards%20Act%2020 06.pdf
5. Government of India, Ministry of Women and Child Development. (2010) http://wcd.nic.in
6. Government of Japan, Ministry of Health, Labour and Welfare (MHLW) (2010)http://www.mhlw.go.jp/english/topics/foodsafety/fhc/02.html (Accessed February 2010).
7. Health Canada (1998) 'Nutraceuticals/Functional Foods and Health Claims on Foods',http://www.hc-sc.gc.ca/fn-an/label-etiquet/claims-reclam/nutra-funct_foodsnutra-fonct_aliment-eng.php # tphp.
8. Health Canada (2004). A regulatory framework for Natural Health Products, http://www.hcsc.gc.ca/dhp-mps/prodnatur/about-apropos/glance-apercu-eng.php.
9. Health Canada (2010). Food and Nutrition–Health Claims, http://www.hc-sc.gc.ca/fn-an/labeletiquet/claims-reclam/index-eng.php.
10. Klimas, M. C. Brethour, and D. Bucknell (2008). International Market Trends Analysis for the Functional Foods and Natural Health Products Industry in the United States,

Australia, the United Kingdom and Japan, Final Report Prepared for Nutri-Net Canada by the George Morris Centre.

11. Kotilainen, L. Rajalahti, R. Ragasa, C. and Pehu E. (2006). Health Enhancing Foods Opportunities for Strengthening the Sector in Developing Countries, The World Bank – Agricultural and Rural Development Discussion Paper 30.
12. Leatherhead Food International (2006). The International Market for Functional Foods: Moving into the Mainstream? Published by: Leatherhead Food Research Association, 2006. ISBN: 1905224249, 9781905224241.
13. Malla Stavroula, Hobbs Jill, Sogah Eric Kofi and Yeung May T (2013). Assessing the functional foods and natural health Products industry: a comparative overview and Literature review. Canadian Agricultural Innovation and Regulation (CAIRN) Network, Canada.
14. Mintel International (2006). Functional Food and Beverages, Mintel International Group Ltd.
15. New Zealand Trade and Enterprise (2007). Market Profile for Functional Foods in the European Market, New Zealand Trade and Enterprise, London.
16. Nutrition Business Journal (NBJ) (2007a). Global Nutrition Industry Report.
17. Nutrition Business Journal (NBJ) (2007b). Healthy Foods Report.
18. Ohama, H. H. Ikeda, and H. Moriyama (2006). Health Foods and Foods with Health Claims in Japan, Toxicology, 221: 95-111.
19. World Health Organization (2002). Globalization, Diets and Non-communicable Diseases, WHO Library Cataloguing-in-Publication Data, ISBN 92 4 159041 6 (NLM classification: QT.
20. http://www.who.int/hpr/NPH/docs/globalization.diet.and.ncds.pdf

CHAPTER 43

Biosensors and Quality Aspects of Functional Foods

Swati Kapoor, Savita Sharma and P S Ranote

Abstract

The use of biosensors is growing in the food industry as a means to ensure food safety and quality because of its versatility in determining biological compounds. It is defined as analytical technique used for detection of analyte that combines a biological component with a physico-chemical detector. The most important contribution of biosensors in the food sectors is to control production processes and ensure food quality and safety by reliable, fast and cost effective procedures. Biosensors have been useful for the evaluation of food composition, particularly in functional food products fortified with phytochemicals, vitamins and minerals.. Biosensors are replacing conventional analytical tools since they offer advantages in size, cost, specificity, rapid response, precision and sensitivity. Future progress in the biosensor technology involves the development of nutraceutical release technology within packaging material where it can be used to trigger the controllable release of bioactive components and targeted delivery of protective agents into the food product. The growing consumer demands for safe and quality food has led the food industries to search for quick and reliable methods for which the applications of the biosensor technique in the field of food processing and quality control are found to be promising.

Keywords: Biosensors, Food quality, Functional foods, Nutraceuticals

Introduction

Functional food is the subject of interest among the consumers due to the health benefits it provide beyond the basic nutrition. Upcoming of the functional foods in the market, makes imperative for the food companies to determine and ensure the content of food additives and the health claims stated on their labels. Moreover, with the increase in the technological advancement, agri-food market has produced increased biological, chemical and physical threats to food products, and a bigger consumer demand on process control, quality and safety of these products. Food contaminants are either formed naturally as by-products of food production process or by human activities carried over from water, air or soil into the food chain (Lavecchia et al., 2010). Some of the examples include acrylamide formation in potato chips during frying, pesticide and insecticide residues in final processed product etc. The other aspect of quality of functional food is to detect and determine the concentration of additional nutrients such as vitamins, minerals, phytochemicals and phytosterols. The evolving food safety and quality issues have made the food laws and standards more stringent to prevent foodborne illness which burdens the food industries to meet these standards and protect consumer's interest. In response to this, the techniques are being explored to make available fast, reliable, low cost control systems to measure the content and quality of food additives and nutrients with health claims to be used by producers and governmental bodies in charge of the legal supervision. The conventional methods for detection of food contamination and composition based on culturing, colony counting, chromatography and immunoassay are tedious, time consuming and costly. These methods often require extraction steps for measurement, well trained operators, and most importantly do not allow continuous on-site monitoring. Thus, biosensor technology offer advantages as alternatives to conventional methods due to their inherent specificity, simplicity, quick response, relative low cost of construction and storage, facility of automation and simple portable equipment construction for a fast analysis and monitoring in platforms of raw material reception, quality control laboratories or some stage during the food processing (Luong et al., 1991).

Biosensor Technology

A biosensor is an analytical device that composed of a biological sensing element with a transducer to produce a signal proportional to the analyte concentration (Nakamura et al., 2003). This signal can result from a change in protons concentration, release or uptake of gases, light emission, absorption, brought about by the metabolism of the target compound by the biological recognition element. The transducer converts this biological signal into a measurable response such as current, potential or absorption of light through electrochemical or optical means, which can be further amplified, processed and stored for later analysis (Mulchandani and Rogers, 1998).

Biosensor consists of two main components:

- **Bioreceptor** or biorecognition element to recognize the target analyte. A bioreceptor can be a tissue, microorganism, organelle, cell, enzyme, antibody, nucleic acid and biomimic.
- **Transducer**, for converting the recognition event into a measurable electrical signal. A transducer may be optical, electrochemical, thermometric, piezoelectric, magnetic and micromechanical or combinations these techniques.

The bioreceptor recognizes the target analyte and the corresponding biological responses are then converted into equivalent electrical signals by the transducer. The amplifier in the biosensor responds to the small input signal from the transducer and delivers a large output signal that contains the essential waveform features of an input signal. The amplified signal is then processed by the signal processor where it can later be stored, displayed and analyzed (Veluswamy et al., 2010).

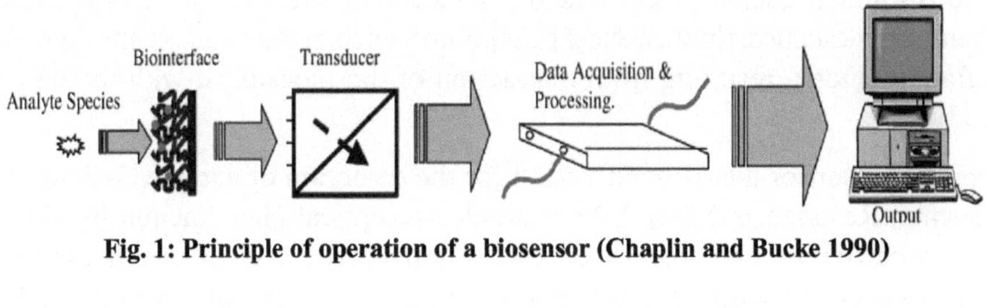

Fig. 1: Principle of operation of a biosensor (Chaplin and Bucke 1990)

Types of Biosensor

Biosensors are basically of three types namely electrochemical sensors, optical sensors and enzymatic biosensors. Bioreceptors that are used in these bosensors are biocatalytic receptors, bioaffinity receptors and hybrid receptors. In biocatalytic receptors the recognition element can be enzyme (mono or multi enzyme), whole cells (microorganisms, such as bacteria, fungi, eukaryotic cells, and yeast), cells organelles and plant or animal tissue slice (Davis et al., 1995). These sensors are generally based on the detection of organic compounds assimilated by the microorganisms or monitoring changes occurring in respiration activity during metabolism (Mello and Kubota, 2002). Bioaffinity receptors employ the use of antigen and antibody where the transducer surface is coated with antigen or antibody. The hybrid receptors such as DNA and

RNA probes have shown promising application in food analysis. In these receptors well defined sequences of single strands of DNA are used as a biological receptor. A DNA probe is added to DNA or RNA from an unknown sample. DNA-based analytical methods seem to be the only method for detecting genetic modifications.

Electrochemical biosensors

An electrochemical biosensor is able to provide specific quantitative analytical information using a biological recognition element (biochemical receptor) which is retained in direct contact with the transduction element (The'venot et al., 2001). Electrochemical biosensors basically rely on a reaction that either produces or consumes electrons and is catalyzed by enzymes. The reaction takes place on electrode surface where the target analyte is detected and the reaction either cause electron transfer across the double layer, producing a current (Amperometric) or can contribute to the double layer potential, producing a voltage (Potentiometric). The current that is produced is directly proportional to the analyte concentration at a fixed potential.

Optical biosensors

The optical biosensors are based on methods such as UV absorption, chemiluminescence, fluorescence/ phosphorescence, reflectance, scattering and refractive index, resulting from interaction of the biocatalyst with the target analyte.

Optical biosensor that is mainly used for the detection of analytes is Surface Plasmon Resonance (SPR). SPR is an electro-optical phenomenon in which energy carried by photons of light can be transferred to electrons in a metal. This transfer of energy creates Plasmon i.e a group of excited electrons on the metal surface (Leonard et al., 2003). Interaction between analytes and bioreceptor changes the chemical properties within the range of the plasmon field which subsequently changes plasmon resonance. These changes can be predicted by interpreting change in the angle of incidence or shift in the wavelength of light absorbed resulting in changing SPR signal.

An SPR biosensor basically consists of a glass slide with a thin gold coating that is mounted on a prism. Light passes through the prism and slide, reflects off the gold and passes back through the prism to a detector. Changes in reflectivity versus angle or wavelength give a signal that is proportional to the volume of analyte bound near the surface. A flow cell allows solutions above the gold surface to be rapidly changed (Fig. 2) (Campbell).

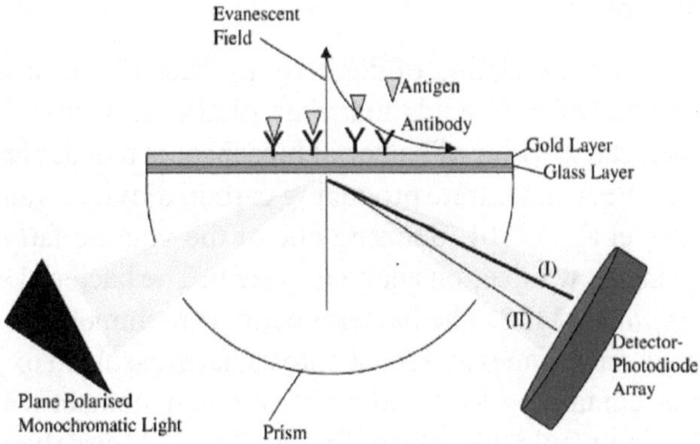

Fig. 2: Surface Plasmon Resonance biosensor (Quinn and Kennedy 1999; Kretschmann 1971)

Enzymatic biosensors

Enzyme based biosensor involves reaction catalyzed by an enzyme in which the union of the substrate is carried out in a concrete region of the enzyme termed the active center, which upon forming the products is recovered and can begin a new reaction cycle (Cock et al., 2009). Oxidoreductases such as glucose oxidase, horseradish peroxidase and alkaline phosphatase (Rogers and Mascini, 1998; Laschi et al., 2000) are the commonly used enzymes, because they are very stable in catalyzing reactions.

Either whole cell from bacteria, fungus, protozoa and higher organisms are being used (Mello and Kubota, 2002) are used or subcellular organelles or tissue that contain more specific complete enzymatic systems are being used (Fig.3). These enzymatic biosensors have been used in the detection of toxic agents such as pesticides, heavy metals or detergents that can inhibit enzymatic systems (D'Souza, 2001).

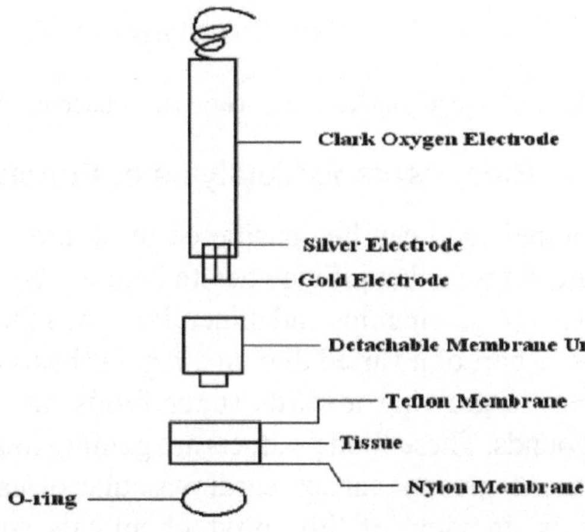

Fig. 3: Representation of biosensors using tissue as a biocomponent (Rekha and Murthy 2010)

Bacterial biosensor

Bacterial biosensors consisting of denitrifying bacteria has been used for the on-line monitoring of fatty acids that are volatile in nature. The underlying principle behind the working of bacterial biosensor is that denitrifying bacteria would consume VFA and nitrate producing carbon dioxide, water and nitrogen gas (Constantin et al., 1996). Benzene one of the volatile fatty acid has been determined using flow injection analysis system. The bacterial strain used was *Pseudomonas putida* ML2. The bacteria cells were immobilised between two cellulose acetate membranes and fixed onto a Clark dissolved oxygen electrode. As the samples containing VFA and nitrate come in contact with the bacteria, the VFA reduction would take place (Fig. 4). The CO_2 dissolved gas produced by the bacterial reaction would then be detected by the CO_2 sensor positioned above the bacteria (Lanyon et al., 2004).

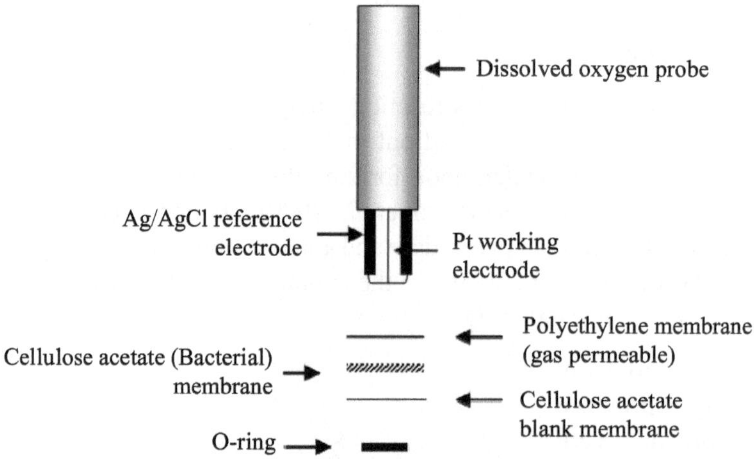

Fig. 4: Diagram representing sequence of bacterial biosensor (Lanyon et al., 2004)

Use of Biosensors for Analysis of Functional Food

Functional food can be considered to be those whole, fortified, enriched or enhanced foods that provide health benefits beyond the provision of essential nutrients (e.g., vitamins and minerals), when they are consumed at efficacious levels as part of a varied diet on a regular basis (Hasler, 2002). Nowadays the concept is growing towards super foods, i.e foods containing antioxidant compounds. These food products are gaining importance due to claimed health benefits to prevent cancer, cardiovascular disease and to protect against free radicals. In view of this, phytochemicals such as carotenoids, phenolic compounds, phytosterols, and ascorbic acid are being incorporated in the food items and is thus, contributing to the development of nutraceutical foods. List of some of the functional food components detected by biosensors is tabulated in Table 1.

Table 1: List of some of the phytochemicals detection by biosensors

Functional component	Food type	Type of Biosensor	References
L-lysine	Milk and Pasta	Amperometric	Curulli et al (1998)
Biotin	Infant formula and milk	Surface Plasmon resonance	Indyk et al (2000)
Ascorbic acid and Polyphenols	Red and white wine	Amperometric	Campanella et al (2004)
Essential fatty acids (linoleic and linolenic acid)	Dairy products and oilseeds	Enzymatic/ Amperometric	Schoemaker et al (1997)
Tocopherols		DNA based electrochemical biosensors	Malpeli et al 1998
Flavonoids (caffeic acid, gallic acid, catechin), anthocyanidins	Complex food samples	Laccase enzyme based biosensor	Gomes and Rebelo (2003)
Folic acid	Vitamin premix, soy and milk based infant formula	SPR	Caselunghe and Lindeberg (2000)
Catechins	Black and Green tea	Tyrosinase based enzymatic biosensor	Nadifiyine et al 2013

Vitamins and minerals

Vitamins and minerals are used to fortify functional foods such as vitamin B_{12} is used in energy drinks, Folic acid to fortify cereal based products, retinol to enhance vitamin A levels in diet, Vitamin D and calcium in milk etc. Vitamin B_{12} is used to fortify energy drinks for sports person to enhance mental performance and reduction of tiredness and fatigue. Detection of Vitamin B_{12} has been made by a dipstick based immune chemiluminescence (immuno-CL) biosensor. This method involves immobilization of vitamin B_{12} antibody on nitrocellulose membrane (NC) followed by treatment with vitamin B_{12} and vitamin B_{12}–alkaline phosphatase conjugate to facilitate the competitive binding. The dipstick was further treated with substrate to generate chemiluminescence. The number of photons generated was inversely proportional to the vitamin B_{12}concentration (Selvakumar and Thakur, 2012).

Folic acid has been used to fortify cereal products for the prevention of pre-natal neural tube defects resulting in spina bifida oranencephaly in new-born babies (Wald, 1991). Optical biosensors have been used to detect folate in breakfast cereals, milk and soy based beverages. Surface Plasmon Resonance technique that has been discussed above is used to obtain signal transduction. In this method a specific interaction takes place between folate binding protein (FBP) and pteroyl L glutamic acid (PGA). The difference between free FBP and associated FBP allows the PGA concentration to be determined which in turn interprets folic acid concentration (Ferancova et al., 2004; Nollet, 2004).

Processing of food leads to degradation of ascorbic acid due to its heat labile nature, therefore, it is commonly added to processed foods as an antioxidant. Conventional methods for the determination of ascorbic acid include titrimetry, spectrophotometry, polarographic methods etc. These chemical methods have low sensitivity, precision and are very elaborate. To alleviate these drawbacks an amperometric principle based biosensor was developed. This biosensor is based on the following reaction

$$2\ \text{Ascorbate} + O_2 \xrightarrow{\text{Ascorbate oxidase}} 2\ \text{Dehydroascorbate} + 2H_2O$$

The reaction results in consumption of oxygen from the solution surrounding the electrode surface which causes a decrease in the voltage response proportional to the ascorbic acid concentration monitored by the amplifier detector system (Rekha and Murthy, 2010).

Biosensor for Phytochemicals

Most of the phytochemials such as phenolic acids (flavoinoids), carotenoids (lycopene, beta-carotene), ascorbic acid are known for there antioxidant power. Therefore, biosensors have been developed to determine the concentration of antioxidant compounds in functional foods.

Polyphenolic compounds are known to have strong correlation with antioxidant activity. Biomediators that are used to evaluate polyphenols concentration are tyrosinase, laccase, horseradish peroxidase (HRP) and horseradish peroxidase/DNA mixtures. Tyrosinase and laccase detects phenolic compounds by using molecular oxygen to catalyze the oxidation of phenolic compounds. Quinone that is formed as a result of this reaction is then electrochemically re-reduced to be determined by electrochemical sensor (Lavecchia et al., 2010).

Bienzymatic biosensors have been used to determine the concentration of glucosinolates in food. Enzymes namely myrosinase and glucose oxidase are used in this system for electrochemical and optical biosensor. Clark type oxygen electrode is used on which a mixture of myrosinase and glu-cose oxidase is immobilized. The enzyme myrosinase catalyzes the hydrolysis of glucosinolates to glucose. Subsequently, the glucose is oxidized by glucose oxidase with the consumption of oxygen. The oxygen level is monitored and related to glucosinolates con-centration (Bi et al., 2008).

Superoxide Dismutase (SOD) biosensor have been developed to determine the concentration of antioxidant compounds. This biosensor is based on the xanthine–xanthine oxidase system to analyze wines, various teas and herbal products such as *chamomile*, dog rose and ginseng. This system is based on the principle where xanthine in solution is oxidized to the superoxide radical and uric acid in the presence of the enzyme xanthine oxidase. The superoxide radical then undergoes disproportion reaction by superoxide dismutase enzyme immobilized on the H_2O_2 electrode that releases oxygen and hydrogen peroxide. The hydrogen peroxide released is oxidized at the anode, i.e. it generates an amperometric signal variation that is proportional to the concentration of superoxide radical in solution. The addition of a sample possessing antioxidant properties produces a decrease in signal strength as the antioxidant species react with the superoxide radical, thus reducing its concentration in solution. There is a consequent decrease in the H_2O_2 released and thus also in the intensity of the amperometric signal (Campanella et al., 1995).

Miscellaneous Uses

Cholesterol

Write about cardiovascular disease. Cholesterol is present in variety of food stuffs namely eggs, dairy products, meat products and others. Efforts are being made by food manufacturers to lower the cholesterol content in the processed food so as to claim the final product healthy and improve its sale ability among health conscious people. Classical method reported use of Libermann-Burchard chemical reaction to determine free and total cholesterol content (Marazuela et al., 1997) which is a complex method and often require more inputs and time. Biosensor technology however has imprinted its success in determination of cholesterol in raw and finished products. Enzymes that are used in the development of cholesterol determining biosensors are cholesterol oxidase and cholesterol esterase (Li et al., 2005; Singh et al., 2004). A simple, fast and economic, cholesterol determining biosensor using the above mentioned enzymes immobilized on polycarbonate membrane attached to the tip of oxygen electrode has been developed by Basu et al., (2007) to determine total cholesterol in food samples. The basic function of the enzyme cholesterol esterase is to convert cholesterol esters present in food to cholesterol and fatty acid. The enzyme cholesterol oxidase then further converts cholesterol and fatty acid into cholest-4 en-3-one and H_2O_2 in the presence of oxygen (Fernandez-Romero et al., 1987). Following this, the amperometric transductor determine the concentration of oxygen consumed or hydrogen peroxide produced which is proportional to the concentration of cholesterol in the food item. In this way, cholesterol can be rapidly determined in the finished foods in the food quality laboratories and save on processing time if further rectification is required.

Fatty acids

Upto date only limited biosensors have been used for detection of fatty acids. Isovaleric acid, one of the important volatile fatty acid (VFAs) has been detected by using piezoelectric quartz crystal sensor. The sensor was shown to respond to a variety of chemical vapors both hydrophobic and hydrophilic achieved by changing the chemical nature of the lipids immobilized on the sensor surface (Yano et al., 1997). In another research, electronic nose has been used for the detection of various volatile fatty acids of interest. Electronic nose which is combination of non-specific conductive chemical sensors and a pattern recognition routine has been used to detect VFAs such as acetic acid, propionic acid and butyric acid can be detected in the headspace of liquid samples in the ppmv range (Maricou et al., 1998). Enzymatic biosensor has also been developed to identify VFAs mainly propionate. The biosensor consists of two recombinant enzymes, propionate coenzyme A (CoA) transferase from *Clostridium propionicum* and acyl-CoA oxidase form *Arabidopsis thaliana*. Production of hydrogen peroxide in the presence of acetyl-CoA, oxygen and VFA substrates was quantified using colorimetric methods (Rajashekhara et al 2006). Electrochemical sensor has also been developed by Zeravik et al., (2010) to determine carboxylic acids such as acetic acid, formic acid etc. in wine. The enzyme, sarcosine oxidase was immobilized on the platinum working electrode by using bovine serum albumin and cross-linking with glutaraldehyde. Short fatty acids that had inhibitory effect on the enzyme activity were then measured by biosensor as an inverse relationship between concentrations of carboxylic acids and enzyme activity signal i.e. higher the amount of carboxylic acids, weaker the enzyme activity signal. The hydrogen peroxide generated during the enzymatic reaction was then measured electrochemically.

Conclusion

Awareness among the consumers regarding the health benefits of functional foods is the primary driver behind the success of the functional food market. People nowadays want to incorporate supplements in there food naturally as a part of there normal diet rather then to take them separately in the form of pills and syrups. Therefore, emphasis has been placed on the safety and scientific analysis aspects. Conventional methods such as spectrophotometric, flurometry, HPLC etc are being used to determine the concentration of vitamins, minerals and phytochemicals in fortified food but these methods are elaborative, less precise and costly. In regard to this biosensor technology is upcoming to examine functional food components and additives. Biosensor offers potential advantages over conventional methods because of there high speed, low cost high precision and senstivity. However, no specific biosensors have been developed till date for detection of individual phytochemical nutrients and continuous progress is being done in this field to evaluate the antioxidant status of individual components.

References

1. Basu, A. K. Chattopadhyay, P. Roychoudhuri, U. and Chakraborty, R. (2007). Development of cholesterol biosensor based on immobilized cholesterol esterase and cholesterol oxidase on oxygen electrode for the determination of total cholesterol in food samples. Bioelectrochemistry, 70:375-379.
2. Bi, X. Wong, W. L. Ji, W. Agarwal, A. Balasubramanian, N. and Yang, K. L. (2008). Development of electrochemical calcium sensors by using silicon nanowires modified with phosphotyrosine. Biosensors Bioelectronics, 23(10):1442 1448.
3. Campanella, L. Favero, G. Sammartino, M. P. and Tomassetti, M. (1994). The effect of organic solvent properties on the response of a tyrosinase enzyme sensor. Talanta, 41:1015–1023.
4. Campbell, C. T. "Surface Plasmon Resonance (SPR) Biosensor Development", http://www.cpac.washington.edu/~campbell/projects/plasmon.pdf.
5. Caselunghe, M.B. Lindeberg, J. (2000). Biosensor-based determination of folic acid in fortified food. Food Chemistry, 70:523-532.
6. Chaplin, M. F. and Bucke, C. (1990). Enzyme technology. 1st edn. Cambridge University Press, England.
7. Cock, L. S. Arenas, A. M. Z. and Aponte, A. A. (2009). Use of enzymatic biosensors as quality indices: a synopsis of present and future trends in the food industry. Chilean Journal of Agriculture Research, 69:270-280.
8. Curulli, A. Kelly, S, O'Sullivan, C. O. Guilbault, G. G. and Palleschi. G. (1998). A new interference-free lysine biosensor using a non-conducting polymer film. Biosensors and Bioelectronics, 13(12):1245–1250.
9. D'Souza, S.F. (2001). Microbial biosensors. Biosensors Bioelectronics, 16: 337-353.
10. Davis, J. Vaughan, D. H. and Cardosi, M. F. (1995). Elements of biosensors construction. Enzyme and Microbial Technology, 17:1030–1035.
11. Ferancova, A. Heilerová, L. Korgová, E. Silhar, S. Stepanek, I. and Labuda, J. (2004). Anti/pro oxidative properties of selected standard chemicals and tea extracts investigated by DNA based electrochemical biosensor. European Food Research Technology, 219(4): 416 420.
12. Fernandez-Romero, J. M. Luque-de-Castro, M. D. and Valcareal, M. (1987). Determination of total cholesterol in serum by flow injection with immobilized enzymes. Clinical Chimistry Acta, 167:97–104.
13. Gomes, S.A.S.S. and Rebelo, M.J.F. (2003). A new laccase biosensor for polyphenols determination. Sensors, 3:166-175.
14. Hasler, C. M. (2002). Functional Foods: Benefits, Concerns and Challenges- A Position Paper from the American Council on Science and Health. The J. Nutr. 132 (12): 3772-3781.

15. Indyk, H. E. Evans, E. A. Caselunghe, M. C. B. Persson, B. S. Finglas, P. M. Woollard, D. C. and Filonzi, E. L. (2000). Determination of biotin and folate in infant formula and milk byoptical biosensor-based immunoassay. Journal of AOAC International, 83(5):1141-1148.
16. Kretschmann, E. (1971). The determination of the optical constants of metals by excitation of surface plasmon resonans. Z Physics, 241: 313–24.
17. Lanyon, Y. H. Marrazza, G. Tothill, I. E. and Mascini, M. (2004ab). Benzene analysis in workplace air using an FIA based bacterial biosensor. Biosensors and Bioelectronics, 20: 2089-2096.
18. Laschi, S. Fránek, M. and Mascini. M. (2000). Screen printed electrochemical immunosensors for PCB detection. Electroanalysis, 12:1293-1298.
19. Lavecchia, T. Tibuzzi, A. and Giardi, M. T. (2010). Biosensors for Functional Food Safety and Analysis. In: Bio Farms for Nutraceuticals: Functional Food and Safety Control by Biosensors. Landes Bioscience and Springer Science+Business Media, New York, USA. Pp 267-281.
20. Leonard, P. Hearty, S. Brennan, J. Dunne, L. Quinn, J. Chakraborty, T. O'Kennedy, R. (2003). Advances in biosensors for detection of pathogens in food and water. Enzyme and Microbial Technology, 32:3–13.
21. Li, G. Lia, J. M. Hu, G. Q. Ma, N. Z. and Wu, P. J. (2005). Study of carbon nanotubes modified biosensor for monitoring total cholesterol in blood. Biosens. Bioelectronics, 20: 2140–2144.
22. Luong, J. H. T. Groom, C. A. and Male, K. B. (1991). Potential role of biosensors in the food and drink industries. Biosensensers Bioelectronics, 6:547–554.
23. Malpeli, G. Folli, C. Cavazzini, D. (1998). Purification and fluorescent titration of cellular retinol binding protein. Springer Protocols. Methods in Molecular Biology, 89:111-122.
24. Marazuela, M. D. Cuesta, B. Moreno-Bondi, M. C. and Quejido, A. (1997). Free cholesterol fiber-optic biosensor for serum samples with simplex optimization. Biosens. Bioelectron. 12: 233–240.
25. Maricou, H. Pereira, D. Verschure, L. Philips, S. and Verstraete, W. (1998). Measurements of some volatile compounds by means of the electronic nose. Water, Air, and Soil Pollution, 107:423-442.
26. Mello, L. D. and Kubota, L. T. (2002). Review of the use of biosensors as analytical tools in the food and drink industries. Food Chemistry, 77:237–256.
27. Mulchandani, A. and Rogers, K.R. (1998). Enzyme and Microbial Biosensors: Techniques and Protocols. Humana Press, Totowa, New Jersey.
28. Nadifiyine, S. Calas-Blanchard, C. Amin, A. and Jean-Louis, M. (2013). Tyrosinase Biosensor Used for the Determination of Catechin Derivatives in Tea: Correlation with HPLC/DAD Method. Food Nutr. Sci. 4: 108-118.
29. Nakamura, H. and Karube, I. (2003). Current research activity in biosensors. Anal Bioanal Chem. 377: 446-468.
30. Nollet, L. M. L. (2004). Handbook of food analysis, 2nd ed., Marcel Dekker Inc. Pp 2138.
31. Quinn, J. G. and O'Kennedy, R. (1999). Transduction platforms and biointerfacial design of biosensors for real-time biomolecular interaction analysis. Analatical Letter, 32(8):1475–1517.
32. Rajashekhara, E. Hosoda, A. Sode, K. Ikenaga, H. and Watanabe, K. (2006). Volatile fatty acid-sensing system involving coenzyme-A transferase. Biotechnology Progress 22: 334-337.
33. Rekha, K. and Murthy, N. (2010). Validation and application of an ascorbate oxidase biosensor for the rapid analysis of vitamin C in food and pharmaceutical samples. Food and Agricultural Immunology, 21 (2):103-111.

34. Rogers, K. R. and Mascini, M. (1998). Biosensors for field analytical monitoring. Field Anal. Chemical Technology, 2:317-331.
35. Schoemaker, M. Feldbrugge, R. Grundig, B. and Spener, F. (1997). The lipoxygenase sensor, a new approach in essential fatty acid determination in foods. Biosen. Bioelectron. 12 (11): 1089-1099.
36. Selvakumar, L. S. and Thakur, M. S. (2012). Dipstick based immunochemiluminescence biosensor for the analysis of vitamin B12 in energy drinks: A novel approach. Analatical Chim. Acta, 722:107- 113.
37. Singh, S. Chaubey, A. and Malhotra, B. D. (2004). Amperometric cholesterol biosensor based on immobilized cholesterol esteresa and cholesterol oxidase on conducting polypyrrole film. Analatical Chim. Acta, 502:229–234.
38. The´venot, D. R. Toth, K., Durst, R. A. and Wilson, G. S. (2001). Electrochemical biosensors: recommended definitions and classification. Biosens Bioelectron, 16:121–131.
39. Velusamy, V. Arshak, K. Korostynska, O. Oliwa, O. and Adley, C. (2010). An overview of foodborne pathogen detection: In the perspective of biosensors. Biotechnology Advances, 28:232-254.
40. Wald, N. (1991). Prevention of neural tube defects: results of the Medical Research Council Vitamin Study. Lancet, 338:131-137.
41. Yano, K. Yoshitake, H. Bornscheuer, U. T. Schmid, R. D. Ikebukuro, K. Yokoyama, K. Masuda, Y. and Karube, I. (1997). Development of a chemical vapour sensor using piezoelectric quartz crystals with coated unusual lipids. Analytica Chimica Acta, 340:41-48.
42. Zeravik, J. Lacina, K. Jilek, M. Vleek, J. and Skladal, P. (2010). Biosensor for determination of carboxylic acids in wines based on the inhibition of sarcosine oxidase. Microchimica Acta , 170:251-256.

CHAPTER 44

Milk Fortification in the Need of Sunshine Vitamin

Taranpreet Singh, Usha Bajwa and Davinder Kaur

Abstract

Vitamin D or "sunshine vitamin" is an essential fat soluble vitamin required for healthy bones. It is responsible for enhancing intestinal absorption of calcium and phosphate. Its deficiency leads to osteoporosis and osteomalacia in adults and rickets and other bone deformities in children. Vitamin D deficiency is being reported to be quite common these days due to modernization accompanied with changes in lifestyle and environmental pollution which limit sun exposure for several individuals. Inadequate sun exposure results in reduced vitamin D synthesis and ultimately poor vitamin D status necessitating compensation through dietary intake. Dietary vitamin D intake is very low in India because of low consumption of vitamin D rich foods, absence of fortification and little use of supplements. It is becoming a major public health problem, needing immediate attention. Fortification of milk and milk products is the safe and effective approach to deal with widespread vitamin D deficiency as milk is consumed and liked equally by all age groups. This paper highlights the importance, sources, recommended allowances and other relevant issues related to vitamin D fortification and its health effects.

Keywords: Milk, Fortification, Vitamin D

Introduction

Vitamin D is a fat-soluble vitamin responsible for enhancing intestinal absorption of calcium and phosphate. It is actually a hormone produced from sterols by the action of ultraviolet light on the skin. Vitamin D plays an important role in the maintenance of healthy bones and teeth along with some essential minerals calcium, phosphorus and magnesium.

Majority of Indian population experience ample sunlight throughout the year. Traditionally, Indians obtained most of their vitamin D through adequate sun exposure. However, seasonal changes, living at high latitude, dark skin pigmentation, aging, modernization accompanied with increased hours spent working indoors and environmental pollution restrict sufficient sun exposure for many. Inadequate sun exposure results in reduced vitamin-D synthesis and ultimately poor vitamin-D status necessitating compensation through dietary intake. Dietary vitamin-D intake is very low in India because of low consumption of vitamin-D rich foods, absence of fortification and little use of supplements. All these factors contribute to poor vitamin-D status which is indicated by low circulating levels of 25-hydroxy vitamin-D and has been found to have a relation with increased risk of chronic diseases, including diabetes mellitus, cancer, autoimmune disorders and osteoporosis (Babu and Calvo, 2010).

An estimated 40 - 90% of adults worldwide are of insufficient Vitamin D status. Rickets, the deforming of the long bones, had been a problem in many countries. Incidence as high as 10% has been reported among breast fed infants and children with little sun exposure. It is therefore very important for infants and children to expose them to sunlight (Wanger and Greer, 2008).

The most important compounds having vitamin D activity are vitamin D_3 (also known as cholecalciferol) and vitamin D_2 (ergocalciferol). Cholecalciferol and ergocalciferol can be ingested from the diet and from supplements. The body can also synthesize vitamin D (cholecalciferol) in the skin, from cholesterol, when exposed to adequate sunshine. Hence it is called “sunshine vitamin” (Institute of Medicine, Washington, US 2010). Vitamin D promotes calcium absorption in the gut and maintains adequate serum calcium and phosphate concentrations to enable normal mineralization of bone and to prevent hypocalcemia. It is also needed for bone growth and bone remodeling. In absence of sufficient vitamin D, bones become thin, brittle, or misshapen. Adequate intake prevents rickets in children and osteomalacia in adults. Together with calcium, vitamin D also helps protect older adults from osteoporosis (Wanger and Greer, 2008).

Sources of Vitamin D

Vitamin D is obtained either through diet or by the sunlight. Its adequacy is represented by combined contribution of cutaneous synthesis and oral ingestion of dietary sources, D_2 from plants and fungi and D_3 from animal sources, fortified foods and supplements.

Most people meet some of their vitamin D needs through exposure to sunlight. Ultraviolet (UV) radiation with a wavelength of 290–320 nanometers penetrates uncovered skin and converts cutaneous 7-dehydrocholesterol to previtamin D_3, which in turn becomes vitamin D_3 (American Academy of Dermatology, 2008). The important factors that affect UV radiation exposure are season, time of day, length of day, cloud cover, smog, skin melanin content, and sunscreen. Complete cloud cover reduces UV energy by 50%; shade (including that produced by severe pollution) reduces it by 60%. UV radiation does not penetrate glass, so exposure to sunshine indoors through a window does not produce vitamin D. Sunscreens with a sun protection factor (SPF) of 8 or more appear to block vitamin D-producing UV rays. It has been suggested by some researchers, that approximately 5–30 minutes of sun exposure between 10 AM and 3 PM at least twice a week to the face, arms, legs, or back without sunscreen usually lead to sufficient vitamin D synthesis. At least 10 % of the body should be exposed to direct sunlight for minimum half hour every day for adequate intake. The moderate use of commercial tanning beds that emit 2–6% UV radiation is also effective. Individuals with limited sun exposure need to include good sources of vitamin D in their diet or take a supplement to achieve recommended levels of intake (Holick, 2005).

According to NIN Hyderabad, under situations of minimal exposure to sunlight, Recommended Daily Allowance of Vitamin D is 400 IU per day (ICMR, 2010). The following points need to be considered:-

- Very few foods of animal origin naturally contain vitamin D.
- It is not present in foods of vegetable origin.
- The major sources of vitamin D are fish liver oils, fatty fish, eggs, butter, cheese, full fat milk powder and milk.
- Some varieties of mushrooms, salami, ham and sausages may also contribute to some vitamin D in the diet.
- Fluid milk and breakfast cereals are the predominant carriers for vitamin D in US, whereas Canada fortifies fluid milk and margarine. Other foods like fruit juices, cheese, butter, tofu and soy milk are fortified with vitamin D and extent of nutrient addition is declared on the label.

- In general, the chief food sources of Vitamin D are fortified milk and fatty fish.

Vitamin D content of some foods is presented in Table 1.

Table 1: Vitamin D content of some foods

Food	Vitamin D I.U./100g
Halibut liver Oil	20,000-4,00,00
Cod Liver Oil	8,00-30,000
Shark Liver Oil	1,200-4,000
Fatty Fish (Sardine, Salmon, Herring)	200-1,200
Egg, hen, whole	50-60
Egg yolk	120-160
Butter	20-60
Ghee	20-60
Milk powder, full fat	15-25
Milk, fresh, whole	2-4

1 microgram (μg) of vitamin D = 40 I.U. (*Source*: N.I.N, Hyderabad)

Vitamin D Toxicity

Vitamin-D intoxication is one of the rarest medical conditions that may be caused either by intentional or inadvertent exposure to excessively high amounts of Vitamin D. Studies have shown that adults can take up to 10,000 IU of Vitamin D a day for at least five months without altering their serum calcium. Vitamin D toxicity may cause hypercalcemia, soft tissue calcification etc. Vitamin D intoxication can contribute to increased risk of death (Cranney et al., 2007).

Despite the importance of the sun for vitamin D synthesis, it is prudent to limit exposure of skin to sunlight. UV radiation is a carcinogen responsible for most of the skin cancers. Lifetime cumulative UV damage to skin is also largely responsible for some age-associated dryness and other cosmetic changes. The American Academy of Dermatology advises that photoprotective measures be taken, including the use of sunscreen, whenever one is exposed to the sun. Assessment of vitamin D requirements cannot address the level of sun exposure because of such public health concerns about skin cancer. Moreover there are no studies to determine whether UV-induced synthesis of vitamin D can occur without increased risk of skin cancer (National Institute of Health, USA, 2011).

Need for Fortification

Early in the 20th century, more than 80% of children in Europe and North America were suffering from rickets. Finding that exposure to ultraviolet radiation or sunlight treated and prevented rickets led to the ultraviolet irradiation of foods including milk. Many countries mandated the fortification of milk with Vitamin D to prevent rickets. The most cost effective way is fortify food with higher levels of Vitamin D along with sensible sun exposure. Because most foods contain only very low amounts of Vitamin D, it is common practice in many countries to fortify certain frequently consumed foods (e.g., baked goods, grain products, milk, yoghurt, cheese, margarine, orange juice, infant foods and some bread). The USFDA has specified that milk should contain 400 - 600 IU/L.

Milk as a Source of Vitamin D

In many countries, the fat from milk is either partially or completely removed to produce reduced-fat or fat-free milks. In such circumstances fat-soluble vitamins, such as vitamins A and D, are removed. However they can be replaced through fortification (Jones and Noelle 2002).

Milk Fortification with Vitamin D

Throughout the world, there are more than 6 billion consumers of milk and milk products. In India, most individuals (vegan and non vegan) consume milk in one or more form (Paneer, Kheer, Cheese etc.). Fortification of milk with vitamin D will help people to meet daily allowance of vitamin D. Addition of vitamin concentrates to milk should be done in a manner that results in vitamin levels that are accurate and agreeable to meet the Recommended Daily Allowance of vitamin D.

The technology to fortify milk is simple. Fortification can be achieved by two ways, ether by batch or continuous method. All the vitamins and minerals that can be added to milk are available in dry powder form. The fat-soluble vitamins are also available in an oily form. Because more than one nutrient is generally added to milk, they are added preferably as a premix, which is a homogenous mixture of the desired amount of fortificants (vitamins and minerals) concentrated in a small amount of the food to be fortified. Premixes ensure the addition of correct amounts and uniform homogenization of the micronutrients in the final product.

Following are guidelines that help processors maintain correct vitamin fortification levels in the fluid milk products

1. *Supplier of Vitamin D*: Vitamin companies/suppliers should be able to provide expert technical support in developing suitable fortification program.

2. *Amount of Vitamin D*: Calculated amount to be added in the milk to prevent the overdose of Vitamin D. Vitamin D must be at levels of not less than 400 IU/l.

3. *Proper Selection of Vitamin Concentrates*: Vitamin Concentrates will be oil based or water dispersible formulations. Most of them contain vitamin D_3. Water soluble formulations may be added before fat standardization. Concentrates commonly used by large dairies are formulated such that 1 ml of vitamin D concentrate fortifies large quantities of homogenized milk at 400 IU Vit D/L. For small batches of milk, more dilute concentrates are accurate. It is desirable to add concentrates at levels slightly higher (e.g., 10-20%) than the calculated usage rate to ensure that milk levels are 100-150% of the target label claim.

4. *Proper storage of vitamin to avoid heat and light losses*: - Excessive exposures to heat, light and air have the potential to shorten vitamin shelf-life. Water dispersible concentrates generally have a shorter shelf-life and should be used more rapidly than oil based formulations.

5. *Follow Standard procedure*: - Standard procedure should be developed to add Vitamin in milk. All vitamin addition must be before the milk is pasteurized.

6. *Testing*: -Product must be tested by a laboratory to ensure proper level of fortification.

Fortification can be done by two ways: - Batch and continuous.

Batch process requires accurate measurement of milk volume as well as accurate measurement of the vitamin concentrate required for that amount of milk, and sufficient mixing time. After adding the appropriate vitamin concentrate, the milk is standardized and allowed sufficient mixing. The fortified milk is tested (to ensure required vitamin quantity/fortification), pasteurized and packed.

Continuous process involves metering pumps which correctly deliver the required amount of concentrate to milk directly into the product flow. Positive pressure pumps (e.g., piston or peristaltic) are required. The best point of metered addition is after standardization, prior to homogenization and pasteurization. It is essential to install check valves to prevent back-flow. Metering pumps are also connected to the control panel so that they turn off during divert flow, to prevent over fortification. This process requires strict control over measuring devices, pumps and other automatic devices controls (Anonymous 2008).

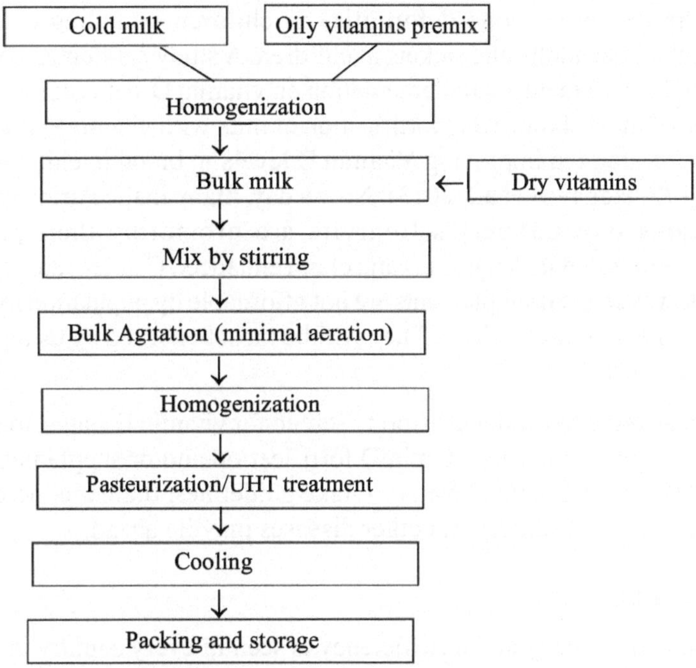

Fig.1: Flowchart for Fortification in Liquid Milk

History of Vitamin D Fortification

Milk fortification began in the first half of this century. Vitamin D fortification of milk in the United Kingdom is reported to have begun in 1923; currently, milk fortification with vitamins A and D is practiced on a voluntary basis. In the United States, the Food and Nutrition Council of the American Medical Association pronounced in 1939 the addition of no more than 400 IU of vitamin D per litre of milk, in the interest of public health. The marked decline in the prevalence of rickets, a vitamin D deficiency disorder, in the USA had been partially attributed to vitamin D fortification of milk. A number of dairies in Ireland fortified milk with vitamin D and calcium following the report of a high prevalence of hypovitaminosis D in the elderly. Vitamin D-fortified milk would be beneficial to populations in whom limited exposure to sunlight reduces vitamin D status, especially in the winter months.

Need of Vitamin D Fortification in India

There is widespread vitamin D deficiency in Indian populations in higher and lower socioeconomic strata, in all age groups, in both genders and in people in various professions. Habitual Indian diets do not provide even 10% of the requirement. Poor vitamin D status in India is accompanied by increased bone disorders and other bone deformities in children including osteoporosis, osteomalacia in adults and rickets in children. A study conducted on Impact of vitamin D fortified milk supplementation on vitamin D status of healthy school children of India showed that fortification of milk with vitamin D is an effective and safe strategy in improving Vitamin D levels in blood in children aged 10-14 years (Khadgawat et al., 2013). Now a day, many malted drink mix making companies like Cadbury's Bourvita are promoting their product by supplementation with Vitamin D and claim enhanced Calcium absorption from milk. However, such supplements are not affordable by population living below poverty line and many others. They will be most benefited by using vitamin D fortified foods.

Without a concerted national effort to screen for vitamin D status, to implement policies or guidelines for vitamin D fortification and/or supplementation and to re-assess recommended dietary intake guidelines, dramatic increase in the number of bone disorders and other diseases may lie ahead.

Conclusion

Vitamin D deficiency and insufficiency is seen in every country in the world. Lack of awareness of the importance of this deficiency is crucial in individual and public health. Fish, cod liver oil and mushrooms exposed to sunlight or ultraviolet radiation are the only natural sources. Excess sun exposure is related to skin cancers. Lifestyle changes reduce outdoor exposure for children and many other age groups. The need for fortifying basic foods with Vitamin D has become now ever more important. Very few foods are fortified with Vitamin D. Even natural foods provide only a small portion to satisfy either a child's or adults Vitamin D requirement. Milk is taken by all age group people regularly. Fortifying it with Vitamin D will meet the RDA of most of the people. As in case of Iodine for Salt, law should be passed by Indian Government to make Vitamin D fortification necessary in the milk.

References

1. Babu, U.S, and Calvo, M.S. (2010). Modern India and the vitamin D dilemma: evidence for the need of a national food fortification program. Molecular Nutrition Food Research, 54(8):1134-47. doi: 10.1002/mnfr.200900480.
2. Institute of Medicine, Food and Nutrition Board (2010). Dietary Reference Intakes for Calcium and Vitamin D. Washington, DC: National Academy Press.
3. Wagner, C. L. and Greer, F. R. (2008). American Academy of Pediatrics Section on Breastfeeding; American Academy of Pediatrics Committee on Nutrition. Prevention of rickets and vitamin D deficiency in infants, children, and adolescents. Pediatrics, 122:1142-1152, PubMed abstract.
4. Holick, M. F. (2005). Photobiology of vitamin D. In: Feldman D, Pike JW, Glorieux FH, eds. Vitamin D, 2ed (1). Burlington, MA: Elsevier.

5. Cranney, C. Horsely, T. O'Donnell, S. Weiler, H. Ooi, D. Atkinson, S. (2007). Effectiveness and safety of vitamin D. Evidence Report/Technology Assessment No. 158, University of Ottawa Evidence-based Practice Center under Contract No. 290-02.0021. AHRQ Publication No. 07-E013.
6. American Academy of Dermatology (2008). Position statement on vitamin D. http://www.aad.org/Forms/Policies/Uploads/PS/PS-VitaminD.pdf
7. National Institution of Health, USA (2011) Dietary Supplement Fact Sheet: Vitamin D http://ods.od.nih.gov/factsheets/VitaminD-HealthProfessional/
8. Jones, Alicia Noelle (2002). Density of Milk. The Physics Factbook. http://hypertextbook.com—/facts/2002/AliciaNoelleJones.shtml
9. Anonymous (2008). Dairy Foods Science Notes: vitamin fortification of fluid milk. Milk Quality Improvement Program, Department of Food Science, Cornell University, USA.
10. Indian Council of Medical Research (2010). Nutritional requirements and recommended dietary allowances for Indians. National Institute of Nutrition, Hyderabad.
11. Khadgawatm, R. Marwaham, R. K. Gargm, M. K. Ramotm, R. Oberoim, A.K. Sreenivasm, V. Gahlot, M.Mehan, N. Mathur, P. and Gupta, N. (2013). Impact of vitamin D fortified milk supplementation on vitamin D status of healthy school children aged 10-14 years. *Osteoporos Int* 24(8):2335-43. doi: 10.1007/s00198-013-2306-9.

Index

C

D

N

O

P

Q

R

S